Number Theory

This is a volume in
PURE AND APPLIED MATHEMATICS

A Series of Monographs and Textbooks

Editors: SAMUEL EILENBERG AND HYMAN BASS

A list of recent titles in this series appears at the end of this volume.

Number Theory

by

Z. I. BOREVICH

and I. R. SHAFAREVICH

Translated by NEWCOMB GREENLEAF for
SCRIPTA TECHNICA, Translators
New York, New York

ACADEMIC PRESS New York San Francisco London
A Subsidiary of Harcourt Brace Jovanovich, Publishers

ACADEMIC PRESS INC.
111 Fifth Avenue, New York, New York 10003

United Kingdom Edition published by
ACADEMIC PRESS, INC. (LONDON) LTD.
24/28 Oval Road, London NW1

LIBRARY OF CONGRESS CATALOG CARD NUMBER: 65-28624

PRINTED IN THE UNITED STATES OF AMERICA

Original Russian Edition: *Teoriya Čisel,*
Moscow, 1964

Translator's Preface

This book was written as a text for the learning of number theory, not as a reference work, and we have attempted to preserve the informal, slow-placed style of the original. The emphasis of the book is on number theory as a living branch of modern mathematics, rather than as a collection of miscellaneous results.

The book should prove accessible to any advanced undergraduate in mathematics, or to any graduate student. The reader should be familiar with the basic concepts of abstract algebra, and should have followed analysis through a standard advanced calculus course. While some results from elementary number theory are occasionally used, a previous course in number theory is certainly not necessary, though the reader without such a course may have a few occasions for consulting a more elementary text.

Almost all of the notation and terminology is standard. The only difficulty arose in the terminology for valuations. Since there does not seem to be any universally adopted terminology in English, we have, after some hesitation, followed that of the authors, which has the advantage of being clear and simple in this context. Thus we reserve the term "valuation" for the case when the value group is the integers. Mappings into the positive reals are called "metrics."

We would like to mention some additional references. The theory of quadratic forms receives a systematic development in the book *Introduction to Quadratic Forms,* by O. T. O'Meara (New York, 1963). In particular, O'Meare presents a proof of the Hasse-Minkowski theorem which does not use the Dirichlet theorem on primes in arithmetic progressions. In Chapter 7 of *Commutative Algebre* (Paris, 1965), Bourbaki gives a complete exposition of the theory of divisors, from a somewhat more abstract standpoint than that found in Chapter 3.

NEWCOMB GREENLEAF

Rochester, New York

Foreword

This book is written for the student in mathematics. Its goal is to give a view of the theory of numbers, of the problems with which this theory deals, and of the methods that are used.

We have avoided that style which gives a systematic development of the apparatus and have used instead a freer style, in which the problems and the methods of solution are closely interwoven. We start from concrete problems in number theory. General theories arise as tools for solving these problems. As a rule, these theories are developed sufficiently far so that the reader can see for himself their strength and beauty, and so that he learns to apply them.

Most of the questions that are examined in this book are connected with the theory of diophantine equations — that is, with the theory of the solutions in integers of equations in several variables. However, we also consider questions of other types; for example, we derive the theorem of Dirichlet on prime numbers in arithmetic progressions and investigate the growth of the number of solutions of congruences.

The methods that we use are primarily algebraic. More precisely, we work with finite field extensions and with metrics on them. However, analytic methods have a considerable place. Chapter 5 is devoted to them, and p-adic analytic functions are used in Chapter 4. Geometric concepts play a considerable role in several spots.

The book does not presuppose a great deal of knowledge on the part of the reader. For reading most of it, two university courses would be completely satisfactory. Some facts on analytic functions are used in the last two chapters.

The necessary prerequisites of an algebraic nature are given in the "Algebraic Supplement" at the end of the book. There the reader will find definitions, results, and some proofs that are used in the book but might not appear in a university course in higher algebra.

This book grew out of a course taught by one of the authors at Moscow University. We would like to thank A. G. Postnikov, who allowed us the use of his notes from this course.

We are also extremely grateful to Dmitri Constantine Faddeev, who made many contributions to this book. He should receive credit for some of the proofs that appear in this book, for example, the new p-adic proof of the theorem of Kummer on the second factor in the number of divisor classes of a cyclotomic field.

Moscow THE AUTHORS

Contents

CHAPTER 1

Congruences

This chapter is devoted to the theory of congruences and to its application to equations in several variables. The connection between congruences and equations is based on the simple remark that if the equation

$$F(x_1, \dots, x_n) = 0, \tag{0.1}$$

where F is a polynomial with integral coefficients, has a solution in integers, then the congruence

$$F(x_1, \dots, x_n) \equiv 0 \pmod{m} \tag{0.2}$$

is solvable for any value of the modulus m. Since the question of the solvability of a congruence can always be decided (if only by trial and error, as there are only finitely many residue classes), we have a sequence of necessary conditions for the solvability of (0.1) in integers.

The question of the sufficiency of these conditions is much more difficult. The assertion that "an equation is solvable if and only if it is solvable as a congruence modulo any integer" is in general false (see, for example, Problem 4), but it is true for certain special classes of equations. In this chapter we shall prove it in the case where F is a quadratic form which satisfies the additional condition, clearly necessary, that (0.1) be solvable in real numbers. (Note that if F is a form, then by the solvability of the equation $F = 0$ we shall understand the existence of a nonzero solution.)

The p-adic numbers, which we shall study and apply to the theory of congruences and equations, will be our basic tool. We now indicate their role. From the elementary theory of numbers it is known that if the congruences

$$F(x_1, \dots, x_n) \equiv 0 \pmod{p_i^{k_i}}$$

are solvable for $i = 1, \ldots, r$, where $p_1, \ldots, p_r$ are distinct primes, then the congruence (0.2) is solvable modulo m, where $m = p_1^{k_1} \ldots p_r^{k_1}$. Thus the solvability of the congruence (0.2) for all m is equivalent to its solvability modulo all powers of primes. We fix a prime p and ask whether the congruence

$$F(x_1, \ldots, x_n) \equiv 0 \pmod{p^k} \tag{0.3}$$

is solvable for all natural numbers k. It was in connection with this problem that Hensel constructed, for each prime p, a new kind of number, calling it p-adic. He showed that the solvability of (0.3) for all k was equivalent to the solvability of (0.1) in p-adic numbers. Hence we may say that the solvability of the congruence (0.2) for all m is equivalent to the solvability of (0.1) in p-adic numbers for all prime numbers p.

Using p-adic numbers, our theorem on quadratic forms then receives the following formulation (its proof will appear in Section 7): If $F(x_1, \ldots, x_n)$ is a quadratic form with integral coefficients, then (0.1) is solvable in integers if and only if it is solvable in p-adic numbers for all p and also in real numbers.

In the formulation of this theorem, called the *Hasse–Minkowski theorem*, and in many other instances, the p-adic numbers occur on equal terms with the real numbers.

If the real numbers are necessary for the study of rational numbers from the standpoint of their size, the p-adic numbers play a completely analogous role in question connected with divisibility by powers of the prime number p. The analogy between real and p-adic numbers can be developed in other ways. It will be shown that the p-adic numbers can be constructed starting from the rational numbers, in exactly the same way that the real numbers are constructed—by adjoining the limits of Cauchy sequences. We shall arrive at different types of numbers by giving different meanings to the notion of convergence.

We make one further remark. If F is a form, then the solvability of (0.1) in integers is equivalent to its solvability in rational numbers. Thus one may speak of rational solvability instead of integral solvability in the Hasse–Minkowski theorem. This obvious remark becomes important when one considers an arbitrary quadratic polynomial F, since the analogous theorem then only holds when one speaks of rational solvability. Hence when we study equations of the second degree, we shall consider not just integral, but also rational solutions.

PROBLEMS

1. Show that the equation $15x^2 - 7y^2 = 9$ has no integral solution.

2. Show that the equation $5x^3 + 11y^3 + 13z^3 = 0$ has no integral solution other than $x = y = z = 0$.

3. Show that an integer of the form $8n+7$ cannot be represented as the sum of three squares.

4. Using the properties of the Legendre symbol, show that the congruence

$$(x^2-13)(x^2-17)(x^2-221) \equiv 0 \pmod{m}$$

is solvable for all m. It is clear that the equation $(x^2-13)(x^2-17)(x^2-221)=0$ has no integral solutions.

5. Show that the equation $a_1x_1 + \cdots + a_nx_n = b$, where $a_1, \ldots, a_n, b$ are integers, is solvable in integers if and only if the corresponding congruence is solvable for all values of the modulus m.

6. Prove the analogous assertion for systems of linear equations.

1. Congruences with Prime Modulus

1.1. *Equivalence of Polynomials*

We first consider congruences with prime modulus p. The residue classes modulo p form a finite field with p elements, and a congruence with modulus p can be considered as an equation in this field. We shall denote the field of residue classes modulo p by Z_p. There exist finite fields other than the various Z_p. All considerations of the next two sections carry over word for word in the general case of any finite field. To do this it is necessary only to replace the number p by the number $q = p^m$ of elements of this field. But we shall confine our attention to the field Z_p and shall use the notation of congruences rather than equations. Only in the construction of the example following Theorem 3 will we need to employ other finite fields.

The field of residue classes modulo a prime (and more generally any finite field) has several properties which distinguish it from the familiar fields of elementary algebra, the fields of rational, real, and complex numbers. Most important in our considerations is the fact that the well-known theorem that polynomials which take equal values for all values of the variables must have equal coefficients is no longer true for this field. For example, by the small Fermat theorem, the polnomials x^p and x take equal values in the field Z_p for all values of the variable x, but their coefficients are unequal. [The following holds for any finite field: If $\alpha_1, \ldots, \alpha_q$ are the elements of the field, then the polynomial $(x-\alpha_1)\cdots(x-\alpha_q)$, which has nonzero coefficients, has value zero for every value for x in the field.]

We write

$$F(x_1, \ldots, x_n) \equiv G(x_1, \ldots, x_n) \pmod{p}$$

and call the polynomials F and G *congruent*, if the coefficients of corresponding terms on the right and left sides are congruent modulo p. If for any set of values $c_1, \ldots, c_n$ we have

$$F(c_1, \ldots, c_n) \equiv G(c_1, \ldots, c_n) \pmod{p},$$

then we write $F \sim G$ and call F and G *equivalent*. It is clear that if $F \equiv G$, then $F \sim G$, but the example of the polynomials x^p and x shows that the converse is, in general, false.

Since, if $F \sim G$, the congruences $F \equiv 0 \pmod{p}$ and $G \equiv 0 \pmod{p}$ have the same solutions, it is natural that in the theory of congruences one needs to be able to replace a polynomial F by a polynomial which is equivalent to it but is possibly in a simpler form. We now return to this problem.

If any variable x_i occurs in the polynomial F to a power not less than p, then using the equivalence $x_i^p \sim x_i$, which follows from the small Fermat theorem, we may replace x_i^p in F by x_i. Since equivalence is preserved under addition and multiplication, we shall obtain a polynomial which is equivalent to F but which contains x_i to a lower degree. This process can be continued until we arrive at an equivalent polynomial which is of degree less than p in each variable x_i. Such a polynomial we call reduced. It is clear that when x_i^p is replaced by x_i, the total degree of F (in all its variables) is not increased. Hence we obtain the following result.

Theorem 1. Every polynomial F is equivalent to a reduced polynomial F^*, whose total degree is not greater than that of F.

We now show that the reduced polynomial equivalent to a given polynomial is uniquely determined.

Theorem 2. If two reduced polynomials are equivalent, then they are congruent.

This theorem is proved in precisely the same way as is the theorem mentioned above on the identity of polynomials, namely, by induction on the number of variables. It clearly suffices to show that if the polynomial F is reduced and $F \sim 0$, then $F \equiv 0 \pmod{p}$.

We consider first the case $n = 1$. If the degree of $F(x)$ is less than p and $F(c) \equiv 0 \pmod{p}$ for all c, then F has more roots than its degree, and this is possible only if all coefficients of F are divisible by p, that is, $F \equiv 0 \pmod{p}$. For arbitrary $n \geqslant 2$, we write F in the form

$$F(x_1, \ldots, x_n) = A_0(x_1, \ldots, x_{n-1}) + A_1(x_1, \ldots, x_{n-1})x_n + \cdots + A_{p-1}(x_1, \ldots, x_{n-1})x_n^{p-1}.$$

Take an arbitrary set of values $x_1 = c_1, \ldots, x_{n-1} = c_{n-1}$, and set $A_0(c_1, \ldots, c_{n-1}) = a_0, \ldots, A_{p-1}(c_1, \ldots, c_{n-1}) = a_{p-1}$. Then

$$F(c_1, \ldots, c_{n-1}, x_n) = a_0 + a_1 x_n + \cdots + a_{p-1} x_n^{p-1}.$$

We have obtained a polynomial in one variable x_n, which is equivalent to zero, since $F \sim 0$. But for polynomials in one variable the theorem has been proved, and therefore the above polynomial must be congruent to zero. Thus

$$A_0(c_1, \dots, c_{n-1}) \equiv 0 \pmod{p},$$
$$\dots\dots\dots\dots\dots\dots\dots$$
$$A_{p-1}(c_1, \dots, c_{n-1}) \equiv 0 \pmod{p},$$

that is, $A_0 \sim 0, \dots, A_{p-1} \sim 0$ (since $c_1, \dots, c_{n-1}$ were arbitrary). Since the polynomials A_i are clearly reduced and depend on $n-1$ variables (and for such polynomials the theorem is true by the induction hypothesis), then

$$A_0 \equiv 0 \pmod{p}, \dots, A_{p-1} \equiv 0 \pmod{p},$$

from which it follows that $F \sim 0 \pmod{p}$.

1.2. *Theorems on the Number of Solutions of Congruences*

From Theorems 1 and 2 we can deduce some corollaries on the number of solutions of congruences.

Theorem 3. If the congruence $F(x_1, \dots, x_n) \equiv 0 \pmod{p}$ has at least one solution, and the total degree of the polynomial F is less than the number of variables, then the congruence has at least two solutions.

Proof. Assume that the polynomial $F(x_1, \dots, x_n)$ with total degree r is such that the congruence $F(x_1, \dots, x_n) \equiv 0 \pmod{p}$ has the unique solution

$$x_1 \equiv a_1 \pmod{p}, \dots, x_n \equiv a_n \pmod{p}.$$

Set $H(x_1, \dots, x_n) = 1 - F(x_1, \dots, x_n)^{p-1}$. By the small Fermat theorem and the assumptions on F we have

$$H(x_1, \dots, x_n) \equiv \begin{cases} 1 & \textit{for} \quad x_1 \equiv a_1, \dots, x_n \equiv a_n \pmod{p}, \\ 0 & \textit{otherwise}. \end{cases}$$

Denote by H^* the reduced polynomial equivalent to H, by Theorem 1. H^* takes the same values as H. But, on the other hand, we can explicitly construct a reduced polynomial taking the same values as H, namely, the polynomial

$$\prod_{i=1}^{n} (1 - (x_i - a_i)^{p-1}).$$

By Theorem 2 we have

$$H^* \equiv \prod_{i=1}^{n} (1 - (x_i - a_i)^{p-1}) \pmod{p}. \tag{1.1}$$

From Theorem 1 it follows that the degree of H^* is not greater than the degree of H, that is, not greater than $r(p-1)$. Thus the degree of the left side of (1.1) is not more than $r(p-1)$ and the degree of the right side equals $n(p-1)$. Hence $n(p-1) \leqslant r(p-1)$, and this proves the assertion for the case $r < n$.

Corollary (Chevalley's Theorem). If $F(x_1, \ldots, x_n)$ is a form of degree less than n, then the congruence

$$F(x_1, \ldots, x_n) \equiv 0 \pmod p$$

has a nonzero solution.

The existence of such a solution follows from Theorem 3, since one solution, namely, zero, always exists in this case.

To complete the picture, we shall show that the inequality $r < n$ cannot be weakened if Chevalley's theorem is to remain valid. We shall construct for every n a form $F(x_1, \ldots, x_n)$ of degree n, such that the congruence

$$F(x_1, \ldots, x_n) \equiv 0 \pmod p \qquad (1.2)$$

has only the zero solution.

We use the fact that for any $n \geqslant 1$ there is a finite field Σ with p^n elements, which contains Z_p as a subfield (see the Supplement, Section 3, Theorem 2). Let $\omega_1, \ldots, \omega_n$ be a basis for the field Σ over Z_p. Consider the linear form $x_1\omega_1 + \cdots + x_n\omega_n$, in which $x_1, \ldots, x_n$ may take arbitrary values in Z_p. Its norm $N_{\Sigma/Z_p}(x_1\omega_1 + \cdots + x_n\omega_n) = \varphi(x_1, \ldots, x_n)$ is clearly a form of degree n in $x_1, \ldots, x_n$ with coefficients in the field Z_p. By the definition of the norm $N(\alpha)$ (Supplement, Section 2.2) of the element $\alpha = x_1\omega_1 + \cdots + x_n\omega_n$ $(x_i \in Z_p)$, it follows that $N(\alpha) = 0$ if and only if $\alpha = 0$, that is, when $x_1 = 0, \ldots, x_n = 0$. Therefore the form φ has the property that the equation $\varphi(x_1, \ldots, x_n) = 0$ has only the zero solution in the field Z_p. Now replace each coefficient of the form φ, which is a residue class modulo p, by any element of this class. We obtain a form $F(x_1, \ldots, x_n)$ with integer coefficients, of degree n in n variables, and for this form F the congruence (1.2) clearly has only the zero solution.

Theorem 3 is a special case of the following fact.

Theorem 4 (Warning's Theorem). The number of solutions of the congruence $F(x_1, \ldots, x_n) \equiv 0 \pmod p$ is divisible by p, provided that the degree of the polynomial $F(x_1, \ldots, x_n)$ is less than n.

Proof. Let the congruence have s solutions $A_i = (a_1^{(i)}, \ldots, a_n^{(i)})$ $i = 1, \ldots, s$. Again set $H = 1 - F^{p-1}$. It is clear that

$$H(X) \equiv \begin{cases} 1 & \text{if } X \equiv A_i \pmod p \qquad (i = 1, \ldots, s), \\ 0 & \textit{otherwise} \end{cases}$$

where X stands for $(x_1, \dots, x_n)$. (Congruence of integer-valued vectors means congruence of their respective components.) For any $A = (a_1, \dots, a_n)$ we form the polynomial

$$D_A(x_1, \dots, x_n) = \prod_{j=1}^{n} (1 - (x_j - a_j)^{p-1}). \tag{1.3}$$

It is clear that

$$D_A(X) \equiv \begin{cases} 1 & \text{for } X \equiv A \pmod p, \\ 0 & \textit{otherwise}. \end{cases} \tag{1.4}$$

Set

$$H^*(x_1, \dots, x_n) = D_{A_1}(x_1, \dots, x_n) + \cdots + D_{A_s}(x_1, \dots, x_n). \tag{1.5}$$

The congruence (1.4) shows that H^* takes the same values as does H for any values of $x_1, \dots, x_n$, that is, $H \sim H^*$. Since each of the polynomials D_{A_i} is reduced, so is H^*, and then by Theorems 1 and 2 the degree of H^* does not exceed the degree of H, which is less than $n(p-1)$. In each D_{A_i} there is just one term of degree $n(p-1)$, namely, the term $(-1)^n(x_1, \dots, x_n)^{p-1}$. Since the degree of H^* is strictly less than $n(p-1)$, the sum of all such terms must vanish, which is possible only if $s \equiv 0 \pmod p$. This is precisely the assertion of Theorem 4.

Theorem 3 follows from the theorem of Warning, since $p \geqslant 2$, and therefore if $s \neq 0$ and $s \equiv 0 \pmod p$, then $s \geqslant 2$.

1.3. *Quadratic Form Modulo a Prime*

We now apply the above results to the case of quadratic forms. The following fact is an immediate corollary of Chevalley's theorem.

Theorem 5. Let $f(x_1, \dots, x_n)$ be a quadratic form with integer coefficients. If $n \geqslant 3$, then the congruence

$$f(x_1, \dots, x_n) \equiv 0 \pmod p$$

has a nonzero solution.

The case of quadratic forms in one variable is trivial [if $a \not\equiv 0 \pmod p$, then the congruence $ax^2 \equiv 0 \pmod p$ has zero as its only solution].

We shall consider the remaining case of binary quadratic forms. We shall assume that $p \neq 2$ (in the case $n = 2, p = 2$, it is easy to examine directly all possible quadratic forms). In this case the form can be written in the form

$$f(x, y) = ax^2 + 2bxy + cy^2.$$

Its discriminant $ac - b^2$ we denote by d.

Theorem 6. The congruence

$$f(x, y) \equiv 0 \pmod{p} \qquad (p \neq 2) \tag{1.6}$$

has a nonzero solution if and only if $-d$ is either divisible by p or is a quadratic residue modulo p.

Proof. It is clear that if two forms f and f_1 are equivalent over the field Z_p (Supplement, Section 1.1), then if the congruence (1.6) has a nonzero solution for one of the forms it has a nonzero solution for both of them. Moreover, in passing from one form to an equivalent form, the discriminant changes by a square nonzero factor from the field Z_p. Hence for the proof of Theorem 6 we may replace the form f by any form equivalent to it. Since any form is equivalent to a diagonal form (Supplement, Section 1, Theorem 3), we may assume that

$$f = ax^2 + cy^2, \qquad d = ac.$$

If $a \equiv 0$ or $c \equiv 0 \pmod{p}$, the theorem is clear. If $ac \not\equiv 0 \pmod{p}$ and (1.6) has a nonzero solution (x_0, y_0), then from the congruence

$$a{x_0}^2 + c{y_0}^2 \equiv 0 \pmod{p}$$

we obtain

$$-ac \equiv \left(\frac{cy_0}{x_0}\right)^2 \pmod{p}$$

[the fraction $w \equiv u/v \pmod{p}$ denotes the result of division in the field Z_p, that is, w is a solution to the congruence $vw \equiv u \pmod{p}$]. Thus $(-d/p) = 1$. On the other hand, if $(-d/p) = 1$ and $-ac \equiv u^2 \pmod{p}$, then we set $(x_0, y_0) = (u, a)$.

PROBLEMS

1. Find the reduced polynomial, modulo p, which is equivalent to the monomial x^k.

2. Construct a cubic form $F(x_1, x_2, x_3)$ for which the congruence

$$F(x_1, x_2, x_3) \equiv 0 \pmod{2}$$

has only the zero solution.

3. Under the assumptions of Warning's theorem, show that the solutions A_i ($i = 1, \dots, s$) satisfy the congruences

$$\sum_{i=1}^{s} a_1^{(i)} \equiv \cdots \equiv \sum_{i=1}^{s} a_n^{(i)} \equiv 0 \pmod{p},$$

provided that $p \neq 2$.

4. Generalize Theorem 4 and Problem 3 to show that

$$\sum_{i=1}^{s} (a_1^{(i)})^k \equiv \cdots \equiv \sum_{i=1}^{s} (a_n^{(i)k} \equiv 0) \pmod{p}$$

for $k = 0, 1, \dots, p-2$.

5. Show that if $F_1(x_1, \dots, x_n), \dots, F_m(x_1, \dots, x_n)$ are polynomials of degrees $r_1, \dots, r_m$ with $r_1 + \cdots + r_m < n$, and the system of congruences

$$\begin{aligned} F_1(x_1, \dots, x_n) &\equiv 0 \pmod p, \\ &\cdots\cdots\cdots \\ F_m(x_1, \dots, x_n) &\equiv 0 \pmod p, \end{aligned} \tag{1.7}$$

has at least one solution, then it has at least two solutions.

6. Show that if the conditions of Problem 5 are fulfilled, then the number of solutions of the system (1.7) is divisible by p.

7. Show that if f is a quadratic form of rank $\geqslant 2$ over the field Z_p and $a \not\equiv 0 \pmod p$ then the congruence

$$f \equiv a \pmod p$$

has a solution.

8. Using Theorems 2 and 3 of Supplement, Section 1, prove that two nonsingular quadratic forms of the same rank over the field Z_p $(p \neq 2)$ are equivalent if and only if the product of their discriminants is a square.

9. Determine the Witt group of classes of quadratic forms over the field Z_p $(p \neq 2)$ (see Problem 5 of Section 1 of the Supplement).

10. Show that the number of nonzero solutions of the congruence $f(x, y) \equiv 0 \pmod p$, where $f(x, y)$ is a quadratic form with discriminant $d \not\equiv 0 \pmod p$, is equal to $(p-1)(1 + (-d/p))$.

11. Using Theorem 7 of Supplement, Section 1, show that if $f(x_1, \dots, x_n)$ is a quadratic form with discriminant $d \not\equiv 0 \pmod p$ and $p \neq 2$, then the number of nonzero solutions of the congruence $f(x_1, \dots, x_n) \equiv 0 \pmod p$ is equal to

$$\begin{aligned} &p^{n-1} - 1 + (p-1)\left(\frac{(-1)^{n/2}d}{p}\right) p^{(n/2)-1} && \text{for } n \text{ even}, \\ &p^{n-1} - 1 && \text{for } n \text{ odd}. \end{aligned}$$

12. Under the assumptions of Problem 11, find the number of solutions to the congruence

$$f(x_1, \dots, x_n) \equiv a \pmod p.$$

2. Trigonometric Sums

2.1. *Congruences and Trigonometric Sums*

In this section (as in the preceding one) we shall consider congruences modulo a prime p, but from a somewhat different point of view. In the theorems of Section 1 we drew conclusions about the number of solutions of congruences, depending on the degrees and the number of variables of the polynomials involved. Here the principal role will be played by the value of the prime modulus p.

We first note that for the equation $F(x_1, \dots, x_n) = 0$ to have a solution, it is necessary that for all m the congruence $F \equiv 0 \pmod m$ have a solution. Even if we limit our considerations to prime values of m, we still have an infinite number of necessary conditions. Clearly, these conditions can be used only if we have a finite method (a method involving a finite number of operations) for verifying them. It can be shown that for a very important class of polynomials such a method (moreover, a very simple one) exists. Namely, for a given polynomial F with integer coefficients from this class the congruence $F \equiv 0 \pmod{p}$ has solutions for all values of p larger than some bound. The polynomials of which we speak are described by the following definition.

Definition. A polynomial $F(x_1, \dots, x_n)$ with rational coefficients is called absolutely irreducible if it cannot be factored in a nontrivial manner in any extension of the field of rational numbers.

The following fundamental theorem holds.

Theorem A. If $F(x_1, \dots, x_n)$ is an absolutely irreducible polynomial with integer coefficients, then the congruence

$$F(x_1, \dots, x_n) \equiv 0 \pmod{p} \tag{2.1}$$

is solvable for all prime numbers p, larger than some bound which depends only on the polynomial F.

An analogous result holds for nonzero solutions when F is homogeneous, and (when the definition of absolute irreducibility is suitably generalized) the result generalizes to systems of congruences.

For $n = 1$ Theorem A is trivial (any polynomial of degree greater than 1 factors in the field of complex numbers and for polynomials of degree 1 the assertion is obvious). But already in the case $n = 2$ its proof requires the use of deep results from algebraic geometry. The first proof of Theorem A for $n = 2$ was given by Weil (A. Weil, Sur les courbes algébriques et les variétés qui s'en deduisent, Act. Sci. Ind. 1041, Paris, Hermann, 1948). The best versions of the proof which have appeared are contained in S. Lang, "Abelian Varieties," Wiley (Interscience), New York, 1959; and A. Mattuck and J. Tate, On an inequality of Castelnuevo-Severi, *Abhandl. Math. Sem. Hamburg* **22**, 195–199 (1959). The transition from $n = 2$ to the general case proved to be much easier. This was done in L. B. Nisnevich, On the number of points of an algebraic variety in a finite prime field, *Dokl. Akad. Nauk SSSR* **99**, No. 1, 17–20 (1954), and in S. Lang and A. Weil, Number of points of varieties in finite fields, *Am. J. Math.* **76**, No. 4, 819–827 (1954).

These last two papers actually contain a result considerably stronger than the assertion of Theorem A. Namely, they show that if the form F is fixed and the prime modulus p varies, then the number N of solutions to the congruence (2.1) becomes arbitrarily large as p increases, and they even give an estimate for the rate of increase of N. Their result can be formulated precisely as follows.

Theorem B. The number $N(F, p)$ of solutions of the congruence (2.1) satisfies the inequality

$$|N(F, p) - p^{n-1}| < C(F)p^{n-1-(1/2)},$$

where the constant $C(F)$ depends only on the polynomial F and not on p.

All known proofs of Theorem A go by way of Theorem B. But the proof of Theorem B demands an algebraic apparatus much more complex than any which we shall use in this book. Therefore we shall not give the proofs of Theorems A and B but instead shall describe a method which can be used to prove these theorems in certain cases, and we shall work out one of these cases.

All our work will be based on the fact that the number of solutions to (2.1) can be given in an explicit formula, or more precisely, can be represented as a sum of certain pth roots of unity. Sums of this type are called *trigonometric*.

We set up the following notations. If $f(x)$ or $f(x_1, \dots, x_n)$ are complex-valued functions whose value depends only on the residue class of the integers $x, x_1, \dots, x_n$ modulo p, then by

$$\sum_x f(x) \qquad \text{and} \qquad \sum_{x_1,\dots,x_n} f(x_1, \dots, x_n)$$

we denote the sums where the values of x and of $x_1, \dots, x_n$ are taken from a full system of residues modulo p, and by

$$\sum_x{}' f(x)$$

the sum where x takes all values from a reduced system of residues.

Let ζ be some fixed primitive pth root of 1. Then it is clear that

$$\sum_x \zeta^{xy} = \begin{cases} p & \text{for} \quad y \equiv 0 \pmod{p}, \\ 0 & \text{for} \quad y \not\equiv 0 \pmod{p}. \end{cases} \tag{2.2}$$

It is this equation which makes it possible to find an explicit formula for the number of solutions of the congruence (2.1).

Consider the sum

$$S = \sum_{x_1,\dots,x_n} \sum_x \zeta^{xF(x_1,\dots,x_n)}.$$

If the values of $x_1, \dots, x_n$ give a solution of (2.1), then, by (2.2),

$$\sum_x \zeta^{xF(x_1,\dots,x_n)} = p.$$

The sum of all such terms entering into S is therefore equal to Np, where N is the number of solutions to the congruence (2.1). If $F(x_1, \dots, x_n) \not\equiv 0 \pmod{p}$, then again, by (2.2),

$$\sum_{x} \zeta^{xF(x_1,\ldots,x_n)} = 0.$$

The sum of all such terms in the formula for S is then zero and we have that $S = Np$. We have thus proved

Theorem 1. The number N of solutions to the congruence (2.1) is given by the formula

$$N = \frac{1}{p} \sum_{x,x_1,\ldots,x_n} \zeta^{xF(x_1,\ldots,x_n)}. \tag{2.3}$$

All terms in which $x \equiv 0 \pmod{p}$ enter into the sum (2.3). Since each such term is equal to 1, and they are p^n in number (each of the variables $x_1, \ldots, x_n$ taking on p different values independently), then

$$N = p^{n-1} + \frac{1}{p} \sum_{x}{}' \sum_{x_1,\ldots,x_n} \zeta^{xF(x_1,\ldots,x_n)}. \tag{2.4}$$

In this form of the formula for N we see a suggestion of Theorem B. The term p^{n-1} is already singled out. We must only show (but this is where all the difficulties lie!) that as p increases the sum of all remaining terms increases in absolute value more slowly than does the principal term p^{n-1}.

2.2. *Sums of Powers*

We now apply the general method of the preceding section to the case when the polynomial F is equal to a sum of powers of the variables, i.e.,

$$F(x_1, \ldots, x_n) = a_1 x_1^{r_1} + \cdots + a_n x_n^{r_n}, \qquad a_i \not\equiv 0 \pmod{p}.$$

We shall assume that $n \geqslant 3$, since for $n = 1$ and $n = 2$ the number of solutions of the congruence $F \equiv 0 \pmod{p}$ can be found by an elementary method.

By formula (2.4) the number N of solutions to the congruence $a_1 x_1^{r_1} + \cdots + a_n x_n^{r_n} \equiv 0 \pmod{p}$ is given by the expression

$$N = p^{n-1} + \frac{1}{p} \sum_{x}{}' \sum_{x_1,\ldots,x_n} \zeta^{x(a_1 x_1^{r_1} + \cdots + a_n x_n^{r_n})},$$

which can be written in the form

$$N = p^{n-1} + \frac{1}{p} \sum_{x}{}' \prod_{i=1}^{n} \sum_{x_i} \zeta^{a_i x x_i^{r_i}}. \tag{2.5}$$

Hence we must investigate sums of the form

$$\sum_{y} \zeta^{ay^r} (a \not\equiv 0 \pmod{p}).$$

Clearly,

$$\sum_{y} \zeta^{ay^r} = \sum_{x} m(x)\zeta^{ax}, \tag{2.6}$$

where $m(x)$ is the number of solutions to the congruence $y^r \equiv x \pmod{p}$. It is clear that $m(0) = 1$. We shall find an explicit formula for $m(x)$ when $x \not\equiv 0 \pmod{p}$.

If g is a primitive root modulo p, then

$$x \equiv g^k \pmod{p}, \tag{2.7}$$

where the exponent k is uniquely determined modulo $p - 1$. Let $y \equiv g^u \pmod{p}$. The congruence $y^r \equiv x \pmod{p}$ is then equivalent to the congruence

$$ru \equiv k \pmod{p-1}. \tag{2.8}$$

By the theory of congruences of the first degree, the congruence (2.8) has $d = (r, p-1)$ solutions in u if d divides k, and otherwise has no solution. Hence

$$m(x) = \begin{cases} d & \text{if } \; k \equiv 0 \pmod{d}, \\ 0 & \text{if } \; k \not\equiv 0 \pmod{d}. \end{cases} \tag{2.9}$$

We shall find another, more convenient formula for $m(x)$. Let ε be a primitive dth root of 1, and for all integers x which are relatively prime to p, we define the functions χ_s $(s = 0, 1, \ldots, d-1)$, by setting

$$\chi_s(x) = \varepsilon^{ks}, \tag{2.10}$$

where k is determined by the congruence (2.7) (since $\varepsilon^{p-1} = 1$ the value of ε^{ks} does not depend on the choice of k). If $k \equiv 0 \pmod{d}$, then $\varepsilon^{ks} = 1$ for all $s = 0, 1, \ldots, d-1$ and hence the sum

$$\sum_{s=0}^{d-1} \chi_s(x)$$

is equal to d. If $k \not\equiv 0 \pmod{d}$, then $\varepsilon^k \neq 1$, and therefore

$$\sum_{s=0}^{d-1} \varepsilon^{ks} = \frac{\varepsilon^{kd} - 1}{\varepsilon^k - 1} = 0.$$

Comparing with (2.9) we obtain (for x not divisible by p) the formula

$$m(x) = \sum_{s=0}^{d-1} \chi_s(x).$$

Using this expression for $m(x)$ we may write the equality (2.6) in the form

$$\sum_{y} \zeta^{ay^r} = 1 + \sum_{x}' \sum_{s=0}^{d-1} \chi_s(x)\zeta^{ax}. \tag{2.11}$$

The functions χ_s, which satisfy

$$\chi_s(xy) = \chi_s(x)\chi_s(y), \tag{2.12}$$

are called multiplicative characters modulo p. We extend them to all values of x by setting $\chi_s(x) = 0$ if p divides x. The property (2.12) obviously still holds after this extension. The character χ_0, which takes the value 1 whenever $p \nmid x$, is called the *unit character*.

We isolate in the sum (2.11) the term corresponding to the unit character. Since

$$1 + \sum_x{}' \zeta^{ax} = \sum_x \zeta^{ax} = 0,$$

we may write (2.11) in the form

$$\sum_y \zeta^{ay^r} = \sum_{s=1}^{d-1} \sum_x \chi_s(x)\zeta^{ax} \tag{2.13}$$

[here we may assume that x runs through a full system of residues modulo p, since $\chi_s(x) = 0$ for $x \equiv 0 \pmod p$)].

Let χ be one of the characters χ_s and a an integer. The expression

$$\sum_x \chi(x)\zeta^{ax}$$

is called a Gaussian sum and is denoted by $\tau_a(\chi)$. Formulas (2.5) and (2.13) allow us to formulate the following theorem.

Theorem 2. Let N be the number of solutions of the congruence

$$a_1x_1^{r_1} + \cdots + a_nx_n^{r_n} \equiv 0 \pmod p), \qquad a_i \not\equiv 0 \pmod p). \tag{2.14}$$

Then

$$N = p^{n-1} + \frac{1}{p}\sum_x{}' \prod_{i=1}^{n} \sum_{s=1}^{d_i-1} \tau_{a_ix}(\chi_{i,s}), \tag{2.15}$$

where $d_i = (r_i, p-1)$ and the character $\chi_{i,s}$ is defined by (2.10) with $d = d_i$.

We note that if at least one of the d_i is equal to 1, i.e., r_i is relatively prime to $p - 1$, then the corresponding interior sum in (2.15) equals zero (as a summation over an empty set of quantities). Hence in this case $N = p^{n-1}$. This, however, was already clear without any computations, since for any values of $x_1, \ldots, x_{i-1}, x_{i+1}, \ldots, x_n$ there is one and only one value for x_i which will satisfy the congruence (2.14).

Theorem 2 is valuable because the absolute value of a Gaussian sum can be precisely computed. In the next section we shall show that

$$|\tau_a(\chi)| = \sqrt{p} \qquad \text{for} \quad a \not\equiv 0 \pmod p) \quad \text{and} \quad \chi \neq \chi_0$$

(see also Problem 8).

We now apply this fact to the result of Theorem 2.

From (2.15) it follows that

$$|N - p^{n-1}| \leqslant \frac{1}{p}\sum_x{}' \prod_{i=1}^{n} \sum_{s=1}^{d_i - 1} |\tau_{a_i x}(\chi_{i,s})|$$

$$= \frac{1}{p}(p-1)\prod_{i=1}^{n}(d_i - 1)p^{1/2} = (p-1)p^{(n/2)-1}\prod_{i=1}^{n}(d_i - 1).$$

We thus obtain the following theorem.

Theorem 3. Let N denote the number of solutions to the congruence

$$a_1 x_1^{r_1} + \cdots + a_n x_n^{r_n} \equiv 0 \pmod{p}.$$

Then for each prime number p which does not divide $a_1, \ldots, a_n$,

$$|N - p^{n-1}| \leqslant C(p-1)p^{(n/2)-1}, \tag{2.16}$$

where $C = (d_1 - 1)\cdots(d_n - 1)$, $d_i = (r_i, p-1)$.

When $n \geqslant 3$ (and we have assumed that this is the case) Theorem 3 implies Theorem B for polynomials of the above type. Indeed

$$|N - p^{n-1}| \leqslant C(p-1)p^{(n/2)-1} \leqslant Cp^{n-1-(1/2)},$$

which is the assertion of Theorem B.

We note in passing that when $n > 3$ the inequality (2.16) is much stronger than that of Theorem B.

Remark. For the proof of Theorem 3 it would suffice, by (2.5), to find a bound for the absolute value of the sum $\sum_x \zeta^{ax^r}$. Such a bound can be found, moreover, by a shorter route, without the use of Gaussian sums (see Problems 9 to 12, for which the authors thank N. M. Korobov). We have chosen the proof involving Gaussian sums because Gaussian sums have many other uses in number theory.

2.3. *The Absolute Value of Gaussian Sums*

Consider the set $\mathfrak{F}$ of all complex functions $f(x)$, defined for rational integers x, and satisfying the condition: $f(x) = f(y)$ if $x \equiv y \pmod{p}$. Since each function $f(x) \in \mathfrak{F}$ is determined by its values on a full system of residues modulo p, $\mathfrak{F}$ is a p-dimensional linear space over the field of complex numbers. We introduce a Hermitian inner product on $\mathfrak{F}$ by setting

$$(f, g) = \frac{1}{p}\sum_x f(x)\overline{g(x)} \qquad (f, g \in \mathfrak{F}).$$

It is easily checked that with respect to this inner product the p functions

$$f_a(x) = \zeta^{-ax} \qquad (a \text{ a residue} \pmod p) \tag{2.17}$$

form an orthonormal basis for $\mathfrak{F}$. Indeed, by (2.2),

$$(f_a, f_{a'}) = \frac{1}{p} \sum_x \zeta^{(a'-a)x} = \begin{cases} 1 & \text{for} \quad a \equiv a' \pmod p, \\ 0 & \text{for} \quad a \not\equiv a' \pmod p. \end{cases}$$

The functions (2.17), which satisfy

$$f_a(x + y) = f_a(x)\, f_a(y),$$

are called additive characters modulo p. We shall find the coordinates of a multiplicative character χ with respect to the basis (2.17). Let

$$\chi = \sum_a \alpha_a f_a. \tag{2.18}$$

Then

$$\alpha_a = (\chi, f_a) = \frac{1}{p} \sum_x \chi(x)\zeta^{ax} = \frac{1}{p}\, \tau_a(\chi). \tag{2.19}$$

We thus see that the Gaussian sums $\tau_a(\chi)$ appear (multiplied by $1/p$) as the coefficients of the multiplicative character χ with respect to the basis of additive characters f_a.

To obtain an important relation between the coordinates α_a [and thus between the Gaussian sums $\tau_a(\chi)$], we multiply the equation

$$\chi(x) = \sum_a \alpha_a f_a(x) \tag{2.20}$$

by $\chi(c)$, where $c \not\equiv 0 \pmod p$, and change the index of summation from a to ac

$$\chi(cx) = \sum_a \chi(c)\alpha_{ac} f_{ac}(x) = \sum_a \chi(c)\alpha_{ac} f_a(cx).$$

Comparing this with (2.20), we obtain

$$\alpha_a = \chi(c)\alpha_{ac}. \tag{2.21}$$

Setting $a = 1$ here and noting that $|\chi(c)| = 1$, we find

$$|\alpha_c| = |\alpha_1| \qquad \text{for} \quad c \not\equiv 0 \pmod p. \tag{2.22}$$

We now assume that the character χ is not the unit character χ_0. Then the number c (relatively prime to p) can be chosen so that $\chi(c) \neq 1$, and in (2.21) with $a = 0$, we find that

$$\alpha_0 = 0. \tag{2.23}$$

We now prove our principal result on the absolute value of Gaussian sums.

Theorem 4. If χ is a multiplicative character modulo p, distinct from the unit character χ_0, and a is an integer relatively prime to χ, then

$$|\tau_a(\chi)| = \sqrt{p}.$$

Proof. We evaluate the inner product (χ, χ) in the space $\mathfrak{F}$. Since $|\chi(x)| = 1$ for $x \not\equiv 0 \pmod{p}$,

$$(\chi, \chi) = \frac{1}{p} \sum_x \chi(x)\overline{\chi(x)} = \frac{p-1}{p}.$$

On the other hand, using (2.18) and considering (2.22) and (2.23) we find

$$(\chi, \chi) = \sum_a |\alpha_a|^2 = (p-1)|\alpha_c|^2.$$

The two results combine to give us

$$|\alpha_c| = \frac{1}{\sqrt{p}} \qquad (c \not\equiv 0 \pmod{p}),$$

from which, by (2.19), the theorem follows.

PROBLEMS

1. Show that $F = x^2 + y^2$, Theorem A (with respect to nonzero solutions) does not hold, and if $F = x^2 - y^2$ Theorem B does not hold. These polynomials, of course, are not absolutely irreducible.

2. Let $\varphi(x)$ be a function, defined for integers x relatively prime to p, and taking nonzero complex values. If $\varphi(x) = \varphi(y)$ when $x \equiv y \pmod{p}$ and $\varphi(xy) = \varphi(x)\varphi(y)$ for all x and y, show that this function is one of the functions $\chi_s(x) = \varepsilon^{ks}$ where ε is a primitive $(p-1)$th root of 1 and k is determined by (7).

3. Show that any complex function $f(x)$ which is nonzero, which depends only on the residue class modulo p of the integer x, and which satisfies

$$f(x+y) = f(x)f(y),$$

has the form $f(x) = \zeta^{tx}$, where t is an integer and ζ is a fixed pth root of 1.

4. Let $p \neq 2$. Show that the character $\chi = \chi_1$, defined by (10) for $d = 2$ (and $s = 1$), coincides with the Legendre symbol

$$\chi(x) = \left(\frac{x}{p}\right).$$

(This character is called the *quadratic character* modulo p.)

5. Let $ab \not\equiv 0 \pmod{p}$ and let χ be the quadratic character modulo $p \neq 2$. For the Gaussian sums $\tau_a(\chi)$ and $\tau_b(\chi)$ prove the relation

$$\tau_a(\chi)\tau_b(\chi) = \left(\frac{-ab}{p}\right) p.$$

6. Under the same conditions show that

$$\sum_x{}' \tau_x(\chi) = 0.$$

7. Solve Problems 10, 11, and 12 of Section 1, using Theorem 2 and the results of Problems 5 and 6.

8. Let χ be an arbitrary multiplicative character modulo p, distinct from χ_0, and let $a \not\equiv 0 \pmod{p}$. Show that

$$|\tau|(\chi)_b{}^2 = \tau_a(\chi)\overline{\tau_a(\chi)} = p.$$

and use this result to give a new proof of Theorem 4.

9. Let $f(x)$ be a polynomial with integer coefficients and let ζ be a primitive mth root of 1. Set $S_a = \Sigma_{x \bmod m}\, \zeta^{af(x)}$. Show that

$$\sum_{a \bmod m} |S_a|^2 = m \sum_{c \bmod m} N(c)^2,$$

where $N(c)$ denotes the number of solutions of the congruence $f(x) \equiv c \pmod{m}$.

10. Denote by ζ a primitive pth root of 1, and set $T_a = \Sigma_x \zeta^{ax^r}$. Show that

$$\sum_a{}' |T_a|^2 = p(p-1)(d-1),$$

where $d = (r, p-1)$.

11. Using the same notations, show that the sums T_a, $a \not\equiv 0 \pmod{p}$, fall into d sets, each with $(p-1)/d$ equal sums. Using this and Problem 10, show that

$$|T_a| < d\sqrt{p}, \qquad a \not\equiv 0 \pmod{p}.$$

12. Using also the fact that $\Sigma_a{}' T_a = 0$, obtain the more precise estimate

$$|T_a| \leqslant (d-1)\sqrt{p}, \qquad a \not\equiv 0 \pmod{p}.$$

[By formula (2.5) this bound gives us another proof of Theorem 3.]

13. Show that the congruence

$$3x^3 + 4y^3 + 5z^3 \equiv 0 \pmod{p}$$

has a nonzero solution for every prime p.

3. *p*-Adic Numbers

3.1. *p-Adic Integers*

We now turn to congruences modulo a power of a prime. We start with an example. Consider the congruence

$$x^2 \equiv 2 \pmod{7^n}$$

modulo a power of the prime 7. For $n = 1$ the congruence has two solutions,

$$x_0 \equiv \pm 3 \pmod{7}. \tag{3.1}$$

Now set $n = 2$. From

$$x^2 \equiv 2 \pmod{7^2} \tag{3.2}$$

it follows that $x^2 \equiv 2 \pmod 7$, and hence any solution of (3.2) must be of the form $x_0 + 7t_1$, where x_0 is a number satisfying the congruence (3.1). We now look for a solution in the form $x_1 = 3 + 7t_1$. (Solutions of the type $-3 + 7t_1$ are found in precisely the same way.) Substituting this expression for x_1 in (3.2), we obtain

$$\begin{aligned} (3 + 7t_1)^2 &\equiv 2 \pmod{7^2}, \\ 9 + 6\cdot 7t_1 + 7^2 t_1^{\,2} &\equiv 2 \pmod{7^2}, \\ 1 + 6t_1 &\equiv 0 \pmod 7, \\ t_1 &= 1 \pmod 7. \end{aligned}$$

We thus have the solution $x_1 \equiv 3 + 7\cdot 1 \pmod{7^2}$. Similarly, when $n = 3$ we have $x_2 = x_1 + 7^2 t_2$ and from the congruence

$$(3 + 7 + 7^2 t_2)^2 \equiv 2 \pmod{7^3}$$

we find that $t_2 \equiv 2 \pmod 7$; that is,

$$x_2 \equiv 3 + 7\cdot 1 + 7^2\cdot 2 \pmod{7^3}.$$

It is easily seen that this process can be continued indefinitely. We obtain a sequence

$$x_0, x_1, \ldots, x_n, \ldots, \tag{3.3}$$

satisfying the conditions

$$\begin{aligned} x_0 &\equiv 3 \pmod 7, \\ x_n &\equiv x_{n-1} \pmod{7^n}, \\ x_n^{\,2} &\equiv 2 \pmod{7^{n+1}}. \end{aligned}$$

The construction of the sequence (3.3) is reminiscent of the process for finding the square root of 2. Indeed, the computation of $\sqrt{2}$ consists of finding a sequence of rational numbers $r_0, r_1, \ldots, r_n, \ldots$, the squares of which converge to 2, for example:

$$|r_n^{\,2} - 2| < \frac{1}{10^n}.$$

In our case we construct a sequence of integers $x_0, x_1, \ldots, x_n, \ldots$, for which $x_n^{\,2} - 2$ is divisible by 7^{n+1}. This analogy becomes more precise if we say that two integers are close (more precisely, p-close, where p is some prime), when their difference is divisible by a sufficiently large power of p. With this concept of closeness we can say that the squares of the numbers in the sequence (3.3) become arbitrarily 7-close to 2 as n increases.

By giving the sequence $\{r_n\}$ we determine the real number $\sqrt{2}$. One might suppose that the sequence (3) also determines a number α, of a different type, such that $\alpha^2 = 2$.

We now note the following fact. If the sequence $\{r_n'\}$ of rational numbers satisfies $|r_n - r_n'| < 1/10^n$ for all n, then its limit is also $\sqrt{2}$. One would naturally assume that a sequence $\{x_n'\}$, for which $x_n \equiv x_n' \pmod{7^{n+1}}$, would determine the same new number α [the new sequence $\{x_n'\}$ clearly, also satisfies $x_n'^2 \equiv 2 \pmod{7^{n+1}}$ and $x_n' \equiv x_{n-1}' \pmod{7^n}$].

These remarks lead to the following definition.

Definition. Let p be some prime number. A sequence of integers

$$\{x_n\} = \{x_0, x_1, \dots, x_n, \dots\},$$

satisfying

$$x_n \equiv x_{n-1} \pmod{p^n} \tag{3.4}$$

for all $n \geqslant 1$, determines an object called a *p-adic integer*. Two sequences $\{x_n\}$ and $\{x_n'\}$ determine the same p-adic integer if and only if

$$x_n \equiv x_n' \pmod{p^{n+1}}$$

for all $n \geqslant 0$.

If the sequence $\{x_n\}$ determines the p-adic integer α, we shall write

$$\{x_n\} \to \alpha.$$

The set of all p-adic integers will be denoted by O_p. To distinguish them from p-adic integers, ordinary integers will be called *rational integers*.

Each rational integer x is associated with a p-adic integer, determined by the sequence $\{x, x, \dots, x, \dots\}$. The p-adic integer corresponding to the rational integer x will also be denoted by x. Two distinct rational integers x and y correspond to distinct p-adic integers. Indeed, if they are equal as p-adic integers, then $x \equiv y \pmod{p^n}$ for all n, which is possible only if $x = y$. Hence we may assume that the set Z of all rational integers is a subset of the set O_p of all p-adic integers.

To clarify the nature of the set O_p, we shall describe a method for choosing, from the set of all possible sequences which determine a given p-adic integer, one standard sequence.

Let a p-adic integer be given by the sequence $\{x_n\}$. Denote the smallest nonnegative integer, congruent to x_n modulo p^{n+1} by $\bar{x}_n$

$$x_n \equiv \bar{x}_n \pmod{p^{n+1}}, \tag{3.5}$$

$$0 \leqslant \bar{x}_n < p^{n+1}. \tag{3.6}$$

The congruence (3.6) shows that

$$\bar{x}_n \equiv x_n \equiv x_{n-1} \equiv \bar{x}_{n-1} \pmod{p^n},$$

so that the sequence $\{\bar{x}_n\}$ determines some p-adic integer, which by (3.5) is the same as that determined by the sequence $\{x_n\}$. A sequence, each term of which satisfies conditions (3.4) and (3.6), will be called *canonical.* Hence we have shown that every p-adic integer is determined by some canonical sequence.

It is easy to see that two distinct canonical sequences determine distinct p-adic integers. If the canonical sequences $\{\bar{x}_n\}$ and $\{\bar{y}_n\}$ determine the same p-adic integer, then from the congruence

$$\bar{x}_n \equiv \bar{y}_n \pmod{p^{n+1}}$$

and the conditions $0 \leqslant \bar{x}_n < p^{n+1}$, $0 \leqslant \bar{y}_n < p^{n+1}$, we obtain $\bar{x}_n = \bar{y}_n$ for all $n \geqslant 0$. Thus the p-adic integers are in one-to-one correspondence with the canonical sequences. From (3.4) it follows that $\bar{x}_{n+1} = \bar{x}_n + a_{n+1}p^{n+1}$, and since $0 \leqslant \bar{x}_{n+1} < p^{n+2}$ and $0 \leqslant \bar{x}_n < p^{n+1}$, we have $0 \leqslant a_{n+1} < p$. Hence every canonical sequence has the form

$$\{a_0, a_0 + a_1 p, a_0 + a_1 p + a_2 p^2, \dots\},$$

where $0 \leqslant a_i < p$. On the other hand, every sequence of this type is a canonical sequence, which determines some p-adic integer. From this it follows that the set of all canonical sequences, and also the set of all p-adic integers, have the cardinality of the continuum.

3.2. *The Ring of p-Adic Integers*

Definition. Let the p-adic integers α and β be determined by the sequences $\{x_n\}$ and $\{y_n\}$. Then the sum (respectively, product) of α and β is the p-adic integer determined by the sequence $\{x_n + y_n\}$ (respectively, $\{x_n y_n\}$).

To verify that this definition makes sense, we must show that the sequences $\{x_n + y_n\}$ and $\{x_n y_n\}$ do indeed determine some p-adic integer, and that this integer depends only on α and β and not on the choice of the sequences which determine them. Both of these assertions are easily verified, and we shall omit the details.

It is now obvious that under these operations the set of p-adic integers becomes a commutative ring, which contains the ring of rational integers as a subring.

Divisibility of p-adic integers is defined as in any commutative ring (see the Supplement, Section 4.1); α divides β if there is a p-adic integer γ such that $\beta = \alpha\gamma$. To investigate the divisibility properties of p-adic integers we must know for which p-adic integers there exists a multiplicative inverse. Such numbers, by Section 4.1 of the Supplement, are called *divisors of unity* or *units.* We shall call them p-adic units.

Theorem 1. A p-adic integer α, which is determined by a sequence $\{x_0, x_1, \ldots, x_n, \ldots\}$, is a unit if and only if $x_0 \not\equiv 0 \pmod{p}$.

Proof. Let α be a unit. Then there is a p-adic integer β such that $\alpha\beta = 1$. If β is determined by the sequence $\{y_n\}$, then the fact that $\alpha\beta = 1$ implies that

$$x_n y_n \equiv 1 \pmod{p^{n+1}}. \tag{3.7}$$

In particular, $x_0 y_0 \equiv 1 \pmod{p}$ and hence $x_0 \not\equiv 0 \pmod{p}$. Conversely, let $x_0 \not\equiv 0 \pmod{p}$. From (3.4) it easily follows that

$$x_n \equiv x_{n-1} \equiv \cdots \equiv x_0 \pmod{p},$$

so that $x_n \not\equiv 0 \pmod{p}$. Consequently, for any n, we may find a y_n such that (3.7) holds. Since $x_n \equiv x_{n-1} \pmod{p^n}$ and $x_{n-n} \equiv x_{n-1} y_{n-1} \pmod{p^n}$, then also $y_n \equiv y_{n-1} \pmod{p^n}$. This means that the sequence $\{y_n\}$ determines some p-adic integer β. Equation (3.7) implies that $\alpha\beta = 1$, which means that α is a unit.

From this theorem it follows that a rational integer a, considered as an element of O_p, is a unit if and only if $a \not\equiv 0 \pmod{p}$. If this condition holds, then a^{-1} belongs to O_p. Hence any rational integer b is divisible by such an a in O_p, that is, any rational number of the form b/a, where a and b are integers and $a \not\equiv 0 \pmod{p}$, belongs to O_p. Rational numbers of this type are called p-integers. They clearly form a ring. We can now formulate the above result as follows:

Corollary. The ring O_p of p-adic integers contains a subring isomorphic to the ring of p-integral rational numbers.

Theorem 2. Every p-adic integer, distinct from zero, has a unique representation in the form

$$\alpha = p^m \varepsilon, \tag{3.8}$$

where ε is a unit of the ring O.

Proof. If α is a unit, then (3.8) holds with $m = 0$. Let $\{x_n\} \to \alpha$, where α is not a unit, so that by Theorem 1, $x_0 \equiv 0 \pmod{p}$. Since $\alpha \neq 0$, the congruence $x_n \equiv 0 \pmod{p^{n+1}}$ does not hold for all n. Let m be the smallest index for which

$$x_m \not\equiv 0 \pmod{p^{m+1}}. \tag{3.9}$$

For any $s \geqslant 0$,

$$x_{m+s} \equiv x_{m-1} \equiv 0 \pmod{p^m}.$$

and therefore the number $y_s = x_{m+s}/p^m$ is an integer. From the congruences

$$p^m y_s - p^m y_{s-1} = x_{m+s} - x_{m+s-1} \equiv 0 \pmod{p^{m+s}},$$

it follows that

$$y_s \equiv y_{s-1} \pmod{p^s}$$

for all $s \geqslant 0$. Thus the sequence $\{y_s\}$ determines some $\varepsilon \in O_p$. Since $y_0 = x_m/p^m \not\equiv 0 \pmod{p}$, ε is a unit by Theorem 1. Finally, from

$$p^m y_s = x_{m+s} \equiv x_s \pmod{p^{s+1}}$$

it follows that $p^m\varepsilon = \alpha$, which is the desired representation.

We assume now that α has another representation $\alpha = p^k\eta$, where $k \geqslant 0$ and η is a unit. If $\{z_s\} \to \eta$, then

$$p^m y_s \equiv p^k z_s \pmod{p^{s+1}} \tag{3.10}$$

for all $s \geqslant 0$, and, by Theorem 1, p never divides y_s or z_s, since ε and η are units. Setting $s = m$ in (3.10), we obtain

$$p^m y_m \equiv p^k z_m \not\equiv 0 \pmod{p^{m+1}},$$

from which we deduce that $k \leqslant m$. By symmetry we also have $m \leqslant k$, i.e., $k = m$. Replacing s by $s + m$ in (3.10) and dividing by p^m we find that

$$y_{m+s} \equiv z_{m+s} \pmod{p^{s+1}}.$$

Since by condition (3.4) $y_{m+s} \equiv y_s \pmod{p^{s+1}}$ and $z_{m+s} \equiv z_s \pmod{p^{s+1}}$, we obtain

$$y_s \equiv z_s \pmod{p^{s+1}}.$$

Since this congruence holds for all $s \geqslant 0$, $\varepsilon = \eta$, and Theorem 2 is proved.

Corollary 1. The p-adic integer α, determined by the sequence $\{x_n\}$, is divisible by p^k if and only if $x_n \equiv 0 \pmod{p^{n+1}}$ for all $n = 0, 1, \ldots, k-1$.

Indeed, we find the exponent m of expression (3.8) as the smallest index m for which (3.9) holds.

Corollary 2. The ring O_p does not have any zero divisors.

If $\alpha \neq 0$ and $\beta \neq 0$, then we have the representations

$$\alpha = p^m\varepsilon, \qquad \beta = p^k\eta,$$

in which ε and η are units. (Thus ε and η have inverses ε^{-1} and η^{-1} in the ring O_p.) If we had $\alpha\beta = 0$, then, multiplying the equation $p^{m+k}\varepsilon\eta = 0$ by $\varepsilon^{-1}\eta^{-1}$, we would obtain $p^{m+k} = 0$, which is impossible.

Definition. The number m in the representation (3.8) of a nonzero p-adic integer α is called the p-adic value, or simply the p-value, of α and is denoted by $v_p(\alpha)$.

In case it is clear which prime p is intended, we shall speak simply of the value, and write $v(\alpha)$. In order that the function $v(\alpha)$ be defined for all p-adic integers, we set $v(0) = \infty$. (This convention is appropriate since 0 is divisible by arbitrarily high powers of p.)

It is easy to verify the following properties of the value function:

$$v(\alpha\beta) = v(\alpha) + v(\beta); \tag{3.11}$$

$$v(\alpha + \beta) = \min(v(\alpha), v(\beta)); \tag{3.12}$$

$$v(\alpha + \beta) = \min(v(\alpha), v(\beta)) \qquad \text{if } v(\alpha) \neq v(\beta). \tag{3.13}$$

The divisibility properties of p-adic integers are concisely expressed in terms of the value function. From Theorem 2 we immediately deduce

Corollary 3. The p-adic integer α is divisible by β if and only if $v(\alpha) \geqslant v(\beta)$.

Thus the arithmetic of the ring O_p is very simple. There is a unique (up to associates) prime element, namely, p. Every nonzero element of O_p is a product of a power of p and a unit.

Finally, we turn to congruences in the ring O_p. Congruence of elements is defined here exactly as it is for rational integers, or, more generally, for elements of any ring (see the Supplement, Section 4.1): $\alpha \equiv \beta \pmod{\gamma}$ means that $\alpha - \beta$ is divisible by γ. If $\gamma = p^n \varepsilon$, where ε is a unit, then any congruence modulo γ is equivalent to a congruence modulo p^n. We thus confine our attention to congruences modulo p^n.

Theorem 3. Any p-adic integer is congruent to a rational integer modulo p^n. Two rational integers are congruent modulo p^n in the ring O_p if and only if they are congruent modulo p^n in the ring Z.

Proof. To prove the first assertion we shall show that if α is a p-adic integer and $\{x_n\}$ is a sequence of rational numbers determining α, then

$$\alpha \equiv x_{n-1} \pmod{p^n}. \tag{3.14}$$

Since x_{n-1} is determined by the sequence $\{x_{n-1}, x_{n-1}, \dots\}$, the sequence $\{x_0 - x_{n-1}, x_1 - x_{n-1}, \dots\}$ determines the number $\alpha - x_{n-1}$. We apply Corollary 1 of Theorem 2 to the p-adic integer $\alpha - x_{n-1}$. We see that the congruence (3.14) is equivalent to the congruence

$$x_k - x_{n-1} \equiv 0 \pmod{p^{k+1}} \qquad (k = 0, 1, \dots, n-1),$$

which is in turn implied by condition (3.4) in the definition of p-adic integers.

We now show that for two rational integers x and y, congruence modulo p in the ring O_p is equivalent to congruence modulo p in the ring Z. Set

$$x - y = p^m a, \qquad a \not\equiv 0 \pmod{p} \tag{3.15}$$

(we assume that $x \neq y$). The congruence

$$x \equiv y \pmod{p^n} \tag{3.16}$$

in the ring Z is equivalent to the condition $n \leqslant m$. On the other hand, (3.15) is a representation of the type (3.8) for the number $x - y$, since a is a p-adic unit. Consequently, $v_p(x - y) = m$, and the condition $n \leqslant m$ can be written in the form $v_p(x - y) \geqslant n$. But this is equivalent to the congruence (3.16) in O_p, since $v(p^n) = n$ (see Corollary 3 of Theorem 2).

Corollary. There are p^n residue classes in O_p modulo p^n.

3.3. *Fractional p-adic Numbers*

Since the ring O_p has no zero divisors (Corollary 2 of Theorem 2), it can be embedded in a field, using the standard construction of a field from an integral domain. Application of this construction to our situation leads to consideration of fractions of the form α/p^k, where α is some p-adic integer, and $k \geqslant 0$. The fractions considered here could more suitably be written as pairs (α, p^k).

Definition. A fraction of the form α/p^k, $\alpha \in O_p$, $k \geqslant 0$, determines a fractional p-adic number, or, more simply, a p-adic number. Two fractions, α/p^k and β/p^m, determine the same p-adic number if and only if $\alpha p^m = \beta p^k$ in O_p.

The set of all p-adic numbers will be denoted by R_p.

A p-adic integer determines an element $\alpha/1 = \alpha/p^0$ in R_p. It is clear that distinct p-adic integers determine distinct elements of R_p. Hence we shall assume that O_p is a subset of the set R_p.

Addition and multiplication are defined in R_p by the rules

$$\frac{\alpha}{p^k} + \frac{\beta}{p^m} = \frac{\alpha p^m + \beta p^k}{p^{k+m}},$$

$$\frac{\alpha}{p^k}\frac{\beta}{p^m} = \frac{\alpha\beta}{p^{k+m}}.$$

It is a simple exercise to verify that the result of these operations does not depend on the choice of fractions to represent the elements of R_p, and that under these operations R_p is turned into a field—the field of all p-adic numbers. It is clear that the field R_p has characteristic zero and thus contains the field of rational numbers.

Theorem 4. Any nonzero p-adic number α is uniquely representable in the form

$$\xi = p^m \varepsilon, \tag{3.17}$$

where m is an integer and ε is a unit of O_p.

Proof. Let $\xi = \alpha/p^k$, $\alpha \in O_p$. By Theorem 2, α can be represented in the form $\alpha = p^l \varepsilon$, $l \geqslant 0$, where ε is a unit of the ring O_p. Thus $\xi = p^m \varepsilon$, where $m = l - k$. The uniqueness of the representation (3.17) follows from the corresponding assertion for p-adic integers, proved in Theorem 2.

The concept of the value of an element, introduced in Section 2, easily generalizes to any p-adic number. We set

$$v_p(\xi) = m,$$

where m is the exponent in (3.17). It is easily seen that properties (3.11), (3.12), and (3.13) of the value automatically carry over to the field R_p. The p-adic number ξ is a p-adic integer if and only if $v_p(\xi) \geqslant 0$.

3.4. *Convergence in the Field of p-Adic Numbers*

In Section 3.1 we noted the analogy between p-adic integers and real numbers, in that both are determined by sequences of rational numbers.

Just as every real number is the limit of any sequence of rational numbers which determines it, it would be natural to conjecture that the same fact should hold for p-adic numbers, if the correct definition of the concept of convergence is given. The definition of limit for real or rational numbers can be based, for example, on the notion of nearness; two real or rational numbers being near if the absolute value of their difference is small. For the definition of convergence for p-adic numbers we thus must decide under what conditions two p-adic numbers are to be considered close to one another.

In the example of the first section, we spoke of the p-nearness of two p-adic integers x and y, meaning by this that the difference of x and y should be divisible by a high power of p. It was under this definition of nearness that the analogy between the definitions of real numbers and of p-adic integers became apparent. If we use the concept of the p-value v_p, then the p-nearness of x and x will be characterized by the value of $v_p(x - y)$. Thus we may speak of two p-adic numbers ξ and η (not necessarily integers) as being near when the value of $v_p(\xi - \eta)$ is sufficiently large. Thus "small" p-adic numbers are characterized by the large value of their p-value.

After these remarks we turn to precise definitions.

Definition. The sequence

$$\{\xi_n\} = \{\xi_0, \xi_1, \dots, \xi_n, \dots\}$$

of p-adic numbers converges to the p-adic number ξ (we denote this by $\lim_{n\to\infty} \xi_n = \xi$ or $\{\xi_n\} \to \xi$) if

$$\lim_{n\to\infty} v_p(\xi_n - \xi) = \infty.$$

A singular feature of this definition (which distinguishes it from the usual definition of convergence for real numbers) is that the convergence of $\{\xi_n\}$ to ξ is determined by the sequence of rational integers $v_p(\xi_n - \xi)$, which must converge to infinity. We can put the definition in a more familiar form if, instead of v_p, we consider another nonnegative real-valued function on the field R_p, which will converge to zero as v_p goes to infinity. Namely, choose some real number ρ, satisfying $0 < \rho < 1$, and set

$$\varphi_p(\xi) = \begin{cases} \rho^{v_p(\xi)} & \text{for } \xi \neq 0, \\ 0 & \text{for } \xi = 0. \end{cases} \tag{3.18}$$

Definition. The function $\varphi_p(\xi)$, $\xi \in R_p$, defined by (3.18), is called a *p-adic metric*. The number $\varphi_p(\xi)$ is called the *p-adic size* of ξ.

As in the case of the value function, we shall sometimes simply call φ_p a *value* and denote it by φ.

Properties (3.11) and (3.12) of the value clearly imply the following properties of the metric:

$$\varphi(\xi\eta) = \varphi(\xi)\varphi(\eta); \tag{3.19}$$

$$\varphi(\xi + \eta) \leqslant \max(\varphi(\xi), \varphi(\eta)). \tag{3.20}$$

From the last inequality we also obtain

$$\varphi(\xi + \eta) \leqslant \varphi(\xi) + \varphi(\eta). \tag{3.21}$$

Properties (3.19) and (3.21) [and also the fact that $\varphi(\xi) > 0$ for $\xi \neq 0$] show that the concept of metric for p-adic numbers is analogous to the concept of absolute value in the field of real (or complex) numbers.

In terms of the valuation φ_p the definition of convergence in the field R_p takes the following form: The sequence $\{\xi_n\}$, $\xi_n \in R_p$, converges to the p-adic number ξ if

$$\lim_{n\to\infty} \varphi_p(\xi_n - \xi) = 0.$$

We may formulate and prove, for the field R_p, general theorems on the limits of sequences, well known in analysis. As an example we shall show that if $\{\xi_n\} \to \xi$ and $\xi \neq 0$, then $\{1/\xi_n\} \to 1/\xi$. First, from some point on, that is, for all $n \geqslant n_0$, we have $v(\xi_n - \xi) > v(\xi)$, from which, by (3.13), $v(\xi_n) = \min(v(\xi_n - \xi), v(\xi)) = v(\xi)$. In particular, $v(\xi_n) \neq \infty$, that is, $\xi_n \neq 0$, so that $1/\xi_n$ makes sense for all $n \geqslant n_0$. Further,

$$v\left(\frac{1}{\xi_n} - \frac{1}{\xi}\right) = v(\xi - \xi_n) - v(\xi_n) - v(\xi) = v(\xi_n - \xi) - 2v(\xi) \to \infty$$

as $n \to \infty$, and our assertion is proved.

Theorem 5. If the p-adic integer α is determined by the sequence $\{x_n\}$ of rational integers, then this sequence converges to α. An arbitrary p-adic number ξ is a limit of a sequence of rational numbers.

Proof. From the congruence (3.14) it follows that $v_p(x_n - \alpha) \geqslant n + 1$. Consequently, $v(x_n - \alpha) \to \infty$ as $n \to \infty$, and this means that $\{x_n\}$ converges to α. Consider now the fractional p-adic number $\xi = \alpha/p^k$. Since

$$v\left(\frac{x_n}{p^k} - \xi\right) = v\left(\frac{x_n - \alpha}{p^k}\right) = v(x_n - \alpha) - k \to \infty$$

as $n \to \infty$, then ξ is the limit of the rational sequence $\{x_n/p^k\}$. The theorem is proved.

From any bounded sequence of real numbers it is possible to choose a convergent subsequence. An analogous property also holds for p-adic numbers.

Definition. The sequence $\{\xi_n\}$ of p-adic numbers is called *bounded* if the numbers $\varphi_p(\xi_n)$ are bounded from above, or equivalently, if the numbers $v_p(\xi_n)$ are bounded from below.

Theorem 6. From any bounded sequence of p-adic numbers (in particular, from any sequence of p-adic integers) it is possible to choose a convergent subsequence.

Proof. We first prove the theorem for a sequence $\{\alpha_n\}$ of p-adic integers. Since in the ring O_p the number of residue classes modulo p is finite (corollary of Theorem 3), there are an infinite number of terms in the sequence $\{\alpha_n\}$ which are congruent modulo p to some rational integer x_0. Choosing all such terms, we obtain a subsequence $\{\alpha_n^{(1)}\}$, all terms of which satisfy the congruence

$$\alpha_n^{(1)} \equiv x_0 \pmod{p}.$$

Analogously, applying the corollary of Theorem 3 to the case $n = 2$, we choose from the sequence $\{\alpha_n^{(1)}\}$ a subsequence with the condition

$$\alpha_n^{(2)} \equiv x_1 \pmod{p^2}$$

where x_1 is some rational integer. Here, clearly, $x_1 \equiv x_0 \pmod{p}$. Continuing this process indefinitely, we obtain for each k a sequence $\{\alpha_n^{(k)}\}$, which is a subsequence of the preceding sequence $\{\alpha_n^{(k-1)}\}$ and for all terms of which the congruence

$$\alpha_n^{(k)} \equiv x_{n-1} \pmod{p^k},$$

holds for some rational integer x_{k-1}. Since $x_k \equiv \alpha_n^{(k+1)} \pmod{p^{k+1}}$ and all $\alpha_n^{(k+1)}$ belong among the $\alpha_n^{(k)}$,

$$x_k \equiv x_{k-1} \pmod{p^k}$$

for all $k \geqslant 1$. Thus the sequence $\{x_n\}$ determines some p-adic integer α. We now take the "diagonal" sequence $\{\alpha_n^{(n)}\}$. It clearly is a subsequence of the initial sequence $\{\alpha_n\}$. We claim that $\{\alpha_n^{(n)}\} \to \alpha$. Indeed, by (3.14) we have $\alpha \equiv x_{n-1} \pmod{p^n}$. But, on the other hand, $\alpha_n^{(n)} \equiv x_{n-1} \pmod{p^n}$ and hence $\alpha_n^{(n)} \equiv \alpha \pmod{p^n}$; that is, $\nu(\alpha_n^{(n)} - \alpha) \geqslant n$. From this it follows that $\nu(\alpha_n^{(n)} - \alpha) \to \infty$ as $n \to \infty$, and thus $\{\alpha_n^{(n)}\}$ converges to α.

We now turn to the proof of the theorem in the general case. If the sequence $\{\xi_n\}$ of p-adic numbers satisfies $\nu(\xi_n) \geqslant -k$ (k some rational integer), then for $\alpha_n = \xi_n p^k$ we have $\nu(\alpha_n) \geqslant 0$. By the above we may extract a convergent subsequence $\{\alpha_{n_i}\}$ from the sequence $\{\alpha_n\}$ of p-adic integers. But then the sequence $\{\xi_{n_i}\} = \{\alpha_{n_i} p^{-k}\}$ is a convergent subsequence of the sequence $\{\xi_n\}$. Theorem 6 is completely proved.

The Cauchy convergence criterion also holds for p-adic numbers: The sequence

$$\{\xi_n\}, \qquad \xi_n \in R_p, \tag{3.22}$$

converges if and only if

$$\lim_{m,n\to\infty} \nu(\xi_m - \xi_n) = \infty. \tag{3.23}$$

The necessity of the condition is clear. For the proof of sufficiency we first note that (3.23) implies that the sequence (3.22) is bounded. Indeed, from (3.23) it follows that there is an n_0, such that $\nu(\xi_m - \xi_{n_0}) \geqslant 0$ for all $m \geqslant n_0$. But then by (3.12) for all $m \geqslant n_0$,

$$\nu(\xi_m) = \nu((\xi_m - \xi_{n_0}) + \xi_{n_0}) \geqslant \min((0, \nu(\xi_{n_0})),$$

and from this it follows that (3.22) is bounded. By Theorem 6 we may extract from (3.22) a convergent subsequence with limit, say ξ. We now show that the sequence (3.22) converges to ξ. Let M be an arbitrarily large number. From (3.23) and the definition of convergence we can find a natural number N so that, first, $\nu(\xi_m - \xi_n) \geqslant M$ for $m, n \geqslant N$, and, second, $\nu(\xi_{n_i} - \xi) \geqslant M$ for $n_i \geqslant N$. Then

$$\nu(\xi_m - \xi) \geqslant \min(\nu(\xi_m - \xi_{n_i}), \qquad \nu(\xi_{n_i} - \xi)) \geqslant M$$

for all $m \geqslant N$. Thus $\lim_{m\to\infty} \nu(\xi_m - \xi) = \infty$, that is, the sequence (3.22) converges.

The principle of convergence proved above can be put in a stronger form. If the sequence (3.22) satisfies (3.23), then it clearly also satisfies

$$\lim_{n\to\infty} v(\xi_{n+1} - \xi_n) = \infty. \tag{3.24}$$

We shall show that, conversely, (3.24) implies (3.23). For if $v(\xi_{n+1} - \xi_n) \geqslant M$ for all $n \geqslant N$, then by (12) from the equation

$$\xi_m - \xi_n = \sum_{i=n}^{m-1} (\xi_{i+1} - \xi_i), \qquad m > n \geqslant N,$$

it follows that

$$v(\xi_m - \xi_n) \geqslant \min_{i=n,\ldots,m-1} v(\xi_{i+1} - \xi_i) \geqslant M,$$

that is, $v(\xi_m - \xi_n) \to \infty$ as $m, n \to \infty$. Thus we have

Theorem 7. For the convergence of the sequence $\{\xi_n\}$ of p-adic numbers, it is necessary and sufficient that $\lim_{n\to\infty} v(\xi_{n+1} - \xi_n) = \infty$.

Having a concept of convergence in the field R_p, we may speak of continuous p-adic functions of p-adic variables. Their definition does not differ at all from the usual one. That is, the function $F(\xi)$ is called *continuous* at $\xi = \xi_0$ if for any sequence $\{\xi_n\}$ which converges to ξ_0, the sequence of values $\{F(\xi_n)\}$ converges to $F(\xi_0)$. A similar definition holds for functions of several variables. Just as in real analysis it is easy to prove the usual theorems on arithmetic operations with continuous p-adic functions. In particular, it is easily verified that polynomials in any number of variables with p-adic coefficients are continuous p-adic functions. This simple fact will be used (Section 5.1) in the future.

To conclude this section we make some remarks on series with p-adic terms.

Definition. If the sequence of partial sums $s_n = \sum_{i=0}^{n} \alpha_i$ of the series

$$\sum_{i=0}^{\infty} \alpha_i = \alpha_0 + \alpha_1 + \cdots + \alpha_n + \cdots \tag{3.25}$$

with p-adic terms converges to the p-adic number α, then we shall say that the series converges and that its sum is α. From Theorem 7 we immediately deduce the following convergence criterion for series.

Theorem 8. In order that the series (3.25) converge, it is necessary and sufficient that the general term converge to zero, that is, that $v(\alpha_n) \to \infty$ as $n \to \infty$.

Convergent p-adic series can clearly be termwise added and subtracted and multiplied by a constant p-adic number. The associativity property of series also holds for them.

Theorem 9. If the terms of a convergent p-adic series are rearranged, its convergence is not affected and its sum does not change.

The simple proof of this theorem is left to the reader.

In analysis it is proved that the property described in Theorem 9, when applied to real numbers, characterizes absolutely convergent series. Thus every convergent p-adic series is "absolutely convergent." From this it follows that convergent p-adic series can be multiplied in the usual manner.

If the p-adic integer α is defined by the canonical sequence $\{a_0, a_0 + a_1 p, a_0 + a_1 p + a_2 p^2, \dots\}$ (Section 3.1), then, by the first assertion of Theorem 5, it will equal the sum of the convergent series

$$a_0 + a_1 p + a_2 p^2 + \cdots + a_n p^n + \cdots, \qquad 0 \leqslant a_n \leqslant p - 1 \ (n = 0, 1, \dots). \tag{3.26}$$

Since distinct canonical sequences determine distinct p-adic integers, the representation of α in the form of the series (3.26) is unique. Conversely, any series of the form (3.26) converges to some p-adic integer.

The representation of p-adic integers in series (3.26) is reminiscent of the expansion of real numbers as infinite decimals.

If we consider the series

$$b_0 + b_1 p + \cdots + b_n p^n + \cdots, \tag{3.27}$$

in which the coefficients are arbitrary rational integers, then it clearly converges [since $\nu(b_n p^n) \geqslant n$], and its sum will equal some p-adic integer α. To obtain the representation (3.26) for this number α, we must successively replace each coefficient in (3.27) by its remainder after division by p, and carry over the quotient at each step to the coefficient of the next term. This observation can be used for computations in the ring O_p. That is, after addition, subtraction, or multiplication of series of the form (3.26) according to the rules for operating with series, we shall obtain a series in the form (3.27), in which the coefficients, in general, will not be the smallest nonnegative residues modulo p. To transform this series into the form (3.26) we need only apply the rule just mentioned. This method of carrying out operations with p-adic integers is easily seen to be analogous to the usual method for operating with real numbers which are expressed as infinite decimals.

From Theorem 1 it easily follows that a p-adic integer, represented in the form of a series (3.26), is a unit in the ring O_p if and only if $a_0 \neq 0$. Along with Theorem 4 this gives us the following result.

Theorem 10. Every nonzero p-adic number ξ is uniquely representable in the form

$$\xi = p^m(a_0 + a_1 p + \cdots + a_n p^n + \cdots), \tag{3.28}$$

where $m = v_p(\xi)$, $1 \leqslant a_0 \leqslant p - 1$, $0 \leqslant a_n \leqslant p - 1$ $(n = 1, 2, \dots)$.

PROBLEMS

1. Set $x_n = 1 + p + \cdots + p^{n-1}$. Show that in the field of p-adic numbers the sequence $\{x_n\}$ converges to $1/(1-p)$.

2. Let $p \neq 2$ and let c be a quadratic residue modulo p. Show that there exist two (distinct) p-adic numbers whose squares equal c.

3. Let c be a rational integer not divisible by p. Show that the sequence $\{c^{p^n}\}$ converges in the field R_p. Show, further, that the limit γ of this sequence satisfies $\gamma \equiv c \pmod{p}$ and $\gamma^{p-1} = 1$.

4. Using the previous problem, show that the polynomial $t^{p-1} - 1$ factors into linear factors over the field R_p.

5. Represent the number -1 in the field of p-adic numbers in a series of the form (3.26).

6. Represent the number $-\frac{2}{3}$ in the form (3.26) in the field of 5-adic numbers.

7. Show that, if $p \neq 2$, there is no pth root of 1 in the field R_p, other than 1.

8. Show that the representation of any nonzero rational number in the form (3.28) has periodic coefficients (from some point on). Conversely, show that any series of the form (3.28), for which the coefficients satisfy $a_{m+k} = a_k$, for all $k \geqslant k_0 (m > 0)$, represents a rational number.

9. Prove the Eisenstein irreducibility criterion for polynomials over the field of p-adic numbers: The polynomial $f(x) = a_0 x^n + a_1 x^{n-1} + \cdots + a_n$ with p-adic integer coefficients is irreducible over the field R_p, if a_0 is not divisible by p, all other coefficients $a_1, \dots, a_n$ are divisible by p, and the constant term a_n is divisible by p but not by p^2.

10. Show that over the field of p-adic numbers there exist finite extensions of arbitrary degree.

11. Show that for distinct primes p and q the fields R_p and R_q are not isomorphic. Further show that no field R_p is isomorphic to the field of real numbers.

12. Show that the field of p-adic numbers has no automorphisms except the identity. (An analogous assertion holds for the field of real numbers.)

4. An Axiomatic Characterization of the Field of p-Adic Numbers

The fields of p-adic numbers are among the basic tools in the theory of numbers. Section 5 will be devoted to applications to some number-theoretic problems. First, we shall make a short detour to clarify the position of the p-adic fields in the general theory of fields.

4.1. *Metric Fields*

We have already remarked several times on the analogy between p-adic and real numbers. In this section we make this analogy precise. Namely, we give a general method for constructing fields, which has as special cases the constructions of the real and p-adic numbers. In the case of real numbers, this method coincides with the construction of Cantor by means of Cauchy sequences of rational numbers.

The generalization of Cantor's method to other fields is based on the following idea. Every concept or construction used in this method can be defined in terms of the concept of convergence of sequences of rational numbers. And this concept is in turn based on that of absolute value. (We say that the sequence $\{r_n\}$ of rational numbers converges to the rational number r if the absolute value of the difference $|r_n - r|$ converges to zero.) Since only certain properties of the absolute value are ever used, we might therefore suspect that, if it is possible to define a function φ from an arbitrary field k to the real numbers which has the same properties as the absolute value function, then the concept of convergence can be defined in k, and by using Cantor's method a new field can be constructed from k.

Definition. Let k be an arbitrary field. A function φ from the field k to the real numbers is called a *metric* of k, if it satisfies the following conditions:

$$(1)\ \varphi(\alpha) > 0 \quad \text{for} \quad \alpha \neq 0,\ \varphi(0) = 0;$$

$$(2)\ \varphi(\alpha + \beta) \leqslant \varphi(\alpha) + \varphi(\beta);$$

$$(3)\ \varphi(\alpha\beta) = \varphi(\alpha)\varphi(\beta).$$

The field k along with the metric given on it is called a *metric field* [and sometimes denoted by (k, φ)]. The following properties of metrics easily follow from the definition:

$$\varphi(\pm 1) = 1;$$

$$\varphi(-\alpha) = \varphi(\alpha);$$

$$\varphi(\alpha - \beta) \leqslant \varphi(\alpha) + \varphi(\beta);$$

$$\varphi(\alpha \pm \beta) \geqslant |\varphi(\alpha) - \varphi(\beta)|;$$

$$\varphi\left(\frac{\alpha}{\beta}\right) = \frac{\varphi(\alpha)}{\varphi(\beta)} \qquad (\beta \neq 0).$$

The following are examples of metrics:

(1) Absolute value in the field of rational numbers.

(2) Absolute value in the field of real numbers.

(3) Modulus or absolute value in the field of complex numbers.

(4) The p-adic metric φ_p (defined in Section 3.4) in the field of p-adic numbers R_p.

(5) The function $\varphi(\alpha)$, defined by $\varphi(0) = 0$, $\varphi(\alpha) = 1$ for $\alpha \neq 0$, for k an arbitrary field. Such a metric is called *trivial.*

If the valuation φ_p of the field R_p is considered only on the rational numbers, then another metric is obtained of the rational field R. This metric, also denoted by φ_p, is called the p-adic metric of R. Its value on the nonzero rational number $x = p^{\nu_p(x)}(a/b)$ (with a and b integers not divisible by p) is clearly given by

$$\varphi_p(x) = \rho^{\nu_p(x)}, \tag{4.1}$$

where ρ is a fixed real number satisfying $0 < \rho < 1$. We shall see below that if Cantor's construction is applied to the field of rational numbers with the p-adic metric (instead of the usual absolute value), then the field R_p will be obtained.

In any field with valuation (k, φ) we define the concept of convergence: the sequence $\{\alpha_n\}$ of elements of k is said to converge to the element $\alpha \in k$, if $\varphi(\alpha_n - \alpha) \to 0$ as $n \to \infty$. In this case we shall say that α is the limit of the sequence $\{\alpha_n\}$, and shall write $\{\alpha_n\} \to \alpha$ or $\alpha = \lim_{n\to\infty} \alpha_n$.

Definition. A sequence $\{\alpha_n\}$ of elements of a metric field with metric φ is called a *Cauchy sequence* if $\varphi(\alpha_n - \alpha_m) \to 0$ as $n, m \to \infty$.

Obviously, any convergent sequence is a Cauchy sequence. For, if $\{\alpha_n\} \to \alpha$, then by the inequality

$$\varphi(\alpha_n - \alpha_m) = \varphi(\alpha_n - \alpha + \alpha - \alpha_m) \leqslant \varphi(\alpha_n - \alpha) + \varphi(\alpha_m - \alpha),$$

$$\varphi(\alpha_n - \alpha_m) \to 0 \qquad [\text{since } \varphi(\alpha_n - \alpha) \to 0 \qquad \text{and} \quad \varphi(\alpha_m - \alpha) \to 0].$$

The converse assertion is valid for some, but not for all, metric fields. It holds for the real and for the p-adic numbers by the Cauchy convergence criterion (see Section 3.4). But it does not hold for the field R of rational numbers, either in the case of the absolute value or in the case of the p-adic metrics.

Definition. A metric field is called *complete* if every Cauchy sequence in it converges.

Cantor's method embeds the noncomplete field of rational numbers (with absolute value as metric) in the complete field of real numbers. It will be shown that such an embedding is possible for any metric field, and the proof of this assertion will consist of an almost verbatim repetition of Cantor's method.

We introduce some terminology. If we say that the metric field (k, φ) is a *subfield* of the metric field (k_1, φ_1), we mean not only that $k \subset k_1$, but also that the metric φ_1 coincides with φ on the field k. Further, a subset of the metric field k will be called *everywhere dense* in k, if every element of k is the limit of some convergent sequence of elements of this subset. Then we have

Theorem 1. For any metric field k there exists a complete metric field $\bar{k}$, which contains k as an everywhere-dense subset.

To formulate the following theorem we need one more definition.

Definition. Let (k_1, φ_1) and (k_2, φ_2) be two isomorphic metric fields. The isomorphism $\sigma : k_1 \to k_2$ is called bicontinuous, or topological, if, for any sequence $\{\alpha_n\}$ of elements of k_1, which converges to the element α under the metric φ_1, the sequence $\{\sigma(\alpha_n)\}$ converges to $\sigma(\alpha)$ under the metric φ_2, and conversely.

Theorem 2. The field $\bar{k}$, given by Theorem 1, is uniquely determined up to a topological isomorphism which leaves fixed all elements of k.

Definition. The field $\bar{k}$, the existence and uniqueness of which is established by Theorems 1 and 2, is called the *completion* of the metric field k.

The field of real numbers is clearly the completion of the field of rational numbers, with the ordinary absolute value as metric. If instead the p-adic metric (4.1) is used with the rational field, then the completion is the field R_p of p-adic numbers. For the second assertion of Theorem 5 of Section 3 shows that R is everywhere dense in R_p, and the Cauchy convergence criterion (Theorem 7 of Section 3) states that R_p is complete. We thus have a new axiomatic description of the field of p-adic numbers: The field of p-adic numbers is the completion of the field of rational numbers under the p-adic metric (4.1).

We now turn to the proofs of Theorems 1 and 2. We shall only sketch the proofs, skipping those parts which are verbatim repetitions of the corresponding arguments in real analysis.

Proof of Theorem 1. We call two Cauchy sequences $\{x_n\}$ and $\{y_n\}$ of elements of the metric field (k, φ) equivalent if $\{x_n - y_n\} \to 0$.

We denote the set of all equivalence classes of Cauchy sequences by $\bar{k}$. In $\bar{k}$ we define the operations of addition and multiplication as follows: if α and β are any two classes and $\{x_n\} \in \alpha$ and $\{y_n\} \in \beta$ are any Cauchy sequences in these classes, then the sum (respectively product) of these classes is the class which contains the sequence $\{x_n + y_n\}$ (respectively $\{x_n y_n\}$). It is easily seen that the sequences $\{x_n + y_n\}$ and $\{x_n y_n\}$ are indeed Cauchy sequences, and that the classes in which they lie do not depend on the choice of sequences $\{x_n\}$ and $\{y_n\}$ from the classes α and β.

It is easily verified that $\bar{k}$ is a ring with unit. Zero and one are the classes containing the sequences $\{0, 0, \ldots\}$ and $\{1, 1, \ldots\}$.

We now show that $\bar{k}$ is a field. If α is a nonzero class, and $\{x_n\}$ is a Cauchy sequence in this class, then, from some point on (say for $n \geqslant n_0$), all x_n are different from zero.

Consider the sequence $\{y_n\}$, defined by

$$y_n = \begin{cases} 1 & \text{for} \quad n < n_0, \\ \dfrac{1}{x_n} & \text{for} \quad n \geqslant n_0. \end{cases}$$

It is easily shown that the sequence $\{y_n\}$ is a Cauchy sequence, and that its class is the inverse of α.

We now introduce a metric on the field $\bar{k}$. We first note that if $\{x_n\}$ is a Cauchy sequence of elements of k, then $\{\varphi(x_n)\}$ is a Cauchy sequence of real numbers. By the completeness of the real field, this sequence converges to a real number, and the limit will not change if we replace the sequence $\{x_n\}$ by an equivalent one. We set $\varphi(\alpha) = \lim_{n\to\infty} \varphi(x_n)$, if α is the class containing the sequence $\{x_n\}$. It is easily shown that the function $\varphi(\alpha)$ satisfies all conditions of being a metric and hence turns $\bar{k}$ into a metric field.

We associate any element a of the field k with that class which contains the sequence $\{a, a, \ldots\}$. This sets up an embedding of metric fields, since, as is easily seen, this isomorphism of k with a subfield of $\bar{k}$ preserves the metric. Identifying each element of k with the corresponding element of $\bar{k}$, we shall consider k to be contained in $\bar{k}$. It is clear that k is everywhere dense in $\bar{k}$; for if α is a class, containing the sequence $\{x_n\}$, then $\{x_n\} \to \alpha$.

We now need only show that $\bar{k}$ is complete. Let $\{\alpha_n\}$ be a Cauchy sequence of elements of $\bar{k}$. Since α_n is the limit of a sequence of elements of the field k, there exists an element $x_n \in k$, such that $\varphi(\alpha_n - x_n) < 1/n$.

The fact that $\{\alpha_n\}$ is a Cauchy sequence implies that the sequence $\{x_n\}$ of elements of k is also a Cauchy sequence. Let α denote the class containing the sequence $\{x_n\}$. It is easily verified that $\{\alpha_n\} \to \alpha$, which completes the proof of Theorem 1.

Proof of Theorem 2. Let $\bar{k}$ and $\bar{k}_1$ be two complete fields containing k as a dense subfield. We shall set up a one-to-one correspondence between k and $\bar{k}_1$, leaving the verification that this is a topological isomorphism to the reader.

Let α be an element of $\bar{k}$, and let $\{x_n\}$ be a sequence of elements of k which converges to α. Since $\{x_n\}$ converges in $\bar{k}$, it is a Cauchy sequence. It remains a Cauchy sequence when regarded as a sequence of elements of k. Since $\bar{k}_1$ is complete, the sequence $\{x_n\}$ converges in $\bar{k}_1$ to some limit, which we denote by α_1. Clearly, if $\{y_n\}$ is another sequence of elements of k which converges to α in $\bar{k}$, then the limit of $\{y_n\}$ in $\bar{k}_1$ will again be α_1. Thus the element α_1

of the field $\bar{k}_1$ is uniquely determined by the element α of the field k. This correspondence, taking α to α_1, is the isomorphism which we need.

4.2. *Metrics of the Field of Rational Numbers*

It is natural to ask now if there exist any completions of the field of rational numbers, other than the real numbers and the p-adic numbers (for all primes p). The answer turns out to be negative; all completions of the rational numbers are of this type. Our immediate goal is the proof of this result.

We may clearly achieve this goal by enumerating all metrics of the rational field R.

In the definition of the p-adic metric φ_p on the field R, we had to choose a real number ρ, satisfying the condition $0 < \rho < 1$ [see (3.1), (3.18)]. Hence we have infinitely many metrics corresponding to the given prime integer p. However, they all clearly give the same conditions for convergence in R, and hence they all lead to the same completion, that is, to the field of p-adic numbers.

We now show that every function of the form

$$\varphi(\chi) = |\chi|^\alpha \tag{4.2}$$

where α is a real number, $0 < \alpha \leqslant 1$, is also a metric of the field R. In the definition of a metric, conditions (1) and (3) are clearly satisfied. Let $|x| \geqslant |y|$, $x \neq 0$. Then

$$|x + y|^\alpha = |x|^\alpha \left|1 + \frac{y}{x}\right|^\alpha \leqslant |x|^\alpha\left(1 + \left|\frac{y}{x}\right|\right)^\alpha$$

$$\leqslant |x|^\alpha\left(1 + \left|\frac{y}{x}\right|\right) \leqslant |x|^\alpha\left(1 + \left|\frac{y}{x}\right|^\alpha\right) = |x|^\alpha + |y|^\alpha,$$

that is, condition (2) is satisfied.

Convergence in R with respect to any metric of the form (4.2) clearly coincides with convergence with respect to the ordinary absolute value, and hence the process of completion under one of these valuations leads again to the real numbers.

Theorem 3 (Ostrowski's Theorem). Every metric of the field of rational numbers is either of the form (4.2), or is a p-adic metric (4.1) for some prime p.

Proof. Let φ be an arbitrary nontrivial metric of the field R. Two cases are possible: Either there is some natural number $a > 1$, for which $\varphi(a) > 1$, or else $\varphi(n) \leqslant 1$ for all natural numbers n. Consider the first case. Since

$$\varphi(n) = \varphi(1 + \cdots + 1) \leqslant \varphi(1) + \cdots + \varphi(1) = n, \tag{4.3}$$

we may set

$$\varphi(a) = a^{\alpha}, \tag{4.4}$$

where α is real and satisfies $0 < \alpha < 1$.

Taking an arbitrary natural number N, we decompose it in powers of a

$$N = x_0 + x_1 a + \cdots + x_{k-1} a^{k-1},$$

where $0 \leqslant x_i \leqslant a - 1$ $(0 \leqslant i \leqslant k - 1)$, $x_{k-1} \geqslant 1$. Hence N satisfies the inequality

$$a^{k-1} \leqslant N < a^k.$$

By the properties of metrics, formulas (4.3) and (4.4) yield

$$\begin{aligned} \varphi(N) &\leqslant \varphi(x_0) + \varphi(x_1)\varphi(a) + \cdots + \varphi(x_{k-1})\varphi(a)^{k-1} \\ &\leqslant (a-1)(1 + a^2 + \cdots + a^{(k-1)\alpha}) \\ &= (a-1)\frac{a^{k\alpha} - 1}{a^{\alpha} - 1} < (a-1)\frac{a^{k\alpha}}{a^{\alpha} - 1} = \frac{(a-1)a^{\alpha}}{a^{\alpha} - 1} a^{(k-1)\alpha} \\ &\leqslant \frac{(a-1)a^{\alpha}}{a^{\alpha} - 1} N^{\alpha} = CN^{\alpha}; \end{aligned}$$

that is,

$$\varphi(N) < CN^{\alpha},$$

where the constant C does not depend on N. Replacing N by N^m in this inequality, for m a natural number, we obtain

$$\varphi(N)^m = \varphi(N^m) < CN^{m\alpha},$$

whence

$$\varphi(N) < \sqrt[m]{C} N^{\alpha}.$$

Letting m tend to infinity, we arrive at

$$\varphi(N) \leqslant N^{\alpha}. \tag{4.5}$$

Now setting $N = a^k - b$, where $0 < b \leqslant a^k - a^{k-1}$, we obtain by condition (2),

$$\varphi(N) \geqslant \varphi(a^k) - \varphi(b) = a^{\alpha k} - \varphi(b).$$

But it is already known that

$$\varphi(b) \leqslant b^{\alpha} \leqslant (a^k - a^{k-1})^{\alpha},$$

and thus

$$\varphi(N) \geqslant a^{\alpha k} - (a^k - a^{k-1})^{\alpha} = \left[1 - \left(1 - \frac{1}{a}\right)^{\alpha}\right] a^{\alpha k} = C_1 a^{\alpha k} > C_1 N^{\alpha},$$

where the constant C_1 does not depend on N. Let m again denote an arbitrary natural number. If N is replaced by N^m in the preceding inequality, then

$$\varphi(N)^m = \varphi(N^m) > C_1 N^{\alpha m},$$

from which

$$\varphi(N) > \sqrt[m]{C_1}\, N^\alpha,$$

and as $m \to \infty$ this yields

$$\varphi(N) \geqslant N^\alpha. \tag{4.6}$$

Comparing (4.5) and (4.6), we see that $\varphi(N) = N^\alpha$ for any natural number N. Now let $x = \pm N_1/N_2$ be an arbitrary rational number, different from zero (N_1 and N_2 are natural numbers). Then

$$\varphi(x) = \varphi\left(\frac{N_1}{N_2}\right) = \frac{\varphi(N_1)}{\varphi(N_2)} = \frac{N_1{}^\alpha}{N_2{}^\alpha} = |x|^\alpha.$$

We have shown that if $\varphi(a) > 1$ for at least one natural number a, then the metric φ is of the form (4.2).

We now turn to the case where

$$\varphi(n) \leqslant 1 \tag{4.7}$$

for all natural n. If for every prime p, we had $\varphi(p) = 1$, then by condition (3) we would also have $\varphi(n) = 1$ for all natural n, and thus also $\varphi(x) = 1$ for all rational $x \neq 0$. But this would contradict the assumption that φ is nontrivial. Thus for some prime p we have $\varphi(p) < 1$. Assume that for some other prime $q \neq p$ we also had $\varphi(q) < 1$. We take exponents k and l so that

$$\varphi(p)^k < \tfrac{1}{2}, \qquad \varphi(q)^l < \tfrac{1}{2}.$$

Since p^k and q^l are relatively prime, there are integers u and v such that $up^k + vq^l = 1$. By (4.7) $\varphi(u) \leqslant 1$ and $\varphi(v) \leqslant 1$, so that

$$1 = \varphi(1) = \varphi(up^k + vq^l) \leqslant \varphi(u)\varphi(p)^k + \varphi(v)\varphi(q)^l < \tfrac{1}{2} + \tfrac{1}{2}.$$

This contradiction shows that there is only one prime p for which

$$\varphi(p) = \rho < 1.$$

Since $\varphi(q) = 1$ for all other prime numbers, $\varphi(a) = 1$ for every integer a which is relatively prime to p. Let $x = p^m(a/b)$ be a nonzero rational number (a and b integers, relatively prime to p). Then

$$\varphi(x) = \varphi(p^m)\frac{\varphi(a)}{\varphi(b)} = \varphi(p^m) = \rho^m.$$

Thus in this case the metric φ coincides with the p-adic metric (4.1).

The proof of Theorem 3 is complete.

PROBLEMS

1. Show that a finite field can have only the trivial metric.

2. Two metrics φ and ψ, defined on the same field k, are called *equivalent* if they define on k the same condition for convergence, that is, if $\varphi(x_n - x) \to 0$ if and only if $\psi(x_n - x) \to 0$. Show that for the equivalence of φ and ψ, it is necessary and sufficient that $\varphi(x) < 1$ if and only if $\psi(x) < 1$ for all $x \in k$.

3. Show that if φ and ψ are equivalent metrics on the field k, then there is a real number δ such that $\varphi(x) = (\psi(x))^{\delta}$ for all $x \in k$.

4. The metric φ, given on the field k, is called *non-Archimedean* if it satisfies not only condition (2) but also the stronger condition

$$\varphi(\alpha + \beta) \leqslant \max(\varphi(\alpha), \varphi(\beta)). \tag{2'}$$

(If this stronger condition fails to hold, then φ is called *Archimedean.*) Show that the metric φ is non-Archimedean if and only if $\varphi(n) \leqslant 1$ for every natural number n (that is, for every multiple of the unit element of k by a natural number).

5. Show that any metric of a field of characteristic p is non-Archimedean.

6. Let k_0 be an arbitrary field, and let $k = k_0(t)$ be the field of all rational functions over k_0. Every nonzero element $u \in k$ can be represented in the form

$$u = t^m \frac{f(t)}{g(t)} \qquad (f(0) \neq 0, g(0) \neq 0),$$

where f and g are polynomials. Show that the function

$$\varphi(u) = \rho^m \quad (0 < \rho < 1), \qquad \varphi(0) = 0, \tag{4.8}$$

is a metric of the field k.

7. Show that the completion of the field $k = k_0(t)$ with respect to the metric (4.8) is isomorphic to the field $k\{t\}$ of formal power series, which consists of all series of the form

$$\sum_{n=m}^{\infty} a_n t^n \qquad (a_n \in k_0)$$

under the usual operations on power series (the integer m may be positive, negative, or zero).

5. Congruences and p-Adic Integers

5.1. *Congruences and Equations in the Ring O_p*

At the beginning of Section 3 we considered the question of the solvability of the congruence $x^2 \equiv 2 \pmod{7^n}$ for $n = 1, 2, \ldots$, and this led us to the concept of a p-adic integer. The close connection between p-adic integers and congruences was already shown in their definition (Section 3.1). This connection is described more fully in the following theorem.

Theorem 1. Let $F(x_1, \dots, x_n)$ be a polynomial whose coefficients are rational integers. The congruence

$$F(x_1, \dots, x_n) \equiv 0 \pmod{p^k} \tag{5.1}$$

is solvable for all $k \geqslant 1$ if and only if the equation

$$F(x_1, \dots, x_n) = 0 \tag{5.2}$$

is solvable in p-adic integers.

Proof. Let (5.2) have the p-adic integral solution $(\alpha_1, \dots, \alpha_n)$. For every k there exist rational integers $x_1^{(k)}, \dots, x_n^{(k)}$ such that

$$\alpha_1 \equiv x_1^{(k)} \pmod{p^k}, \dots, \alpha_n \equiv x_n^{(k)} \pmod{p^k}. \tag{5.3}$$

From this we obtain

$$F(x_1^{(k)}, \dots, x_n^{(k)}) \equiv F(\alpha_1, \dots, \alpha_n) \equiv 0 \pmod{p^k};$$

that is, $(x_1^{(k)}, \dots, x_n^{(k)})$ is a solution of the congruence (5.1).

Now assume that (5.1) has the solution $(x_1^{(k)}, \dots, x_n^{(k)})$ for each k. Select from the sequence $\{x_1^{(k)}\}$ of rational integers a p-adically converging subsequence $\{x_1^{(k_i)}\}$ (Theorem 6, Section 3). From the sequence $\{x_2^{(k_i)}\}$ select again a convergent subsequence. Repeating this process n times, we arrive at a sequence of natural numbers $\{l_1, l_2, \dots\}$, such that each of the sequences $\{x_i^{(l_1)}, x_i^{(l_2)}, \dots\}$ is p-adically convergent. Let

$$\lim_{m\to\infty} x_i^{(l_m)} = \alpha_i .$$

It will be shown that $(\alpha_1, \dots, \alpha_n)$ is a solution of (5.2). Since the polynomial $F(x_1, \dots, x_n)$ is a continuous function,

$$F(\alpha_1, \dots, \alpha_n) = \lim_{m\to\infty} F(x_1^{(l_m)}, \dots, x_n^{(l_m)}).$$

On the other hand, by the choice of the subsequence $(x_1^{(l_m)}, \dots, x_n^{(l_m)})$,

$$F(x_1^{(l_m)}, \dots, x_n^{(l_m)}) \equiv 0 \pmod{p^{l_m}},$$

so that $\lim_{m\to\infty} F(x_1^{(l_m)}, \dots, x_n^{(l_m)}) = 0$. Thus $F(\alpha_1, \dots, \alpha_n) = 0$, and the theorem is proved.

Consider now the case when $F(x_1, \dots, x_n)$ is a form. Assume that the equation $F(x_1, \dots, x_n) = 0$ has a nonzero solution $(\bar{\alpha}_1, \dots, \bar{\alpha}_n)$ in p-adic integers. Set $m = \min(\nu_p(\bar{\alpha}_1), \dots, \nu_p(\bar{\alpha}_n))$. Then each $\bar{\alpha}_i$ is represented in the form

$$\bar{\alpha}_i = p^m \alpha_i \qquad (i = 1, \dots, n),$$

where all α_i are integers and at least one of them is not divisible by p. Clearly, $(\alpha_1, \dots, \alpha_n)$ is also a solution of the equation $F(x_1, \dots, x_n) = 0$. The numbers

$(x_1^{(k)}, \dots, x_n^{(k)})$, satisfying (5.3), then give, as we have seen, a solution of (5.1), not all terms of which are divisible by p.

Conversely, assume that (5.1) with F homogeneous, has for each k a solution $(x_1^{(k)}, \dots, x_n^{(k)})$ in which at least one of the numbers $x_i^{(k)}$ is not divisible by p. Clearly, for some index $i = i_0$ there will be an infinite number of values of m for which $x_{i_0}^{(m)}$ is not divisible by p. Therefore the sequence $\{l_1, l_2, \dots\}$ can be chosen so that all $x_{i_0}^{(l_m)}$ are not divisible by p. But then from $\alpha_{i_0} = \lim x_{i_0}^{(l_m)}$ it follows that α_{i_0} is not divisible by p, and a fortiori $\alpha_{i_0} \neq 0$. Thus we have proved the following theorem.

Theorem 2. Let $F(x_1, \dots, x_n)$ be a form whose coefficients are rational integers. The equation $F(x_1, \dots, x_n) = 0$ has a nontrivial solution in the ring O_p if and only if for every m the congruence $F(x_1, \dots, x_n) \equiv 0 \pmod{p^m}$ has a solution in which not all terms are divisible by p.

It is clear that in Theorems 1 and 2 F may be a polynomial whose coefficients are p-adic integers.

5.2. *On the Solvability of Some Congruences*

By Theorem 1, we can solve (5.2) in p-adic integers provided we can solve an infinite sequence of congruences (5.1). It is generally difficult to tell when we may limit our consideration to only a finite number of these. Here we shall consider a special case.

Theorem 3. Let $F(x_1, \dots, x_n)$ be a polynomial whose coefficients are p-adic integers. Let $\gamma_1, \dots, \gamma_n$ be p-adic integers such that for some i $(1 \leqslant i \leqslant n)$ we have

$$F(\gamma_1, \dots, \gamma_n) \equiv 0 \pmod{p^{2\delta+1}},$$

$$\frac{\partial F}{\partial x_i}(\gamma_1, \dots, \gamma_n) \equiv 0 \pmod{p^{\delta}},$$

$$\frac{\partial F}{\partial x_i}(\gamma_1, \dots, \gamma_n) \not\equiv 0 \pmod{p^{\delta+1}}$$

(δ is a nonnegative rational integer). Then there exist p-adic integers $\theta_1, \dots, \theta_n$, such that

$$F(\theta_1, \dots, \theta_n) = 0$$

and

$$\theta_1 \equiv \gamma_1 \pmod{p^{\delta+1}}, \dots, \theta_n \equiv \gamma_n \pmod{p^{\delta+1}}.$$

Proof. Consider the polynomial $f(x) = F(\gamma_1, \dots, \gamma_{i-1}, x, \gamma_{i+1}, \dots, \gamma_n)$. To prove the theorem it suffices to find a p-adic integer α, for which $f(\alpha) = 0$ and $\alpha \equiv \gamma_i \pmod{p^{\delta+1}}$ (if such an α is found, then set $\theta_j = \gamma_j$ for $j \neq i$, and $\theta_i = \alpha$). Let $\gamma_i = \gamma$. We construct a sequence

$$\alpha_0, \alpha_1, \ldots, \alpha_m, \ldots \tag{5.3'}$$

of p-adic integers, congruent to γ modulo $p^{\delta+1}$, such that

$$f(\alpha_m) \equiv 0 \pmod{p^{2\delta+1+m}} \tag{5.4}$$

for all $m \geqslant 0$. For $m = 0$ set $\alpha_0 = \gamma$. Assume that for some $m \geqslant 1$ the p-adic integers $\alpha_0, \ldots, \alpha_{m-1}$, satisfying the above requirements, have already been found. In particular, $\alpha_{m-1} \equiv \gamma \pmod{p^{\delta+1}}$ and $f(\alpha_{m-1}) \equiv 0 \pmod{p^{2\delta+m}}$. Expand the polynomial $f(x)$ in powers of $x - \alpha_{m-1}$:

$$f(x) = \beta_0 + \beta_1(x - \alpha_{m-1}) + \beta_2(x - \alpha_{m-1})^2 + \cdots \qquad (\beta_i \in O_p).$$

By the induction assumption $\beta_0 = f(\alpha_{m-1}) = p^{2\delta+m}A$, where A is a p-adic integer. Further, since $\alpha_{m-1} \equiv \gamma \pmod{p^{\delta+1}}$, then $\beta_1 = f'(\alpha_{m-1}) = p^{\delta}B$, where B is not divisible by p in O_p. Setting $x = \alpha_{m-1} + \xi p^{m+\delta}$, we obtain

$$f(\alpha_{m-1} + \xi p^{m+\delta}) = p^{2\delta+m}(A + B\xi) + \beta_2 p^{2\delta+2m}\xi^2 + \cdots$$

We now choose a value $\xi = \xi_0 \in O_p$ so that $A + B\xi_0 \equiv 0 \pmod{p}$ [this is possible since $B \not\equiv 0 \pmod{p}$]. Noting that $k\delta + km \geqslant 2\delta + 1 + m$ for $k \geqslant 2$, we have

$$f(\alpha_{m-1} + \xi_0 p^{m+\delta}) \equiv 0 \pmod{p^{2\delta+1+m}}.$$

Thus we may set $\alpha_m = \alpha_{m-1} + \xi_0 p^{m+\delta}$. Since $m + \delta \geqslant \delta + 1$, $\alpha_m \equiv \gamma \pmod{p^{\delta+1}}$. By our construction $v_p(\alpha_m - \alpha_{m-1}) \geqslant m + \delta$, and thus the sequence (5.3′) converges. Denote its limit by α. Clearly, $\alpha \equiv \gamma \pmod{p^{\delta+1}}$. From (5.4) it follows that $\lim_{m\to\infty} f(\alpha_m) = 0$; on the other hand, by the continuity of the polynomial f, $\lim_{m\to\infty} f(\alpha_m) = f(\alpha)$. Thus $f(\alpha) = 0$.

Corollary. If the polynomial $F(x_1, \ldots, x_n)$ has p-adic integers as coefficients and for some i $(1 \leqslant i \leqslant n)$ the p-adic integers $\gamma_1, \ldots, \gamma_n$ satisfy

$$F(\gamma_1, \ldots, \gamma_n) \equiv 0 \pmod{p},$$

$$F'_{x_i}(\gamma_1, \ldots, \gamma_n) \not\equiv 0 \pmod{p},$$

then there exist p-adic integers $\theta_1, \ldots, \theta_n$ such that

$$F(\theta_1, \ldots, \theta_n) = 0$$

and

$$\theta_1 \equiv \gamma_1 \pmod{p}, \ldots, \theta_n \equiv \gamma_n \pmod{p}.$$

Thus a solution $(c_1, \ldots, c_n)$ to the congruence $F(x_1, \ldots, x_n) \equiv 0 \pmod{p}$ can be extended to a solution of the equation $F(x_1, \ldots, x_n) = 0$ in the ring O_p, provided that at least one of the following congruences does not hold:

$$F'_{x_1}(c_1, \dots, c_n) \equiv 0 \pmod{p};$$

$$\dots\dots\dots\dots\dots\dots$$

$$F'_{x_n}(c_1, \dots, c_n) \equiv 0 \pmod{p}.$$

This last assertion has an important application to the question which we dealt with at the beginning of Section 2. There we noted that to show directly that the congruence

$$F(x_1, \dots, x_n) \equiv 0 \pmod{m}$$

is solvable for all m involves the verification of an infinite number of conditions. In the case where the modulus is prime, Theorems A and B of Section 2.1 allow the possibility of an effective verification, in that they show that a direct verification is only necessary for a finite number of primes. Now we can say something about the case of arbitrary moduli. As we have already noted, it suffices to consider moduli which are powers of a prime, and for moduli having the form p^k ($k = 1, 2, \dots$), the solvability of the congruence (5.1) is equivalent to the solvability of the equation $F = 0$ in the ring of p-adic integers.

Using Theorems A and B of Section 2.1 (which we have not proved), and also Theorem 3, we prove the following result.

Theorem C. If $F(x_1, \dots, x_n)$ is an absolutely irreducible polynomial with rational integer coefficients, then the equation $F(x_1, \dots, x_n) = 0$ is solvable in the ring O_p of p-adic integers for all prime numbers p greater than some bound which depends only on the polynomial F.

Hence, for all but a finite number of primes p, the congruence

$$F(x_1, \dots, x_n) \equiv 0 \pmod{p^k} \tag{5.5}$$

is solvable for all k.

Theorem C thus reduces the question of the solvability of the congruence (5.5) for all p to the question of the solvability of the equation $F = 0$ in the ring O_p for a finite number of primes p. We shall not deal here with the question of the solvability of the equation $F = 0$ in the ring O_p for these finitely many p (for the case of quadratic polynomials this will be done in Section 6).

The idea of the proof of Theorem C is very simple: Using the estimate of Theorem B for the number of solutions of the congruence (2.1), we shall show that the number of solutions to this congruence is greater, for sufficiently large p, than the number of solutions to the system of congruences

$$\begin{aligned} F(x_1, \dots, x_n) &\equiv 0 \pmod p, \\ F'_{x_n}(x_1, \dots, x_n) &\equiv 0 \pmod p. \end{aligned} \tag{5.6}$$

To do this we need another estimate for the number of solutions of a congruence.

Lemma. If not all coefficients of the polynomial $F(x_1, \dots, x_n)$ are divisible by p, then the number $N(p)$ of solutions to the congruence

$$F(x_1, \dots, x_n) \equiv 0 \pmod p \tag{5.7}$$

satisfies the inequality

$$N(p) \leqslant Lp^{n-1}, \tag{5.8}$$

in which the constant L is equal to the total degree of F.

We prove the lemma by induction on n. For $n = 1$ it follows from the fact that the number of roots of a nonzero polynomial in the field Z_p cannot exceed its degree.

If $n > 1$, consider $F(x_1, \dots, x_n)$ as a polynomial in $x_1, \dots, x_{n-1}$, the coefficients of which are polynomials in x_n. Denote by $f(x_n)$ the greatest common divisor of these coefficients modulo p. Then

$$F(x_1, \dots, x_n) \equiv f(x_n)F_1(x_1, \dots, x_n) \pmod p,$$

where for any a the polynomial $F_1(x_1, \dots, x_{n-1}, a)$ is not identically congruent to zero modulo p. Let l and L_1 be the degrees of f and F_1, respectively. It is clear that f and F_1 can be chosen so that $l + L_1 \leqslant L$. We can now bound the number of solutions $(c_1, \dots, c_n)$ to the congruence (5.7) by considering the different values for x_n in these solutions. Consider first those solutions for which

$$f(c_n) \equiv 0 \pmod p. \tag{5.9}$$

If (5.9) is fulfilled, then for any choice of $c_1, \dots, c_{n-1}$ we obtain a solution of (5.7). Since the numbers of values of c_n, modulo p, is at most l, then the number of solutions of (5.7), for which (5.9) holds, is at most lp^{n-1}. Consider now solutions for which $f(c_n) \not\equiv 0 \pmod p$. All such solutions clearly satisfy the congruence $F_1(x_1, \dots, x_n) \equiv 0 \pmod p$. Since $F(x_1, \dots, x_{n-1}, c_n)$ is not identically congruent to zero modulo p, then by the induction hypothesis the number $N(p, c_n)$ of solutions of the congruence $F(x_1, \dots, x_{n-1}, c_n) \equiv 0 \pmod p$ satisfies the inequality $N(p, c_n) \leqslant L_1 p^{n-2}$. Since c_n takes not more than p values, the total number of solutions of this type does not exceed $L_1 p^{n-1}$. Thus the total number of solutions of (5.7) does not exceed $lp^{n-1} + L_1 p^{n-1} \leqslant Lp^{n-1}$, which is what was to be proved.

Proof of Theorem C. We may, of course, assume that the polynomial F

actually depends on the variable x_n. Consider F as a polynomial in x_n with coefficients which are polynomials in $x_1, \dots, x_{n-1}$. Since F is absolutely irreducible, it follows that the discriminant $D_{x_n}(x_1, \dots, x_{n-1})$ of the polynomial F, considered as a polynomial in x_n, is a polynomial in $x_1, \dots, x_{n-1}$ which is not identically zero, since otherwise F would be divisible by the square of some polynomial. Consider a prime number p, which does not divide all coefficients of $D_{x_n}(x_1, \dots, x_{n-1})$, and let $N_1(p)$ be the number of solutions of (5.6). If $(c_1, \dots, c_n)$ is a solution of (5.6), then c_n is a common root, modulo p, of the polynomials $F(c_1, \dots, c_{n-1}, x)$ and $F'_{x_n}(c_1, \dots, c_{n-1}, x_n)$ and therefore

$$D_{x_n}(c_1, \dots, c_{n-1}) \equiv 0 \pmod{p}.$$

By the lemma, the number of solutions to this congruence does not exceed $K_1 p^{n-2}$, where K_1 is some constant which depends only on the polynomial F. For given $c_1, \dots, c_{n-1}$ the values of c_n are determined by the congruence

$$F(c_1, \dots, c_{n-1}, x_n) \equiv 0 \pmod{p}$$

and therefore there are at most m of them, where m is the degree of the polynomial F in x_n. Thus the number $N_1(p)$ of solutions to system (5.6) does not exceed Kp^{n-2}, where $K = mK_1$. We now show that for sufficiently large p, the number $N(p)$ solutions to the congruence (5.7) is larger than the number $N_1(p)$ of solutions to system (5.6). Indeed, by Theorem B,

$$N(p) > p^{n-1} - Cp^{n-1-(1/2)},$$

and we have just shown that $N_1(p) < Kp^{n-2}$. Thus

$$N(p) - N_1(p) > p^{n-1} - Cp^{n-1-(1/2)} - Kp^{n-2} = p^{n-2}(p - Cp^{1/2} - K),$$

which means that $N(p) > N_1(p)$ for sufficiently large p. Thus, for sufficiently large p, the congruence $F \equiv 0 \pmod{p}$ has a solution $(\gamma_1, \dots, \gamma_n)$ for which

$$\frac{\partial F}{\partial x_n}(\gamma_1, \dots, \gamma_n) \not\equiv 0 \pmod{p}.$$

By the corollary of Theorem 3, this proves that the equation $F = 0$ has a solution in the ring O_p for all p, larger than some given constant.

PROBLEMS

1. Show that if m and p are relatively prime, then any p-adic unit ε, satisfying the congruence $\varepsilon \equiv 1 \pmod{p}$, is an mth power in R_p.

2. Let $m = p^\delta m_0$, $(m_0, p) = 1$, and let $\varepsilon \equiv 1 \pmod{p^{2\delta+1}}$. Show that the p-adic unit ε is an mth power in R_p.

3. If $p \neq 2$ and the p-adic integers α and β are not divisible by p, show that the solvability of the congruence $\alpha x^p \equiv \beta \pmod{p^2}$ implies the solvability of the equation $\alpha x^2 = \beta$ in the field R_p.

4. Assume that the coefficients ε_i in the form $G = \varepsilon_1 x_1{}^p + \cdots + \varepsilon_n x_n{}^p$ are p-adic units ($p \neq 2$). Show that if the congruence $G \equiv 0 \pmod{p^2}$ has a solution in which at least one of the variables is not divisible by p, then the equation $G = 0$ has a nonzero solution in the field R_p.

5. Let all coefficients of the form $G = \alpha_1 x_1{}^p + \cdots + \alpha_n x_n{}^p$ be p-adic integers which are divisible at most by the $(p-1)$th power of p. If the congruence $G \equiv 0 \pmod{p^{p+2}}$ has a solution in which not all variables are divisible by p, show that the equation $G = 0$ has a nonzero solution in the field R_p. [If $p \neq 2$, it suffices to have a solution to the congruence $G \equiv 0 \pmod{p^{p+1}}$.]

6. Let the quadratic form $F = \alpha_1 x_1{}^2 + \cdots + \alpha_n x_n{}^2$ have coefficients which are p-adic integers ($p \neq 2$) not divisible by p. Show that if the congruence $F \equiv 0 \pmod{p^2}$ has a solution in which not all values of the variables are divisible by p, then the equation $F = 0$ has a nonzero solution in the field R_p.

7. If the form $F = \alpha_1 x_1{}^m + \cdots + \alpha_n x_n{}^m$ has coefficients which are nonzero p-adic integers, set $r = \nu_p(m)$, $s = \max(\nu_p(\alpha_1), \ldots, \nu_p(\alpha_n))$ and $N = 2(r+s)+1$. Show that the equation $F = 0$ has a nonzero solution in the field R_p if and only if the congruence $F \equiv 0 \pmod{p^N}$ has a solution in which not all values of the variables are divisible by p.

8. Show that the form $3x^3 + 4y^3 + 5z^3$ represents zero in the field R_p for all p (see Problem 13, Section 2).

9. Let the polynomial $F(x_1, \ldots, x_n)$ have coefficients in O_p and, denote by c_m ($m \geqslant 0$) the number of solutions to the congruence $F(x_1, \ldots, x_n) \equiv 0 \pmod{p^m}$. Consider the series $\varphi(t) = \sum_{m=0}^{\infty} c_m t^m$. It has been conjectured that the series $\varphi(t)$, called the Poincaré series of the polynomial F, represents a rational function of t. Find the Poincaré series for the polynomial $F = \varepsilon_1 x_1{}^2 + \cdots + \varepsilon_n x_n{}^2$, where ε_i is a p-adic unit, and check that the function $\varphi(t)$ is rational.

10. Find the Poincaré series for a polynomial $F(x_1, \ldots, x_n)$ with p-adic integral coefficients, which satisfies the condition that for any solution of the congruence $F \equiv 0 \pmod{p}$, $\partial F/\partial x_i \not\equiv 0 \pmod{p}$ for some $i = 1, \ldots, n$.

11. Compute the Poincaré series for the polynomial $F(x, y) = x^2 - y^3$.

6. Quadratic Forms with p-Adic Coefficients

In this and the next section we shall apply the theory of p-adic numbers which we have developed to the investigation of the simplest types of equations. We shall consider the problem of the representation of p-adic and rational numbers by quadratic forms. The algebraic preliminaries that we shall need on the properties of quadratic forms over arbitrary fields are given in the Supplement, Section 1.

6.1. *Squares in the Field of p-Adic Numbers*

For the study of quadratic forms over a given field it is important to know which of the elements of the field are squares. Therefore we first turn to the study of squares in the field R_p of p-adic numbers. We know (Section 3, Theorem 4) that every nonzero p-adic number α can be represented uniquely in the form $\alpha = p^m\varepsilon$, where ε is a p-adic unit (that is, ε is a unit in the ring O_p of p-adic integers). If α is the square of the p-adic number $\gamma = p^k\varepsilon_0$, then $m = 2k$ and $\varepsilon = \varepsilon_0{}^2$. To determine all squares of the field R_p, we must thus determine which units of O_p are squares.

Theorem 1. Let $p \neq 2$. In order that the p-adic unit

$$\varepsilon = c_0 + c_1 p + c_2 p^2 + \cdots \qquad (0 \leqslant c_i < p,\ c_0 \neq 0) \tag{6.1}$$

be a square, it is necessary and sufficient that the integer c_0 be a quadratic residue modulo p.

Proof. If $\varepsilon = \eta^2$ and $\eta \equiv b \pmod p$ (b a rational integer), then $c_0 \equiv b^2 \pmod p$. Conversely, if $c_0 \equiv b^2 \pmod p$, let $F(x) = x^2 - \varepsilon$. We have $F(b) \equiv 0 \pmod p$ and $F'(b) = 2b \not\equiv 0 \pmod p$. By the corollary of Theorem 3 of Section 5 there is a $\eta \in O_p$ such that $F(\eta) = 0$ and $\eta \equiv b \pmod p$. Thus $\varepsilon = \eta^2$, and the theorem is proved.

Corollary 1. If $p \neq 2$, any p-adic unit which is congruent to 1 modulo p is a square in R_p.

Corollary 2. If $p \neq 2$, the index $(R_p{}^* : R_p{}^{*2})$ of the subgroup of squares $R_p{}^{*2}$ in the multiplicative group of the field R_p is equal to 4.

For if ε is not a square, then the quotient of any pair of numbers from $1, \varepsilon, p, p\varepsilon$ is not a square in R_p. But any nonzero p-adic number can be represented as the product of one of the numbers $1, \varepsilon, p, p\varepsilon$ with some square.

If $p \neq 2$ and the unit ε is given by (6.1), set

$$\left(\frac{\varepsilon}{p}\right) = \begin{cases} +1 & \text{if } \varepsilon \text{ is a square in } R, \\ -1 & \text{otherwise.} \end{cases}$$

By Theorem 1 we have

$$\left(\frac{\varepsilon}{p}\right) = \left(\frac{c_0}{p}\right),$$

where (c_0/p) is the Legendre symbol. If ε is a rational integer relatively prime to p, then the symbol (ε/p) which we have defined clearly coincides with the Legendre symbol. It is easily seen that for p-adic units ε and η we have

$$\left(\frac{\varepsilon\eta}{p}\right) = \left(\frac{\varepsilon}{p}\right)\left(\frac{\eta}{p}\right).$$

We turn to the case $p = 2$.

Theorem 2. In order that the 2-adic unit ε be a square (in the field R_2), it is necessary and sufficient that $\varepsilon \equiv 1 \pmod 8$.

Proof. The necessity follows from the fact that the square of an odd integer is always congruent to 1 modulo 8. To prove sufficiency, set $F(x) = x^2 - \varepsilon$ and apply Theorem 3 of Section 5, taking $\delta = 1$ and $\gamma = 1$. Since $F(1) \equiv 0 \pmod 8$ and $F'(1) = 2 \not\equiv 0 \pmod 4$, the theorem implies that there is an $\eta \equiv 1 \pmod 4$, such that $F(\eta) = 0$; that is, $\varepsilon = \eta^2$.

Corollary $(R_2{}^* : R_2{}^{*2}) = 8$, where $R_2{}^{*2}$ is the subgroup of squares of the multiplicative group $R_2{}^*$ of the field of 2-adic numbers.

By the above theorem the reduced system of residues modulo 8, namely, 1, 3, 5, 7, forms a system of coset representatives for the subgroup of squares in the group of all 2-adic units. If we also take the products $2\cdot 1, 2\cdot 3, 2\cdot 5, 2\cdot 7$, then we obtain a full system of coset representatives for the subgroup $R_2{}^{*2}$ of the group $R_2{}^*$.

6.2. *Representation of Zero by p-Adic Quadratic Forms*

As is the case in any field, a nonsingular quadratic form over the field R_p can be put in the form

$$\alpha_1 x_1{}^2 + \cdots + \alpha_n x_n{}^2 \qquad (\alpha_i \neq 0)$$

after a linear change of variables (see the Supplement, Section 1.1). If $\alpha_i = p^{2k_i}\varepsilon_i$ or $\alpha_i = p^{2k_i+1}\varepsilon_i$ (ε_i a unit in O_p), then after the substitution $p^{k_i}x_i = y_i$ we obtain a form in which all coefficients are p-adic integers which are divisible at most by the first power of p. Thus any nonsingular quadratic form over the field R_p is equivalent to a form

$$F = F_0 + pF_1 = \varepsilon_1 x_1{}^2 + \cdots + \varepsilon_r x_r{}^2 + p(\varepsilon_{r+1} x_{r+1}{}^2 + \cdots + \varepsilon_n x_n{}^2), \qquad (6.2)$$

where the ε_i are p-adic units.

While considering the question of the representation of zero, we may assume that $r \geqslant n - r$. The form pF is clearly equivalent to the form $F_1 + pF_0$. Since F and pF simultaneously represent zero, we may take the form $F_1 + pF_0$ instead of $F_0 + pF_1$.

We first consider the case $p \neq 2$.

Theorem 3. Let $p \neq 2$ and $0 < r < n$. The form (6.2) represents zero in the field R_p if and only if at least one of the forms F_0 or F_1 represents zero.

Proof. Let the form (6.2) represent zero:

$$\varepsilon_1 \xi_1{}^2 + \cdots + \varepsilon_r \xi_r{}^2 + p(\varepsilon_{r+1}\xi_{r+1}{}^2 + \cdots + \varepsilon_n \xi_n{}^2) = 0. \qquad (6.3)$$

We may assume that all ξ_i are integers and that at least one of them is not divisible by p. If not all $\xi_1, \dots, \xi_r$ are divisible by p, say $\xi_1 \not\equiv 0 \pmod p$, then, considering (6.3) modulo p, we have

$$F_0(\xi_1, \dots, \xi_r) \equiv 0 \pmod p,$$

$$\frac{\partial F_0}{\partial x_1}(\xi_1, \dots, \xi_r) = 2\varepsilon_1\xi_1 \not\equiv 0 \pmod p.$$

By the corollary of Theorem 3 of Section 5, the form F_0 represents zero. Assume now that $\xi_1, \dots, \xi_r$ are all divisible by p, so that $\varepsilon_1\xi_1^2 + \cdots + \varepsilon_r\xi_r^2 \equiv$ (mod p^2). We consider (6.3) modulo p^2. Dividing this congruence by p, we obtain

$$F_1(\xi_{r+1}, \dots, \xi_n) \equiv 0 \pmod p,$$

where at least one of $\xi_{r+1}, \dots, \xi_n$ is not divisible by p. Again applying the corollary of Theorem 3 of Section 5, we conclude that in this case the form F_1 represents zero. Since the sufficiency of the condition is obvious, Theorem 3 is proved. The following corollaries are immediate.

Corollary 1. If $\varepsilon_1, \dots, \varepsilon_r$ are p-adic units and $p \neq 2$, then the form $f = \varepsilon_1 x_1^2 + \cdots + \varepsilon_r x_r^2$ represents zero in R_p if and only if the congruence $f(x_1, \dots, x_r) \equiv 0 \pmod p$ has a nontrivial solution in O_p.

Corollary 2. If we also assume that $r \geqslant 3$, then the form $f(x_1, \dots, x_r)$ always represents zero in R_p.

For by Theorem 5 of Section 1, the congruence $f(x_1, \dots, x_r) \equiv 0 \pmod p$ has a nontrivial solution.

In the proof of Theorem 3 the equality (6.3) was not actually used; we used only the congruences $F \equiv 0 \pmod p$ and $F \equiv 0 \pmod{p^2}$. Thus the solvability of the second of these congruences already implies that one of the forms F_0 or F_1, and hence F, represents zero. Hence we have

Corollary 3. If $p \neq 2$ the form (6.2) represents zero if and only if the congruence $F \equiv 0 \pmod{p^2}$ has a solution in which not all variables are divisible by p.

We now consider quadratic forms over the field of 2-adic numbers. In this case Theorem 3 and all its corollaries are false. For example, if $f = x_1^2 + x_2^2 + x_3^2 + x_4^2$, then the equation $f = 0$ has no nontrivial solution (since the congruence $f \equiv 0 \pmod 8$ has no solution in integers, at least one of which is odd). But we shall see that the form $f + 2x_5^2$ does represent zero in R_2 (Theorem 5).

Theorem 4. The form (6.2) (with $p = 2$) represents zero in the field of 2-adic numbers if and only if the congruence $F \equiv 0 \pmod{16}$ has a solution in which at least one of the variables is odd.

Proof. Let $F(\xi_1, \ldots, \xi_n) \equiv 0 \pmod{16}$, where not all of the 2-adic integers ξ_i are divisible by 2. We first assume that $\xi_i \not\equiv 0 \pmod 2$ for some $i \leqslant r$, say $\xi_1 \not\equiv 0 \pmod 2$. Since $F(\xi_1, \ldots, \xi_n) \equiv 0 \pmod 8$ and $(\partial F/\partial x_1)(\xi_1, \ldots, \xi_n) = 2\varepsilon_1\xi_1 \not\equiv 0 \pmod 4$, by Theorem 3 of Section 5, the form F represents zero. If $\xi_1, \ldots, \xi_r$ are all divisible by 2, set $\xi_i = 2\eta_i$ $(1 \leqslant i \leqslant r)$, where η_i is a 2-adic integer. Divide the congruence

$$4\sum_{i=1}^{r} \varepsilon_i\eta_i^2 + 2\sum_{i=r+1}^{n} \varepsilon_i\xi_i^2 \equiv 0 \pmod{16}$$

by 2 to obtain

$$\sum_{i=r+1}^{n} \varepsilon_i\xi_i^2 + 2\sum_{i=1}^{r} \varepsilon_i\eta_i^2 \equiv 0 \pmod 8,$$

where at least one of $\xi_{r+1}, \ldots, \xi_n$ is not divisible by 2. As above it follows from this congruence that the form $F_1 + 2F_0$ represents zero. Since the forms F and $2F$ represent zero simultaneously, the sufficiency of the condition is proved. The converse is obvious.

In the course of the proof we have obtained the following result.

Corollary. If the congruence $F \equiv 0 \pmod 8$, where F is given by (2) with $p = 2$, has a solution in which at least one of the variables $x_1, \ldots, x_r$ takes an odd value, then this form represents zero in the field R_2.

Theorem 5. Any quadratic form over the field R_p of p-adic numbers in five or more variables always represents zero.

Proof. We may assume that our form is (6.2) with $r \geqslant n - r$. Since $n \geqslant 5$, then $r \geqslant 3$. Corollary 2 of Theorem 3 then implies that the form F_0, and hence also the form F, represents zero. The theorem is proved if $p \neq 2$.

Let $p = 2$. If $n - r > 0$, consider the "partial" form $f = \varepsilon_1 x_1^2 + \varepsilon_2 x_2^2 + \varepsilon_3 x_3^2 + 2\varepsilon_n x_n^2$. We claim that such a form always represents zero in R_2. Since $\varepsilon_1 + \varepsilon_2 = 2\alpha$ (α a 2-adic integer), then $\varepsilon_1 + \varepsilon_2 + 2\varepsilon_n\alpha^2 \equiv 2\alpha + 2\alpha^2 = 2\alpha(1 + \alpha) \equiv 0 \pmod 4$, that is, $\varepsilon_1 + \varepsilon_2 + 2\varepsilon_n\alpha^2 = 4\beta$, where β is a 2-adic integer. Setting $x_1 = x_2 = 1$, $x_3 = 2\beta$, $x_n = \alpha$, we have

$$\varepsilon_1 \cdot 1^2 + \varepsilon_2 \cdot 1^2 + \varepsilon_3(2\beta)^2 + 2\varepsilon_n\alpha^2 \equiv 4\beta + 4\beta^2 \equiv 0 \pmod 8.$$

By the corollary of Theorem 4 the form f represents zero. But then F also represents zero. In the case $n = r$, we take as a partial form $f = \varepsilon_1 x_1^2 + \varepsilon_2 x_2^2$

$+ \varepsilon_3 x_3{}^2 + \varepsilon_4 x_4{}^2 + \varepsilon_5 x_5{}^2$. If $\varepsilon_1 + \varepsilon_2 \equiv \varepsilon_3 + \varepsilon_4 \equiv 2 \pmod 4$ then set $x_1 = x_2 = x_3 = x_4 = 1$, and if, say, $\varepsilon_1 + \varepsilon_2 \equiv 0 \pmod 4$, set $x_1 = x_2 = 1$, $x_3 = x_4 = 0$. In general we get $\varepsilon_1 x_1{}^2 + \varepsilon_2 x_2{}^2 + \varepsilon_3 x_3{}^2 + \varepsilon_4 x_4{}^2 = 4\gamma$, where γ is a 2-adic integer. Set $x_2 = 2\gamma$, and then

$$f \equiv 4\gamma + 4\gamma^2 \equiv 0 \pmod 8).$$

We complete the proof by applying the corollary of Theorem 4. Theorem 5 is completely proved.

By Theorem 6 of Section 1 of the Supplement, Theorem 5 implies the following corollary.

Corollary 1. Any nonsingular quadratic form in four or more variables over the field R_p represents all p-adic numbers.

Corollary 2. Let $F(x_1, \dots, x_n)$ be a nonsingular quadratic form whose coefficients are rational integers. If $n \geqslant 5$, then for any m the congruence $F(x_1, \dots, x_n) \equiv 0 \pmod m$ has a nontrivial solution.

Indeed, since the form F represents zero in R_p, then for any $s \geqslant 1$, the congruence $F \equiv 0 \pmod{p^s}$ has a solution in which at least one variable is not divisible by p.

6.3. *Binary Forms*

Binary quadratic forms form an important special case of the general theory. We consider the question of the representation of numbers of the field R_p by the quadratic form

$$x^2 - \alpha y^2, \qquad \alpha \neq 0, \qquad \alpha \in R_p. \tag{6.4}$$

(Any nonsingular binary form can be put in this form by a change of variables and by multiplying the form by some p-adic number.)

Let H_α denote the set of all nonzero p-adic numbers represented by the form (6.4). This set has the surprising property of being a group under multiplication. Indeed, if $\beta = x^2 - \alpha y^2$, $\beta_1 = x_1{}^2 - \alpha y_1{}^2$, then a simple computation shows that

$$\beta\beta_1 = (xx_1 + \alpha yy_1)^2 - \alpha(xy_1 + yx_1)^2,$$

$$\beta^{-1} = \left(\frac{x}{\beta}\right)^2 - \alpha\left(\frac{y}{\beta}\right)^2.$$

Another proof of this fact can be given, using the quadratic extension $R_p(\sqrt{\alpha})$ of the field R_p (assuming that α is not a square in R_p). The equation $\beta = x^2 - \alpha y^2$ simply says that β is the norm of the number $\xi = x + y\sqrt{\alpha}$ of the field $R_p(\sqrt{\alpha})$. But if $\beta = N(\xi)$ and $\beta_1 = N(\xi_1)$, then $\beta\beta_1 = N(\xi\xi_1)$ and $\beta^{-1} = N(\xi^{-1})$.

If α is a square in R, then the form (6.4) represents zero, and hence represents all numbers of R_p. Hence in this case H_α coincides with the entire multiplicative group R_p^* of the field R_p.

Since the form (6.4) represents all squares of the field R_p (set $y = 0$), then $R_p^{*2} \subset H_\alpha$. By the corollaries to Theorems 1 and 2 the index $(R_p^* : R_p^{*2})$ is finite, so that the group H_α has finite index in R_p^*.

Theorem 6. If the number $\alpha \in R_p^*$ is not a square, then $(R_p^* : H_a) = 2$.

Proof. First note that the form (6.4) represents the p-adic number β if and only if the form

$$\alpha x^2 + \beta y^2 - z^2 \tag{6.5}$$

represents zero (Theorem 6 of the Supplement, Section 1). The representability of zero will not be changed if α and β are multiplied by squares. Hence we may assume that α and β are taken from some fixed system of coset representatives of R_p^{*2} in R_p^*.

First, let $p \neq 2$. We claim that $H_\alpha \neq R_p^{*2}$. This is clear if $-\alpha$ is not a square (since $-\alpha \in H_\alpha$). If $-\alpha$ is a square, then the form $x^2 - \alpha y^2$ is equivalent to the form $x^2 + y^2$, which represents all p-adic units (Corollary 2 of Theorem 3), so that H_α does not coincide with R_p^{*2}. Further, we claim that H_α does not coincide with R_p^* (assuming, of course, that $\alpha \notin R_p^{*2}$). If ε is a nonsquare p-adic unit, then we may assume that α is ε, p, or $p\varepsilon$. But by Theorem 3 (and Theorem 10 of the Supplement, Section 1) the form (6.5) does not represent zero if $\alpha = \varepsilon$, $\beta = p$, or when $\alpha = p$ or $p\varepsilon$ and $\beta = \varepsilon$. Thus $H_\alpha \neq R_p^*$. Since $R_p^{*2} \subset H_\alpha \subset R_p^*$, the index $(R_p^* : H_\alpha)$ must divide the index $(R_p^* : R_p^{*2}) = 4$ (by Corollary 2 of Theorem 1). But we have shown that the index is neither 4 nor 1, so that $(R_p^* : H_\alpha) = 2$ and Theorem 6 is proved in the case $p \neq 2$.

Now let $p = 2$. In this case we have $(R_2^* : R_2^{*2}) = 8$, and as coset representatives we may take the numbers 1, 3, 5, 7, $2\cdot1$, $2\cdot3$, $2\cdot5$, $2\cdot7$. We shall therefore assume that α and β, in the form (6.5), are taken from this set. We thus need to check which of these forms represents zero in R. The answer is given in the following table, in which a "+" sign denotes that for the corresponding α and β the form (6.5) represents zero in R, and an empty square denotes that the form does not represent zero.

α \ β	1	3	5	7	2·1	2·3	2·5	2·7
1	+	+	+	+	+	+	+	+
3	+		+			+		+
5	+	+	+	+				
7	+		+		+		+	
2·1	+			+	+			+
2·3	+	+					+	+
2·5	+			+		+	+	
2·7	+	+			+	+		

[Since the form (6.5) is symmetric in α and β, the table is symmetric about its main diagonal.] We see that in each row except the first one the "+" occurs in four columns. This means that for each nonsquare $\alpha \in R_2{}^*$, the form (6.4) represents precisely four cosets of the subgroup $R_2{}^{*2}$. Thus $(H_\alpha : R_2{}^{*2}) = 4$ and since $(R_2{}^* : R_2{}^{*2}) = 8$ (corollary of Theorem 2), then $(R_2{}^* : H_\alpha) = 2$.

We use the results of Section 6.2 to verify the table. Let $\alpha = 2\varepsilon$, $\beta = 2\eta$, where ε and η are 2-adic units, and let

$$2\varepsilon x^2 + 2\eta y^2 - z^2 = 0. \tag{6.6}$$

We may assume that x, y and z are integers, not all divisible by 2. It is clear that $z \equiv 0 \pmod 2$, and that neither x nor y are divisible by 2 [otherwise the left side of (6.6) would not be divisible by 4]. Setting $z = 2t$, we put (6.6) in the form

$$\varepsilon x^2 + \eta y^2 - 2t^2 = 0.$$

This equation, by the corollary of Theorem 4, is equivalent to the corresponding congruence modulo 8 (with x and y odd). Since $x^2 \equiv y^2 \equiv 1 \pmod 8$, and either $2t^2 \equiv 2 \pmod 8$ or $2t^2 \equiv 0 \pmod 8$, then (6.6) is solvable if and only if one of the following holds:

$$\varepsilon + \eta \equiv 2 \pmod 8; \qquad \varepsilon + \eta \equiv 0 \pmod 8.$$

Let now $\alpha = 2\varepsilon$, $\beta = \eta$. In the equation $2\varepsilon x^2 + \eta y^2 - z^2 = 0$ (with x, y and z 2-adic integers not all divisible by 2) we obtain, by similar reasoning, $y \not\equiv 0$

(mod 2) and $z \not\equiv 0 \pmod 2$. Hence (again by the corollary of Theorem 4), this equation can be satisfied if and only if we have one of the congruences:

$$2\varepsilon + \eta \equiv 1 \pmod 8; \qquad \eta \equiv 1 \pmod 8; \tag{6.7}$$

which correspond to the cases $2 \nmid x$ and $2 \mid x$.

Only the case $\alpha = \varepsilon$, $\beta = \eta$ remains. If in the equation $\varepsilon x^2 + \eta y^2 - z^2 = 0$ the p-adic integers x, y, and z are not all divisible by 2, then precisely one of them is divisible by 2 and the other two are not. If $z \equiv 0 \pmod 2$ then $\varepsilon x^2 + \eta y^2 \equiv \varepsilon + \eta \equiv 0 \pmod 4$, so that either $\varepsilon \equiv 1 \pmod 4$ or $\eta \equiv 1 \pmod 4$. If $z \not\equiv 0 \pmod 2$, then $\varepsilon x^2 + \eta y^2 \equiv 1 \pmod 4$, and since precisely one of the numbers x and y is divisible by 2, we again find that at least one of the congruences

$$\varepsilon \equiv 1 \pmod 4, \qquad \eta \equiv 1 \pmod 4 \tag{6.8}$$

holds. Conversely, assume, say, that $\varepsilon \equiv 1 \pmod 4$. Then the congruence $\varepsilon x^2 + \eta y^2 - z^2 \equiv 0 \pmod 8$ is satisfied by $x = 1, y = 0, z = 1$ if $\varepsilon \equiv 1 \pmod 8$, and by $x = 1$, $y = 2$, $z = 1$ if $\varepsilon \equiv 5 \pmod 8$, and this means that the form $\varepsilon x^2 + \eta y^2 - z^2$ represents zero.

This ends the verification of the table and hence the proof of Theorem 6.

From Theorem 6 it follows that if α is a nonzero p-adic number which is not a square, then the factor group R_p^*/H_α is a cyclic group of order 2. We can thus establish an isomorphism between this factor group and the group $\{1, -1\}$ of square roots of 1. The unique isomorphism between R_p^*/H_α and $\{1, -1\}$ sends the subgroup H_α to the number $+1$, and the coset βH_α, distinct from H_α, to the number -1. It will be easier for us to deal with the homomorphism of the group R_p^* onto the group $\{1, -1\}$ with kernel H_α, since then we will have a function on R_p^* (and not on the factor group R_p^*/H_α).

Definition. For any pair $\alpha \neq 0$, $\beta \neq 0$ of p-adic numbers, we define the symbol (α, β) to be equal to $+1$ or to -1, depending on whether the form $\alpha x^2 + \beta y^2 - z^2$ represents zero in the field R_p^* or not. The symbol (α, β) is called the *Hilbert symbol.*

It follows immediately from the definition that if α is a square, then $(\alpha, \beta) = 1$ for all β. If $\alpha \notin R_p^{*2}$, then $(\alpha, \beta) = 1$ if and only if $\beta \in H_\alpha$. Thus for any $\alpha \neq 0$, the mapping $\beta \to (\alpha, \beta)$ is a homomorphism of the group R_p^* to the group $\{1, -1\}$ with kernel H_α. In other words,

$$(\alpha, \beta_1\beta_2) = (\alpha, \beta_1)(\alpha, \beta_2). \tag{6.9}$$

Further, the definition of the symbol (α, β) depends on the solvability of (6.5), which is symmetric in α and β, so that

$$(\beta, \alpha) = (\alpha, \beta), \tag{6.10}$$

from which, by (6.9),

$$(\alpha_1\alpha_2, \beta) = (\alpha_1, \beta)(\alpha_2, \beta). \tag{6.11}$$

We note that

$$(\alpha, -\alpha) = 1 \tag{6.12}$$

for any $\alpha \in R_p{}^*$ (since the equation $\alpha x^2 - \alpha y^2 - z = 0$ has the solution $x = y = 1$, $z = 0$), and thus, by (6.9),

$$(\alpha, \alpha) = (\alpha, -1). \tag{6.13}$$

Using (6.9) to (6.13) the computation of (α, β) in the general case is reduced to the computation of (p, ε) and (ε, η), where ε and η are p-adic units. Indeed, if $\alpha = p^k\varepsilon$, $\beta = p^l\eta$, then from these formulas we obtain

$$(p^k\varepsilon, p^l\eta) = (p, p)^{kl}(\varepsilon, p)^l(p, \eta)^k(\varepsilon, \eta) = (p, \varepsilon^l\eta^k(-1)^{kl})(\varepsilon, \eta).$$

We now compute (p, ε) and (ε, η). If $p \neq 2$, then by Theorem 3 the form $px^2 + \varepsilon y^2 - z^2$ represents zero if and only if $\varepsilon y^2 - z^2$ represents zero, that is, if and only if the unit ε is a square. Thus $(p, \varepsilon) = (\varepsilon/p)$ for $p \neq 2$ (see Section 6.1). By Corollary 2 of Theorem 3, the form $\varepsilon x^2 + \eta y^2 - z^2$ always represents zero, and thus $(\varepsilon, \eta) = +1$ for any p-adic units ε and η $(p \neq 2)$.

If $p = 2$, the values of the symbols $(2, \varepsilon)$ and (ε, η) have already essentially been found in the proof of Theorem 6. For by (6.7), with $\varepsilon = 1$, the form $2x^2 + \eta y^2 - z^2$ represents zero if and only if $\eta \equiv \pm 1 \pmod 8$. Hence $(2, \eta) = (-1)^{(\eta^2-1)/8}$. Further, the form $\varepsilon x^2 + \eta y^2 - z^2$ represents zero if and only if one of the congruences of (6.8) is fulfilled. Thus $(\varepsilon, \eta) = (-1)^{[(\varepsilon-1)/2][(\eta-1)/2]}$. Summing up, we have

Theorem 7. The values of the Hilbert symbols (p, ε) and (ε, η) for p-adic units ε and η are given by the formulas

$$(p, \varepsilon) = \left(\frac{\varepsilon}{p}\right), \quad (\varepsilon, \eta) = 1 \qquad \text{for} \quad p \neq 2,$$

$$(2, \varepsilon) = (-1)^{(\varepsilon^2-1)/8}, \quad (\varepsilon, \eta) = (-1)^{[(\varepsilon-1)/2][(\eta-1)/2]} \qquad \text{for } p = 2.$$

6.4. *Equivalence of Binary Forms*

The Hilbert symbol allows us to give explicit conditions for the equivalence of two binary quadratic forms over the field R_p. Let $f(x, y)$ and $g(x, y)$ be two binary nonsingular quadratic forms over R_p with determinants $\delta(f)$ and $\delta(g)$. For f and g to be equivalent, it is necessary that $\delta(f)$ and $\delta(g)$ differ by a factor which lies in $R_p{}^{*2}$ (Theorem 1 of the Supplement, Section 1). To formulate another necessary condition for equivalence, which, along with the above one will be sufficient, we need the following fact.

Theorem 8. Let the binary form f have determinant $\delta \neq 0$. Then the Hilbert symbol $(\alpha, -\delta)$ takes the same value for all nonzero p-adic numbers α represented by f.

Proof. Let α and α' be two nonzero p-adic numbers represented by the form f. By Theorem 2 of Section 1 of the Supplement the form f is equivalent to a form f_1 of the type $\alpha x^2 + \beta y^2$. Since α' is also represented by f, then $\alpha' = \alpha x_0{}^2 + \beta y_0{}^2$, so that $\alpha\alpha' - \alpha\beta y_0{}^2 - (\alpha x_0)^2 = 0$. Hence the form $\alpha\alpha' x^2 - \alpha\beta y^2 - z^2$ represents zero, so that $(\alpha\alpha', -\alpha\beta) = 1$. But $\alpha\beta$ differs from δ by a square factor, so that $(\alpha\alpha', -\delta) = 1$, and thus by $(\alpha, -\delta) = (\alpha', -\delta)$, which proves the theorem.

Theorem 8 implies that the binary form f has a new invariant, and we set

$$e(f) = (\alpha, -\delta(f)),$$

where α is any nonzero p-adic number which is represented by f.

Theorem 9. Let f and g be two nonsingular binary quadratic forms over the field R_p. f and g are equivalent if and only if both of the following conditions hold:

$$(1)\ \delta(f) = \delta(g)\gamma^2, \qquad \gamma \in R_p{}^*;$$

$$(2)\ e(f) = e(g).$$

Proof. The necessity of both conditions is clear. To prove sufficiency we first show that the two forms represent the same p-adic numbers. Let the number $\gamma \in R_p{}^*$ be represented by the form g. Letting $f = \alpha x^2 + \beta y^2$, we have

$$(\alpha, -\alpha\beta) = e(f) = e(g) = (\gamma, -\delta(g)) = (\gamma, -\alpha\beta),$$

by which

$$(\gamma\alpha^{-1}, -\alpha\beta) = 1.$$

By the definition of the Hilbert symbol this means that we can solve the equation

$$\gamma\alpha^{-1}x^2 - \alpha\beta y^2 - z^2 = 0$$

in nonzero x, y, and z. But then

$$\gamma = \alpha\left(\frac{z}{x}\right)^2 + \beta\left(\frac{\alpha y}{x}\right)^2,$$

that is, γ is represented by the form f. The equivalence of f and g now follows from Theorem 11 of Section 1 of the Supplement.

6.5. *Remarks on Forms of Higher Degree*

Theorem 5 on quadratic forms over the field R_p is one of a class of theorems in number theory which run as follows: "All is well as long as the number of variables is sufficiently large." In this case "well" means that the quadratic form represents zero over the field of p-adic numbers, and "sufficiently large" means that the number of variables is at least five. It would be most interesting to observe this phenomenon also in the case of forms of higher degree over the field of p-adic numbers.

The precise formulation is this. For any natural number r there is a number $N(r)$ such that any form of degree r over the field R_p represents zero, provided that the number of variables exceeds $N(r)$. We note that there is no reason to believe, a priori, that any such number $N(r)$ exists, but Brauer showed that it does. However, his bound is rather large [R. Brauer, "A note on systems of homogeneous algebraic equations," *Bull. Am. Math. Soc.* **51** (1945) pp. 749-755]. For $r = 2$, Theorem 5 shows that $N(r) = r^2$. For $r = 3$, Demyanov and Lewis showed that $N(r) = r^2$ also; that is, any cubic form over the field of p-adic numbers in at least 10 variables represents zero [V. B. Demyanov, "On cubic forms over discrete normed fields," *Dokl. Akad. Nauk SSSR* **74** (1950) pp. 889-891; D. J. Lewis, "Cubic homogeneous polynomials over p-adic fields," *Ann. Math.* **56** (1952) pp. 473-478]. It was believed for some time that $N(r) = r^2$ is true in general, but recently, G. Terjanian found a counterexample.

One may also consider systems of equations

$$\begin{gathered} F_1(x_1, \ldots, x_m) = 0 \\ \cdots\cdots\cdots\cdots \\ F_k(x_1, \ldots, x_m) = 0 \end{gathered} \tag{6.14}$$

in which $F_1, \ldots, F_k$ are forms with p-adic coefficients of degrees $r_1, \ldots, r_k$. In the case of two quadratic forms, with $m \geqq 9$, the solvability of the system was shown by Demyanov [a simple proof of this result of Demyanov was given in B. J. Birch, D. J. Lewis, and T. G. Murphy, "Simultaneous quadratic forms," *Am. J. Math.* **84** (1962) pp. 110-115]. A general method is known which shows how to get solutions for systems (6.14) when m is sufficiently large compared to $r_1, \ldots, r_k$ provided one knows the function $N(r)$ mentioned in the preceding paragraph [see, for instance, S. Lang, "On quasi-algebraic closure," *Ann. Math.* **55** (1952) pp. 373-390].

Finally, it is easily shown that the hypothesized value for $N(r)$ is the best possible, that is, for any r there is a form of degree r in r^2 variables which does not represent zero over the field of p-adic numbers. We give an example of such a form. Recall that in Section 2.1 we constructed a form $F(x_1, \ldots, x_n)$

of degree n in n variables such that the congruence

$$F(x_1, \dots, x_n) \equiv 0 \pmod{p}$$

had only the zero solution:

$$x_1 \equiv 0 \pmod{p}, \dots, x_n \equiv 0 \pmod{p}. \tag{6.15}$$

Set

$$\Phi(x_1, \dots, x_{n^2}) = F(x_1, \dots, x_n) + pF(x_{n+1}, \dots, x_{2n}) + \cdots + p^{n-1}F(x_{n^2-n+1}, \dots, x_{n^2})$$

We shall show that the form Φ does not represent zero in the field of p-adic numbers. Assume the contrary, that is, assume that the equation

$$\Phi(x_1, \dots, x_{n^2}) = 0 \tag{6.16}$$

has a nonzero solution. Since Φ is homogeneous we may assume that all variables are integers and that at least one of them is not divisible by p. Considering as a congruence modulo p, we obtain that $F(x_1, \dots, x_n) \equiv 0 \pmod{p}$, from which it follows by (6.15) that $x_1 = px_1', \dots, x_n = px_n'$. Equation (6.16) then takes the form

$$p^n F(x_1', \dots, x_n') + pF(x_{n+1}, \dots, x_{2n}) + \cdots + p^{n-1}F(x_{n^2-n+1}, \dots, x_{n^2}) = 0$$

or, after dividing by p,

$$F(x_{n+1}, \dots, x_{2n}) + \cdots + p^{n-2}F(x_{n^2-n+1}, \dots, x_{n^2}) + p^{n-1}F(x_1', \dots, x_n') = 0.$$

As in the previous step, we obtain here that $x_{n+1}, \dots, x_{2n}$ are divisible by p. Repeating this process n times, we obtain that $x_1, \dots, x_{n^2}$ are divisible by p, which is a contradiction.

PROBLEMS

1. Verify the following properties of the Hilbert symbol:

(1) $\qquad (\alpha, 1-\alpha) = +1, \ \alpha \neq 1;$

(2) $\qquad (\alpha, \beta) = (\gamma, -\alpha\beta), \ \gamma = \alpha\xi^2 + \beta\eta^2 \neq 0;$

(3) $\qquad (\alpha\gamma, \beta\gamma) = (\alpha, \beta)(\gamma, -\alpha\beta).$

2. Let $f = \alpha_1 x_1^2 + \cdots + \alpha_n x_n^2$ ($\alpha_i \in R_p^*$) be a quadratic form, and define the Hasse symbol by the formula

$$c_p(f) = (-1, -1) \prod_{1 \leqslant i \leqslant j \leqslant n} (\alpha_i, \alpha_j)$$

Show that

$$c_p(\alpha x^2 \dot{+} f) = c_p(f)(\alpha, -\delta),$$

$$c_p(\alpha x^2 \dot{+} \beta y^2 \dot{+} f) = c_p(f)(\alpha\beta, -\delta)(\alpha, \beta)$$

(δ is the determinant of the form f).

3. Let the form $f = \alpha_1 x_1^2 + \cdots + \alpha_n x_n^2$ with p-adic coefficients represent the number $\gamma \neq 0$ of R_p. Show that there is a representation $\gamma = \alpha_1 \xi_1^2 + \cdots + \alpha_n \xi_n^2$ ($\xi_i \in R_p$) such that all "partial sums" $\gamma_k = \alpha_1 \xi_1^2 + \cdots + \alpha_k \xi_k^2 (1 \leqslant k \leqslant n)$ are nonzero. (Use Theorems 5 and 8 of Section 1 of the Supplement.)

4. Using the same notation, show that the form f is equivalent to a diagonal form $g = \gamma y_1^2 + \beta_2 y_2^2 + \cdots + \beta_n y_n^2$, for which $c_p(g) = c_p(f)$. (First show that the form $\alpha x^2 + \beta y^2$ is transformed by the substitution $x = \mu X - \nu\beta Y$, $y = \nu X + \mu\alpha Y$ $(\alpha\mu^2 + \beta\nu^2 = \gamma \neq 0)$ into $\gamma X^2 + \alpha\beta\gamma Y^2$, and $(\alpha, \beta) = (\gamma, \alpha\beta\gamma)$.)

5. Show, by induction on the number of variables, that equivalent diagonal nonsingular quadratic forms over the field R_p have the same Hasse symbol (use Theorem 4 of Section 1 of the Supplement). The Hasse symbol can thus be defined for arbitrary nonsingular quadratic forms: If the form f is equivalent to the diagonal form f_0, set $c_p(f) = c_p(f_0)$.

6. Let f_1 and f_2 be two quadratic forms over the field R_p with determinants $\delta_1 \neq 0$ and $\delta_2 \neq 0$. Show that

$$c_p(f_1 \dot{+} f_2) = c_p(f_1)c_p(f_2)(-1, -1)(\delta_1, \delta_2).$$

7. Let f be a nonsingular quadratic form over the field R_p, with δ its discriminant and α a nonzero number from R_p. Show that

$$c_p(\alpha f) = \begin{cases} c_p(f)(\alpha, (-1)^{(n+1)/2} & \text{if } n \text{ is odd,} \\ c_p(f)(\alpha, (-1)^{n/2}\delta), & \text{if } n \text{ is even.} \end{cases}$$

8. Show that a nonsingular quadratic form in three variables over the field R_p represents zero if and only if $c_p(f) = +1$.

9. Let f be a nonsingular quadratic form in four variables over the field R_p with determinant δ. Show that f does not represent zero in R if and only if δ is a square in R_p and $c_p(f) = -1$.

10. Let f be a nonsingular quadratic form in n variables over R_p with determinant δ. Show that f represents the nonzero p-adic number α if and only if one of the following holds:

(a) $n = 1$ and $\alpha\beta$ is a square in R_p.
(b) $n = 2$ and $c_p(f) = (-\alpha, -\delta)$.
(c) $n = 3$, $-\alpha\delta$ is a square in R_p and $c_p(f) = 1$.
(d) $n = 3$ and $-\alpha\delta$ is not a square in R_p.
(e) $n \geqslant 4$.

11. Give the conditions under which a nonsingular quadratic form over the field R_p does not represent zero (nontrivially), but otherwise does represent all p-adic numbers.

12. In which p-adic fields does the form $2x^2 - 15y^2 + 14z^2$ fail to represent zero?

13. Which 5-adic numbers are represented by the form $2x^2 + 5y^2$?

14. Let f and f' be nonsingular quadratic forms in n variables over the field R_p with determinants δ and δ'. Show that f and f' are equivalent if and only if $c_p(f) = c_p(f')$ and $\delta = \delta'\alpha^2$ ($\alpha \in R_p$).

7. Rational Quadratic Forms

7.1. *The Hasse–Minkowski Theorem*

In this section we shall give the proof of one of the most important results of number theory—the so-called *Hasse–Minkowski theorem*—of which we have already spoken at the beginning of this chapter.

Theorem 1 (Hasse–Minkowski). A quadratic form with rational coefficients represents zero in the field of rational numbers if and only if it represents zero in the field of real numbers and in all fields of p-adic numbers (for all primes p).

The proof of this theorem depends essentially on the number n of variables of the quadratic form. For $n = 1$ the assertion of the theorem is trivial. In the case $n = 2$ the proof is very simple. If the binary rational quadratic form f with discriminant $d \neq 0$ represents zero in the field of real numbers, then $-d > 0$ (see Theorem 10 of Section 1 of the Supplement); hence $-d = p_1^{k_1} \cdots p_s^{k_s}$, where the p_i are distinct primes. If now f represents zero in the field R_p, then since $-d$ is a square in R_{p_i}, the exponent k_i must be even ($i = 1, \ldots, s$). But in this case $-d$ is a square in the field of rational numbers and hence f represents zero in R.

The proof of the theorem for $n \geqslant 3$ is rather difficult. The various cases which occur will be analyzed in the following paragraphs. But first we make some preliminary remarks.

We may assume that the coefficients of the quadratic form $f(x_1, \ldots, x_n)$ are rational integers (if not, multiply it by the least common multiple of the denominators of the coefficients). It is clear that the equation

$$f(x_1, \ldots, x_n) = 0 \tag{7.1}$$

can be solved in the field of rational numbers R (or in the field of p-adic numbers R_p) if and only if it can be solved in the ring Z of integers (respectively, in the ring O_p of p-adic integers). Further, (7.1) is solvable in real numbers if and only if the form f is indefinite. Hence, by Theorem 2 of Section 5, we may formulate the Hasse–Minkowski theorem as follows:

Equation (7.1) is solvable in rational integers if and only if the form f is indefinite and for any modulus p the congruence

$$f(x_1, \ldots, x_n) \equiv 0 \pmod{p^m}$$

has a solution in which at least one of the variables is not divisible by p. By Theorem 5 of Section 6 any form in five or more variables represents zero in the field of p-adic numbers. Hence, for such forms the Hasse–Minkowski theorem reads: In order that a nonsingular rational quadratic form in

$n \geqslant 5$ variables represent zero in the field of rational numbers, it is necessary and sufficient that it be indefinite.

Thus the conditions for solvability in p-adic fields actually need only be verified for $n = 3$ and $n = 4$. For these values of n the Hasse–Minkowski theorem gives us an effective criterion for the solvability of (7.1). Indeed, if the form f is given by $f = \sum a_i x_i^2$, then Corollary 2 of Theorem 3, Section 6, implies that, for any odd prime p which does not provide any of the a_i, the form f with $n \geqslant 3$ always represents zero in R. Thus only a finite number of primes p actually need be considered. For each of these p the theorems of Section 6 decide the question of the representation of zero by f in R_p.

By Theorem 6, Section 1, of the Supplement, Theorem 1 implies the following.

Corollary. A nonsingular quadratic form with rational coefficients represents the rational number a if and only if it represents a in the field of real numbers and in the field of p-adic numbers for all primes p.

7.2. *Forms with Three Variables*

We now turn to the proof of the Hasse–Minkowski theorem, treating the case $n = 3$ in this section. For forms in three variables Theorem 1 was proved (in somewhat different terminology) by Legendre. The formulation of Legendre is given in Problem 1.

Let the form be given as $a_1x^2 + a_2y^2 + a_3z^2$. Since the form is indefinite, not all the coefficients a_1, a_2, a_3 have the same sign. Multiplying (if necessary) the form by -1, we may assume that two coefficients are positive and one negative. We may assume that a_1, a_2, a_3 are integers, square-free and relatively prime (they can be divided by their greatest common divisor). Further, if, say, a_1 and a_2 have a common prime factor p, then, multiplying the form by p and taking px and py as new variables, we obtain a form with coefficients a_1/p, a_2/p, pa_3. Repeating this process as necessary, we arrive at a form

$$ax^2 + by^2 - cz^2, \tag{7.2}$$

whose coefficients are positive integers a, b, and c which are pairwise relatively prime (and square-free).

Let p be some odd prime divisor of the number c. Since by assumption the form (7.2) represents zero in R_p, by Theorem 3 of Section 6 and Corollary 1 of that theorem, the congruence $ax^2 + by^2 \equiv 0 \pmod{p}$ has a nontrivial solution, say (x_0, y_0). Then the form $ax^2 + by^2$ factors, modulo p, into linear factors:

$$ax^2 + by^2 \equiv ay_0^{-2}(xy_0 + yx_0)(xy_0 - yx_0) \pmod{p}.$$

The same also holds for the form (7.2), so that we have

$$ax^2 + by^2 - cz^2 \equiv L^{(p)}(x, y, z)M^{(p)}(x, y, z) \pmod{p}, \tag{7.3}$$

where $L^{(p)}$ and $M^{(p)}$ are integral linear forms. An analogous congruence also holds for the odd prime divisors of the coefficients a and b, and also for the prime 2, since

$$ax^2 + by^2 - cz^2 \equiv (ax + by - cz)^2 \pmod{2}.$$

We find linear forms $L(x, y, z)$ and $M(x, y, z)$ such that

$$L(x, y, z) \equiv L^{(p)}(x, y, z) \pmod{p},$$

$$M(x, y, z) \equiv M^{(p)}(x, y, z) \pmod{p}$$

for all prime divisors p of the coefficients a, b, and c. The congruence (7.3) shows that

$$ax^2 + by^2 - cz^2 \equiv L(x, y, z)M(x, y, z) \pmod{abc}. \tag{7.4}$$

We shall give integer values to the variables x, y, and z, satisfying the inequalities

$$0 \leqslant x < \sqrt{bc}, \quad 0 \leqslant y < \sqrt{ac}, \quad 0 \leqslant z < \sqrt{ab}. \tag{7.5}$$

If we exclude from consideration the case $a = b = c = 1$ (the assertion of the theorem is obvious for the form $x^2 + y^2 - z^2$, since it represents zero in any field), then, since a, b, c are pairwise relatively prime, the numbers $\sqrt{ac}$, $\sqrt{bc}$, and $\sqrt{ab}$ will not all be integers. Hence the number of triples (x, y, z) satisfying (7.5) will be strictly greater than $\sqrt{ab}\cdot\sqrt{bc}\cdot\sqrt{ac} = abc$. Since the number of triples (x, y, z) is greater than the number of residues modulo abc, there are two distinct triples (x_1, y_1, z_1) and (x_2, y_2, z_2) such that

$$L(x_1, y_1, z_1) \equiv L(x_2, y_2, z_2) \pmod{abc}.$$

The linearity of L implies that

$$L(x_0, y_0, z_0) \equiv 0 \pmod{abc},$$

where

$$x_0 = x_1 - x_2, \quad y_0 = y_1 - y_2, \quad z_0 = z_1 - z_2.$$

From (7.4) it follows that

$$a{x_0}^2 + b{y_0}^2 - c{z_0}^2 \equiv 0 \pmod{abc}. \tag{7.6}$$

Since the triples (x_1, y_1, z_1) and (x_2, y_2, z_2) satisfy (7.5),

$$|x_0| < \sqrt{bc}, \quad |y_0| < \sqrt{ac}, \quad |z_0| < \sqrt{ab},$$

so that

$$-abc < a{x_0}^2 + b{y_0}^2 - c{z_0}^2 < 2abc. \tag{7.7}$$

The inequality (7.7) and the congruence (7.6) combine to give either

$$ax_0^2 + by_0^2 - cz_0^2 = 0, \tag{7.8}$$

or

$$ax_0^2 + by_0^2 - cz_0^2 = abc. \tag{7.9}$$

In the first case we have a nontrivial representation of zero by the form (7.2), which is what was required. In the second case we rely on the following lemma.

Lemma 1. If the form (7.2) represents abc, then it also represents zero.

Let x_0, y_0, z_0 satisfy (7.9). It is easily seen that then

$$a(x_0z_0 + by_0)^2 + b(y_0z_0 - ax_0)^2 - c(z_0^2 + ab)^2 = 0. \tag{7.10}$$

If $z_0^2 + ab \neq 0$, then this equality proves the lemma. If $-ab = z_0^2$, then the form $ax^2 + by^2$ represents zero (Theorem 10 of Section 1 of the Supplement). But then (7.2) also represents zero, so that the lemma is proved.

This proof is very short, but is based on the computation involved in (7.10). We shall give another proof which uses more general methods. If bc is a square, then the form $by^2 - cz^2$, and hence also (7.2), represents zero. Assume that bc is not a square. It will be shown that in this case the representability of zero by (7.2) is equivalent to ac being the norm of some element from the field $R(\sqrt{bc})$. Indeed, from (7.8) (where we may assume that $x \neq 0$) it follows that

$$ac = \left(\frac{cz_0}{x_0}\right)^2 - bc\left(\frac{y_0}{x_0}\right)^2 = N\left(\frac{cz_0}{x_0} + \frac{y_0}{x_0}\sqrt{bc}\right).$$

Conversely, if $ac = N(u + v\sqrt{bc})$, then

$$ac^2 + b(cv)^2 - cu^2 = 0.$$

Assume now that (7.9) holds. Multiplying it by c, we obtain either

$$ac(x_0^2 - bc) = (cz_0)^2 - bcy_0^2$$

or

$$acN(\alpha) = N(\beta),$$

where $\alpha = x_0 + \sqrt{bc}$, $\beta = cz_0 + y_0\sqrt{bc}$. But then

$$ac = N(\gamma),\ \gamma = \frac{\beta}{\alpha} \in R(\sqrt{bc}),$$

and this, as we have seen, means that (7.2) represents zero in R.

We now note the following fact. In the proof of Theorem 1 for the case of three variables we have never used the fact that the form (7.2) represents zero over the field of 2-adic numbers. Hence, from the solvability of (7.2) in the field of real numbers and also in the field of p-adic numbers for all odd p it follows that (7.2) is solvable in the field R_2. It will be shown that an analogous result holds also for any other field R_q. Namely, if the rational quadratic form in three variables represents zero in the field of real numbers and also in all fields R_p, with the possible exception of the field R_q, then it represents zero in the field R_q (and hence, by what has been proved, also in the field R).

We shall try to explain the cause of this phenomenon. Consider the conditions for the representability of zero by the form

$$ax^2 + by^2 - z^2 \tag{7.11}$$

in all fields R_p and in the field of real numbers (here a and b are arbitrary nonzero rational numbers; hence any nonsingular rational quadratic form in three variables can, after change of variables and multiplication by some rational number, be put into this form). By Section 3.6, the condition for the representability of zero in the field of p-adic numbers can be expressed as

$$\left(\frac{a, b}{p}\right) = 1, \tag{7.12}$$

where $(a, b/p)$ is the Hilbert symbol in the field R_p. For rational a and b we use the notation $(a, b/p)$ for the Hilbert symbol (a, b) to denote the field in which it is being considered. This change in notation is necessary because we will now be considering the Hilbert symbol simultaneously in different fields.

As for the real numbers, the form (7.11) clearly represents zero if and only if at least one of the numbers a, b is positive. To write this condition in the form (7.12), we carry over the results of Section 6.3 to the field of real numbers. We first agree on the following notation. All p-adic fields R_p and the field of real numbers together comprise all completions of the field R of rational numbers (Section 4.2). The fields R_p are in one-to-one correspondence with the rational primes p. To extend this correspondence to the field of real numbers, we introduce the symbol ∞, which we call the infinite prime, and we say that the real numbers are the completion of the field R with respect to the infinite prime. An ordinary prime p, in contrast, is called a *finite prime.* By analogy with the notation R_p for the p-adic fields, we denote the field of real numbers by R_∞.

For any α from the multiplicative group $R_\infty{}^*$ of the field R_∞, we consider the form

$$x^2 - \alpha y^2 \tag{7.13}$$

and by H_α we denote the set of all $\beta \in R_\infty{}^*$, represented by this form. If $\alpha > 0$, that is, $\alpha \in R_\infty{}^{*2}$, then the form (7.13) represents all real numbers, and thus $H_\alpha = R_\infty{}^*$. If $\alpha < 0$, that is, α is not a square, then the form (7.13) represents only positive numbers, and therefore as in Theorem 6 of Section 6, we have

$$(R_\infty{}^*:H_\alpha) = 2. \tag{7.14}$$

For α, $\beta \in R_\infty{}^*$ we set (α, β) equal to $+1$ or -1 depending on whether the form represents β or not, and it follows from the above that the symbol (α, β) will have all the properties (6.9) to (6.13). In analogy with Theorem 7 of Section 6 by which the Hilbert symbol was computed for the p-adic fields, we have here the much simpler relations

$$\begin{aligned} (\alpha, \beta) &= +1, \qquad \text{if } \alpha > 0 \text{ or } \beta > 0, \\ (\alpha, \beta) &= -1, \qquad \text{if } \alpha < 0 \text{ and } \beta < 0. \end{aligned} \tag{7.15}$$

For rational a and b we denote the value of the Hilbert symbol in the field R_∞ by $(a, b/\infty)$.

Using the Hilbert symbols $(a, b/p)$ we can now reformulate Theorem 1 for forms in three variables as follows.

The form $ax^2 + by^2 - z^2$ with nonzero rational coefficients a and b represents zero in the field of rational numbers if and only if for all p (including $p = \infty$)

$$\left(\frac{a, b}{p}\right) = 1. \tag{7.16}$$

For any nonzero rational numbers a and b the symbol $(a, b/p)$ differs from $+1$ only for finitely many values of p. Indeed, if p is not equal to 2 or ∞ and if p does not enter into the factorizations of a and b into prime powers (which means that a and b are p-adic units), then, by Corollary 2 of Theorem 3 of Section 6, the form (7.11) represents zero in R_p and thus for all such p the symbol $(a, b/p)$ equals $+1$. Besides this, it will now be shown that the value of the symbol $(a, b/p)$ for fixed a and b is subject to one further condition. Namely, the number of p (including $p = \infty$) for which $(a, b/p) = -1$ is always even. Another way of expressing this fact is to say that

$$\prod_p \left(\frac{a, b}{p}\right) = 1, \tag{7.17}$$

where p runs through all prime numbers and the symbol ∞. For the formal infinite product on the left contains only a finite number of terms different from $+1$, so the product will be 1 if and only if the number of p for which $(a, b/p) = -1$ is even.

We now prove (7.17). Factoring a and b into prime powers and using formulas (6.9) to (6.13) (also valid, as mentioned, for $p = \infty$), we easily reduce the proof of the general formula (7.17) to the proof of the following special cases:

(1) $a = -1, b = -1$.
(2) $a = q, b = -1$ (q a prime).
(3) $a = q, b = q'$ (q and q' distinct primes).

By Theorem 7 of Section 6 and (7.15), we have

$$\prod_p \left(\frac{-1, -1}{p}\right) = \left(\frac{-1, -1}{2}\right)\left(\frac{-1, -1}{\infty}\right) = (-1)\cdot(-1) = 1,$$

$$\prod_p \left(\frac{2, -1}{p}\right) = \left(\frac{2, -1}{2}\right)\left(\frac{2, -1}{\infty}\right) = 1 \cdot 1 = 1,$$

$$\prod_p \left(\frac{q, -1}{p}\right) = \left(\frac{q, -1}{q}\right)\left(\frac{q, -1}{2}\right) = \left(\frac{-1}{q}\right)(-1)^{\frac{q-1}{2}\cdot\frac{-1-1}{2}} = 1,$$

$$\prod_p \left(\frac{2, q}{p}\right) = \left(\frac{2, q}{q}\right)\left(\frac{2, q}{2}\right) = \left(\frac{2}{q}\right)(-1)^{\frac{q^2-1}{8}} = 1,$$

$$\prod_p \left(\frac{q, q'}{p}\right) = \left(\frac{q, q'}{q}\right)\left(\frac{q, q'}{q'}\right)\left(\frac{q, q'}{2}\right) = \left(\frac{q'}{q}\right)\left(\frac{q}{q'}\right)(-1)^{\frac{q'-1}{2}\cdot\frac{q-1}{2}} = 1.$$

These computations, in which q and q' denote distinct odd primes, prove the relation (7.17).

Note that in the proof of (7.17) we have used the quadratic reciprocity law of Gauss. On the other hand, knowing the explicit formulas for the Hilbert symbol (Theorem 7, Section 6), we can deduce all parts of the law of quadratic reciprocity from the formula (7.17). Thus (7.17) is equivalent to Gauss' reciprocity law.

Assume now that the form (7.11) represents zero in all fields R_p, except perhaps for R_q. From the equality (7.17), along with the fact that $(a, b/p) = 1$ for all $p \neq q$, we deduce that $(a, b/q) = 1$. In other words we have the following assertion.

Lemma 2. If a rational quadratic form in three variables represents zero in all fields R_p (p running through all prime numbers and the symbol ∞), except possibly for R_q, then it also represents zero in R_q.

7.3. *Forms in Four Variables*

We shall assume that our form is given by

$$a_1 x_1^2 + a_2 x_2^2 + a_3 x_3^2 + a_4 x_4^2, \tag{7.18}$$

where all a_i are square-free integers. Since the form is indefinite we may assume that $a_1 > 0$ and $a_4 < 0$. Along with the form (7.18) we consider the forms

$$g = a_1 x_1^2 + a_2 x_2^2 \qquad \text{and} \qquad h = -a_3 x_3^2 - a_4 x_4^2.$$

The idea of the proof of the Hasse–Minkowski theorem is as follows. Using the fact that the form (7.18) represents zero in the fields R_p, we shall show that there is a rational integer $a \neq 0$ which is simultaneously rationally represented by the forms g and h. This immediately gives us a rational representation of zero by the form (7.18).

Let $p_1, \ldots, p_s$ be all distinct odd primes dividing the coefficients a_1, a_2, a_3, a_4. For each of the preceding primes and also for $p = 2$, choose a representation of zero,

$$a_1 \xi_1^2 + a_2 \xi_2^2 + a_3 \xi_3^2 + a_4 \xi_4^2 = 0,$$

in the field R_p for which all $\xi_i \neq 0$ (see Theorem 8, Section 1, of the Supplement) and set

$$b_p = a_1 \xi_1^2 + a_2 \xi_2^2 = -a_3 \xi_3^2 - a_4 \xi_4^2.$$

Our representation can be chosen so that each b_p is a nonzero p-adic integer divisible at most by the first power of p (if $b_p = 0$, then the forms f and g represent zero in R_p and hence by Theorem 5 of Section 1 of the Supplement they represent all numbers of R_p).

Consider the system of congruences

$$\begin{aligned} a &\equiv b_2 \pmod{16}, \\ a &\equiv b_{p_1} \pmod{p_1^2}, \\ &\cdots\cdots\cdots\cdots \\ a &\equiv b_{p_s} \pmod{p_s^2}. \end{aligned} \tag{7.19}$$

A rational number a, satisfying these congruences, is uniquely determined modulo $m = 16 p_1^2 \cdots p_s^2$. Since b_{p_i} is divisible by at most the first power of p_i, $b_{p_i} a^{-1}$ is a p-adic unit, and

$$b_{p_i} a^{-1} \equiv 1 \pmod{p_i}.$$

By Corollary 1, Theorem 1, of Section 6 the quantity $b_{p_i} a^{-1}$ is a square in the field R_p. Analogously, since b_2 is not divisible by any higher power of 2 than the first, $b_2 a^{-1} \equiv 1 \pmod 8$, and therefore (Theorem 2 of Section 6) $b_2 a^{-1}$ is a square in R_2.

From the fact that b_p and a differ by a square factor in R_p, it follows that for all $p = 2, p_1, \ldots, p_s$ the forms

$$-a x_0^2 + g \qquad \text{and} \qquad -a x_0^2 + h \tag{7.20}$$

represent zero in R_p. If a is chosen to be positive, then since $a_1 > 0$ and

$-a_4 > 0$ the forms (7.20) represent zero in the field of real numbers. Finally, if p is different from $2, p_1, \ldots, p_s$ and does not divide a; that is, if p is odd and does not divide the coefficients of the form (7.20), then, by Corollary 2 of Theorem 3 of Section 6, these forms represent zero in R_p. If, in addition to $2, p, \ldots, p$, there were at most one more prime q dividing the integer a, then we could apply Lemma 2 and conclude (using the Hasse–Minkowski theorem for forms in three variables) that the forms (7.20) represent zero in the field of rational numbers. In such a case we would have the representations

$$a = a_1c_1^2 + a_2c_2^2, \qquad a = -a_3c_3^2 - a_4c_4^2$$

with rational c_i, from which

$$a_1c_1^2 + a_2c_2^2 + a_3c_3^2 + a_4c_4^2 = 0,$$

and the Hasse–Minkowski theorem would be proved for forms in four variables. We will now show that a number a, satisfying the congruence (7.19) and possessing the desired additional property, always can be found. To do this we shall have to apply the theorem of Dirichlet on prime numbers in arithmetic progressions, which we shall prove in Chapter 5, Section 3.2.† Dirichlet's theorem asserts that if the increment and first term of an infinite arithmetic progression are relatively prime, then the progression contains an infinite number of primes. Let $a^* > 0$ be any number satisfying (7.19). Let d denote the greatest common divisor of a^* and m. Since a^*/d and m/d are relatively prime, Dirichlet's theorem implies that there is an integer $k \geqslant 0$ such that $a^*/d + km/d = q$ is prime. As a we may then take

$$a = a^* + km = dq.$$

Since all the divisors of d are among $2, p_1, \ldots, p_s$, this choice of a allows us to finish the proof of Theorem 1 for forms in four variables.

7.4. *Forms in Five and More Variables*

Let an indefinite rational quadratic form in five variables be given by

$$a_1x_1^2 + a_2x_2^2 + a_3x_3^2 + a_4x_4^2 + a_5x_5^2, \tag{7.21}$$

where all a_i are square-free integers. We can assume that $a_1 > 0$ and $a_5 < 0$. Set

$$g = a_1x_1^2 + a_2x_2^2, \qquad h = -a_3x_3^2 - a_4x_4^2 - a_5x_5^2.$$

Reasoning precisely as in the case $n = 4$, we use Dirichlet's theorem to find a rational integer $a > 0$ which is represented by the forms g and h in the field of real numbers and in all the fields R_p, with the possible exception of R_q, where q is some prime number which does not divide the coefficients a_i. We claim that the forms g and h represent a in the field R_q. For the form g this is established exactly as before, using Lemma 2. The form h represents zero in R (Corollary 2 of Theorem 3, Section 6), and thus represents all numbers in R_q (Theorem 5 of Section 1 of the Supplement). By the Corollary to the Hasse–Minkowski theorem (see the end of Section 7.1), which has already been proved for forms in two and three variables, we find that the forms g and h represent a in the field of rational numbers. As before it easily follows that the form (7.21) admits a rational representation of zero.

For the proof of Theorem 1 in the case $n > 5$, we simply note that any indefinite quadratic form f, after being diagonalized, is easily represented as $f = f_0 \dotplus f_1$, where f_0 is an indefinite form in five variables. We have proved that f_0 represents zero in the field of rational numbers, and hence so does f. The Hasse–Minkowski theorem is completely proved.

7.5. *Rational Equivalence*

The Hasse–Minkowski theorem allows us to solve another important question on rational quadratic forms, the question of their equivalence.

Theorem 2. In order that two nonsingular quadratic forms with rational coefficients be equivalent over the field of rational numbers, it is necessary and sufficient that they be equivalent over the field of real numbers and over all the p-adic fields R_p.

Proof. The necessity of the condition is clear. The proof of sufficiency is carried out by induction on the number of variables. Let $n = 1$. The forms ax^2 and bx^2 are equivalent over any field provided only that a/b is a square in that field. But, if a/b is a square in the real field and also in all p-adic fields, then as we saw in Section 7.1, a/b is a square in the field R of rational numbers. Hence for $n = 1$ Theorem 2 holds.

Now let $n > 1$. Let $a \neq 0$ be a rational number represented by the form f (in the field R). Since equivalent forms represent the same numbers, the form g represents a in the real field and also in all fields R_p. By the corollary to the Hasse–Minkowski theorem, the form g represents a in the field R. Applying Theorem 2 of the Supplement, Section 1, we obtain

$$f \sim ax^2 \dotplus f_1, \quad g \sim ax^2 \dotplus g_1,$$

where f and g are quadratic forms in $n - 1$ variables over the field R (the sign $\sim$ denotes equivalence over R). Since the forms f and g are equivalent in the fields R_p, it follows (Supplement, Section 1, Theorem 4) that the forms f_1 and

g_1 are also equivalent in all these fields. By the induction hypothesis, f_1 and g_1 are equivalent in the rational field R. Then f and g are also equivalent in R, and Theorem 2 is proved.

As an example we consider the question of the equivalence of binary quadratic forms.

The determinant $d(f)$ of a nonsingular rational quadratic form has a unique representation as

$$d(f) = d_0(f)c^2,$$

where $d_0(f)$ is a square-free integer. When we pass to an equivalent form the value of $d_0(f)$ does not change (Supplement, Section 1, Theorem 1) and thus it is an invariant of the equivalence class of rationally equivalent forms. Let a be any nonzero rational number represented by the nonsingular binary form f. For each prime p (including $p = \infty$) set

$$e_p(f) = \left(\frac{a, -d(f)}{p}\right).$$

By Theorem 8 of Section 6 (which clearly also holds for the real field R_p), the value of $e_p(f)$ does not depend on the choice of a. It is consequently also an invariant of f under rational equivalence.

Combining Theorem 2 with Theorem 9 of Section 6 (which also holds for R_∞), we obtain the following criterion for rational equivalence of binary quadratic forms.

Theorem 3. Two binary quadratic forms f and g are rationally equivalent if and only if

$$d_0(f) = d_0(g) \quad \text{and} \quad e_p(f) = e_p(g) \qquad \text{for all } p.$$

Note that while, formally, an infinite number of invariants appear, their number is actually finite, since $e_p(f) = +1$ for all but a finite number of p.

7.6. *Remarks on Forms of Higher Degree*

As was done for forms with p-adic coefficients in relation to Theorem 5, Section 6, it would be interesting to include the Hasse–Minkowski theorem, or its corollary for $n \geqslant 5$, in a system of more general results, or at least hypotheses, concerning forms of higher degree.

It is natural to ask first if the analog of the Hasse–Minkowski theorem for forms of higher degree is true; that is, if a form represents zero in all p-adic fields and in the real field, does it represent zero in the rationals? It is easy to construct examples which disprove this hypothesis. For instance, if q, l, q', l' are distinct primes such that $(l/q) = -1$ and $(l'/q') = -1$ and the form

$x^2 + qy^2 - lz^2$ represents zero in the field R_2, then the form in four variables

$$(x^2 + qy^2 - lz^2)(x^2 + q'y^2 - l'z^2) \tag{7.22}$$

represents zero in all fields R_p and in the field of real numbers, but fails to represent zero in the field of rational numbers. Indeed, in the field R_2 the first factor represents zero by hypothesis. If p is odd and different from q and l, then the first factor represents zero in R_p by Corollary 2, Theorem 3, Section 6. As for q and l, the second factor represents zero in R_q and R_l for the same reason. However, neither factor represents zero in R, since the first factor fails to represent zero in R_q and the second in $R_{q'}$ [since $(l/q) = -1$ and $(l'/q') = -1$]. As a numerical example of (7.22) consider

$$(x^2 + 3y^2 - 17z^2)(x^2 + 5y^2 - 7z^2).$$

This example may perhaps appear somewhat artificial, since the form (7.22) is reducible, and the cause of this phenomenon may lie in its reducibility. Selmer gave a simple example free from this deficiency [E. S. Selmer, The diophantine equation $ax^3 + by^3 + cz^3 = 0$, *Acta. Math.* **85**, 203–362 (1951)]. He showed that the form $3x^3 + 4y^3 + 5z^3$ represents zero in every p-adic field R_p and in the real field, but does not represent zero in the field of rational numbers. The fact that this form represents zero in all fields R_p is easily shown (see Problem 8, Section 5). But the nonrepresentability of zero over the rational numbers is a more delicate question (see Problem 23, Section 7, Chapter 3).

The analog of the Hasse-Minkowski theorem for forms of higher degree is not even true when the number of variables is large. For example, the form

$$({x_1}^2 + \cdots + {x_n}^2)^2 - 2({y_1}^2 + \cdots + {y_n}^2)^2$$

with $n \geqslant 5$ represents zero in all p-adic fields and in the real field, but does not represent zero in the field of rational numbers for any n. The same also holds for the form

$$3({x_1}^2 + \cdots + {x_n}^2)^3 + 4({y_1}^2 + \cdots + {y_n}^2)^3 - 5({z_1}^2 + \cdots + {z_n}^2)^3,$$

which, unlike the previous example, is absolutely irreducible.

In the preceding examples both forms had even degree. Analogous examples for forms of odd degree have never been found. Hence it is possible that the analog of the Hasse-Minkowski theorem holds for forms of odd degree in sufficiently many variables. Since Brauer's theorem says that forms in sufficiently many variables represent zero in all p-adic fields (see Section 6.5), we are led to the following hypothesis: A rational form of odd degree in sufficiently many variables represents zero rationally.

This hypothesis was proved by Birch [B. J. Birch, "Homogeneous forms of odd degree in a large number of variables," *Mathematika* **4** (1957) pp.

102-105], who showed that forms of odd degree represent zero in the field of rational numbers provided that the number of variables is sufficiently large compared to the degree.

PROBLEMS

1. Prove the following theorem of Legendre: If a, b, and c are rational integers, pairwise relatively prime, square-free, and not all of the same sign, then the equation

$$ax^2 + by^2 + cz^2 = 0$$

is solvable in rational numbers if and only if the congruences

$$x^2 \equiv -bc \pmod{a},$$

$$x^2 \equiv -ca \pmod{b},$$

$$x^2 \equiv -ab \pmod{c},$$

are all solvable.

2. Do either of the forms $3x^2 + 5y^2 - 7z^2$, $3x^2 - 5y^2 - 7z^2$, represent zero in the field of rational numbers?

3. Which prime integers are represented by the forms $x^2 + y^2$, $x^2 + 5y^2$, $x^2 - 5y^2$?

4. Give a description of the set of all rational numbers represented by the form $2x^2 - 5y^2$.

5. Which rational numbers are represented by the form $2x^2 - 6y^2 + 15z^2$?

6. Let f be a nonsingular quadratic form over the field of rational numbers, with the number of variables not equal to 4. Show that f represents zero if and only if it represents all rational numbers.

7. For which rational integers a does the form $x^2 + 2y^2 - az^2$ represent zero rationally?

8. Find all solutions of the equation $x^2 + y^2 - 2z^2 = 0$ in rational numbers.

9. Which of the forms

$$x^2 - 2y^2 + 5z^2, \quad x^2 - y^2 + 10z^2, \quad 3x^2 - y^2 + 30z^2$$

are equivalent over the field of rational numbers?

10. Let a and b be square-free rational integers with $|a| > |b|$. If the form $ax^2 + by^2 - z^2$ represents zero in all p-adic fields, show that there are rational integers a and c, such that

$$aa_1 = c^2 - b, \quad |a_1| < |a|.$$

(The equation $aa_1 + b - c^2 = 0$ shows that the form $aa_1x^2 + by^2 - z^2$ represents zero rationally.)

11. By consideration of forms $ax^2 + by^2 - z^2$, where a and b are square-free integers, prove the Hasse–Minkowski theorem for forms in three variables by induction on $m = \max(|a|, |b|)$ (use Problem 10 and Problem 3, Section 1, the Supplement).

CHAPTER 2

Representation of Numbers by Decomposable Forms

In Chapter 1 we considered questions dealing with the existence and determination of rational solutions to equations. This chapter deals with the same questions, but only with respect to integral solutions. We consider a simple example.

The problem consists in finding all integral solutions to the equation

$$x^2 - 2y^2 = 7. \tag{0.1}$$

We may assume that $x > 0$, $y > 0$ (the remaining solutions are obtained by change of sign). This equation has the solutions (3, 1) and (5, 3). From these two solutions we can obtain an infinite number of others by the following method: if (x, y) is a solution of (0.1), then $(3x + 4y, 2x + 3y)$ is also a solution, as is shown by substitution. Starting from the solution $(x_0, y_0) = (3, 1)$, we thus obtain an infinite sequence of solutions (x_n, y_n), determined by the recursion formula

$$\begin{aligned} x_{n+1} &= 3x_n + 4y_n, \\ y_{n+1} &= 2x_n + 3y_n. \end{aligned} \tag{0.2}$$

Starting from the solution $(x_0', y_0') = (5, 3)$ we use the same formula to obtain another infinite sequence of solutions (x_0', y_0'). It can be shown that these two sequences exhaust all solutions to (0.1) with $x > 0$ and $y > 0$.

This completely elementary solution of (0.1) was obtained by computation. We can connect it with some general concepts and lay the groundwork for future generalizations.

We note that the form $x^2 - 2y^2$ is irreducible over the field R of rational numbers, but in the extension field $R(\sqrt{2})$ it can be factored into linear factors $(x + y\sqrt{2})(x - y\sqrt{2})$. If we use the concept of the norm for the extension $R(\sqrt{2})/R$ (Supplement, Section 2.2), then (0.1) can be written in the form

$$N(\xi) = N(x + y\sqrt{2}) = 7. \tag{0.3}$$

The problem then is to find in the field $R(\sqrt{2})$ all numbers $\xi = x + y\sqrt{2}$, where x and y are rational integers, whose norms are equal to 7. If the norm of the number $\varepsilon = u + v\sqrt{2}$ (u and v rational integers) equals 1, then by the multiplicativity of the norm, if ξ is a solution of (0.3) so are all numbers of the form $\xi\varepsilon^n$. Since $N(3 + 2\sqrt{2}) = 1$, we may take ε to be $3 + 2\sqrt{2}$. The passage from ξ to $\xi\varepsilon$ corresponds to that from (x, y) to $(3x + 4y, 2x + 3y)$. The two infinite sequences given by (0.2) now take the form

$$\begin{aligned} x_n + y_n\sqrt{2} &= (3 + \sqrt{2})(3 + 2\sqrt{2})^n \\ x_n' + y_n'\sqrt{2} &= (5 + 3\sqrt{2})(3 + 2\sqrt{2})^n \end{aligned} \qquad n \geqslant 0.$$

The possibility of obtaining an infinite number of solutions of (0.1) from one solution thus depends on the existence of a number $\varepsilon = u + v\sqrt{2}$ with integral u and v for which $N(\varepsilon) = 1$. In turn, the question of the existence of such numbers is connected to the basic concepts of the theory of algebraic numbers. Consider the set of all numbers of the form $x + y\sqrt{2}$, where x and y are rational integers. It is easily checked that this set, which we denote by $\mathfrak{O}$, forms a ring. In the arithmetic of this ring a major role is naturally played by the units, that is, those numbers $\alpha \in \mathfrak{O}$ such that $\alpha^{-1} \in \mathfrak{O}$ also. It is easily shown that α is a unit in $\mathfrak{O}$ if and only if $N(\alpha) = \pm 1$. This indicates the deeper significance of those numbers $\varepsilon \in \mathfrak{O}$ whose norm is 1; along with the numbers of norm -1, they form all units of the ring $\mathfrak{O}$.

In this chapter we consider the general theory, of which (0.1) is one of the simplest examples. Our success with the equation (0.1) was based on the fact that the form $x^2 - 2y^2$ is irreducible over the rational numbers and factors into linear factors over the field $R(\sqrt{2})$, allowing the equation to be written in the form (0.3). Our general theory will deal with forms which factor, in some extension of the field of rational numbers, into a product of linear forms.

Although our goal is the investigation of equations in which the coefficients and values of the variables are integers, we shall find it necessary to consider the more general case of forms with rational coefficients. The values of the variables will always be assumed to be integers.

1. Decomposable Forms

1.1. *Integral Equivalence of Forms*

Definition. Two forms $F(x_1, \ldots, x_m)$ and $G(y_1, \ldots, y_l)$ of the same degree with rational coefficients are called *integrally equivalent* if each can be obtained from the other by a linear change of variables with rational integer coefficients.

For example, the forms $x^2 + 7y^2 + z^2 - 6xy - 2xz + 6yz$ and $2u^2 - v^2$ are equivalent, since the linear substitutions

$$x = 3v, \quad y = u + v, \quad z = -u + v, \qquad u = -x + 2y + z, \quad v = x - y - z$$

take one into the other. In the case of forms which depend on the same number of variables, this is equivalent to saying that one of the forms can be transformed into the other by a linear change of variables with unimodular matrix (that is, an integral square matrix with determinant equal to ± 1).

If the forms F and G are equivalent, then, knowing all integral solutions of the equation $F = a$, we can obtain all integral solutions of the equation $G = a$, and conversely. Hence if we are interested in integral solutions of an equation of the form $F = a$, we may take instead of the form F any form which is equivalent to it.

Lemma 1. Any form of degree n is equivalent to a form in which the nth power of one of the variables occurs with nonzero coefficient.

Let $F(x_1, \ldots, x_m)$ be a form of degree n. We shall show that there exist rational integers $a_2, \ldots, a_m$, so that

$$F(1, a_2, \ldots, a_m) \neq 0.$$

The proof goes by induction on m. If $m = 1$, the form F is given by $Ax_1{}^n$, where $A \neq 0$, so that $F(1) \neq 0$. Assume that the lemma has already been proved for any form in $m - 1$ variables ($m \geqslant 2$). Write F as

$$F = G_0 x_m{}^n + G_1 x_m{}^{n-1} + \cdots + G_n,$$

where G_k $(0 \leqslant k \leqslant n)$ is either zero or is a form of degree k in the variables $x_1, \ldots, x_{m-1}$ (we say that a form is of degree zero if it is a nonzero constant). All the G_k are not zero, since F, as a form of degree n, has at least one nonzero coefficient. By the induction assumption there exist integers $a_2, \ldots, a_{m-1}$ such that $G_k(1, a_2, \ldots, a_{m-1}) \neq 0$ for at least one k. Since the polynomial $F(1, a_2, \ldots, a_{m-1}, x_m)$ in the single variable x_m is not identically zero, we may choose the value of a_m distinct from its roots and thus can obtain

$$F(1, a_2, \ldots, a_m) \neq 0.$$

We now make the following change of variables:

$$x_1 = y_1;$$
$$x_2 = a_2 y_1 + y_2;$$
$$\cdots\cdots\cdots$$
$$x_m = a_m y_1 + y_m.$$

After this transformation the form F becomes

$$G(y_1, \ldots, y_m) = F(y_1, a_2 y_1 + y_2, \ldots, a_m y_1 + y_m).$$

Since the matrix of the transformation is integral and has determinant 1, the forms F and G are equivalent, and the coefficient of $y_1{}^n$ is

$$G(1, 0, \ldots, 0) = F(1, a_2, \ldots, a_m),$$

which is nonzero. Lemma 1 is proved.

1.2. *Construction of Decomposable Forms*

Definition. The form $F(x_1, \ldots, x_m)$ with coefficients in the field of rational numbers is called *decomposable* if it factors into linear factors in some extension Ω/R.

An example of a decomposable form is the form

$$F(x, y) = a_0 x^n + a_1 x^{n-1} y + \cdots + a_n y^n$$

in two variables ($a_0 \neq 0$). Indeed, if Ω is a splitting field for the polynomial $F(x, 1)$ and $\alpha_1, \ldots, \alpha_n$ are its roots, then in Ω we have the factorization

$$F(x, y) = a_0(x - \alpha_1 y) \ldots (x - \alpha_n y).$$

Among the nonsingular quadratic forms considered in Chapter 1, the only decomposable forms are those in one or two variables (Problem 1).

It is clear that if F is decomposable, then so are all forms equivalent to F.

In the definition of decomposable forms no mention was made of the nature of the field Ω, in which the form factors into linear terms. We shall now show that Ω may always be taken to be a finite extension of R. The basic tools here are the results from the theory of finite extensions of fields. The results which we shall need are collected in the Supplement, Section 2.

Definition. A finite extension field of the field of rational numbers is called an *algebraic number field*, and its elements are called *algebraic numbers.*

Theorem 1. Any rational decomposable form factors into linear terms in some algebraic number field.

Proof. By Lemma 1 we may assume that we are given

$$F = (\alpha_{11}x_1 + \cdots + \alpha_{1m}x_m)\cdots(\alpha_{n1}x_1 + \cdots + \alpha_{nm}x_m) \qquad (\alpha_{ij} \in \Omega),$$

in which the coefficient of $x_1{}^n$ is nonzero. Since in this case the coefficients α_{i1} $(1 \leqslant i \leqslant n)$ are all nonzero, we may set

$$F = A(x_1 + \beta_{12}x_2 + \cdots + \beta_{1m}x_m)\cdots(x_1 + \beta_{n2}x_2 + \cdots + \beta_{nm}x_m), \quad (1.1)$$

where $A = \alpha_{11}\cdots\alpha_{n1}$ and $\beta_{ij} = \alpha_{ij}\alpha_{i1}{}^{-1}$. The number A is rational because it is the coefficient of $x_1{}^n$. For some fixed j $(2 \leqslant j \leqslant n)$ we set $x_j = 1$, and we set all remaining variables, except x_1, equal to zero. Then

$$F(x_1, 0, \ldots, 1, \ldots, 0) = A(x_1 + \beta_{1j})\cdots(x_1 + \beta_{nj}).$$

Since on the left there is a polynomial (of degree n) with rational coefficients, it follows that the β_{ij} are algebraic numbers. Let L denote the subfield of Ω generated over R by all β_{ij}. The extension L/R is clearly finite (Supplement, Section 2.1); that is, L is an algebraic number field.

From now on we shall consider only forms which are irreducible over the field of rational numbers, since for such forms the question of integral representation of rational numbers is of greatest interest. We now give a method for constructing irreducible decomposable forms.

Let K be any algebraic number field of degree n, and let θ be a primitive element for K over R, so that $K = R(\theta)$ (Supplement, Section 2.3). The minimum polynomial $\varphi(t)$ of the number θ over the field R has degree n. Construct an extension L over K in which $\varphi(t)$ factors completely,

$$\varphi(t) = (t - \theta^{(1)})\cdots(t - \theta^{(n)}) \qquad (\theta^{(1)} = \theta)$$

we may assume that $L = R(\theta^{(1)}, \ldots, \theta^{(n)})$. For any number $\alpha = f(\theta) \in K$ [$f(t)$ a polynomial with rational coefficients] we set

$$\alpha^{(i)} = f(\theta^{(i)}) \in R(\theta^{(i)}) \subset L.$$

Then the norm $N(\alpha) = N_{K/R}(\alpha)$ satisfies

$$N(\alpha) = \alpha^{(1)}\alpha^{(2)}\cdots\alpha^{(n)}$$

(Supplement, Section 2.3).

Now let $\mu_1, \ldots, \mu_m$ be any set of nonzero elements of K. These numbers determine a form

$$F(x_1, \ldots, x_m) = \prod_{i=1}^{n}(x_1\mu_1{}^{(i)} + \cdots + x_m\mu_m{}^{(i)}). \quad (1.2)$$

Since $\mu_k{}^{(i)} = f_k(\theta^{(i)})$ [$1 \leqslant k \leqslant m$, $f_k(t)$ a polynomial with rational coefficients], the coefficients of the form (1.2) are symmetric functions in $\theta^{(1)}, \ldots, \theta^{(n)}$, which means that they are rational functions of the coefficients of the polynomial $\varphi(t)$. Hence the form (1.2) has rational coefficients. If we substitute arbitrary rational numbers for the variables $x_1, \ldots, x_m$, then, since

$$x_1\mu_1^{(i)} + \cdots + x_m\mu_m^{(i)} = (x_1\mu_1 + \cdots + x_m\mu_m)^{(i)},$$

the product (1.2) will be the norm of the number $x_1\mu_1 + \cdots + x_m\mu_m$ (with respect to the extension K/R). Hence (1.2) can more simply be written

$$F(x_1, \ldots, x_m) = N(x_1\mu_1 + \cdots + x_m\mu_m). \tag{1.3}$$

The form (1.2) will not always be irreducible. For example, if in the field $R(\sqrt{2}, \sqrt{3})$ we take $\mu_1 = \sqrt{2}$, $\mu_2 = \sqrt{3}$, then the corresponding form will be $(2x_1^2 - 3x_2^2)^2$. However, we have the following theorem.

Theorem 2. If the numbers $\mu_2, \ldots, \mu_m$ generate the field K, i.e., $K = R(\mu_2, \ldots, \mu_m)$, then the form

$$F(x_1, \ldots, x_m) = N(x_1 + x_2\mu_2 + \cdots + x_m\mu_m) \tag{1.4}$$

is irreducible (over the field of rational numbers). Conversely, every irreducible decomposable form is equivalent to some constant multiple of a form of the type (1.4).

Proof. Assume that

$$F = GH,$$

where the forms G and H have rational coefficients. Since factorization in polynomial rings is unique (up to constant factors), each of the linear forms

$$L_i = x_1 + x_2\mu_2^{(i)} + \cdots + x_m\mu_m^{(i)}$$

must divide either G or H. Let $L_1 = x_1 + x_2\mu_2 + \cdots + x_m\mu_m$ divide G; that is,

$$G = L_1 M_1.$$

In this last equation, replace all coefficients by their images under the isomorphism $\alpha \to \alpha^{(i)}$ of the field $K = R(\theta)$ onto the field $R(\theta^{(i)})$. Since the coefficients of the form G are rational, we obtain

$$G = L_i M_i,$$

which means that L_i divides G for all $i = 1, \ldots, n$ [$n = (K : R)$]. Note that the isomorphism $\alpha \to \alpha^{(i)}$, $\alpha \in R(\mu_2, \ldots, \mu_m)$ is completely determined by the images $\mu_2^{(i)}, \ldots, \mu_m^{(i)}$ of the numbers $\mu_2, \ldots, \mu_m$. From this it follows that the sets of numbers $\mu_2^{(i)}, \ldots, \mu_m^{(i)}$ $(1 \leqslant i \leqslant n)$ are pairwise-distinct (since the isomorphisms $\alpha \to \alpha^{(i)}$ are pairwise-distinct), which means that the forms $L_1, \ldots, L_n$ are pairwise-distinct. Since the coefficient of x_1 in each form L_i

is equal to 1, these forms are pairwise-nonproportional. Using again the uniqueness of factorization, we conclude that G is divisible by the product $L_1 \cdots L_n$; that is, G is divisible by F. Hence H is a constant and the first assertion of the theorem is proved.

We prove now the second assertion. Let $F^*(x_1, \ldots, x_m)$ be any irreducible decomposable form of degree n. By Lemma 1 we may assume that the coefficient of $x_1{}^n$ is nonzero, so that F^* will have a factorization of the type (1.1), where β_{ij} are some algebraic numbers. Set $\beta_{1j} = \mu_j$ $(2 \leqslant j \leqslant m)$ and consider the field $K = R(\mu_2, \ldots, \mu_m)$, whose degree we denote by r. By what has been proved, the form

$$F = N(x_1 + x_2\mu_2 + \cdots + x_m\mu_m)$$

is irreducible, and one of its linear factors, $L_1 = x_1 + x_2\mu_2 + \cdots + x_m\mu_m$, is a divisor of the form F^*. Replacing all coefficients in the equation $F^* = L_1M_1$ by their images under the isomorphism $\alpha \to \alpha^{(i)}$ $(\alpha \in K,\ 1 \leqslant i \leqslant r)$, we obtain $F^* = L_iM_i$. We have already seen that the forms $L_1, \ldots, L_r$ are pairwise-nonproportional, so that F^* is divisible by their product $L_1 \cdots L_r$, which coincides with F. Since F is irreducible, $F = AF$, where A is a constant, and Theorem 2 is proved. (In the process we have also proved that $r = n$.)

1.3. *Modules*

It is clear that the question of integral solutions to the equation $F(x_1, \ldots, x_m) = a$, where F is given by (1.3), reduces to the determination of all numbers ξ in the field K which can be represented in the form

$$\xi = x_1\mu_1 + \cdots + x_m\mu_m \tag{1.5}$$

with $x_1, \ldots, x_m$ rational integers, and for which $N(\xi) = a$. It is thus natural to study the set of all numbers of the form (1.5).

Definition. Let K be an algebraic number field, and let $\mu_1, \ldots, \mu_m$ be an arbitrary finite set of elements of K. The set M of all linear combinations

$$c_1\mu_1 + \cdots + c_m\mu_m$$

with rational integer coefficients c_i $(1 \leqslant i \leqslant m)$ is called a *module* in K. The numbers $\mu_1, \ldots, \mu_m$ are called *generators* for the module M.

A given module M can be generated by many different sets. If $\mu_1, \ldots, \mu_m$ is a set of generators for the module M, we write $M = \{\mu_1, \ldots, \mu_m\}$.

We consider how the form (1.3) changes if, instead of $\mu_1, \ldots, \mu_m$ we take another set of numbers $\rho_1, \ldots, \rho_l$, generating the same module M. We have

$$\rho_j = \sum_{k=1}^{m} c_{jk}\mu_k \qquad (1 \leqslant j \leqslant l)$$

with rational integers c_{jk}. Let

$$G(y_1, \dots, y_l) = N(y_1\rho_1 + \cdots + y_l\rho_l).$$

Since

$$\sum_{j=1}^{l} y_j\rho_j = \sum_{k=1}^{m} \left(\sum_{j=1}^{l} c_{jk}y_j\right)\mu_k,$$

then the linear substitution

$$x_k = \sum_{j=1}^{l} c_{jk}y_j \qquad (1 \leqslant k \leqslant m)$$

takes the form F into the form G. As the sets of generators μ_k and ρ_j of the module M play a symmetric role, there is also an integral linear change of variables which takes G into F. This means that different systems of generators for the same module M correspond to equivalent forms; that is, with each module M of the field K is associated a uniquely determined class of equivalent decomposable forms.

For each module $M = \{\mu_1, \dots, \mu_m\}$ and each number $\alpha \in K$, we denote by αM the set of all products $\alpha\xi$, where ξ is any element of M. It is clear that αM coincides with the set of all integral linear combinations of the numbers $\alpha\mu_1, \dots, \alpha\mu_m$; that is, $M = \{\alpha\mu_1, \dots, \alpha\mu_m\}$.

Definition. Two modules M and M_1 in the algebraic number field K are called *similar* if $M_1 = \alpha M$ for some $\alpha \neq 0$ in K.

The forms associated to similar modules M and αM differ only by a constant multiple, equal to $N(\alpha)$. Hence if we are considering forms only up to constant multiples, we may replace the module M by any module similar to it, and in particular we may assume that one of the generators of the module, say μ_1, equals 1.

We may now formulate our problem on the representation of numbers by irreducible decomposable form as follows. If the form F is given by

$$F(x_1, \dots, x_m) = AN(x_1\mu_1 + \cdots + x_m\mu_m)$$

(for suitable choice of the field K), then the finding of all integral solutions to the equation $F(x_1, \dots, x_m) = a$ is equivalent to the finding in the module $M = \{\mu_1, \dots, \mu_m\}$ of all numbers α, such that $N(\alpha)$ equals the rational number a/A. Hence in the future we shall start from the problem of finding in a given module all numbers with given norm. We have seen that this is equivalent to finding all numbers in the similar module μM with norm $N(\mu)a/A$. Hence we may replace the given module by any similar module whenever such replacement is helpful.

If the degree of the algebraic number field K equals n, then any module M of the field K contains at most n linearly independent numbers (over R).

Definition. Let K be an algebraic number field of degree n; Let M be a module in K. If M contains n linearly independent elements (over the field of rational numbers), then it is called *full*, otherwise *nonfull*. The forms connected with the module are correspondingly called full or nonfull.

For example, if the rational integer d is not a cube, then the numbers 1, $\sqrt[3]{d}$, $\sqrt[3]{d^2}$ form a basis for the field $R(\sqrt[8]{d})$ over R, and thus the form

$$N(x + y\sqrt[3]{d} + z\sqrt[3]{d^2}) = x^3 + dy^3 + d^2z^3 - 3dxyz$$

is full. As an example of a nonfull form, take

$$N(x + y\sqrt[3]{d}) = x^3 + dy^3.$$

If $\{1, \mu_2, \dots, \mu_m\}$ is a full module of the field K, then $K = R(\mu_2, \dots, \mu_m)$. By Theorem 1 it follows that any full form is irreducible.

The problem of the representation of numbers by nonfull irreducible forms is very difficult, and at this time there is little satisfactory general theory. A particular case will be considered in Chapter 4.

The problem of the representation of rational numbers by full forms is much easier and is essentially solved. We shall deal with it in this chapter. This problem, as we have noted, is equivalent to the problem of finding in a fixed full module of an algebraic number field K all numbers with given norm.

PROBLEMS

1. Show that a rational quadratic form is decomposable if and only if its rank is $\leqslant 2$.

2. Show that a form connected with an arbitrary module of an algebraic number field K is a constant multiple of a power of an irreducible form.

3. Show that in the field of rational numbers R any module has the form aZ, where $a \in R$ (Z is the ring of rational integers).

2. Full Modules and Their Rings of Coefficients

2.1. *Bases of Modules*

Definition. A system $\alpha_1, \dots, \alpha_m$ of generators of the module M is called a *basis* for M if it is linearly independent over the ring of integers, that is, if the equation

$$a_1\alpha_1 + \cdots + a_m\alpha_m = 0, \qquad (a_i \in Z),$$

occurs only when all a_i are zero.

It is clear that if $\alpha_1, \ldots, \alpha_m$ is a basis for the module M, then any $\alpha \in M$ has a unique representation in the form

$$\alpha = c_1\alpha_1 + \cdots + c_m\alpha_m, \qquad (c_i \in Z). \tag{2.1}$$

We now show that any module has a basis. The proof of this does not depend on the fact that the module consists of numbers from some algebraic number field, but only on the fact that the module is a finitely generated Abelian group under addition, containing no elements of finite order. Therefore we shall prove the result we need in the theory of Abelian groups. We use the following terminology. A system of elements $\alpha_1, \ldots, \alpha_m$ of an Abelian group M (whose operation is written additively) is called a system of generators if every $\alpha \in M$ can be represented in the form (2.1). In this case we write: $M = \{\alpha_1, \ldots, \alpha_m\}$. If this system satisfies the above definition, it is called a basis for M.

Theorem 1. If an Abelian group without elements of finite order possesses a finite system of generators, then it possesses a basis.

Proof. Let $\alpha_1, \ldots, \alpha_s$ be an arbitrary set of generators of the group M, First, note that if any integral multiple of one generator is added to another. the resulting system is also a system of generators. Let, for instance, $\alpha_1' = \alpha_1 + k\alpha_2$. Then for any $\alpha \in M$ we have

$$\alpha = c_1\alpha_1 + c_2\alpha_2 + \cdots + c_s\alpha_s = c_1\alpha_1' + (c_2 - kc_1)\alpha_2 + \cdots + c_s\alpha_s,$$

where all coefficients are integers, which means that $M = \{\alpha_1', \alpha_2, \ldots \alpha_s\}$.

If the elements $\alpha_1, \ldots, \alpha_s$ are linearly independent, they form a basis of M. Assume that they are linearly dependent, that is, that

$$c_1\alpha_1 + c_2\alpha_2 + \cdots + c_s\alpha_s = 0 \tag{2.2}$$

for some set of integers c, not all zero. Choose among the nonzero coefficients c the smallest one in absolute value. Let it be c_1. Assume that not all coefficients c_i are divisible by c_1, say, $c_2 = c_1 q + c'$, where $0 < c' < |c_1|$. If we pass to the new set of generators

$$\alpha_1' = \alpha_1 + q\alpha_2, \alpha_2, \ldots, \alpha_s,$$

then the relation (2.2) takes the form

$$c_1\alpha_1' + c'\alpha_2 + \cdots + c_s\alpha_s = 0,$$

and in this relation the coefficient $c' > 0$ appears, which is less than c. Thus, if for the generators $\alpha_1, \ldots, \alpha_s$ we have a nontrivial relation (2.2), in which the nonzero coefficient of smallest absolute value does not divide all remaining coefficients, then we can construct another system of generators for which we also have a nontrivial relation with integer coefficients in which the nonzero coefficient of smallest absolute value is smaller (in absolute value) than the analogous quantity in the first system. Hence after a finite number of such transformations, we arrive at a new system of generators $\beta_1, \dots, \beta_s$, for which we have the dependence

$$k_1\beta_1 + k_2\beta_2 + \cdots + k_s\beta_s = 0 \tag{2.3}$$

with integer coefficients k_i, where one of the coefficients, say, k_1, is a divisor of all the others. Dividing the relation (2.3) by k_1 (this can be done since we have assumed that M contains no elements of finite order), we obtain

$$\beta_1 + l_2\beta_2 + \cdots + l_s\beta_s = 0 \tag{2.4}$$

with integers $l_2, \dots, l_s$. From (2.4) it follows that β_1 can be dropped from the system of generators; that is, $M = \{\beta_2, \dots, \beta_s\}$.

We have shown that if some system of generators of M is linearly dependent, then we can construct a new system with fewer generators. After carrying out this procedure several times, we must arrive at a system of generators which is linearly independent, that is, a basis for the group M.

Corollary. Any module in an algebraic number field K has a basis.

The number of elements m in any basis of the module M is equal to the maximal number of linearly independent (over R) elements in M. Hence this number will be the same for all bases. It is called the *rank* of the module M. The rank of the module consisting only of zero is set equal to zero.

Let $\omega_1, \dots, \omega_m$ and $\omega_1', \dots, \omega_m'$ be any two bases of a module M of rank m. It is clear that the matrix of transition C from the first to the second basis is integral. By symmetry the transition matrix from the second basis to the first, that is, C^{-1}, is also integral. Consequently, det $C = \pm 1$. We thus obtain that any transition matrix from one basis of a module of rank m to another is unimodular of rank m.

If the degree of the field K over R is equal to n, then the rank of any module of K does not exceed n. It is clear that the rank of a module is equal to n if and only if it is a full module. Nonfull modules are thus characterized by having rank less than n, the degree of the the field.

Any system of generators of a module of rank m contains not less than m elements. It follows that among the forms associated with a given module there are forms in m variables and there are no forms in less than m variables. A full form of degree n could thus be characterized as an irreducible decomposable form which is not equivalent to a form in less than n variables.

Theorem 2. Let M be a finitely generated Abelian group without elements of finite order and let N be a subgroup. Then N has a finite set of generators and hence a basis. For any basis $\omega_1, \dots, \omega_m$ of the group M (for some ordering of this basis) there is a basis for N of the form

$$\begin{aligned}
\eta_1 &= c_{11}\omega_1 + c_{12}\omega_2 + \cdots + c_{1k}\omega_k + \cdots + c_{1m}\omega_m, \\
\eta_2 &= \qquad\qquad c_{22}\omega_2 + \cdots + c_{2k}\omega_k + \cdots + c_{2m}\omega_m, \\
&\cdots\cdots\cdots\cdots\cdots\cdots\cdots\cdots \\
\eta_k &= \qquad\qquad\qquad\qquad\qquad c_{kk}\omega_k + \cdots + c_{km}\omega_m,
\end{aligned}$$

where the c_{ij} are integers with $c_{ii} > 0$, $k \leqslant m$.

Proof. The theorem will be proved by induction on the rank m of the group M, that is, on the number of elements of the basis of M. The case $m = 0$ is trivial. Let $m \geqslant 1$. If N consists only of zero, then $k = 0$ and the theorem is valid. If $\alpha \in N$, $\alpha \neq 0$, then

$$\alpha = c_1\omega_1 + \cdots + c_m\omega_m, \tag{2.5}$$

where at least one of the coefficients c_i is not zero. By reordering the basis we may assume that $c_1 \neq 0$. If $c_1 < 0$ then the coefficient of ω_1 in $-\alpha$ will be positive. Among all elements of the subgroup N choose that element

$$\eta_1 = c_{11}\omega_1 + c_{12}\omega_2 + \cdots + c_{1m}\omega_m,$$

in which the coefficient $c_{11} > 0$ of ω_1 is smallest. We now claim that for any $\alpha \in N$ the coefficient c_{11} will be divisible by c_{11}. Indeed, if $c_1 = c_{11}q + c'$, $0 < c' < c$ (q an integer), then for the element $\alpha - q\eta$, we have

$$\alpha - q\eta_1 = c'\omega_1 + c_2{}'\omega_2 + \cdots + c_m{}'\omega_m,$$

so by the minimality of c_{11} it follows that $c' = 0$. Consider now in M the subgroup $M_0 = \{\omega_2, \dots, \omega_m\}$. Since the intersection $N \cap M_0$ is a subgroup of the group M_0, by the induction hypothesis $N \cap M_0$ has a basis of the type

$$\begin{aligned}
\eta_2 &= c_{22}\omega_2 + c_{23}\omega_3 + \cdots + c_{2k}\omega_k + \cdots + c_{2m}\omega_m, \\
\eta_3 &= \qquad\qquad c_{33}\omega_3 + \cdots + c_{3k}\omega_k + \cdots + c_{3m}\omega_m, \\
&\cdots\cdots\cdots\cdots\cdots\cdots\cdots\cdots \\
\eta_k &= \qquad\qquad\qquad\qquad\qquad c_{kk}\omega_k + \cdots + c_{km}\omega_m,
\end{aligned}$$

where c_{ij} are integers, $c_{ii} > 0$, $k - 1 \leqslant m - 1$ (for suitable ordering of the basis elements $\omega_2, \dots, \omega_n$). We assert that N consists of all integral linear combinations of the elements $\eta_1, \eta_2, \dots, \eta_k$. Let α be an arbitrary element of N. If we write α in the form (2.5), then since we have shown that $c_1 = c_{11}q_1$, with q_1 an integer,

$$\alpha - q_1\eta_1 = c_2{}'\omega_2 + \cdots + c_m{}'\omega_m,$$

which lies in the intersection $M_0 \cap N$. By the induction assumption we have

$$\alpha - q_1\eta_1 = q_2\eta_2 + \cdots + q_k\eta_k,$$

where the q_i are integers, so that $\alpha = q_1\eta_1 + \cdots + q_k\eta_k$. We have thus shown that $N = \{\eta_1, \eta_2, \ldots, \eta_k\}$. The generators $\eta_1, \ldots, \eta_k$, as is easily seen, are linearly independent over Z, which means that they form a basis for N of the required type.

The proof of Theorem 2 essentially reproduces Gauss' method for eliminating variables in systems of linear equations. The only difference is that in our case the coefficients lie not in a field but in the ring of integers.

Corollary. Any subgroup N of a module M in an algebraic number field K is also a module (a submodule of the module M).

2.2. *Coefficient Rings*

Definition. A number α of the algebraic number field K is called a *coefficient* of the full module M of the field K if $\alpha M \subset M$, that is, if for any $\xi \in M$ the product $\alpha\xi$ also belongs to M.

The set $\mathfrak{O}_M$ of all coefficients of a module M forms a ring. For if α and β belong to $\mathfrak{O}_M$, then for any $\xi \in M$ we have $(\alpha - \beta)\xi = \alpha\xi - \beta\xi \in M$ and $(\alpha\beta)\xi = \alpha(\beta\xi) \in M$; that is, $\alpha - \beta \in \mathfrak{O}_M$ and $\alpha\beta \in \mathfrak{O}_M$. The ring $\mathfrak{O}_M$ is called the *ring of coefficients* of the full module M. Since $1 \in \mathfrak{O}_M$, $\mathfrak{O}_M$ is a ring with unit.

To ascertain whether a given number $\alpha \in K$ lies in the ring $\mathfrak{O}_M$, it is not necessary to check for all $\xi \in M$ whether the product $\alpha\xi$ lies in M or not. It suffices to check this only for any basis $\mu_1, \ldots, \mu_n$ of the module M. Indeed, if $\alpha\mu_i \in M$ for all $i = 1, \ldots, n$, then for $\xi = c_1\mu_1 + \cdots + c_n\mu_n \in M$ we have

$$\alpha\xi = c_1(\alpha\mu_1) + \cdots + c_n(\alpha\mu_n) \in M.$$

We now show that the coefficient ring $\mathfrak{O}_M$ is a full module in K. Let γ be an arbitrary nonzero element of M. Since $\alpha\gamma \in M$ for any $\alpha \in \mathfrak{O}_M$, then $\gamma\mathfrak{O}_M \subset M$. The set of all numbers $\gamma\mathfrak{O}_M$ is clearly a group under addition and thus by the Corollary of Theorem 2, $\gamma\mathfrak{O}_M$ is a module. But then $\mathfrak{O}_M = \gamma^{-1}(\gamma\mathfrak{O}_M)$ is also a module. We now need to show that this module is full. Let α be any nonzero element of K and denote by c a common denominator for all rational numbers a_{ij}, determined by

$$\alpha\mu_i = \sum_{j=1}^{n} a_{ij}\mu_j \qquad (1 \leqslant i \leqslant n). \tag{2.6}$$

Since the products ca_{ij} are integers, $c\alpha\mu_i \in M$ and thus $c\alpha \in \mathfrak{O}_M$. If we now take an arbitrary basis $\alpha_1, \ldots, \alpha_n$ for the field K, then by what has just been proved for some rational integers $c_1, \ldots, c_n$, the products $c_1\alpha_1, \ldots, c_n\alpha_n$ will all belong to $\mathfrak{O}_M$. We thus see that $\mathfrak{O}_M$ contains n linearly independent numbers, and this means that $\mathfrak{O}_M$ is a full module.

Definition. A full module in the field of algebraic numbers K which contains the number 1 and is a ring is called an *order* of the field K.

Using this definition we can formulate our result as follows.

Theorem 3. The coefficient ring for any full module of the algebraic number field K is an order of this field.

The converse also holds: Any order $\mathfrak{O}$ of the field K is the coefficient ring for some full module, for example, for itself (since $1 \in \mathfrak{O}$, $\alpha\mathfrak{O} \subset \mathfrak{O}$, if and only if $\alpha \in \mathfrak{O}$).

For any number $\gamma \neq 0$ of K the condition $\alpha\xi \in M$ is equivalent to the condition $\alpha(\gamma\xi) \in \gamma M$ (here $\xi \in M$). It follows that the similar modules M and γM have the same coefficient rings; that is,

$$\mathfrak{O}_{\gamma M} = \mathfrak{O}_M .$$

Let $\mu_1, \ldots, \mu_n$ be a basis for the module M, and $\omega_1, \ldots, \omega_n$ a basis for its coefficient ring $\mathfrak{O}_M$. For each $i = 1, \ldots, n$, we have

$$\mu_i = \sum_{j=1}^{n} b_{ij}\omega_j ,$$

where the b_{ij} are rational numbers. If b is a common denominator for all the coefficients b_{ij}, then the number $b\mu_i$ will be an integral linear combination of the basis elements of $\mathfrak{O}_M$; that is, $b\mu_i$ will lie in $\mathfrak{O}_M$. The module bM thus satisfies $bM \subset \mathfrak{O}_M$.

We summarize these results.

Lemma 1. The coefficient rings of similar full modules coincide. Every full module is similar to a module contained in its coefficient ring.

2.3. *Units*

Consider the problem of integral representation of rational numbers by full decomposable forms. In Section 1.3 we saw that this problem reduces to the determination in a full module M of all numbers μ, for which

$$N(\mu) = a. \tag{2.7}$$

For any ω of the coefficient ring $\mathfrak{O} = \mathfrak{O}_M$, the product $\omega\mu$ lies in M and by the multiplicativity of the norm,

$$N(\omega\mu) = N(\omega)a.$$

If $N(\omega) = 1$, then (2.7) still holds, with μ replaced by $\omega\mu$. Thus the coefficients

ω of norm 1 allow us to obtain a whole class of new solutions of (2.7) from one solution. This fact is the foundation of the method of solving (2.7) which we are going to describe.

We shall show that the coefficients $\omega \in \mathfrak{O}$, with $N(\omega) = 1$, are contained in the set of elements ε of the ring $\mathfrak{O}$ for which ε^{-1} also belongs to $\mathfrak{O}$. Such numbers ε are called the *units of the ring* $\mathfrak{O}$ (Supplement, Section 4.1). Since the inclusions $\varepsilon M \subset M$ and $\varepsilon^{-1} M \subset M$ are equivalent to the equality $\varepsilon M = M$, the units of the ring $\mathfrak{O}_M$ can be characterized as being those elements $\alpha \in K$ for which $\alpha M = M$.

Lemma 2. If the number α belongs to the order $\mathfrak{O}$, then its characteristic and minimum polynomials have integer coefficients. In particular, the norm $N(\alpha) = N_{K/R}(\alpha)$ and trace $Sp(\alpha) = Sp_{K/R}(\alpha)$ are rational integers.

Proof. Let the order $\mathfrak{O}$ be the coefficient ring of the module $M = \{\mu_1, \dots, \mu_n\}$ (for example, we may take $M = \mathfrak{O}$). If $\alpha \in \mathfrak{O}$, then in (2.6) the coefficients a_{ij} are integers, from which it follows that the characteristic polynomial of the number α (with respect to the extension K/R) has integer coefficients. The remaining assertions of the lemma are now obvious.

Theorem 4. Let $\mathfrak{O}$ be an arbitrary order of the algebraic number field K. In order that the number $\varepsilon \in \mathfrak{O}$ be a unit of the ring $\mathfrak{O}$, it is necessary and sufficient that $N(\varepsilon) = \pm 1$.

Proof. We first show that for any $\alpha \neq 0$ of $\mathfrak{O}$ the norm $N(\alpha)$ is divisible (in the ring $\mathfrak{O}$) by α. By Lemma 2 the characteristic polynomial $\varphi(t) = t^n + c_1 t^{n-1} + \cdots + c_n$ of the number α has integer coefficients. Since $\varphi(\alpha) = 0$, then $N(\alpha)/\alpha$ lies in $\mathfrak{O}$, which means that $N(\alpha)$ is divisible by α.

Now if $N(\alpha) = \pm 1$, then 1 is divisible by α; that is, α is a unit of the ring $\mathfrak{O}$. Conversely, if ε is a unit of the ring $\mathfrak{O}$, so that $\varepsilon\varepsilon' = 1$ for some $\varepsilon' \in \mathfrak{O}$, then since $N(\varepsilon)$ and $N(\varepsilon')$ are integers, the equation $N(\varepsilon)N(\varepsilon') = 1$ implies that $N(\varepsilon) = \pm 1$. Theorem 4 is proved.

To find all coefficients $\omega \varepsilon \mathfrak{O}$ with $N(\omega) = 1$, we thus must determine all units of the ring $\mathfrak{O}$, and then isolate those units with norm $+1$. Two numbers μ_1 and μ_2 of the full module M are called *associates* if their quotient $\mu_1/\mu_2 = \varepsilon$ is a unit of the coefficient ring $\mathfrak{O} = \mathfrak{O}_M$. It is clear that if $M = \mathfrak{O}$, then this concept coincides with the usual notion of associates in commutative rings with unit (Supplement, Section 4.1). This relation induces an equivalence relation on the set of all solutions to (2.7), and therefore the set of all solutions to (2.7) is divided into equivalence classes of associate solutions. If μ_1 and μ_2 are two associate solutions, then $\mu_1 = \mu_2\varepsilon$, where ε is a unit of the ring $\mathfrak{O}$ with $N(\varepsilon) = 1$. Conversely, if ε is any unit of norm 1 and μ is a solution of (2.7), then $\mu\varepsilon$ is also a solution of (2.7) and μ and $\mu\varepsilon$ are associates. Thus all solutions from a given class of associate solutions are obtained by multiplying one solution by all units with norm 1. We now show that the number of such classes of solutions is finite.

Theorem 5. An order $\mathfrak{O}$ contains only a finite number of nonassociate elements of given norm.

Proof. Let $\omega_1, \dots, \omega_n$ be a basis of the order $\mathfrak{O}$ and let $c > 1$ be an arbitrary natural number.

Using the general definition of the Supplement, Section 4.1, we say that two numbers α and β of $\mathfrak{O}$ are congruent modulo c if their difference $\alpha - \beta$ is divisible by c (in the ring $\mathfrak{O}$). It is clear that any $\alpha \in \mathfrak{O}$ is congruent to a unique number of the form

$$x_1\omega_1 + \cdots + x_n\omega_n, \quad 0 \leqslant x_i < c \qquad (1 \leqslant i \leqslant n).$$

Hence $\mathfrak{O}$ contains c^n congruence classes modulo c. Let the numbers α and β belong to the same congruence class and satisfy $|N(\alpha)| = |N(\beta)| = c$. The equation $\alpha - \beta = c\gamma$, $\gamma \in \mathfrak{O}$, implies that $\alpha/\beta = 1 \pm [N(\beta)/\beta]\gamma \in \mathfrak{O}$ [since $N(\beta)/\beta \in \mathfrak{O}$; see the start of the proof of Theorem 4], and analogously $\beta/\alpha = 1 \pm [N(\alpha)/\alpha]\gamma \in \mathfrak{O}$. Thus the numbers α and β divide one another, which means that they are associates in the ring $\mathfrak{O}$. This proves that $\mathfrak{O}$ can contain only a finite number (not greater than c^n) of pairwise-nonassociate elements whose norm in absolute value is equal to c.

Corollary. A full module M of the field K contains only a finite number of pairwise-nonassociate elements with given norm.

Indeed, if $\mathfrak{O}$ is the coefficient ring of the module M, then for some natural number b the module bM is contained in $\mathfrak{O}$. If $\gamma_1, \dots, \gamma_k$ are pairwise-nonassociate elements of M with norm c, then the numbers $b\gamma_1, \dots, b\gamma_k$ of $\mathfrak{O}$ have norm $b^n c$ and are pairwise-nonassociate in $\mathfrak{O}$. Thus the number k cannot be arbitrarily large.

Remark. The proof of Theorem 5 shows that in the ring $\mathfrak{O}$ (and also in the module M) there is a finite set of numbers with given norm c such that any number of $\mathfrak{O}$ (or of M) with the same norm c is associate with one of these. However the proof is noneffective, that is, it does not allow us to find these numbers, although it does give an effective bound on their number.

Our basic problem of finding all solutions to (2.7) thus splits into the following two problems:

(1) Find all units ε in the coefficient ring $\mathfrak{O}_M$ with norm $N(\varepsilon) = 1$.
(2) Find numbers $\mu_1, \dots, \mu_k$ in M with norm a such that they are pairwise-

nonassociate and such that any $\mu \in M$ with norm a is associative to one of them, that is, $\mu = \mu_i \varepsilon$, where $1 \leqslant i \leqslant k$ and ε is a unit of the coefficient ring $\mathfrak{O}_M$.

If these two problems are solved, then we will have solved the problem of integral representation of rational numbers by full decomposable forms.

2.4. *Maximal Orders*

The concept of an order leads naturally to the question of the relationship between different orders in a given algebraic number field K. In this section we show that among the various orders of the field K there is one maximal one which contains all other orders. Lemma 2 shows that the minimum polynomial of any number in any order has integer coefficients. We shall see below that the maximal order of an algebraic number field K coincides with the set $\tilde{\mathfrak{O}}$ of all numbers of K whose minimum polynomial has integer coefficients. We first prove the following lemmas.

Lemma 3. If $\alpha \in \tilde{\mathfrak{O}}$, that is, if the minimum polynomial $t^m + c_1 t^{m-1} + \cdots + c_M$ of the number α has integer coefficients, then the module $M = \{1, \alpha, \ldots, \alpha^{m-1}\}$ is a ring.

Proof. It is clearly sufficient to show that any power α^k ($k \geqslant 0$) of the number α lies in M. For $k \leqslant m - 1$ this is true by the definition of M. Further, $\alpha^m = -c_1\alpha^{m-1} - \cdots - c_M$ with integers c_i, so that $\alpha^m \in M$. Let $k > m$, and assume that it is already proved that $\alpha^{k-1} \in M$; that is, $\alpha^{k-1} = a_1\alpha^{m-1} + \cdots + a_m$ with integers a_i. Then

$$\alpha^k = \alpha\alpha^{k-1} = a_1\alpha^m + a_2\alpha^{m-1} + \cdots + a_m\alpha.$$

Since all terms on the right lie in M, α^k also belongs to M. Lemma 3 is proved.

Lemma 4. If $\mathfrak{O}$ is any order of the field K and $\alpha \in \tilde{\mathfrak{O}}$, then the ring $\mathfrak{O}[\alpha]$, consisting of all polynomials in α with coefficients from $\mathfrak{O}$, also is an order of the field K.

Proof. Since $\mathfrak{O} \subset \mathfrak{O}[\alpha]$, the ring $\mathfrak{O}[\alpha]$ contains $n = (K: R)$ linearly independent numbers over R. We thus need only show that $\mathfrak{O}[\alpha]$ is a module (that is, that it is finitely generated). Let $\omega_1, \ldots, \omega_n$ be a basis for the order $\mathfrak{O}$. By Lemma 3, any power α^k ($k \geqslant 0$) can be represented in the form $a_0 + a_1\alpha + \cdots + a_{m-1}\alpha^{m-1}$ with integers a_i, where m is the degree of the minimum polynomial of the number α. From this it easily follows that any number of $\mathfrak{O}[\alpha]$ can be represented as an integral linear combination of the products $\omega_i\alpha^j$ ($1 \leqslant i \leqslant n$, $0 \leqslant j \leqslant m - 1$), and this means that $\mathfrak{O}[\alpha]$ is a module.

Repeated application of Lemma 4 gives us the following.

Corollary. If $\mathfrak{O}$ is an order and $\alpha_1, \ldots, \alpha_p$ are numbers of $\tilde{\mathfrak{O}}$, then the ring $\mathfrak{O}[\alpha_1, \ldots, \alpha_p]$ of all polynomials in $\alpha_1, \ldots, \alpha_p$ with coefficients in $\mathfrak{O}$ is also an order.

Theorem 6. The set of all numbers of the algebraic number field K whose minimum polynomial has integer coefficients is the maximal order of the field K.

Proof. Let $\mathfrak{O}$ be any order of the field K and let α and β be arbitrary numbers of $\tilde{\mathfrak{O}}$. By the Corollary of Lemma 4 the ring $\mathfrak{O}[\alpha, \beta]$ is an order, and hence it is contained in $\tilde{\mathfrak{O}}$ (Lemma 2). But then the difference $\alpha - \beta$ and the product $\alpha\beta$ are also contained in $\tilde{\mathfrak{O}}$. This proves that $\tilde{\mathfrak{O}}$ is a ring. Since $\mathfrak{O} \subset \tilde{\mathfrak{O}}$, $\tilde{\mathfrak{O}}$ contains n linearly independent numbers. We thus need only show that $\tilde{\mathfrak{O}}$ is a module.

Let $\omega_1, \ldots, \omega_n$ be any basis of the order $\mathfrak{O}$, and let $\omega_1^*, \ldots, \omega_n^*$ be the dual basis to it in the field K (Supplement, Section 2.3). We shall show that the ring $\tilde{\mathfrak{O}}$ is contained in the module $\mathfrak{O}^* = \{\omega_1^*, \ldots, \omega_n^*\}$. Let α be any element of the ring $\tilde{\mathfrak{O}}$. Represent it in the form

$$\alpha = c_1\omega_1^* + \cdots + c_n\omega_n^*$$

with rational c_i. Multiplying by ω_i and taking the trace, we obtain

$$c_i = \operatorname{Sp} \alpha\omega_i \qquad (1 \leqslant i \leqslant n)$$

(we are using here the fact that $\operatorname{Sp} \omega_i\omega_i^* = 1$ and $\operatorname{Sp} \omega_i\omega_j^* = 0$ for $i \neq j$). All products $a\omega_i$ are contained in the order $\mathfrak{O}[\alpha]$, and therefore by Lemma 2 all numbers c_i are integers, and this means that $\alpha \in \mathfrak{O}^*$. Thus $\tilde{\mathfrak{O}} \subset \mathfrak{O}^*$. By the Corollary of Theorem 2 we now conclude that $\tilde{\mathfrak{O}}$ is a module, and Theorem 6 is proved.

The proof which we have given that $\tilde{\mathfrak{O}}$ is a ring is of very general character; that is, it remains valid (with insignificant changes) also in the general theory of commutative rings without zero divisors. The corresponding concepts in the general case are given in Section 4 of the Supplement. Using the terminology introduced there we can say that the maximal order of an algebraic number field K is the integral closure of the ring Z of rational integers in the field K. Here the maximal order $\tilde{\mathfrak{O}}$ will frequently be called the *ring of integers* of K, and any number in $\tilde{\mathfrak{O}}$ will be called an *integer* of K.

The units of the maximal order $\tilde{\mathfrak{O}}$ are also called the *units of the algebraic number field K.*

2.5. *The Discriminant of a Full Module*

Let $\mu_1, \ldots, \mu_n$ and $\mu_1', \ldots, \mu_n'$ be two bases for the full module M of the algebraic number field K. We have seen (Section 2.1) that the transition matrix from one basis to the other is unimodular (that is, it is an integral matrix with determinant ± 1). It follows that the discriminants $D(\mu_1, \dots, \mu_n)$ and $D(\mu_1', \dots, \mu_n')$ are equal [Supplement, Eq. (2.12)]. All bases of the module thus have the same discriminant. This common value is clearly a rational number, and it is called the *discriminant of the module M*.

Every order of the field K is a full module in K. Hence we may speak of the discriminant of an order. Since the trace of any number in an order is an integer, the discriminant of an order will always be a rational integer (the same holds for any full module contained in $\tilde{\mathfrak{D}}$).

A basis of the maximal order $\tilde{\mathfrak{D}}$ of the algebraic number field K is frequently called a *fundamental basis of* K, and its discriminant is called the *discriminant of the field* K. The discriminant of an algebraic number field is a very important arithmetic invariant and will play a key role in many questions.

PROBLEMS

1. Let $\omega_1, \omega_2, \omega_3$ be linearly independent numbers of the algebraic number field K. Show that the set of numbers of the form $a\omega_1 + b\omega_2 + c\omega_3$, where the rational integers a, b, c satisfy $2a + 3b + 5c = 0$, forms a module in K, and find its basis.

2. Find the coefficient ring of the module $\{2, \sqrt{2}/2\}$ in the field $R(\sqrt{2})$. Show that the module $\{1, \sqrt{2}\}$ is the maximal order of the field $R(\sqrt{2})$.

3. Show that the field of rational numbers contains only one order, the ring of rational integers.

4. Show that in the order $\{1, \sqrt[3]{2}, \sqrt[3]{4}\}$ of the field $R(\sqrt[3]{2})$ every number with norm 2 is an associate of $\sqrt[3]{2}$.

5. Show that the intersection of two full modules is again a full module.

6. Show that any module of an algebraic number field which is a ring is contained in the maximal order.

7. Let $M = \{\alpha_1, \dots, \alpha_n\}$ and $N = \{\beta_1, \dots, \beta_n\}$ be two full modules of the field K. The module generated by the products $\alpha_i\beta_j$ $(1 \leqslant i, j \leqslant n)$ does not depend on the choice of the bases α_i and β_j. It is called the *product of the modules M and N* and is denoted by MN. Show that the coefficient rings of the modules M and N are contained in the coefficient ring of their product MN.

8. Let M be a full module contained in the maximal order $\tilde{\mathfrak{D}}$ of the algebraic number field K. Show that if the discriminant of the module M is not divisible by the square of any integer other than 1, then M coincides with $\tilde{\mathfrak{D}}$.

9. Let θ be a primitive element of the algebraic number field K of degree n, with θ contained in the maximal order. Show that if the discriminant of the minimum polynomial of the number θ is not divisible by any square, then the numbers $1, \theta, \dots, \theta^{n-1}$ form a fundamental basis of the field K.

10. Find a fundamental basis and the discriminant of the field $R(\sqrt[3]{2})$.

11. Find a fundamental basis and the discriminant of the field $R(\rho)$, where ρ is a root of the equation $x^3 - x - 1 = 0$.

12. Let M be a full module of the algebraic number field K. Show that the set M^* of all $\xi \in K$, for which $\operatorname{Sp} \alpha\xi \in Z$ for all $\alpha \in M$, is also a full module of the field K. The module M^* is called the *dual of the module* M. Show that if $\mu_1, \ldots, \mu_n$ is a basis of M, then the dual basis $\mu_1^*, \ldots, \mu_n^*$ of the field K (with respect to R) is a basis of M^*.

13. Show that $(M^*)^* = M$; that is, the dual of the module M^* coincides with M.

14. Show that the dual modules M and M^* have the same coefficient ring.

15. Show that for full modules M_1 and M_2 the inclusions $M_1 \subset M_2$ and $M_1^* \supset M_2^*$ are equivalent.

16. Let θ be a primitive element of the algebraic number field K of degree n, with θ contained in the maximal order $\tilde{\mathfrak{O}}$, and let $f(t)$ be the minimum polynomial of θ over R. Show that for the module $M = \{1, \theta, \ldots, \theta^{n-1}\}$ (which is clearly an order), the dual module M^* coincides with $(1/f'(\theta))M$.

17. Let M be a full module in K with coefficient ring $\mathfrak{O}$. Show that the product MM^* (see Problem 7) coincides with $\mathfrak{O}^*$.

18. Let $M = \{4, \theta, \theta^2\}$ be a module in the field $R(\theta)$, where $\theta^3 = 2$. Show that the coefficient ring of M is the order $\{1, 2\theta, 2\theta^2\}$, while that of the module $M^2 = \{2, 2\theta, \theta^2\}$ is the maximal order $\{1, \theta, \theta^2\}$.

19. The polynomial $t^n + a_1 t^{n-1} + \cdots + a_n$ with rational integer coefficients is called an *Eisenstein polynomial* with respect to the prime number p if all the coefficients $a_1, \ldots, a_n$ are divisible by p, and the constant term a_n, while divisible by p, is not divisible by p^2. Show that if a primitive element θ of an algebraic number field K of degree n is a root of an Eisenstein polynomial with respect to p, then

$$N(c_0 + c_1\theta + \cdots + c_{n-1}\theta^{n-1}) \equiv c_0^n \pmod{p}$$

for any rational integers $c_0, c_1, \ldots, c_{n-1}$.

20. If θ is a primitive element of the algebraic number field K and θ lies in $\tilde{\mathfrak{O}}$, then the index of the order $\{1, \theta, \ldots, \theta^{n-1}\}$ in the maximal order is called the *index of the number* θ. Show that if θ is a root of an Eisenstein polynomial with respect to the prime p, then p does not divide the index of θ.

21. Show that each of the three cubic fields

$$\begin{aligned} K_1 &= R(\theta), & \theta^3 - 18\theta - 6 &= 0,\\ K_2 &= R(\theta), & \theta^3 - 36\theta - 78 &= 0,\\ K^3 &= R(\theta), & \theta^3 - 54\theta - 150 &= 0, \end{aligned}$$

have as a fundamental basis $1, \theta, \theta^2$. Verify that all three fields have the same discriminant, namely, $22356 = 23 \cdot 2^2 \cdot 3^5$. (It follows from Problem 14, Section 7, Chapter 3, that the three fields K_1, K_2, K_3 are distinct.)

22. Show that a fundamental basis for the field $R(\theta)$, $\theta^3 - \theta - 4 = 0$, is given by $1, \theta, (\theta + \theta^2)/2$.

23. Let a and b be relatively prime natural numbers which are square-free. Set $k = ab$ if $a^2 - b^2 \equiv 0 \pmod 9$, and $k = 3ab$ if $a^2 - b^2 \not\equiv 0 \pmod 9$. Show that the discriminant of the field $R(\sqrt[3]{ab^2})$ is $D = -3k^2$.

24. Show that the numbers $1, \sqrt[3]{6}, (\sqrt[3]{6})^2$ form a fundamental basis for the field $R(\sqrt[3]{6})$.

3. Geometric Methods

Two problems were formulated at the end of Section 2.3 (to which we were led by the question of the representation of numbers by full decomposable forms) whose solution requires the introduction of new concepts of a geometric character. At the base of these concepts is a method of representing algebraic numbers as points in n-dimensional space, analogous to the well-known planar representation of the complex numbers.

3.1. *Geometric Representation of Algebraic Numbers*

If the algebraic number field K is of degree n over the rational numbers R, then there are precisely n distinct isomorphisms of this field into the field C of all complex numbers.

Definition. If the image of the field K under the isomorphism $\sigma : K \to C$ is contained in the real numbers, then the isomorphism σ is called *real*, and, if this does not hold, it is called *complex*.

Thus for the cubic field $K = R(\theta)$, where $\theta^3 = 2$, the isomorphism $R(\theta) \to R(\sqrt[3]{2})$, for which $\theta \to \sqrt[3]{2}$, is real (by $\sqrt[3]{2}$ we understand here the real root). The two other isomorphisms $R(\theta) \to R(\varepsilon\sqrt[3]{2})$ and $R(\theta) \to R(\varepsilon^2\sqrt[3]{2})$ [$\varepsilon = \cos(2\pi/3) + i\sin(2\pi/3)$] are complex. If d is a nonsquare rational number, then for the field $R(\theta)$, $\theta^2 = d$, both isomorphisms are real if $d > 0$, and both are complex if $d < 0$. In general, if θ is a primitive element of the arbitrary algebraic number field K, which is a root of the irreducible polynomial $\varphi(t)$ over R, and if $\theta_1, \ldots, \theta_n$ are the roots of $\varphi(t)$ in the field C, then the isomorphism

$$K = R(\theta) \to R(\theta_i) \subset C, \qquad (\theta \to \theta_i) \tag{3.1}$$

will be real if the root θ_i is real, and complex otherwise.

If $\gamma = x + yi$ is any complex number (x and y real) we denote by γ the complex conjugate $x - yi$.

Let $\sigma : K \to C$ be a complex isomorphism. The mapping $\bar{\sigma} : K \to C$, defined by

$$\bar{\sigma}(\alpha) = \overline{\sigma(\alpha)}, \qquad (\alpha \in K),$$

is also a complex isomorphism of K into C. This isomorphism $\bar{\sigma}$ is called conjugate to σ. Since $\bar{\sigma} \neq \sigma$ and $\bar{\bar{\sigma}} = \sigma$, the set of all complex isomorphisms of K into C is divided into pairs of conjugate isomorphisms. In particular, the number of complex isomorphisms is always even. Two complex isomorphisms of the form (3.1) are conjugate if and only if the corresponding roots θ_i and θ_j are complex-conjugate numbers.

Assume that among the isomorphisms of K into C there are s real ones $\sigma_1, \ldots, \sigma_s$ and $2t$ complex ones, so that $s + 2t = n = (K : R)$. From each pair of complex-conjugate isomorphisms, choose one. Denote this set of isomorphisms by $\sigma_{s+1}, \dots, \sigma_{s+t}$. The set of all isomorphisms of K into C then takes the form

$$\sigma_1, \dots, \sigma_s, \sigma_{s+1}, \bar{\sigma}_{s+1}, \dots, \sigma_{s+t}, \bar{\sigma}_{s+t}.$$

In the future we shall always assume the isomorphisms to be enumerated in this manner. Clearly there exist fields with no real isomorphisms ($s = 0$) or no complex isomorphisms ($t = 0$).

Consider the set $\mathfrak{L}^{s,t}$ of all rows of the form

$$x = (x_1, \dots, x_s; x_{s+1}, \dots, x_{s+t}), \tag{3.2}$$

in which the first s components, $x_1, \dots, x_s$, are real, and the remaining ones, $x_{s+1}, \dots, x_{s+t}$, are complex numbers. Addition and multiplication of these rows, as well as scalar multiplication by a real number, are defined componentwise. Under these operations $\mathfrak{L}^{s,t}$ becomes a commutative ring with unit $(1, \dots, 1)$ and at the same time a real linear space. The rows (3.2) will be called *vectors* or *points of the space* $\mathfrak{L}^{s,t}$.

As a basis of $\mathfrak{L}^{s,t}$ (over the field of real numbers) we may clearly take the vectors

$$\left.\begin{matrix}(1, \dots, 0; 0, \dots, 0)\\ \cdots\cdots\cdots\\ (0, \dots, 1; 0, \dots, 0)\end{matrix}\right\} s,$$

$$\left.\begin{matrix}(0, \dots, 0; 1, \dots, 0)\\ (0, \dots, 0; i, \dots, 0)\\ \cdots\cdots\cdots\\ (0, \dots, 0; 0, \dots, 1)\\ (0, \dots, 0; 0, \dots, i)\end{matrix}\right\} 2t. \tag{3.3}$$

Thus the dimension of the space $\mathfrak{L}^{s,t}$ over R equals $n = s + 2t$. If we set

$$x_{s+j} = y_j + iz_j \qquad (j = 1, \dots, t),$$

then the vector (3.2) will have coordinates

$$(x_1, \dots, x_s; y_1, z_1 \dots, y_t, z_t) \tag{3.4}$$

with respect to the basis (3.3).

In cases where $\mathfrak{L}^{s,t}$ is being considered as an n-dimensional real linear space, we shall also denote it by $\mathfrak{R}^n$.

Fix some point x in $\mathfrak{L}^{s,t}$. The transformation $x' \to xx'$ ($x' \in \mathfrak{L}^{s,t}$), that is, multiplication of an arbitrary point of $\mathfrak{L}^{s,t}$ by x, is clearly a linear transformation of the real space $\mathfrak{L}^{s,t} = \mathfrak{R}^n$. In terms of the basis (3.3), the matrix of this transformation is seen to be

$$\begin{pmatrix} x_1 & & & & & & & & \\ & \ddots & & & & & & & \\ & & x_s & & & & & & \\ & & & y_1 & -z_1 & & & & \\ & & & z_1 & y_1 & & & & \\ & & & & & \ddots & & & \\ & & & & & & & y_t & -z_t \\ & & & & & & & z_t & y_t \end{pmatrix},$$

where all other entries are zero. The determinant of this matrix is

$$x_1 \dots x_s(y_1{}^2 + z_1{}^2) \cdots (y_t{}^2 + z_t{}^2) = x_1 \cdots x_s|x_{s+1}|^2 \cdots |x_{s+t}|^2.$$

This suggests the following definition. By the norm $N(x)$ of any point $x = (x_1, \dots, x_{s+t}) \in \mathfrak{L}^{s,t}$, we shall understand the expression

$$N(x) = x_1 \cdots x_s|x_{s+1}|^2 \cdots |x_{s+t}|^2.$$

Thus we have just shown that the norm $N(x)$ of a point x can be defined as the determinant of the matrix of the linear transformation $x' \to xx'$.

The norm is clearly multiplicative:

$$N(xx') = N(x)N(x').$$

We now turn to the representation of numbers of the field K by points of the space $\mathfrak{L}^{s,t}$. Each number α of K will be made to correspond to the point

$$x(\alpha) = (\sigma_1(\alpha), \dots, \sigma_s(\alpha); \sigma_{s+1}(\alpha), \dots, \sigma_{s+t}(\alpha)) \tag{3.5}$$

of $\mathfrak{L}^{s,t}$. This point is the geometric representation of the number α.

If α and β are different numbers of K, then for any $k = 1, \dots, s + t$, the numbers $\sigma_k(\alpha)$ and $\sigma_k(\beta)$ are distinct, and therefore $x(\alpha) \neq x(\beta)$. Thus the embedding

$$\alpha \to x(\alpha) \qquad (\alpha \in K)$$

is one-to-one. (Of course it is not a mapping "onto"; that is, not every point of $\mathfrak{L}^{s,t}$ is the image of some number of the field K.)

Since $\sigma_k(\alpha + \beta) = \sigma_k(\alpha) + \sigma_k(\beta)$ and $\sigma_k(\alpha\beta) = \sigma_k(\alpha)\sigma_k(\beta)$,

$$x(\alpha + \beta) = x(\alpha) + x(\beta), \tag{3.6}$$

$$x(\alpha\beta) = x(\alpha)x(\beta); \tag{3.7}$$

that is, if numbers of K are added or multiplied, the corresponding points are also added or multiplied. Further, if a is a rational number, then $\sigma_k(a\alpha) = \sigma_k(a)\sigma_k(\alpha) = a\sigma_k(\alpha)$, so that

$$x(a\alpha) = ax(\alpha). \tag{3.8}$$

Thus by Section 2.3 of the Supplement we have

$$\begin{aligned} N(\alpha) &= N_{K/R}(\alpha) \\ &= \sigma_1(\alpha)\cdots\sigma_s(\alpha)\sigma_{s+1}(\alpha)\bar{\sigma}_{s+1}(\alpha)\cdots\sigma_{s+t}(\alpha)\bar{\sigma}_{s+t}(\alpha) \\ &= \sigma_1(\alpha)\cdots\sigma_s(\alpha)|\sigma_{s+1}(\alpha)|^2\cdots|\sigma_{s+t}(\alpha)|^2, \end{aligned}$$

so that the norm $N(x(\alpha))$ of the point $x(\alpha)$ coincides with the norm $N(\alpha)$ of the number α:

$$N(x(\alpha)) = N(\alpha), \qquad (\alpha \in K).$$

We consider two simple examples. If d is a positive rational number which is not a square, then for the real quadratic field $R(\theta)$, $\theta^2 = d$, the geometric representation of the number $\alpha = a + b\theta$ (a and b rational) will be the point $x(\alpha) = (a + b\sqrt{d}, a - b\sqrt{d})$. In the case of the imaginary quadratic field $R(\eta)$, $\eta^2 = -d$, the representation of the number $\alpha = a + b\eta$ will be the point in the complex plane with coordinates $(a, b\sqrt{d})$ [the basis (3.3) in this case consists of the numbers 1 and i].

We shall show that if $\alpha_1, \ldots, \alpha_n$ is any basis of the field K (over R), then the corresponding vectors $x(\alpha_1), \ldots, x(\alpha_n)$ of $\mathfrak{L}^{s,t} = \mathfrak{R}^n$ are linearly independent (over the reals). For this set

$$\sigma_k(\alpha_l) = x_k^{(l)} \qquad (1 \leqslant k \leqslant s),$$

$$\sigma_{s+j}(\alpha_l) = y_j^{(l)} + iz_j^{(l)} \qquad (1 \leqslant j \leqslant t).$$

Since the vectors

$$x(\alpha_l) = (x_1^{(l)}, \ldots, x_s^{(l)}; y_1^{(l)} + iz_1^{(l)}, \ldots, y_t^{(l)} + iz_t^{(l)})$$

with the basis (3.3) have the coordinates

$$(x_1^{(l)}, \ldots, x_s^{(l)}, y_1^{(l)}, z_1^{(l)}, \ldots, y_t^{(l)}, z_t^{(l)}),$$

then to prove our assertion we need only show that the determinant

$$d = \begin{vmatrix} x_1^{(1)}\cdots x_s^{(1)} & y_1^{(1)} & z_1^{(1)}\cdots y_t^{(1)} & z_t^{(1)} \\ \cdots & \cdots & \cdots & \cdots \\ x_1^{(n)}\cdots x_s^{(n)} & y_1^{(n)} & z_1^{(n)}\cdots y_t^{(n)} & z_t^{(n)} \end{vmatrix}$$

is nonzero. Consider instead the determinant

$$d^* = \begin{vmatrix} x_1^{(1)}\cdots x_s^{(1)} & y_1^{(1)} + iz_1^{(1)} & y_1^{(1)} - iz_1^{(1)}\cdots \\ \cdots & \cdots & \cdots \\ x_1^{(n)}\cdots x_s^{(n)} & y_1^{(n)} + iz_1^{(n)} & y_1^{(n)} - iz_1^{(n)}\cdots \end{vmatrix},$$

which can also be written in the form

$$d^* = \begin{vmatrix} \sigma_1(\alpha_1) \cdots \sigma_s(\alpha_1) & \sigma_{s+1}(\alpha_1) & \bar{\sigma}_{s+1}(\alpha_1) \cdots \\ \cdots & \cdots & \cdots \\ \sigma_1(\alpha_n) \cdots \sigma_s(\alpha_n) & \sigma_{s+1}(\alpha_n) & \bar{\sigma}_{s+1}(\alpha_n) \cdots \end{vmatrix}.$$

In the determinant d add to the column number $s + 1$ the succeeding column and then take the 2 outside the determinant sign. Then subtract this new column from the succeeding one, and take the $-i$ outside the determinant. Performing this operation on each succeeding pair of columns, we wind up with the equation

$$d^* = (-2i)^t\, d. \tag{3.9}$$

In Section 2.3 of the Supplement it is shown that

$$d^{*2} = D, \tag{3.10}$$

where $D = D(\alpha_1, \ldots, \alpha_n)$ is the discriminant of the basis $\alpha_1, \ldots, \alpha_n$ (with respect to the extension K/R). Since $D \neq 0$, it follows from (3.9) and (3.10) that the determinant d also is nonzero.

Assume now that $\alpha_1, \ldots, \alpha_n$ is a basis for the full module M of the field K. From (3.6) and (3.8) it follows that if $\alpha = a_1\alpha_1 + \cdots + a_n\alpha_n$ is in M ($a_1, \ldots, a_n$ rational integers), then the geometric representation of α in $\mathfrak{L}^{s,t}$ will be the vector $x(\alpha) = a_1 x(\alpha_1) + \cdots + a_n x(\alpha_n)$. We thus have the following result.

Theorem 1. Let K be a number field of degree $n = s + 2t$, with $M = \{\alpha_1, \ldots, \alpha_n\}$ a full module in K. Under the geometric representation of numbers of K by points of the space $\mathfrak{R}^n$, the numbers of M are represented by the set of all integral linear combinations of the n linearly independent (in the space $\mathfrak{R}^n$) vectors $x(\alpha_1), \ldots, x(\alpha_n)$.

3.2. *Lattices*

The geometric study of full modules is based on the fact established in Theorem 1. We therefore consider sets of vectors in $\mathfrak{R}^n$ of the above type, without necessarily assuming that they represent the numbers of some full module.

Definition. Let $e_1, \ldots, e_m$, $m \leqslant n$, be a linearly independent set of vectors in $\mathfrak{R}^n$. The set $\mathfrak{M}$ of all vectors of the form

$$a_1 e_1 + \cdots + a_m e_m,$$

where the a_i independently take on all rational integral values, is called an m-dimensional lattice in $\mathfrak{R}^n$, and the vectors $e_1, \ldots, e_m$ are called a *basis* of this lattice. If $m = n$, the lattice is called *full*; otherwise it is called *nonfull*.

Theorem 1 thus states that the geometric representation of the numbers of a full module is some full lattice.

It is easily seen that two linearly independent sets $e_1, \dots, e_m$ and $f_1, \dots, f_m$ determine the same lattice if and only if they are connected by a unimodular transformation, that is, if

$$f_i = \sum_{j=1}^{m} c_{ij} e_j \qquad (1 \leqslant i \leqslant m),$$

where (c_j) is an integral matrix with determinant ± 1.

The more detailed study of lattices is based on the metric properties of the space $\mathfrak{R}^n$. We introduce an inner product on $\mathfrak{L}^{s,t} = \mathfrak{R}^n$ by taking the vectors (3.3) to be an orthonormal basis. If the vectors x and x' have the coordinates $(x_1, \dots, x_n)$ and $(x_1', \dots, x_n')$ with respect to this basis, then the inner product of x and x', (x, x') is given by the formula

$$(x, x') = x_1 x_1' + \cdots + x_n x_n'.$$

The length of a vector x will be denoted by $\|x\|$.

Let r be a positive real number. The set of all points x with coordinates $(x_1, \dots, x_n)$ [with respect to the basis (3.3)], for which

$$\|x\| = \sqrt{x_1^2 + \cdots + x_n^2} < r,$$

will be denoted by $U(r)$. The set $U(r)$ is called the (open) *ball of radius r* with center at the origin.

A set of points in $\mathfrak{R}^n$ is called *bounded* if it is contained in some ball $U(r)$.

A set of points is called discrete if for every $r > 0$ there are only a finite number of points of the set in the ball $U(r)$.

Lemma 1. The set of points of any lattice $\mathfrak{M}$ in $\mathfrak{R}^n$ is discrete.

Proof. Since any nonfull lattice can be embedded in a full lattice (in many ways), it suffices only to consider a full lattice $\mathfrak{M}$. Let $e_1, \dots, e_n$ be any basis for $\mathfrak{M}$. The conditions

$$(x, e_2) = 0, \dots, (x, e_n) = 0$$

give us a system of $n - 1$ homogeneous linear equations in n unknowns. Since such a system has a nonzero solution, there is a nonzero vector x which is orthogonal to the vectors $e_2, \dots, e_n$. If we also had $(x, e_1) = 0$, then the vector x would be orthogonal to all vectors of the space $\mathfrak{R}^n$, which is impossible. Hence $(x, e_1) \neq 0$. The vector $f_1 = [1/(x, e_1)]\, x$ will also be orthogonal to all the vectors $e_2, \dots, e_n$, and $(f_1, e_1) = 1$. In this manner, for every i, $1 \leqslant i \leqslant n$, we can choose a vector f_i, for which

$$(f_i, e_j) = \begin{cases} 1 & \text{if } j = i, \\ 0 & \text{if } j \neq i. \end{cases}$$

Assume now that the vector $z = a_1 e_1 + \cdots + a_n e_n$ of $\mathfrak{M}$ (a_i rational integers) lies in the ball $U(r)$; that is, $\|z\| < r$. Since $a_k = (z, f_k)$, by the Cauchy-Schwartz inequality we have

$$|a_k| = |(z, f_k)| \leqslant \|z\| \cdot \|f_k\| < r\|f_k\|,$$

where $r\|f_k\|$ does not depend on z. Thus there are only a finite number of possibilities for the integers a_k, so that the set of all $z \in \mathfrak{M}$ for which $\|z\| < r$ is finite. Lemma 1 is proved.

Let X be some set of points of the space $\mathfrak{R}^n$ and z a point of $\mathfrak{R}^n$. The set of all points of the form $x + z$, where x is in X, is called the *translate of the set X* by the vector z and is denoted by $X + z$.

Definition. Let $e_1, \ldots, e_m$ be any basis for the lattice $\mathfrak{M}$. The set T of points of the form

$$\alpha_1 e_1 + \cdots + \alpha_m e_m,$$

where $\alpha_1, \ldots, \alpha_m$ independently take on all real values satisfying $0 \leqslant \alpha_i < 1$, is called a *fundamental parallelepiped* of the lattice $\mathfrak{M}$.

A fundamental parallelepiped is not uniquely determined by the lattice; it depends on the choice of basis.

Lemma 2. If T is a fundamental parallelepiped of the full lattice $\mathfrak{M}$, then the sets

$$T_z = T + z,$$

where z runs through all points of the lattice $\mathfrak{M}$, are pairwise-disjoint and fill the entire space $\mathfrak{R}^n$.

Proof. Let $e_1, \ldots, e_n$ be the basis of $\mathfrak{M}$ used to construct the parallelepiped T. We must show that every point $x = x_1 e_1 + \cdots + x_n e_n$ of $\mathfrak{R}^n$ lies in one and only one set T_z. For each i write the real number x_i in the form $x_i = k_i + \alpha_i$, where k_i is a rational integer and α_i satisfies the condition $0 \leqslant \alpha_i < 1$. Setting $z = k_1 e_1 + \cdots + k_n e_n$ and $u = \alpha_1 e_1 + \cdots + \alpha_n e_n$, we have

$$x = u + z \qquad (u \in T,\ z \in \mathfrak{M}),$$

which means that $x \in T_z$. Now if $x \in T_{z'}$, that is, $x = u' + z'$ ($u' \in T$, $z' \in \mathfrak{M}$), then by comparing the coefficients of e_i in the equation $u + z = u' + z'$ we easily obtain $z = z'$. Lemma 2 is proved.

Lemma 3. For any real number $r > 0$ there are only a finite number of sets T_z (see the notation of Lemma 2) which intersect the ball $U(r)$.

Proof. Let $e_1, \ldots, e_n$ be the basis of the lattice used to construct the parallelepiped T. If we set $d = \|e_1\| + \cdots + \|e_n\|$, then for any vector $u = \alpha_1 e_1 + \cdots + \alpha_n e_n \in T$, we will have

$$\|u\| \leqslant \|\alpha_1 e_1\| + \cdots + \|\alpha_n e_n\| = \alpha_1 \|e_1\| + \cdots + \alpha_n \|e_n\| < d.$$

Assume that the set T_z ($z \in \mathfrak{M}$) intersects $U(r)$. This means that for some vector $x = u + z$, where $u \in T$, $z \in \mathfrak{M}$, we have $\|x\| < r$. Since $z = x - u$,

$$\|z\| \leqslant \|x\| + \|-u\| < r + d,$$

that is, the point z is contained in the ball $U(r + d)$. By Lemma 1 there are only finitely many such points, and Lemma 3 is proved.

It is clear that the vectors of a lattice form a group under the operation of vector addition. In other words, every lattice is a subgroup of the additive group $\mathfrak{R}^n$. Lemma 1 shows that all subgroups are not lattices. We now show that the property of lattices which was established in that lemma characterizes lattices among all subgroups of the group $\mathfrak{R}^n$.

Lemma 4. A subgroup $\mathfrak{M}$ of the group $\mathfrak{R}^n$, the points of which are discrete, is a lattice.

Proof. Denote by $\mathfrak{G}$ the smallest linear subspace of the space $\mathfrak{R}^n$ which contains the set $\mathfrak{M}$, and by m the dimension of $\mathfrak{G}$. We can then choose m vectors $e_1, \ldots, e_m$ in $\mathfrak{M}$ which form a basis for the subspace $\mathfrak{G}$. Denote by $\mathfrak{M}_0$ the lattice with basis $e_1, \ldots, e_m$. Clearly $\mathfrak{M}_0 \subset \mathfrak{M}$. We shall show that the index $(\mathfrak{M} : \mathfrak{M}_0)$ is finite. Indeed, we may represent any vector x of $\mathfrak{M}$ (even any vector of $\mathfrak{G}$) in the form

$$x = u + z, \tag{3.11}$$

where $z \in \mathfrak{M}_0$ and u lies in the fundamental parallelepiped T of the lattice $\mathfrak{M}_0$, constructed with the basis $e_1, \ldots, e_m$. Since $x \in \mathfrak{M}$ and $z \in \mathfrak{M}_0 \subset \mathfrak{M}$, and since $\mathfrak{M}$ is a group, $u \in \mathfrak{M}$. But T is a bounded set, and since $\mathfrak{M}$ is discrete, T can contain only a finite number of vectors of $\mathfrak{M}$. This shows that the number of vectors u which occur in (3.11) for all $x \in \mathfrak{M}$ is finite, which means that the index $(\mathfrak{M} : \mathfrak{M}_0)$ is finite. Let $(\mathfrak{M} : \mathfrak{M}_0) = j$. Since the order of any element of the factor group $\mathfrak{M}/\mathfrak{M}_0$ is a divisor of j, then $jx \in \mathfrak{M}_0$ for all $x \in \mathfrak{M}$, which means that x is a linear combination, with integer coefficients, of $(1/j)e_1, \ldots, (1/j)e_m$. The group $\mathfrak{M}$ is thus contained in the lattice $\mathfrak{M}^*$ with basis $(1/j)e_1, \ldots, (1/j)e_m$. Applying Theorem 2 of Section 2, we see that the subgroup $\mathfrak{M}$ of the group $\mathfrak{M}^*$ must possess a basis of $l \leqslant m$ vectors $f_1, \ldots, f_l$. To show that $\mathfrak{M}$ is a lattice, we need only verify that the vectors $f_1, \ldots, f_l$ are linearly independent over the real numbers. But this follows from the fact that the m linearly independent (over the reals) vectors $e_1, \ldots, e_m$ are linear combinations of the f_i (since $\mathfrak{M}_0 \subseteq \mathfrak{M}$). Lemma 4 is proved.

3.3. *The Logarithmic Space*

Along with the above geometric representation of the numbers of the field K, in which the operation of multiplication of numbers was represented by the operation of multiplication of vectors in $\mathfrak{R}^n$, we must consider another geometric representation, in which the operation of multiplication also has a simple interpretation.

Let there be s real and $2t$ complex isomorphisms of the algebraic number field K into the field C of complex numbers. We shall assume that these isomorphisms are indexed as in Section 3.1.

Consider the real linear space $\mathfrak{R}^{s+t}$ of dimension $s + t$, consisting of rows $(\lambda_1, \ldots, \lambda_{s+t})$ with real components. If $x \in \mathfrak{L}^{s,t}$ is of the form (3.2), with all components different from zero, set

$$\begin{aligned} l_k(x) &= \ln|x_k| && \text{for } k = 1, \ldots, s, \\ l_{s+j}(x) &= \ln|x_{s+j}|^2 && \text{for } j = 1, \ldots, t. \end{aligned} \tag{3.12}$$

We associate to each such point x of $\mathfrak{L}^{s,t}$ the vector

$$l(x) = (l_1(x), \ldots, l_{s+t}(x)) \tag{3.13}$$

of the space $\mathfrak{R}^{s+t}$. If x and x' are any points of $\mathfrak{L}^{s,t}$ with nonzero components, then

$$l_k(xx') = l_k(x) + l_k(x') \qquad (1 \leqslant k \leqslant s + t),$$

so that

$$l(xx') = l(x) + l(x'). \tag{3.14}$$

The collection of all points $x \in \mathfrak{L}^{s,t}$ of the form (3.2) with nonzero components [that is, for which $N(x) \neq 0$] form a group under componentwise multiplication. Equation (3.14) shows that the mapping $x \to l(x)$ is a homomorphism of this multiplicative group onto the additive group of the vector space $\mathfrak{R}^{s+t}$.

Comparing (3.12) with the definition of the norm $N(x)$ of a point $x \in \mathfrak{L}^{s,t}$, we easily see that

$$\sum_{k=1}^{s+t} l_k(x) = \ln|N(x)|. \tag{3.15}$$

If α is a nonzero number from the field K, set

$$l(\alpha) = l(x(\alpha)),$$

where $x(\alpha)$ is the representation of the number α in the space $\mathfrak{L}^{s,t}$ described in Section 3.1. From (3.5), (3.12), and (3.13) we see that the vector $l(\alpha)$ has the form

$$l(\alpha) = (\ln|\sigma_1(\alpha)|, \ldots, \ln|\sigma_s(\alpha)|, \ln|\sigma_{s+1}(\alpha)|^2, \ldots, \ln|\sigma_{s+t}(\alpha)|^2).$$

We call the vector $l(\alpha) \in \mathfrak{R}^{s+t}$ the logarithmic representation of the nonzero number $\alpha \in K$, and call the space $\mathfrak{R}^{s+t}$ the logarithmic space of the field K.

From (3.7) and (3.14) it follows that

$$l(\alpha\beta) = l(\alpha) + l(\beta) \qquad (\alpha \neq 0, \beta \neq 0). \tag{3.16}$$

The mapping $\alpha \to l(\alpha)$ is thus a homomorphism of the multiplicative group of the field K into the group of vectors of the space $\mathfrak{R}^{s+t}$. In particular it follows that

$$l(\alpha^{-1}) = -l(\alpha), \qquad (\alpha \neq 0).$$

The sum of the components

$$l_k(\alpha) = l_k(x(\alpha)), \qquad (1 \leqslant k \leqslant s+t),$$

of the vector $l(\alpha)$ is given by the formula

$$\sum_{k=1}^{s+t} l_k(\alpha) = \ln|N(\alpha)|. \tag{3.17}$$

Indeed, the sum on the left is the logarithm of the absolute value of the product

$$\sigma_1(\alpha) \cdots \sigma_s(\alpha)\sigma_{s+1}(\alpha)\overline{\sigma_{s+1}(\alpha)} \cdots \sigma_{s+t}(\alpha)\overline{\sigma_{s+t}(\alpha)},$$

and this product (Supplement, Section 2.3) equals the norm $N(\alpha)$ (with respect to the extension K/R).

This proof of (3.17) [which does not rely on (3.15)] shows why the definition (3.12) of the components $l_k(x)$ of the vector $l(x)$ distinguished between real and complex isomorphisms. The component $l_{s+j}(x)$ corresponds not to one, but to the two complex-conjugate isomorphisms σ_{s+j} and $\bar{\sigma}_{s+j}$.

3.4. *Geometric Representation of Units*

Let $\mathfrak{D}$ be some fixed order of the field K. In the logarithmic space $\mathfrak{R}^{s+t}$ consider the set of all vectors $l(\varepsilon)$, where ε is a unit in the ring $\mathfrak{D}$. The mapping $\varepsilon \to l(\varepsilon)$ is not one-to-one. For if the unit $\eta \in \mathfrak{D}$ is a root of 1, that is, if $\eta^m = 1$ for some natural number m, then $|\sigma_k(\eta)| = 1$ for all $k = 1, \ldots, s+t$, so that $l(\eta)$ is the zero vector. Thus all roots of 1 (and the order $\mathfrak{D}$ contains at least two: $+1$ and -1) are mapped to the zero vector. In order to use the mapping $\varepsilon \to l(\varepsilon)$ to clarify the structure of the group of units, we must answer the following two questions:

(1) Which units $\varepsilon \in \mathfrak{D}$ are mapped to the zero vector?

(2) What is the form of the set of all vectors $l(\varepsilon)$?

We start with the first question. Denote by W the set of all numbers $\alpha \in \mathfrak{D}$ for which $l(\alpha) = 0$. By (3.16) the product of any two numbers of W again lies in W. Since the condition $l(\alpha) = 0$ is equivalent to

$$|\sigma_k(\alpha)| = 1 \qquad (1 \leqslant k \leqslant s+t),$$

the set of all points $x(\alpha) \in \mathfrak{R}^n = \mathfrak{L}^{s,t}$ for all $\alpha \in W$ is bounded, that is, it is contained in some ball $U(r)$. Applying Lemma 1 we find that the set W is finite. For an arbitrary number $\alpha \in W$ consider its powers $1, \alpha, \ldots, \alpha^k, \ldots$. Since these powers are contained in W, there is some equality $\alpha^k = \alpha^l$, $l > k$. Setting $m = l - k$, we obtain $\alpha^m = 1$. Thus all numbers of W are roots of 1, and this means that W is a finite group, contained in the group of units of the ring $\mathfrak{O}$.

Since the group W contains a subgroup of order 2 (consisting of $+1$ and -1), then it has even order. Further, all finite subgroups of the multiplicative group of a field are cyclic (Supplement, Section 3), and therefore W is cyclic.

We therefore have the following answer to the first question.

Theorem 2. The units of the order $\mathfrak{O}$, for which $l(\varepsilon)$ is the zero vector, form a finite cyclic group of even order. This group consists of all roots of 1 contained in $\mathfrak{O}$.

We therefore turn to the second question, that is, we shall seek to clarify the structure of the set $\mathfrak{E}$ in $\mathfrak{R}^{s+t}$ which consists of all vectors $l(\varepsilon)$, where ε is a unit of the ring $\mathfrak{O}$.

By Theorem 4 of Section 2 the norm of any unit ε of $\mathfrak{O}$ equals ± 1, and therefore $\ln|N(\varepsilon)| = 0$. By (3.17) we therefore have

$$\sum_{k=1}^{s+t} l_k(\varepsilon) = 0. \tag{3.18}$$

This means that all points $l(\varepsilon)$ belong to the subspace $\mathfrak{L} \subset \mathfrak{R}^{s+t}$, consisting of all points $(\lambda_1, \ldots, \lambda_{s+t}) \in \mathfrak{R}^{s+t}$, for which $\lambda_1 + \cdots + \lambda_{s+t} = 0$. The dimension of the subspace $\mathfrak{L}$ clearly equals $s + t - 1$.

We shall show that $\mathfrak{E}$ is a lattice. Since $\mathfrak{E}$ is a subgroup of the additive group of the vector space $\mathfrak{R}^{s+t}$, by Lemma 4 it suffices to prove that the set $\mathfrak{E}$ is discrete. (As an orthonormal basis in $\mathfrak{R}^{s+t}$ we take those vectors with one component equal to 1 and the rest equal to 0.) Let r be any positive real number and let $\|l(\varepsilon)\| < r$. Since

$$l_k(\varepsilon) \leqslant |l_k(\varepsilon)| \leqslant \|l(\varepsilon)\|, \text{ then } l_k(\varepsilon) < r \quad (1 \leqslant k \leqslant s+t),$$

which means that

$$|\sigma_k(\varepsilon)| < e^r \qquad (k = 1, \ldots, s),$$

$$|\sigma_{s+j}(\varepsilon)|^2 < e^r \qquad (j = 1, \ldots, t).$$

From this it follows that the set of points $x(\varepsilon)$ in $\mathfrak{R}^n$, where ε runs through all units of $\mathfrak{O}$ for which $\|l(\varepsilon)\| < r$, is bounded. But since the vectors $x(\alpha) \in \mathfrak{R}^n$ for all $\alpha \in \mathfrak{O}$ form a lattice (Theorem 1), it follows from Lemma 1 that the number of such ε is finite. Thus the number of vectors $l(\varepsilon)$ such that $\|l(\varepsilon)\| < r$ is also finite and this means that the set $\mathfrak{E}$ is discrete.

Since the lattice $\mathfrak{E}$ is contained in the subspace $\mathfrak{L}$, its dimension does not exceed $s + t - 1$.

We have thus proved the following fact.

Theorem 3. The set of all points $l(\varepsilon)$, where ε is a unit of the order $\mathfrak{D}$, forms a lattice $\mathfrak{E}$ in the logarithmic space $\mathfrak{L}$ of dimension $r \leqslant s + t - 1$.

3.5. *Partial Results on the Group of Units*

Theorems 2 and 3, which were derived from very simple geometric considerations, contain much important information on the structure of the group of units of the order $\mathfrak{D}$. From these theorems it follows that in $\mathfrak{D}$ there exist units $\varepsilon_1, \dots, \varepsilon_r$, $r \leqslant s + t - 1$, such that every unit ε is uniquely representable in the form

$$\varepsilon = \zeta \varepsilon_1{}^{a_1} \cdots \varepsilon_r{}^{a_r}, \tag{3.19}$$

where $a_1, \dots, a_r$ are rational integers and ζ is some root of 1 contained in $\mathfrak{D}$. In other words, the group of units of the order $\mathfrak{D}$ is the product of a finite group and r infinite cyclic groups.

To prove this assertion we take any basis for the lattice $\mathfrak{E}$, say, $l(\varepsilon_1), \dots, l(\varepsilon_r)$, and will show that the units $\varepsilon_1, \dots, \varepsilon_r$ satisfy the desired properties. Let ε be an arbitrary unit of the ring $\mathfrak{D}$. Since $l(\varepsilon) \in \mathfrak{E}$, then

$$l(\varepsilon) = a_1 l(\varepsilon_1) + \cdots + a_r l(\varepsilon_r),$$

where a_i are rational integers. Consider the unit:

$$\zeta = \varepsilon \varepsilon_1{}^{-a_1} \cdots \varepsilon_r{}^{-a_r}.$$

By formula (3.16) we have $l(\zeta) = l(\varepsilon) - a_1 l(\varepsilon_1) - \cdots - a_r l(\varepsilon_r)$, and then by Theorem 2 ζ is a root of 1. Hence ε has a representation (3.19). We now prove that this representation is unique. Assume that we also have $\varepsilon = \zeta' \varepsilon_1{}^{b_1} \cdots \varepsilon_r{}^{b_r}$. Since the vectors $l(\varepsilon_1), \dots, l(\varepsilon_r)$ are linearly independent, it follows from $l(\varepsilon) = b_1 l(\varepsilon_1) + \cdots + b_r l(\varepsilon_r)$ that $a_1 = b_1, \dots, a_r = b_r$. But then also $\zeta = \zeta'$, and our assertion is completely proved.

There remains the open question of the precise value of the number r, since we have proved only that it does not exceed $s + t - 1$. Using the methods we have relied on so far, we cannot even guarantee that $r > 0$ (when $s + t - 1 > 0$). In Section 4 we shall show that actually $r = s + t - 1$. But this will be an existence theorem; it will establish the existence of $s + t - 1$ independent units. It is therefore not surprising that its proof will require some new concepts.

By Theorem 3 the remaining assertion is equivalent to the fact that the dimension of the lattice $\mathfrak{E}$, which represents the units of the order $\mathfrak{O}$ in the logarithmic space, is of dimension equal to $s + t - 1$.

PROBLEMS

1. Show that the set of all images $x(\alpha) \in \mathfrak{R}^n$ of numbers α of the algebraic number field K of degree n is an everywhere-dense subset of the space $\mathfrak{R}^n$.

2. Assume that $s \neq 0$, that is, that there is an isomorphism of the field K into the field of real numbers. Show that the group of roots of 1 contained in K consists only of two numbers: $+1$ and -1. (This condition always holds when the degree of K is odd.)

3. Determine all roots of 1 which can be contained in an algebraic number field of degree 4.

4. Find all units of the field $R(\sqrt{3})$.

5. Show that in the field $R(\sqrt[3]{2})$ any unit has the form $\pm(1 + \sqrt[3]{2})^k$.

6. Let the algebraic number field K contain a complex root of 1. Show that then the norm of any $\alpha \neq 0$ of K is positive.

4. The Group of Units

4.1. *A Criterion for the Fullness of a Lattice*

In this section we finish our investigation of the structure of the group of units of an order in an algebraic number field. The basic problem, which we are going to solve, has already been considered at the end of the preceding section. It is to prove that the lattice $\mathfrak{E}$, the vectors of which represent the units of the order $\mathfrak{O}$ in the logarithmic representation, has dimension $s + t - 1$ (we preserve all notations from Section 3).

The lattice $\mathfrak{E}$ lies in the space $\mathfrak{R}^{s+t}$ and is contained in the subspace $\mathfrak{L}$, which consists of all points $(\lambda_1, \ldots, \lambda_{s+t})$ for which $\lambda_1 + \cdots + \lambda_{s+t} = 0$. Since the dimension of $\mathfrak{L}$ is equal to $s + t - 1$, we need to show that $\mathfrak{E}$ is a full lattice in the space $\mathfrak{L}$. This will be proved in Section 4.3, using the following criterion for the fullness of a lattice.

Theorem 1. A lattice $\mathfrak{M}$ in a linear space $\mathfrak{L}$ is full if and only if there exists a bounded set U in $\mathfrak{L}$, such that the translates of U by all vectors of $\mathfrak{M}$ occupy the whole space $\mathfrak{L}$ (they may intersect).

Proof. If the lattice $\mathfrak{M}$ is full, then as U we may take any of its fundamental parallelepipeds. Lemma 2 of Section 3 now implies that the translates of a fundamental parallelepiped by all points of the lattice fill the entire space (it is clear that a fundamental parallelepiped is bounded). Assume now that the lattice $\mathfrak{M}$ is not full, and let U be an arbitrary bounded set in $\mathfrak{L}$. We shall show that in this case the translates of U by all vectors of $\mathfrak{M}$ cannot fill the entire space $\mathfrak{L}$. Since U is bounded, there exists a real number $r > 0$ such that $\|u\| < r$ for all $u \in U$. Let $\mathfrak{L}'$ denote the subspace generated by the vectors of $\mathfrak{M}$. Since the lattice $\mathfrak{M}$ is not full, $\mathfrak{L}'$ is a proper subspace of $\mathfrak{L}$, and therefore there exists in $\mathfrak{L}$ vectors of arbitrary length which are orthogonal to the subspace $\mathfrak{L}'$ (and hence to all vectors in $\mathfrak{M}$). We claim that any such vector y, for which $\|y\| \geqslant r$, cannot lie within a translate of U by a vector of $\mathfrak{M}$. For suppose that such a y (orthogonal to $\mathfrak{L}'$) is contained in some translate, say, $y = u + z$, where $u \in U$, $z \in \mathfrak{M}$. By the Cauchy-Schwarz inequality

$$\|y\|^2 = (y, y) = (y, u) \leqslant \|y\| \; \|u\| < r\|y\|,$$

so that $\|y\| < r$. Theorem 1 is proved. (The geometric meaning of the proof is that all translates of U by vectors of the nonfull lattice $\mathfrak{M}$ lie in the strip consisting of all points at distance less than r from the subspace $\mathfrak{L}'$).

Remark. In topological terms, the fullness of the lattice $\mathfrak{M}$ in the space $\mathfrak{L}$ is equivalent to the compactness of the factor group $\mathfrak{L}/\mathfrak{M}$ ($\mathfrak{L}$ being considered as a topological group under addition).

4.2. *Minkowski's Lemma*

Our proof of the existence of $s + t - 1$ independent units will be based on a simple geometric fact which has many applications in number theory. The formulation and proof of this assertion (Theorem 3) use the concept of volume in n-dimensional space and some of its properties.

The volume $v(X)$ of the set X in the n-dimensional space $\mathfrak{R}^n$ can be defined as the multiple integral

$$v(X) = \int_{(X)} \cdots \int dx_1 \, dx_2 \cdots dx_n,$$

carried out over the set X. [Here we sometimes deviate from (3.4) and denote the coordinates of the point x in $\mathfrak{R}^n$ by $(x_1, \ldots, x_n)$.] We shall not enter into the question of conditions under which the volume exists. In the cases which we shall consider, the set X will be given by some inequalities of a very simple type, and the question of the existence of the volume can be decided by elementary considerations. We list some simple properties of volume, easily verified from properties of the integral.(We assume that all volumes considered exist.)

(1) If X is contained in X', then

$$v(X) \leqslant v(X').$$

(2) If the sets X and X' do not intersect, then

$$v(X \cup X') = v(X) + v(X').$$

(3) Any translate of a set has the same volume; that is,

$$v(X + z) = v(X).$$

(4) Let α be a positive real number. Let αX denote the set of all points of the form αx, where x runs through all points of X. (The set αX is called the *expansion of* X by α.) Then

$$v(\alpha X) = \alpha^n v(X).$$

We now compute the volume of a fundamental parallelepiped T of the full lattice $\mathfrak{M}$ in $\mathfrak{R}^n$, which is constructed from some basis $e_1, \dots, e_n$. Let

$$e_j = (a_{1j}, \dots, a_{nj}) \qquad (1 \leqslant j \leqslant n).$$

We shall show that then

$$v(T) = |\det(a_{ij})|. \tag{4.1}$$

In the integral

$$v(T) = \underset{(T)}{\int \cdots \int} dx_1 \cdots dx_n$$

we change variables by the formula

$$x_i = \sum_{j=1}^{n} a_{ij} x_j' \qquad (1 \leqslant i \leqslant n).$$

The Jacobian of this transformation is $\det(a_{ij})$, which is nonzero since the vectors $e_1, \dots, e_n$ are linearly independent. Under this transformation the set T is taken to the set T_0, consisting of all points $(x_1', \dots, x_n')$, for which $0 \leqslant x_i' < 1$ $(i = 1, \dots, n)$, so that

$$\begin{aligned} v(T) &= \underset{(T_0)}{\int \cdots \int} |\det(a_{ij})| \, dx_1' \cdots dx_n' \\ &= |\det(a_{ij})| \int_0^1 \cdots \int_0^1 dx_1' \cdots dx_n' = |\det(a_{ij})|, \end{aligned}$$

and (4.1) is proved.

Let the mapping $x \to x'$ give a nonsingular linear transformation of the space $\mathfrak{R}^n$ into itself. The lattice $\mathfrak{M}$ is taken by this transformation into some lattice $\mathfrak{M}'$ (clearly full), and its fundamental parallelepiped T is taken to a fundamental parallelepiped T' of the lattice $\mathfrak{M}'$. It is clear that the parallelepiped T' will be constructed from the images $e_1', \dots, e_n'$ of the vectors of basis $e_1, \dots, e_n$. If $e_j' = (b_{1j}, \dots, b_{nj})$ $(1 \leqslant j \leqslant n)$, then the volume $v(T') = |\det(b_{ij})|$. Let $C = (c_{ij})$ denote the matrix of the linear transformation $x \to x'$

with respect to the basis $e_1, \ldots, e_n$, so that

$$e_j' = \sum_{i=1}^{n} c_{ij} e_i \qquad (1 \leqslant j \leqslant n).$$

It is easily seen that $b_{ij} = \sum_{s=1}^{n} a_{is} c_{sj}$; that is, the matrix (b_{ij}) is the product of (a_{ij}) and (c_{ij}), which means that

$$v(T') = v(T) \cdot |\det C|. \tag{4.2}$$

Now assume that $e_1, \ldots, e_n$ and $e_1', \ldots, e_n'$ are two bases of the same lattice $\mathfrak{M}$. Since these bases are obtained from one another by unimodular transformations (that is, C has integer entries and $\det C = \pm 1$), it follows from (4.2) that $v(T') = v(T)$. This shows that the volume of a fundamental parallelepiped of a lattice depends only on the lattice itself and not on the choice of basis.

Combining (4.1) with (3.9) and (3.10) we obtain the following strengthening of Theorem 1, Section 3.

Theorem 2. Let K be a number field of degree $n = s + 2t$ and let M be a full module with discriminant D. Under the geometric representation of numbers of K by points of $\mathfrak{L}^{s,t} = \mathfrak{R}^n$, the points which represent the numbers of M form a full lattice with the volume of a fundamental parallelepiped equal to $2^{-t}\sqrt{|D|}$.

To formulate the basic proposition of this section we need two more geometric concepts.

A set $X \subset \mathfrak{R}^n$ is called centrally symmetric if whenever $x \in X$, then also $-x \in X$.

A set X is called convex if for any two points $x \in X$ and $x' \in X$, all points of the form $\alpha x + (1 - \alpha)x'$, where α is a real number satisfying $0 \leqslant \alpha \leqslant 1$, lie in X. In other words, X is convex if the entire line segment connecting any two points of X lies in X.

Theorem 3 (Minkowski's Lemma on Convex Bodies). Let $\mathfrak{M}$ be a full lattice in the n-dimensional space $\mathfrak{R}^n$, with the volume of a fundamental parallelepiped of $\mathfrak{M}$ given by Δ, and let X be a bounded, centrally symmetric, convex set with volume $v(X)$. If $v(X) > 2^n\Delta$, then the set X contains at least one nonzero point of the lattice $\mathfrak{M}$.

Proof. We base our proof on the following (intuitively obvious) assertion: If a bounded set $Y \subset \mathfrak{R}^n$ has the property that its translates $Y_z = Y + z$ by vectors $z \in \mathfrak{M}$ are pairwise-nonintersecting, then $v(Y) \leqslant \Delta$. To prove this assertion we consider some fundamental parallelepiped T of the lattice $\mathfrak{M}$ and look at the intersections $Y \cap T_{-z}$ of the set Y with all translates $T_{-z} = T - z$

of the parallelepiped T. It is clear that

$$v(Y) = \sum_{z \in \mathfrak{M}} v(Y \cap T_{-z}).$$

(Although this sum is formally infinite, it contains only a finite number of nonzero terms, since the set Y is bounded and thus intersects only a finite number of the T_{-z}; see Lemma 3 of Section 3.) The translate of the set $Y \cap T_{-z}$ by the vector z is clearly equal to $Y_z \cap T$, and therefore $v(Y \cap T_{-z}) = v(Y_z \cap T)$, so that

$$v(Y) = \sum_{z \in \mathfrak{M}} v(Y_z \cap T).$$

If the translates Y_z are pairwise-nonintersecting, then the intersections $Y_z \cap T$ are also pairwise-nonintersecting, and since they are all contained in T, the sum on the right of the above equation cannot be more than $v(T)$. Hence $v(Y) \leqslant v(T)$, and our assertion is proved.

Consider now the set $\frac{1}{2}X$ (obtained from X by contracting by a factor of 2). From the assumptions of the theorem it follows that $v(\frac{1}{2}X) = (1/2^n)v(X) > \Delta$. If all translates $\frac{1}{2}X + z$ by vectors $z \in \mathfrak{M}$ were pairwise-nonintersecting, then we would necessarily have $v(\frac{1}{2}X) \leqslant \Delta$, which is not the case. Hence for two distinct vectors z_1 and z_2 of $\mathfrak{M}$, the sets $\frac{1}{2}X + z_1$ and $\frac{1}{2}X + z_2$ have a common point:

$$\tfrac{1}{2}x' + z_1 = \tfrac{1}{2}x'' + z_2 \qquad (x', x'' \in X).$$

We write this in the form

$$z_1 - z_2 = \tfrac{1}{2}x'' - \tfrac{1}{2}x'.$$

Since the set X is centrally symmetric, $-x' \in X$, and since it is convex,

$$\tfrac{1}{2}x'' - \tfrac{1}{2}x' = \tfrac{1}{2}x'' + \tfrac{1}{2}(-x') \in X.$$

Thus the nonzero point $z_1 - z_2$ of $\mathfrak{M}$ lies in the set X and the theorem is proved.

From the first part of the proof we may draw the following obvious corollary (which will be used in Section 5).

Lemma 1. If the union of all translates of the set Y by vectors of the lattice $\mathfrak{M}$ completely fills the space $\mathfrak{R}^n$, then $v(Y) \geqslant \Delta$.

For in this case the intersections $Y_z \cap T$ completely fill the fundamental parallelepiped T (possibly overlapping) and therefore

$$v(Y) = \sum_{z \in \mathfrak{M}} v(Y_z \cap T) \geqslant v(T) = \Delta.$$

To investigate the group of units we shall apply Minkowski's lemma to lattices in the space $\mathfrak{L}^{s,t}$, and as the set X we shall take all points x of the form (3.2) for which

$$|x_1| < c_1, \dots, |x_s| < c_s; |x_{s+1}|^2 < c_{s+1}, \dots, |x_{s+t}|^2 < c_{s+t},$$

where $c_1, \dots, c_{s+t}$ are positive real numbers. The convexity and central symmetry of this set X are clear. We compute its volume. Using (3.4) for the coordinates of x, we obtain

$$v(X) = \int_{-c_1}^{c_1} dx_1 \cdots \int_{-c_s}^{c_s} dx_s \iint\limits_{y_1^2+z_1^2<c_{s+1}} dy_1\, dz_1 \cdots \iint\limits_{y_t^2+z_t^2<c_{s+t}} dy_t\, dz_t = 2^s \pi^t \prod_{i=1}^{s+t} c_i .$$

Applying Minkowski's lemma to this set gives us the following result (to which we shall refer in the future).

Theorem 4. If the volume of a fundamental parallelepiped of the full lattice $\mathfrak{M}$ in the space $\mathfrak{L}^{s,t}$ is Δ, and if the positive real numbers $c_1, \dots, c_{s+t}$ satisfy $\prod_{i=1}^{s+t} c_i > (4/\pi)^t \Delta$, then there is a nonzero vector $x = (x_1, \dots, x_{s+t})$ in the lattice $\mathfrak{M}$ for which

$$|x_1| < c_1, \dots, |x_s| < c_s; |x_{s+1}|^2 < c_{s+1}, \dots, |x_{s+t}|^2 < c_{s+t}. \tag{4.3}$$

4.3. *The Structure of the Group of Units*

We can now completely solve the question of the structure of the group of units of an arbitrary order.

Theorem 5 (Dirichlet's Theorem). Let $\mathfrak{O}$ be any order of the algebraic number field K of degree $n = s + 2t$. Then there exist units $\varepsilon_1, \dots, \varepsilon_r$, $r = s + t - 1$, such that every unit $\varepsilon \in \mathfrak{O}$ has a unique representation in the form

$$\varepsilon = \zeta \varepsilon_1^{a_1} \cdots \varepsilon_r^{a_r},$$

where $a_1, \dots, a_r$ are rational integers and ζ is some root of 1 contained in $\mathfrak{O}$.

Proof. We have already remarked that we need only prove that that lattice $\mathfrak{E}$, which is the image of the units of $\mathfrak{O}$, is full in the space $\mathfrak{L}$ (which is of dimension $s + t - 1$). From Theorem 1 we see that it will suffice to show that there is a bounded set U in $\mathfrak{L}$ such that the translates of U by all vectors of $\mathfrak{E}$ cover the entire space $\mathfrak{L}$. We rephrase this last assertion in terms of the space $\mathfrak{L}^{s,t}$.

It is clear that any point $(\lambda_1, \dots, \lambda_{s+t})$ of $\mathfrak{L}$ (and also of $\mathfrak{R}^{s+t}$) is the image of some point x of $\mathfrak{L}^{s,t}$ under the mapping $x \to l(x)$. It follows immediately from (3.15) that the image of a point $x \in \mathfrak{L}^{s,t}$ (with nonzero components) under the logarithmic mapping lies in the subspace $\mathfrak{L}$ of $\mathfrak{R}^{s+t}$ if and only if $|N(x)| = 1$.

Denote by S the set of all $x \in \mathfrak{L}^{s,t}$ for which $|N(x)| = 1$. If X_0 is an arbitrary bounded subset of S, then its image $l(X_0)$ is also bounded. For if the point $x = (x_1, \dots, x_{s+t})$ has norm ± 1 and satisfies

$$|x_k| < C \qquad (1 \leqslant k \leqslant s), \qquad |x_{s+j}|^2 < C \qquad (1 \leqslant j \leqslant t);$$

then $l_k(x) < \ln C$ for all $k = 1, \dots, s+t$ and thus

$$l_k(x) = -\sum_{i \neq k} l_i(x) > -(s+t-1)C,$$

so that $l(X_0)$ is bounded. Since the norm is multiplicative, if $x \in S$ and $X_0 \subset S$, then the product $X_0 x$ is also contained in S. In particular, for any unit ε of the order $\mathfrak{O}$, we have $X_0 x(\varepsilon) \subset S$ [since $N(x(\varepsilon)) = N(\varepsilon) = \pm 1$]. If the products $X_0 x(\varepsilon)$ for all units ε completely cover S, then the translates $l(X_0) + l(\varepsilon)$ clearly cover the entire space $\mathfrak{L}$. We have thus shown that to prove Theorem 5 it suffices to find in S a bounded subset X_0 whose "multiplicative translates" $X_0 x(\varepsilon)$ completely cover S.

Let y be any point of S and let $\mathfrak{M}$ be the lattice of points in $\mathfrak{L}^{s,t}$ corresponding to the numbers of the order $\mathfrak{O}$. We map the space $\mathfrak{L}^{s,t}$ into itself by the linear transformation $x \to yx$ $(x \in \mathfrak{L}^{s,t})$. In Section 3.1 we saw that the determinant of the matrix of this transformation was equal to $N(y)$, that is, to ± 1. Hence, by (4.2), the volumes of fundamental parallelepipeds for the lattices $\mathfrak{M}$ and $y\mathfrak{M}$ are the same. Denote this volume by Δ.

Choose positive real numbers $c_1, \dots, c_{s+t}$ so that

$$Q = c_1 \cdots c_{s+t} > \left(\frac{4}{\pi}\right)^t \Delta,$$

and denote by X the set of all points $x \in \mathfrak{L}^{s,t}$ for which the inequality (4.3) holds. Theorem 4 implies that there is a nonzero point $x = yx(\alpha)$ $(\alpha \in \mathfrak{O}, \alpha \neq 0)$ contained in X. Since $N(x) = N(y)N(\alpha) = \pm N(\alpha)$ and $|N(x)| < c_1 \cdots c_{s+t} = Q$, $|N(\alpha)| < Q$. By Theorem 5 of Section 2 there are only finitely many pairwise-nonassociate numbers in the order $\mathfrak{O}$ whose norms have absolute value less than Q. Fix some set $\alpha_1, \dots, \alpha_N$ of nonzero numbers of $\mathfrak{O}$ such that any nonzero number of $\mathfrak{O}$, whose norm has absolute value less than Q, is associate with one of these. Then for some i $(1 \leqslant i \leqslant N)$ we have $\alpha\varepsilon = \alpha_i$, where ε is a unit in $\mathfrak{O}$. Then y can be represented as

$$y = xx(\alpha_i^{-1})x(\varepsilon). \tag{4.4}$$

Set

$$X_0 = S \cap \left(\bigcup_{i=1}^{N} Xx(\alpha_i^{-1}) \right). \tag{4.5}$$

Since X is bounded, each of the sets $Xx(\alpha_i^{-1})$ is also bounded and thus X_0 is bounded. Further, the choice of the numbers $c_1, \ldots, c_{s+t}$ which determine the set X and of the numbers $\alpha_1, \ldots, \alpha_N$ did not depend on the point y, and therefore the set (4.5) is completely determined by the order $\mathfrak{D}$. Since y and $x(\varepsilon)$ lie in S, then by (4.4) the point $xx(\alpha_i^{-1})$ also lies in S, which means that it belongs to X_0. Equation (4.4) thus shows that the point y of S, which we chose arbitrarily, is contained in the set $X_0x(\varepsilon)$. Thus S is completely covered by all these sets (for all units ε), and this proves Theorem 5.

As we remarked in Section 3.5, Dirichlet's theorem implies that the group of units in any order of an algebraic number field of degree $n = s + 2t$ is the product of a finite group with $s + t - 1$ infinite cyclic groups.

If $s + t = 1$ (and this is the case only for the field of rational numbers and for imaginary quadratic fields), then $r = 0$. In this case the lattice $\mathfrak{E}$ consists only of the zero vector and the group of units of the order $\mathfrak{D}$ is just the finite group of roots of 1.

The units $\varepsilon_1, \ldots, \varepsilon_r$, whose existence is established by Dirichlet's theorem, are called fundamental units of the order $\mathfrak{D}$. From Section 3.5 it is clear that a set of units $\varepsilon_1, \ldots, \varepsilon_r$ is fundamental if and only if the vectors $l(\varepsilon_1), \ldots, l(\varepsilon_r)$ form a basis for the lattice $\mathfrak{E}$. From this it easily follows that the units

$$\varepsilon_i' = \zeta_i \varepsilon_1^{a_{i1}} \cdots \varepsilon_r^{a_{ir}} \qquad (1 \leqslant i \leqslant r)$$

(where ζ_i is an arbitrary root of 1 contained in $\mathfrak{D}$) will be a fundamental set of units if and only if the matrix (a_{ij}) is unimodular.

Remark. The proof of Dirichlet's theorem is not effective in that it does not give an algorithm for finding some set of fundamental units for the order $\mathfrak{D}$. This was caused by our use of the full system of nonassociate numbers $\alpha_1, \ldots, \alpha_N$ whose norm is less than Q. The existence of such a system was established in a noneffective manner. We return to the question of effectiveness in Section 5.

Dirichlet's theorem also holds (as does Theorem 2 of Section 3) for the maximal order $\tilde{\mathfrak{D}}$ of the field K. Fundamental units for the maximal order $\tilde{\mathfrak{D}}$ are also called *fundamental units for the algebraic number field K.*

4.4. *The Regulator*

By the construction of Sections 3.3 and 3.4, there is associated to each order $\mathfrak{D}$ of an algebraic number field K of degree $r = s + 2t$ a lattice $\mathfrak{E}$ of dimension $r = s + t - 1$ in the subspace $\mathfrak{L} \subset \mathfrak{R}^{s+t}$. The volume v of a fundamental parallelepiped of this lattice does not depend on the choice of basis and thus is completely determined by the lattice $\mathfrak{E}$. We now compute this volume. Let T_0 be a fundamental parallelepiped of the lattice $\mathfrak{E}$ constructed from the basis $l(\varepsilon_1), \ldots, l(\varepsilon_r)$ (here $\varepsilon_1, \ldots, \varepsilon_r$ is a system of fundamental units of the order $\mathfrak{D}$). The vector $l_0 = 1/\sqrt{s+t}\,(1, \ldots, 1)$ in $\mathfrak{R}^{s+t}$ is clearly orthogonal to the subspace $\mathfrak{L}$ and has unit length. Then the r-dimensional volume $v = v(T_0)$ equals the $(s+t)$-dimensional volume of the parallelepiped T determined by the vectors $l_0, l(\varepsilon_1), \ldots, l(\varepsilon_r)$. By (4.1) the volume v is equal to the absolute value of the determinant of the matrix whose rows are the components of these vectors. If we now add all other columns to the ith column and use (3.18), we may expand the determinant by the ith column and obtain

$$v = \sqrt{s+t}\,R,$$

where R is the absolute value of any of the minors of order r of the matrix

$$\begin{pmatrix} l_1(\varepsilon_1) \cdots l_{s+t}(\varepsilon_1) \\ \cdot\ \cdot\ \cdot\ \cdot\ \cdot\ \cdot\ \cdot\ \cdot\ \cdot \\ l_1(\varepsilon_r) \cdots l_{s+t}(\varepsilon_r) \end{pmatrix}. \tag{4.6}$$

It thus follows that all rth-order minors of (4.6) have the same absolute value, which is, moreover, independent of the choice of the system of fundamental units $\varepsilon_1, \ldots, \varepsilon_r$. The number R (as well as v) thus depends only on the order $\mathfrak{D}$. It is called the *regulator of the order* $\mathfrak{D}$.

The regulator of the maximal order $\tilde{\mathfrak{D}}$ is called the *regulator of the algebraic number field K*. (For the field of rational numbers and for imaginary quadratic fields the regulator is by definition equal to 1.)

PROBLEMS

1. Show that the inequality $v(X) > 2^n\Delta$ in Minkowski's lemma cannot be weakened. To do this construct a convex bounded centrally symmetric set X with volume $v(X) = 2^n\Delta$ which does not contain any nonzero point of the lattice.

2. Let a be a positive real number. Show that the volume of the set $X \subset \mathfrak{L}^{s,t}$, consisting of all points x for which

$$|x_1| + \cdots + |x_s| + 2\sqrt{y_1^2 + z_1^2} + \cdots + 2\sqrt{y_t^2 + z_t^2} < a$$

[in the coordinates (3.4)], equals

$$v(X) = 2^s \left(\frac{\pi}{2}\right)^t \frac{1}{n!} a^n.$$

Check that the set X is bounded, centrally symmetric, and convex.

3. Let a and b be natural numbers which are not squares. Show that fundamental units for the order $\{1, \sqrt{a}\}$ of the field $R(\sqrt{a})$ are also fundamental units for the order $\{1, \sqrt{a}, \sqrt{-b}, \sqrt{a}\sqrt{-b}\}$ in the field $R(\sqrt{a}, \sqrt{-b})$.

4. Show that the group of units of an arbitrary order is a subgroup of finite index in the group of units of the maximal order $\tilde{\mathfrak{D}}$.

5. Let the units $\eta_1, \dots, \eta_r$ $(r = s + t - 1)$ of the order $\mathfrak{D}$ be such that the vectors $l(\eta_1), \dots, l(\eta_r)$ are linearly independent. Show that the group of all units of the form $\eta_1^{c_1} \cdots \eta_n^{c_r}$ is a subgroup of finite index of the group of all units of the order $\mathfrak{D}$.

6. Let $c_1, \dots, c_n$ be positive real numbers and let (a_{ij}) be a nonsingular real matrix of order n. Show that if $c_1 \cdots c_n > d = |\det(a_{ij})|$, then there exist rational integers $x_1, \dots, x_n$, not all zero, such that

$$\left|\sum_{j=1}^{n} a_{ij}x_j\right| < c_i \qquad (i = 1, \dots, n).$$

[*Hint*: Verify that the set of all points $(x_1, \dots, x_n)$ in $\mathfrak{R}^n$ which satisfy the above inequality is bounded, centrally symmetric, and convex, and has volume $(1/d)2^n c_1 \cdots c^n$. Apply Minkowski's lemma on convex bodies.]

7. Let $a_{ij}(1 \leqslant i \leqslant k,\ 1 \leqslant j \leqslant n)$ be rational integers and let m_i $(1 \leqslant i \leqslant k)$ be natural numbers. Show that the set of all integral points $(x_1, \dots, x_n)$ in $\mathfrak{R}^n$ for which

$$\sum_{j=1}^{n} a_{ij}x_j \equiv 0 \pmod{m_i} \qquad (1 \leqslant i \leqslant k),$$

forms a full lattice, with the volume of a fundamental parallelepiped $\leqslant m_1 \cdots m_k$.

8. Let a, b, c be nonzero rational integers, pairwise relatively prime, and square-free. Then $|abc| = 2^\lambda p_1 \cdots p_s$ (p_i are odd primes, λ is 0 or 1). Assume that the form $ax^2 + by^2 + cz^2$ represents zero in all p-adic fields. Show that there exist integral linear forms in three variables, $L_1, \dots, L_s, L', L''$ such that whenever u, v, w are integers satisfying

$$\begin{aligned} L_i(u, v, w) &\equiv 0 \pmod{p_i}, \qquad (1 \leqslant i \leqslant s),\\ L'(u, v, w) &\equiv 0 \pmod{2^{1+\lambda}}, \\ L''(u, v, w) &\equiv 0 \pmod{2}, \end{aligned} \qquad (*)$$

then

$$au^2 + bv^2 + cw^2 \equiv 0 \pmod{4|abc|}.$$

9. Under the same assumptions as in Problem 8, let $\mathfrak{M}$ denote the lattice of integral points $(u, v, w) \in \mathfrak{R}^3$ which satisfy $(*)$. By Problem 7 the volume of a fundamental parallelepiped of the lattice $\mathfrak{M}$ does not exceed $4|abc|$. Let X denote the ellipsoid of points which satisfy

$$|a|x^2 + |b|y^2 + |c|z^2 < 4|abc|,$$

the volume of which is easily computed to be $(32/3)\pi|abc|$. Apply the Minkowski lemma on convex bodies to the lattice $\mathfrak{M}$ and the ellipsoid X to prove that the form $ax^2 + by^2 + cz^2$ has a rational zero. (In this proof of the Hasse–Minkowski theorem for forms in three variables the fact that the form is indefinite is not used.)

5. The Solution of the Problem of the Representation of Rational Numbers by Full Decomposable Forms

5.1. *Units with Norm* $+1$

In Section 2.3 we saw that to solve the problem of finding all numbers in a given full module with certain norm it was necessary to find all units ε of the coefficient ring $\mathfrak{O}$ for which $N(\varepsilon) = +1$. The set of all such units clearly forms a group. We now study the structure of this group.

We first assume that the degree n of the field K is odd. In this case the ring $\mathfrak{O}$ only contains two roots of 1, namely, ± 1 (Problem 2 of Section 3). If for some unit $\varepsilon \in \mathfrak{O}$ we have $N(\varepsilon) = -1$, then

$$N(-\varepsilon) = N(-1)N(\varepsilon) = (-1)^n(-1) = 1.$$

Let $\varepsilon_1, \dots, \varepsilon_r$ $(r = s + t - 1)$ be any system of fundamental units of the ring $\mathfrak{O}$. Suppose that among the ε_i there are some units with norm -1. Replacing each such unit by $-\varepsilon_i$, we obtain a new system of fundamental units $\eta_1, \dots, \eta_r$ with $N(\eta_i) = 1$ for $i = 1, \dots, r$. The norm of an arbitrary unit $\varepsilon = \pm\eta_1{}^{a_1} \cdots \eta_r{}^{a_r}$ will then equal $N(\pm 1) = (\pm 1)^n = \pm 1$. Hence all units $\varepsilon \in \mathfrak{O}$ for which $N(\varepsilon) = 1$ have the form

$$\varepsilon = \eta_1{}^{a_1} \cdots \eta_r{}^{a_r} \qquad (a_i \in Z).$$

Now let n be an even number. We shall show that in this case any root of 1 contained in K has norm $+1$. This certainly holds for the roots ± 1. If K contains a complex root ζ of 1, then $s = 0$, and this means that the set of all isomorphisms of K into the field of complex numbers is divided into pairs of complex-conjugate isomorphisms. If σ and $\bar{\sigma}$ are complex-conjugate isomorphisms, then $\sigma(\zeta)\bar{\sigma}(\zeta) = |\sigma(\zeta)|^2 = 1$. By the results of Section 2.3 of the Supplement, this means that $N(\zeta) = 1$, and our assertion is proved.

Again let $\varepsilon_1, \dots, \varepsilon_r$ be any system of fundamental units of the ring $\mathfrak{O}$. If $N(\varepsilon_i) = 1$ for $i = 1, \dots, r$, then the norm of any unit of the ring $\mathfrak{O}$ is $+1$. Assume that

$$N(\varepsilon_1) = 1, \dots, N(\varepsilon_k) = 1, \qquad N(\varepsilon_{k+1}) = -1, \dots, N(\varepsilon_r) = -1,$$

where $k < r$. Setting

$$\eta_1 = \varepsilon_1, \dots, \eta_k = \varepsilon_k, \qquad \eta_{k+1} = \varepsilon_{k+1}\varepsilon_r, \dots, \eta_{r-1} = \varepsilon_{r-1}\varepsilon_r,$$

we obtain a new system of fundamental units $\eta_1, \dots, \eta_{r-1}, \varepsilon_r$, where $N(\eta_i) = 1$ $(1 \leqslant i \leqslant r-1)$. Now let $\varepsilon = \zeta\eta_1{}^{a_1} \cdots \eta_{r-1}^{a_{r-1}}\varepsilon_r{}^b$ $(a_1, \dots, a_{r-1}, b \in Z)$ be any unit. Since $N(\varepsilon) = (-1)^b$, then $N(\varepsilon) = +1$ if and only if the exponent b is even, that is, $b = 2a$. We thus find that if n is even, any unit $\varepsilon \in \mathfrak{O}$ with norm $+1$ has the form

$$\varepsilon = \zeta\eta_1{}^{a_1} \cdots \eta_{r-1}^{a_{r-1}}\eta_r{}^{a_r} \qquad (a_i \in Z),$$

where $\eta_r = \varepsilon_r{}^2$, and ζ is any root of 1 contained in $\mathfrak{O}$.

Hence if we have found a system of fundamental units in the order $\mathfrak{O}$, then we can also find all units with norm $+1$.

5.2. *The General Form for Solutions of the Equation* $N(\mu) = a$

When we combine the results of Section 5.1 with the Corollary of Theorem 5 of Section 2, we obtain the following result, which gives a complete characterization of the set of all solutions to (2.7).

Theorem 1. Let M be a full module in the algebraic number field K of degree $n = s + 2t$, let $\mathfrak{O}$ be its coefficient ring, and let a be a nonzero rational number. In the order $\mathfrak{O}$ there exist units $\eta_1, \dots, \eta_r$ ($r = s + t - 1$) with norm $+1$, and in the module M there is a finite set (possibly empty) of numbers $\mu_1, \dots, \mu_k$ with norm a, such that every solution $\mu \in M$ of the equation

$$N(\mu) = a \tag{5.1}$$

has a unique representation in the form

$$\mu = \mu_i \eta_1{}^{a_1} \cdots \eta_r{}^{a_r} \qquad \text{for } n \text{ odd,}$$

$$\mu = \mu_i \zeta \eta_1{}^{a_1} \cdots \eta_r{}^{a_r} \qquad \text{for } n \text{ even.}$$

Here μ_i is one of the numbers $\mu_1, \dots, \mu_k$, ζ is a root of 1, and $a_1, \dots, a_r$ are rational integers.

In the case of even n, if we take as a new system of number μ_i the set of all products $\mu_i\zeta$, we obtain a representation which has the same form as that for odd n.

In any order of an imaginary quadratic field there are only finitely many units (since $r = s + t - 1 = 0$). Hence in this case (5.1) has only a finite number of solutions. If K is not an imaginary quadratic field (and, of course, not the field of rational numbers), then $r > 0$ and hence (5.1) either has no solution or has infinitely many.

Remark. Theorem 1 shows us the structure of the set of all solutions of (5.1), but it does not give an effective means for finding these solutions. For the practical solution of (5.1) we must find an effective method for finding a system of fundamental units for the order $\mathfrak{O}$, and a method for finding a full set of pairwise-nonassociate numbers $\mu_1, \dots, \mu_k$ in the module M with given norm. In the following parts of this section we shall show that both of these problems can actually be solved in a finite number of steps. It must be said, however, that the general effective methods to be described for finding fundamental units and numbers with given norm in a module are very ill-suited to actual computation, in view of the very large amount of unnecessary computation. Our goal is only to show that in principle it is possible to carry out these constructions in a finite number of steps. In any given case, by using other considerations and examining the particular behavior of the special case, it is usually possible to find a much shorter route. In Section 5.3, by way of example, we shall give a simple method for solving our problems in the case of quadratic fields.

5.3. *The Effective Construction of a System of Fundamental Units*

Let $\sigma_1, \ldots, \sigma_n$ denote all isomorphisms of the algebraic number field K into the field of complex numbers.

Lemma 1. Let $c_1, \ldots, c_n$ be arbitrary positive real numbers. In any full module M of the field K there exist only finitely many numbers α for which

$$|\sigma_1(\alpha)| < c_1, \ldots, |\sigma_n(\alpha)| < c_n, \tag{5.2}$$

and all such numbers can be effectively located.

Proof. Take any basis $\alpha_1, \ldots, \alpha_n$ in M (if the module M is given by a set of generators which is not a basis, by following the proof of Theorem 1 of Section 2 a basis may be constructed in a finite number of steps). Any number of M can then be represented in the form

$$\alpha = a_1\alpha_1 + \cdots + a_n\alpha_n \tag{5.3}$$

with rational integers a_j. Let $\alpha_1^*, \ldots, \alpha_n^*$ be the dual basis to $\alpha_1, \ldots, \alpha_n$ in the field K (see Section 2.3 of the Supplement) and take a real number $A > 0$ for which

$$|\sigma_i(\alpha_j^*)| \leqslant A \tag{5.4}$$

for all i and j. Multiplying (5.3) by α_j^* and taking the trace, we obtain

$$a_j = \operatorname{Sp} \alpha\alpha_j^* = \sum_{i=1}^{n} \sigma_i(\alpha)\sigma_i(\alpha_j^*).$$

Now if $\alpha \in M$ satisfies (5.2), then by (5.4) the coefficients a_j satisfy

$$|a_j| \leqslant A \sum_{i=1}^{n} |\sigma_i(\alpha)| < A \sum_{i=1}^{n} c_i. \tag{5.5}$$

Hence there are only finitely many possibilities for the a_j. By testing all such numbers we easily find those which satisfy (5.2).

Until the end of this section we shall use all concepts and notations of the preceding two sections.

The possibility of effectively finding a system of fundamental units for an arbitrary order of an algebraic number field is based on the following theorem.

Theorem 2. Let $\mathfrak{O}$ be any order of the algebraic number field K. A number $\rho > 0$ can be found such that the ball of radius ρ in the logarithmic space

$\mathfrak{R}^{s+t}$ must contain a basis for the lattice $\mathfrak{E}$ (which represents the units of the order $\mathfrak{O}$).

We show that this theorem does actually give us a method for constructing a system of fundamental units for the order $\mathfrak{O}$. If for the unit $\varepsilon \in \mathfrak{O}$, $l(\varepsilon)$ is contained in the ball of radius ρ, then

$$|\sigma_k(\varepsilon)| < e^{\rho} \quad (1 \leqslant k \leqslant s), \qquad |\sigma_{s+j}(\varepsilon)| < e^{\rho/2} \quad (1 \leqslant j \leqslant t). \tag{5.6}$$

By Lemma 1 the number of units $\varepsilon \in \mathfrak{O}$ satisfying this condition is finite and they can actually be found (to determine which numbers of the order $\mathfrak{O}$ are units, use Theorem 4 of Section 2). From this collection of units form all possible systems $\varepsilon_1, \ldots, \varepsilon_r$, where $r = s + t - 1$, for which the vectors $l(\varepsilon_1), \ldots, l(\varepsilon_r)$ are linearly independent. By Theorem 2 at least one of these systems will be a system of fundamental units of the order $\mathfrak{O}$. For each such system we compute the volume of the fundamental parallelepiped determined by the vectors $l(\varepsilon_1), \ldots, l(\varepsilon_r)$. Hence that system for which this volume is smallest will be a system of fundamental units.

Theorem 2 will follow trivially from the following two lemmas, when applied to the lattice $\mathfrak{E}$. Note that we can always enumerate all vectors of the lattice $\mathfrak{E}$ which lie in any bounded set. For if the coordinates of the point $l(\varepsilon)$ are bounded, then we have a bound of the type (5.6) on the unit ε, and by Lemma 1 we may explicitly find all such units. In general, we shall say that a lattice $\mathfrak{M}$ is effectively given if ther^ is an algorithm for locating all points of the lattice in any bounded set.

Lemma 2. Let $\mathfrak{M}$ be a full lattice in $\mathfrak{R}^m$ which is effectively given, and let Δ be the volume of its fundamental parallelepiped. Then a number ρ can be found such that the ball of radius ρ contains a basis for $\mathfrak{M}$.

Proof. If $m = 1$, then we can set $\rho = 2\Delta$. In general, we will prove the lemma by induction on m. Take any bounded convex centrally symmetric set in $\mathfrak{R}^m$ with volume greater than $2^m\Delta$. By Minkowski's lemma (Section 4.2) this set will contain a nonzero vector of the lattice $\mathfrak{M}$. Let u be any such vector with $u \neq nx$ for $x \in \mathfrak{M}$, with $n > 1$ an integer. Let $\mathfrak{L}'$ denote the subspace orthogonal to the vector u and let $\mathfrak{M}'$ be the projection of the lattice $\mathfrak{M}$ into $\mathfrak{L}'$. If $x' \in \mathfrak{M}'$, then for some $x \in \mathfrak{M}$ we have $x = \xi u + x'$, where ξ is a real number. For any integer k the vector $x - ku$ also belongs to $\mathfrak{M}$, so we may choose the vector x in $\mathfrak{M}$ (with given projection x') so that $|\xi| \leqslant \frac{1}{2}$. For such an x we shall have

$$\|x\|^2 = \xi^2\|u\|^2 + \|x'\|^2 \leqslant \tfrac{1}{4}\|u\|^2 + \|x'\|^2.$$

This inequality shows that the set of vectors $x' \in \mathfrak{M}'$ in some bounded region are the projections of vectors x from some bounded region of $\mathfrak{R}^m$, so that the lattice $\mathfrak{M}'$ is also effectively given. If $u_2, \ldots, u_m$ are vectors in $\mathfrak{M}$ such that the projections $u_2', \ldots, u_m'$ form a basis for $\mathfrak{M}'$, then the set $u, u_2, \ldots, u_m$ is easily seen to be a basis for $\mathfrak{M}$. Hence the volume of a fundamental parallelepiped of the lattice $\mathfrak{M}'$ is $\Delta/\|u\|$, which we can compute explicitly. By the induction hypothesis we can find a number ρ' such that $\mathfrak{M}'$ has a basis $u_2', \ldots, u_m'$ for which $\|u_i'\| < \rho'$ $(i = 2, \ldots, m)$. But we have already shown that then the vectors $u_2, \ldots, u_m$ in $\mathfrak{M}$ can be chosen so that

$$\|u_i\| < (\tfrac{1}{4}\|u\|^2 + \rho'^2)^{1/2}.$$

Thus in the ball of radius

$$\rho = \max(\|u\| + 1, (\tfrac{1}{4}\|u\|^2 + \rho'^2)^{1/2})$$

there necessarily exists a basis $u, u_2, \ldots, u_m$ for the lattice $\mathfrak{M}$, and this completes the proof of Lemma 2.

To complete the proof of Theorem 2 we now need only find a bound for the volume of a fundamental parallelepiped of the lattice $\mathfrak{E}$.

Lemma 3. If v is the volume of a fundamental parallelepiped of the lattice $\mathfrak{E}$, then

$$v \leqslant C(\ln Q)^{s+t-1}N \leqslant C(\ln Q)^{s+t-1}\sum_{a=1}^{[Q]} a^n,$$

where $Q = (2/\pi)^t\sqrt{|D|} + 1$ (D is the discriminant of the order $\mathfrak{O}$), N is the number of pairwise-nonassociate numbers of $\mathfrak{O}$ for which $|N(\alpha)| \leqslant Q$, and C is some constant which depends only on $s + t$.

Proof. We use the notations of the proof of Theorem 5 of Section 4. The real numbers $c_1, \ldots, c_{s+t}$ are chosen so that

$$c_1 \cdots c_{s+t} = \left(\frac{4}{\pi}\right)^t \Delta + 1 = \left(\frac{2}{\pi}\right)^t \sqrt{|D|} + 1 = Q.$$

Since the set of all translates of the set $l(X_0)$ by the vectors of the lattice $\mathfrak{E}$ completely covers $\mathfrak{L}$, by Lemma 1 of Section 4 we have

$$v \leqslant v(l(X_0)).$$

Let U_i $(i = 1, \ldots, N)$ denote the intersection of the set $l(X) - l(\alpha_i)$ with the subspace $\mathfrak{L}$. By (4.5) the sets U_i cover $l(X_0)$, so that

$$v \leqslant \sum_{i=1}^{N} v(U_i). \tag{5.7}$$

We now compute the volume $v(U_i)$. The intersection U of the set $l(X) - l(\alpha)$

with the subspace $\mathfrak{L}$ consists of all points $(\lambda_1, \dots, \lambda_{s+t}) \in \mathfrak{R}^{s+t}$ for which

$$\begin{gathered} \lambda_1 + \cdots + \lambda_{s+t} = 0, \\ \lambda_k < \ln c_k - l_k(\alpha) \qquad (1 \leqslant k \leqslant s+t). \end{gathered} \tag{5.8}$$

Set $|N(\alpha)| = a$ (so that $\Sigma\, l_k(\alpha) = \ln a$) and translate the set U by the vector $(\lambda_1{}^*, \dots, \lambda_{s+t}{}^*) \in \mathfrak{L}$ with components

$$\lambda_k{}^* = -\ln c_k + l_k(\alpha) + \frac{1}{s+t} \ln \frac{Q}{a}.$$

Under this translation the set U is carried into a set U^* with the same volume, in which, by (5.8),

$$\lambda_k < \frac{1}{s+t} \ln \frac{Q}{a} \qquad (1 \leqslant k \leqslant s+t).$$

Let U_0 denote the set of points in $\mathfrak{L}$ for which

$$\lambda_k < 1 \qquad (1 \leqslant k \leqslant s+t),$$

and let C_0 denote the volume of U_0. The constant C_0 clearly depends only on $s+t$. Since U^* is obtained from U_0 by multiplying by the factor $1/(s+t) \ln (Q/a)$, then

$$v(U^*) = \left(\frac{1}{s+t} \ln \frac{Q}{a}\right)^{s+t-1} v(U_0),$$

so that

$$v(U) = C_0\left(\frac{1}{s+t} \ln \frac{Q}{a}\right)^{s+t-1} \tag{5.9}$$

We now return to the inequality (5.7). For each $i = 1, \dots, N$ we have $1 \leqslant |N(\alpha_i)| \leqslant [Q]$. Further, we saw in the proof of Theorem 5 of Section 2 that the ring $\mathfrak{D}$ contains at most a^n pairwise-nonassociate numbers with norm in absolute value a. Combining these facts with (5.7) and (5.8), we obtain for v the estimate indicated in the lemma.

5.4. *The Numbers in a Module with Given Norm*

We now turn to the question of the construction in a module of a full set of pairwise-nonassociate elements with given norm.

In the coefficient ring $\mathfrak{D}$ of the full module M we fix some system of fundamental units $\varepsilon_1, \dots, \varepsilon_r$. The vectors $l(\varepsilon_1), \dots, l(\varepsilon_r)$, along with the vector $l_0 = (1, \dots, 1)$, form a basis for the logarithmic space $\mathfrak{R}^{s+t}$. Hence for any $\mu \in M$, the vector $l(\mu)$ can be represented in the form

$$l(\mu) = \xi l_0 + \sum_{i=1}^{r} \xi_i l(\varepsilon_i) \tag{5.10}$$

with real coefficients $\xi, \xi_1, \ldots, \xi_r$. By formulas (3.17) and (3.18) the coefficient ξ is given by

$$\xi = \frac{1}{s+t} \ln |N(\mu)|.$$

Each real number ξ_i can be represented in the form $\xi_i = k_i + \gamma_i$, where k_i is an integer and $|\gamma_i| \leqslant \frac{1}{2}$. If $\mu' = \mu\varepsilon_1^{-k_1} \cdots \varepsilon_r^{-k_r}$, then μ and μ' are associates, and (5.10) takes the form

$$l(\mu') = \frac{\ln a}{s+t} l_0 + \gamma_1 l(\varepsilon_1) + \cdots + \gamma_r l(\varepsilon_r),$$

where $a = |N(\mu)| = |N(\mu')|$. We have thus found a bounded set in $\mathfrak{R}^{s+t}$ such that for any $\mu \in M$ with $|N(\mu)| = a$, μ has an associate μ' with the logarithmic representation of μ' lying in the bounded set. We hence have a bound of the type (5.2) for the number μ'. By Lemma 1 we can enumerate all numbers of M satisfying this bound. We now pick from this finite set all numbers with the specified norm, and from this latter set pick one number in each equivalence class of associates. In this way we obtain a set $\mu_1, \ldots, \mu_k$ of pairwise-nonassociate numbers with given norm such that any number of M with this norm is associate with one of the μ_i. The results of this section give us a method for finding, in a finite number of steps, all numbers in a given full module with specified norm (or of establishing the nonexistence of such numbers). This gives a final solution to the problem of integral representation of rational numbers by full decomposable forms.

PROBLEMS

1. Let d be a rational integer which is square-free and divisible by at least one prime of the form $4k+3$. Show that any unit of the order $\{1, \sqrt{d}\}$ in the field $R(\sqrt{d})$ has norm $+1$.

2. Show that $5 + 2\sqrt{6}$ is a fundamental unit for the maximal order in the field $R(\sqrt{6})$.

3. Find all integral solutions to the equation

$$3x^2 - 4y^2 = 11.$$

4. Show that in the cubic field $R(\theta)$, $\theta^3 = 6$, the number $\varepsilon = 1 - 6\theta + 3\theta^2$ is a fundamental unit.

6. Classes of Modules

In view of the role played by the concept of a full module, it is important to investigate the structure of the set of all full modules of a given algebraic number field K. The number of all such modules is clearly infinite. But modules which are similar (Section 1.3) have many properties in common. We have seen that similar modules have the same coefficient ring (Lemma 1 of Section 2) and that the problems of finding numbers with given norm in similar modules are equivalent (Section 1.3). In view of this it is natural to collect similar modules in equivalence classes and to investigate the set of all equivalence classes of similar modules. In this section we shall show that the set of equivalence classes of similar modules with a given order $\mathfrak{O}$ of the algebraic number field K as coefficient ring is a finite set. This result, along with the theorem of Dirichlet on the group of units, is one of the most fundamental results in the theory of algebraic numbers. Its proof depends, as does the proof of the theorem on units, on the lemma of Minkowski on convex bodies. Another very important tool will be the concept of the norm of a module.

6.1. *The Norm of a Module*

Let M be an arbitrary full module in the algebraic number field K of degree n and let $\mathfrak{O}$ denote its coefficient ring. Pick bases $\omega_1, \dots, \omega_n$ for $\mathfrak{O}$ and $\mu_1, \dots, \mu_n$ for M. The transition matrix $A = (a_{ij})$ from the first basis to the second, that is, the matrix defined by

$$\mu_j = \sum_{i=1}^{n} a_{ij}\omega_i \qquad (1 \leqslant j \leqslant n,\ a_{ij} \in R), \tag{6.1}$$

depends on the module M and the choice of the bases ω_i and μ_j. Let $\omega_1', \dots, \omega_n'$ and $\mu_1', \dots, \mu_n'$ be other bases for the modules $\mathfrak{O}$ and M and let $\mu_j' = \sum^n{}_{i=1} a_{ij}'\omega_i'$ $(a_{ij}' \in R)$. The matrix $A_1 = (a_{ij}')$ is related to the matrix A by the relation

$$A_1 = CAD, \tag{6.2}$$

where $C = (c_{ij})$ and $D = (d_{ij})$ are integral unimodular matrices satisfying

$$\omega_j = \sum_{i=1}^{n} c_{ij}\omega_i', \quad \mu_j' = \sum_{i=1}^{n} d_{ij}\mu_i \qquad (c_{ij}, d_{ij} \in R)$$

(we know that the transition matrix from one basis of a module to another is unimodular). Thus the module M has as invariants any functions of the matrix A which remain unchanged when A is replaced by A_1 according to (6.2). The collection of all such numbers is the set of "rational invariant factors" of the matrix A. We consider the simplest of these, the absolute value of the determinant det A. Its invariance is evident:

$$|\det A_1| = |\det C| \cdot |\det A| \cdot |\det D| = |\det A|.$$

Definition. Let M be a full module in K with coefficient ring $\mathfrak{O}$. The absolute value of the determinant of the transition matrix from a basis of the ring $\mathfrak{D}$ to a basis of the module M is called the *norm of the module* M and is denoted by $N(M)$.

By (2.12) of the Supplement, the discriminants $D = D(\mu_1, \dots, \mu_n)$ and $D_0 = D(\omega_1, \dots, \omega_n)$ of the bases μ_i and ω_i (that is, the discriminants of the modules M and $\mathfrak{D}$, see Section 2.5) are connected by the relation $D = D_0 (\det A)^2$. The concept of the norm allows us to write this formula

$$D = D_0 N(M)^2. \tag{6.3}$$

If a module is contained in its coefficient ring, then the matrix (a_{ij}), determined by (6.1), is integral, and therefore the norm of the module is an integer. The value of this integer is clarified by the following theorem.

Theorem 1. If the full module M is contained in its coefficient ring $\mathfrak{D}$, then its norm $N(M)$ equals the index $(\mathfrak{D} : M)$.

This theorem is an immediate corollary of the following lemma.

Lemma 1. If M_0 is a torsion-free Abelian group of rank n, and M is a subgroup which is also of rank n, then the index $(M_0 : M)$ is finite and equals the absolute value of the determinant of the transition matrix from any basis of M_0 to any basis of M.

Proof. Let $\omega_1, \dots, \omega_n$ be any basis of M_0. By Theorem 2 of Section 2 there is a basis $\eta_1, \dots, \eta_n$ for the subgroup M of the form

$$\begin{aligned} \eta_1 &= c_{11}\omega_1 + c_{12}\omega_2 + \cdots + c_{1n}\omega_n, \\ \eta_2 &= \qquad\qquad c_{22}\omega_2 + \cdots + c_{2n}\omega_n, \\ &\cdots\cdots\cdots\cdots\cdots\cdots \\ \eta_n &= \qquad\qquad\qquad\qquad\qquad c_{nn}\omega_n, \end{aligned}$$

where the c_{ij} are rational integers and $c_{ii} > 0$ $(1 \leqslant i \leqslant n)$. It is clear that $|\det A|$ does not depend on the choice of the bases for M and M_0 and that

$$|\det A| = c_{11}c_{22}\cdots c_{nn}.$$

We consider the elements

$$x_1\omega_1 + \cdots + x_n\omega_n, \qquad 0 \leqslant x_i < c_{ii} \qquad (1 \leqslant i \leqslant n) \tag{6.4}$$

and will show that they form a complete system of coset representatives for the subgroup M of the group M_0. Let $\alpha = a_1\omega_1 + \cdots + a_n\omega_n$ be an arbitrary element of M_0. Dividing a_1 by c_{11} we get $a_1 = c_{11}q_1 + x_1$, $0 \leqslant x_1 < c_{11}$. Then

$$\alpha - q_1\eta_1 - x_1\omega_1 = a_2'\omega_2 + \cdots + a_n'\omega_n.$$

Dividing a_2' by c_{22} yields $a_2' = c_{22}q_2 + x_2$, $0 \leqslant x_2 < c_{22}$, so that

$$\alpha - q_1\eta_1 - q_2\eta_2 - x_1\omega_1 - x_2\omega_2 = a_3''\omega_3 + \cdots + a_n''\omega_n.$$

Continuing this process n times we arrive at

$$\alpha - q_1\eta_1 - \cdots - q_n\eta_n - x_1\omega_1 - \cdots - x_n\omega_n = 0,$$

where q_i and x_i are rational integers with $0 \leqslant x_i < c_{ii}$. Since $q_1\eta_1 + \cdots + q_n\eta_n$ belongs to M, α and the element $x_1\omega_1 + \cdots + x_n\omega_n$ of the form (6.4) lies in the same coset of the subgroup M. This means that every coset of M in M_0 has a representative of the form (6.4). We now need to show that the various elements of the form (6.4) lie in distinct cosets of M in M_0. Asserting the converse, we assume that the difference of two distinct elements $x_1\omega_1 + \cdots + x_n\omega_n$ and $x_1'\omega_1 + \cdots + x_n'\omega_n$ of the form (6.4) lies in M. Letting s denote the smallest index $(1 \leqslant s \leqslant n)$ for which $x \neq x'$, we obtain

$$(x_s - x_s')\omega_s + \cdots + (x_n - x_n')\omega_n = b_1\eta_1 + \cdots + b_n\eta_n$$

with integral b_i. Substituting their expressions in terms of the ω_i for $\eta_1, \ldots, \eta_n$, and equating the coefficients of the various ω_i on both sides of the equation, we easily find that $b_1 = 0, \ldots, b_{s-1} = 0$, and that $c_{ss}b_s = x_s - x_s'$. But the latter equation is impossible for integral b_s, since $0 < |x_s - x_s'| < c_{ss}$. Thus the elements of the form (6.4) form a complete system of coset representatives for M in M_0. Since there are $c_{11}c_{22}\cdots c_{nn} = |\det A|$ of them, Lemma 1 and Theorem 1 are proved.

Theorem 2. The norms of the similar modules M and αM are connected by the relation

$$N(\alpha M) = |N(\alpha)|N(M).$$

In particular, if a module is similar to the order $\mathfrak{O}$, then

$$N(\alpha\mathfrak{O}) = |N(\alpha)|.$$

Proof. If $\mu_1, \ldots, \mu_n$ is a basis for M, then we may take $\alpha\mu_1, \ldots, \alpha\mu_n$ as a basis for αM. The norm $N(\alpha)$ of the number α is the determinant of the transition matrix from the basis μ_1 to the basis $\alpha\mu_i$ (see Section 2.2 of the Supplement). By Lemma 1 of Section 2 the modules M and αM have the same coefficient ring $\mathfrak{O}$. Let A and A_1 denote the transition matrices from some base of the ring $\mathfrak{O}$ to the bases μ_i and $\alpha\mu_i$, respectively. Then $A_1 = AC$ and we obtain

$$(N\alpha M) = |\det A_1| = |\det A| \cdot |\det C| = N(M)|N(\alpha)|.$$

The second assertion of the theorem follows from the fact that $N(\mathfrak{O}) = 1$.

6.2. *Finiteness of the Number of Classes*

We now turn to the proof of the basic theorem of this section. Its proof will rely on two lemmas.

Lemma 2. If M_1 is a full module in the field K and M_2 is any full submodule of M_1, then there are only finitely many intermediate submodules M (that is, modules satisfying $M_1 \supset M \supset M_2$).

Proof. Let $\xi_1, \dots, \xi_s$, $s = (M_1 : M_2)$, be any system of coset representatives of M_2 in M_1. If $\alpha_1, \dots, \alpha_n$ is a basis of M_2, then every element $\theta \in M_1$ has a unique representation $\theta = \xi_k + c_1\alpha_1 + \cdots + c_n\alpha_n$, where ξ_k is one of these representatives, and $c_1, \dots, c_n$ are rational integers. Let $\theta_1, \dots, \theta_n$ be a basis of the intermediate module M. Then each θ_j has a representation $\theta_j = \xi_{kj} + c_{1j}\alpha_1 + \cdots + c_{nj}\alpha_n$ with integral c_{ij}. Therefore,

$$M = \{\theta_1, \dots, \theta_n\} = \{\theta_1, \dots, \theta_n, \alpha_1, \dots, \alpha_n\} = \{\xi_{k_1}, \dots, \xi_{k_n}, \alpha_1, \dots, \alpha_n\}.$$

Since there are only finitely many possibilities for the set $\xi_{k_1}, \dots, \xi_{k_n}$ there are only finitely many possibilities for the intermediate module M.

Corollary. If M_0 is any full module in the field K and r is any natural number, there are only finitely many full modules in K which contain M_0 such that $(M : M_0) = r$.

For by the finiteness of the factor group M/M_0 we have $rM \subset M_0$, and hence $(1/r)M_0 \supset M \supset M_0$.

Lemma 3. Let K be an algebraic number field of degree $n = s + 2t$ and let M be a full module in K with discriminant D. Then there exists a nonzero number α in M whose norm satisfies

$$|N(\alpha)| \leqslant \left(\frac{2}{\pi}\right)^t \sqrt{|D|}. \tag{6.5}$$

Proof. We take positive real numbers $c_1, \dots, c_{s+t}$ so that

$$c_1 \cdots c_{s+t} = \left(\frac{2}{\pi}\right)^t \sqrt{|D|} + \varepsilon, \tag{6.6}$$

where ε is an arbitrary positive real number. From Theorems 2 and 4 of Section 4 it follows that there exists a number $\alpha \neq 0$ in M satisfying

$$|\sigma_k(\alpha)| < c_k \quad (1 \leqslant k \leqslant s), \qquad |\sigma_{s+j}(\alpha)|^2 < c_{s+j} \quad (1 \leqslant j \leqslant t).$$

The norm

$$N(\alpha) = \sigma_1(\alpha) \cdots \sigma_s(\alpha)|\sigma_{s+1}(\alpha)|^2 \cdots |\sigma_{s+t}(\alpha)|^2$$

of such a number must have absolute value not exceeding (6.6). Since this is true for arbitrarily small ε, there must be a nonzero number of M which satisfies the inequality (6.5).

Theorem 3. If $\mathfrak{O}$ is any order of the algebraic number field K, there are only finitely many equivalence classes of similar modules which have $\mathfrak{O}$ as their coefficient ring.

Proof. Let M be any module which has $\mathfrak{O}$ as coefficient ring. Let D denote the discriminant of the module M and D_0 the discriminant of the order $\mathfrak{O}$. We take a nonzero number in M which satisfies (6.5). Using (6.3) we may write (6.5) in the form

$$|N(\alpha)| \leqslant \left(\frac{2}{\pi}\right)^t N(M)\sqrt{|D_0|}.$$

Since $\alpha\mathfrak{O} \subset M$, then $\mathfrak{O} \subset (1/\alpha)M$. By Lemma 1 and the definition of the norm of a module we have

$$\left(\frac{1}{\alpha}M:\mathfrak{O}\right) = N\left(\frac{1}{\alpha}M\right)^{-1} = \frac{|N(\alpha)|}{N(M)} \leqslant \left(\frac{2}{\pi}\right)^t \sqrt{|D_0|}.$$

This proves that every class of similar modules with coefficient ring $\mathfrak{O}$ contains a module M' for which

$$M' \supset \mathfrak{O}, \qquad (M':\mathfrak{O}) \leqslant \left(\frac{2}{\pi}\right)^t \sqrt{|D_0|}. \tag{6.7}$$

By the Corollary of Lemma 2 there are only finitely many such modules M' satisfying (6.7). Hence the number of classes of modules with $\mathfrak{O}$ as coefficient ring is finite, and Theorem 3 is proved.

Remark. If M_1 and M_2 are any two full modules of the algebraic number field K, we may effectively determine whether or not they are similar. To do this we first determine their coefficient rings. If these are different, then M_1 and M_2 are not similar. Suppose that M_1 and M_2 have the same coefficient ring $\mathfrak{O}$. Replacing, if necessary, one of our modules by a module similar to it, we may assume that $M_1 \supset M_2$. We compute the index $(M_1 : M_2) = a$. If $\alpha M_1 = M_2$, then $\alpha \in \mathfrak{O}$ and $|N(\alpha)| = a$. Therefore we find a full set of pairwise-nonassociate numbers $\alpha_1, \dots, \alpha_k$ in the ring $\mathfrak{O}$ whose norms are equal in absolute value to a (by Section 5.4 such a system can be effectively computed). If α is any number of the ring $\mathfrak{O}$ for which $|N(\alpha)| = a$, then α is associate with some α_i, and $\alpha M_1 = \alpha_i M_1$. We therefore compare the modules M_2 and $\alpha_i M_1$ $(1 \leqslant i \leqslant k)$. The modules M_1 and M_2 will be similar if and only if the module M_2 coincides with one of the $\alpha_i M_1$.

PROBLEMS

1. Show that any algebraic number field other than the field of rational numbers contains an infinite number of orders. (Hence the number of equivalence classes of similar modules corresponding to all possible orders is infinite.)

2. Use Problem 2 of Section 4 to show that in a full module M with discriminant D there is a number $\alpha \neq 0$ for which

$$|N(\alpha)| \leqslant \left(\frac{4}{\pi}\right)^t \frac{n!}{n^n} \sqrt{|D|}$$

($n = s + 2t$ being the degree of the algebraic number field).

3. Use Stirling's formula

$$n! = \sqrt{2\pi n} \left(\frac{n}{e}\right)^n e^{\theta/12n} \qquad (0 < \theta < 1)$$

and Problem 2 to show that the discriminant D_0 of an algebraic number field K with degree $n = s + 2t$ satisfies

$$|D_0| > \left(\frac{\pi}{4}\right)^{2t} \frac{1}{2\pi n} e^{2n-(1/6n)}.$$

Thus as n increases, the discriminants of algebraic number fields of degree n converge to infinity.

4. Show that the discriminant of any algebraic number field of degree $n > 1$ is not equal to ± 1 (Minkowski's theorem).

5. Show that there exist only finitely many algebraic number fields with given discriminant (Hermite's theorem).

Remark. By Problem 3 it suffices to show that there exist only finitely many fields K with fixed degree $n = s + 2t$ whose discriminant is a given value D_0. In the space $\mathfrak{R}^n$ [consisting of all points $(x_1, \ldots, x_s, y_1, z_1, \ldots, y_t, z_t)$] consider the set X defined in the case $s > 0$ by

$$|x_1| < \sqrt{|D_0| + 1}, \quad |x_k| < 1 \qquad (2 \leqslant k \leqslant s),$$
$$y_j^2 + z_j^2 < 1 \qquad (1 \leqslant j \leqslant t),$$

and in the case $s = 0$ by

$$|y_1| < \tfrac{1}{2}, \quad |z_1| < \sqrt{|D_0| + 1}, \quad y_j^2 + z_j^2 < 1 \qquad (2 \leqslant j \leqslant t).$$

Applying Minkowski's lemma on convex bodies to the set X and to the lattice representing the numbers of the maximal order $\tilde{\mathfrak{O}}$, deduce that K contains a primitive element $\theta \in \tilde{\mathfrak{O}}$ whose minimum polynomial has bounded coefficients.

7. Representation of Numbers by Binary Quadratic Forms

In this section we make a more detailed study of the questions of this chapter in the case of binary quadratic forms. Since any irreducible rational form $ax^2 + bxy + cy^2$ decomposes into linear factors in some quadratic field, our problem is connected with the study of full modules and their coefficient rings in quadratic fields.

7.1. *Quadratic Fields*

We call any extension of the rational field of degree 2 a quadratic field. We first describe this simplest class of algebraic number fields.

Let $d \neq 1$ be a square-free rational integer (positive or negative). Since the polynomial $t^2 - d$ is irreducible over the rationals, the field $R(\theta)$, obtained from R by adjoining a root θ of this polynomial, is of degree 2 over R; $R(\theta)$ is a quadratic field. We shall denote it $R(\sqrt{d})$.

It is easily seen that any quadratic field K is of this type. We prove this. If α lies in K and is not rational, then clearly $K = R(\alpha)$. The minimum polynomial for α over R will have degree 2, so for some rational p and q we have $\alpha^2 + p\alpha + q = 0$. Set $\beta = \alpha + (p/2)$; then $\beta^2 = (p^2/4) - q$. The rational number $(p^2/4) - q$ can be represented in the form $c^2 d$, where d is a square-free integer. It is clear that $d \neq 1$, since otherwise β and also α would be rational. If $\theta = \beta/c$, then $\theta^2 = d$ and $K = R(\theta)$; that is, $K = R(\sqrt{d})$.

We now show that for distinct d (not equal to 1 and square-free), the fields $R(\sqrt{d})$ are distinct. For if $R(\sqrt{d'}) = R(\sqrt{d})$, then

$$\sqrt{d'} = x + y\sqrt{d}$$

for some rational x and y, so that

$$d' = x^2 + dy^2 + 2xy\sqrt{d}$$

and consequently

$$d' = x^2 + dy^2, \qquad 2xy = 0.$$

If $y = 0$, then $d' = x^2$, which is impossible. If $x = 0$, then $d' = dy^2$, which means that $d' = d$.

We have shown that there is a one-to-one correspondence between quadratic fields and square-free rational integers $d \neq 1$.

7.2. *Orders in Quadratic Fields*

Any number of the field $R(\sqrt{d})$ has the form

$$\alpha = x + y\sqrt{d},$$

where x and y are rational. Since the characteristic polynomial of α equals

$$t^2 - 2xt + x^2 - dy^2,$$

then α will lie in the maximal order $\tilde{\mathfrak{D}}$ of the field $R(\sqrt{d})$ if and only if

$2x = \mathrm{Sp}(\alpha)$ and $x^2 - dy^2 = N(\alpha)$ are rational integers. Set $2x = m$. Since $(m^2/4) - dy^2$ will be an integer and d is square-free, the denominator of the rational number y (in reduced form) must be 2; that is, $y = n/2$ with integral n. Clearly, $N(\alpha) = (m^2/4) - d(n^2/4)$ will be integral only if

$$m^2 - dn^2 \equiv 0 \pmod 4. \tag{7.1}$$

The solvability of this congruence depends on the residue class of d modulo 4. Since d is square-free, $d \not\equiv 0 \pmod 4$, and we have the three possibilities

$$d \equiv 1 \pmod 4, \qquad d \equiv 2 \pmod 4, \qquad d \equiv 3 \pmod 4.$$

If $d \equiv 1 \pmod 4$, then the congruence (7.1) takes the form $m^2 \equiv n^2 \pmod 4$, which is equivalent to $m \equiv n \pmod 2$; that is, $m = n + 2$, and we obtain

$$\alpha = \frac{m}{2} + \frac{n}{2}\sqrt{d} = l + n\,\frac{1 + \sqrt{d}}{2}$$

with integral l and n. Hence in this case, as a basis for the maximal order $\tilde{\mathfrak{O}}$ (that is, as a fundamental basis for the field $R(\sqrt{d})$, see the end of Section 2), we may take the numbers 1 and $\omega = (1 + \sqrt{d})/2$.

Now let $d \equiv 2 \pmod 4$ or $d \equiv 3 \pmod 4$. If the congruence (7.1) had a solution with n odd, then from $d \equiv m^2 \pmod 4$ it would follow that $d \equiv 0 \pmod 4$ for m even and $d \equiv 1 \pmod 4$ for m odd. But this contradicts our assumptions. If n is even, then from the congruence $m^2 \equiv 0 \pmod 4$ we find that m is even. Thus in these cases the number $x + y\sqrt{d}$ belongs to the maximal order $\tilde{\mathfrak{O}}$ of the field $R(\sqrt{d})$ only when $x = m/2$ and $y = n/2$ are integers. As a basis for the order $\tilde{\mathfrak{O}}$ we may thus take the numbers 1 and $\omega = \sqrt{d}$.

In the future, when we speak of a basis for the maximal order of the field $R(\sqrt{d})$, we shall always have in mind the basis $1, \omega$, where $\omega = (1 + \sqrt{d})/2$ for $d \equiv 1 \pmod 4$ and $\omega = \sqrt{d}$ for $d \equiv 2, 3 \pmod 4$.

Now consider an arbitrary order $\mathfrak{O}$ of the field $R(\sqrt{d})$. Since $\mathfrak{O}$ is contained in the maximal order $\tilde{\mathfrak{O}}$ (see Section 2.4), then all numbers of $\mathfrak{O}$ have the form $x + y\omega$ with integral x and y. We choose from these numbers one for which y takes its smallest positive value, say, $a + f\omega$. Since a, being a rational integer, is contained in $\mathfrak{O}$, then $f\omega \in \mathfrak{O}$. It is then clear that for any $x + y\omega \in \mathfrak{O}$, the coefficient y is divisible by f, and hence $\mathfrak{O} = \{1, f\omega\}$. Conversely, by Lemma 3 of Section 2, for any natural number f the module $\{1, f\omega\}$ is a ring and hence is an order in the field $R(\sqrt{d})$. Since for distinct natural numbers f the orders $\{1, f\omega\}$ are distinct, we obtain the following fact: The set of all orders of a quadratic field is in one-to-one correspondence with the set of all natural numbers.

In the future we shall denote the order $\{1, f\omega\}$ by $\mathfrak{O}_f$. It is easily seen that the number f equals the index of the order $\mathfrak{O}_f$ in the maximal order $\tilde{\mathfrak{O}} = \mathfrak{O}_1 = \{1, \omega\}$. Thus an order of a quadratic field is completely determined by its index in the maximal order.

We now turn to the computation of the discriminant D_f of the order $\mathfrak{O}_f$. We first assume that $d \equiv 1 \pmod 4$. Since $\operatorname{Sp}\sqrt{d} = 0$,

$$\operatorname{Sp}\omega = \operatorname{Sp}\left(\frac{1+\sqrt{d}}{2}\right) = 1,$$

$$\operatorname{Sp}\omega^2 = \operatorname{Sp}\left(\frac{d+1}{4} + \frac{\sqrt{d}}{2}\right) = \frac{d+1}{2}$$

and hence

$$D_f = \begin{vmatrix} \operatorname{Sp} 1 & \operatorname{Sp} f\omega \\ \operatorname{Sp} f\omega & \operatorname{Sp} f^2\omega^2 \end{vmatrix} = \begin{vmatrix} 2 & f \\ f & f^2 \dfrac{d+1}{2} \end{vmatrix} = f^2 d.$$

Now if $d \equiv 2, 3 \pmod 4$, then

$$D_f = \begin{vmatrix} \operatorname{Sp} 1 & \operatorname{Sp} f\sqrt{d} \\ \operatorname{Sp} f\sqrt{d} & \operatorname{Sp} f^2 d \end{vmatrix} = \begin{vmatrix} 2 & 0 \\ 0 & 2f^2 d \end{vmatrix} = f^2 \cdot 4d.$$

These formulas for D_f show that each order of a quadratic field is uniquely determined by its discriminant.

The results of this section are summarized in the following theorem.

Theorem 1. Let d be a square-free rational integer, $d \neq 1$. As a basis for the maximal order $\tilde{\mathfrak{O}}$ of the quadratic field $R(\sqrt{d})$ we may take the numbers 1 and ω, where $\omega = (1+\sqrt{d})/2$ when $d \equiv 1 \pmod 4$ and $\omega = \sqrt{d}$ for $d \equiv 2, 3 \pmod 4$. The discriminant D_1 of the maximal order $\tilde{\mathfrak{O}}$ [that is, the discriminant of the field $R(\sqrt{d})$] equals d in the first case and $4d$ in the second case. Any order $\mathfrak{O}$ of the field $R(\sqrt{d})$ is of the form $\mathfrak{O}_f = \{1, f\omega\}$, where f is the index $(\tilde{\mathfrak{O}} : \mathfrak{O})$. The discriminant of the order $\mathfrak{O}_f$ equals $D_1 f^2$.

7.3. *Units*

Since any number of the order $\mathfrak{O}_f$ is represented in the form $x + yf\omega$ with x and y rational integers, then by Theorem 4 of Section 2 we shall find all units in $\mathfrak{O}_f$ if we solve the equation

$$N(x + yf\omega) = \pm 1, \tag{7.2}$$

that is, the equation

$$x^2 + fxy + f^2 \frac{1-d}{4} y^2 = \pm 1 \tag{7.3}$$

for $d \equiv 1 \pmod 4$ and the equation

$$x^2 - df^2 y^2 = \pm 1 \tag{7.4}$$

for $d \equiv 2, 3 \pmod 4$.

For an imaginary quadratic field $s = 0$, $t = 1$, $r = s + t - 1 = 0$, so that the group of units of any order of such a field is finite and consists of roots of 1. This fact also follows from a direct examination of (7.3) and (7.4), which only have a finite number of integral solutions when $d < 0$. When $d = -1, f = 1$, (7.4) has the four solutions: $x = \pm 1, y = 0$; $x = 0, y = \pm 1$, which correspond to the numbers ± 1, $\pm i$, which are the fourth roots of 1. When $d = -3$, $f = 1$, (7.3) has six solutions: $x = \pm 1, y = 0$; $x = 0, y = \pm 1$; $x = 1, y = -1$; $x = -1$, $y = 1$, which correspond to the numbers ± 1, $\pm \frac{1}{2} \pm (i\sqrt{3}/2)$, which are the sixth roots of 1. For all remaining orders of imaginary quadratic number fields (7.3) or (7.4) have only two solutions: $x = \pm 1$, $y = 0$; that is, ± 1 are the only units.

The case of real quadratic fields is more complicated. For the field $R(\sqrt{d})$ with $d > 0$ we have $s = 2$, $t = 0$, and hence $r = 1$, so all units of the order $\mathfrak{O}_f$ have the form $\pm \varepsilon^n$, where ε is a fundamental unit of the order $\mathfrak{O}_f$. Our problem is thus to determine the fundamental unit ε. Along with ε the numbers $1/\varepsilon$, $-\varepsilon$, and $-1/\varepsilon$ will also be fundamental units. We may thus assume that $\varepsilon > 1$. It is clear that the condition $\varepsilon > 1$ determines the fundamental unit ε uniquely.

Let $\eta > 1$ be any unit of $\mathfrak{O}_f$. We shall show that in the representation $\eta = x + yf\omega$ the coefficients x and y are positive (for $d = 5, f = 1$, it is possible that $x = 0$). For any $\alpha \in F(\sqrt{d})$ we denote by α' its conjugate, that is, the image of α under the automorphism $\sqrt{d} \to -\sqrt{d}$ of the field $R(\sqrt{d})$. It is easy to check that $\omega - \omega' > 0$. Since $N(\eta) = \eta\eta' = \pm 1$, then η' equals either $1/\eta$ or $-1/\eta$; and in both cases $\eta - \eta' > 0$; that is, $yf(\omega - \omega') > 0$, and hence $y > 0$. Further, since $|\eta'| = |x + yf\omega'| < 1$ and $f\omega' < -1$ with the exception of the case $d = 5$, $f = 1$, we have $x > 0$ [if $d = 5$, $f = 1$, then $-1 < f\omega' = (1 - \sqrt{5})/2 < 0$ and we obtain $x \geqslant 0$].

Let $\varepsilon > 1$ be a fundamental unit of the order $\mathfrak{O}_f$. For the unit $\varepsilon^n = x_1 + y_1 f\omega$ with natural number n we have $x_1 > x$ and $y_1 > y$. Hence to find the fundamental unit $\varepsilon > 1$, we must find the integral solution of (7.2) with smallest positive values for x and y. Using the results of Section 5.3 we may find an upper bound C for the desired values of x and y and then find them after a finite number of steps.

We now show that the number of steps in the location of a fundamental unit can be significantly reduced if a basic result from the theory of continued fractions is employed. We speak of the theorem which asserts that if for the real number $\xi > 0$ as well as the relatively prime natural numbers x and y we have

$$\left|\frac{x}{y} - \xi\right| < \frac{1}{2y^2},$$

then x/y is necessarily one of the convergents in the continued fraction expansion of the number ξ.

By (7.2),

$$\left|\frac{x}{y} + f\omega'\right| = \frac{1}{y(x + yf\omega)}.$$

If $d \equiv 1 \pmod 4$, then except in the case $d = 5, f = 1$, we have

$$\left|\frac{x}{y} - f\frac{\sqrt{d} - 1}{2}\right| = 1 \Big/ \left[y^2\left(\frac{x}{y} + f\frac{\sqrt{d} + 1}{2}\right)\right] < \frac{1}{2y^2}$$

[since $x/y > 0$ and $f(\sqrt{d} + 1)/2 > 2$]. If $d \equiv 2, 3 \pmod 4$, then since $x^2 = fdy^2 \pm 1 \geqslant dy^2 - 1 \geqslant y^2(d-1)$ and $d \geqslant 2$,

$$\left|\frac{x}{y} - f\sqrt{d}\right| = \frac{1}{y(x + yf\sqrt{d})} \leqslant \frac{1}{y^2(\sqrt{d-1} + \sqrt{d})} < \frac{1}{2y^2}.$$

By the theorem mentioned above, the reduced fraction x/y is one of the convergents in the continued fraction expansion of the number $-f\omega'$. To find the smallest positive solution to (7.2) we therefore need only test those numbers that occur as numerators and denominators of the convergents of the continued fraction expansion for $-f\omega'$ (and that do not exceed the previously computed constant C). The practical computation is expediently carried out as follows. Find the sequence of entries q_k of the continued fraction expansion of $-f\omega'$ and let P_k and Q_k denote the numerators and denominators of the corresponding convergents. Continue the computations until a stage is reached at which $N(P_k + \omega f Q_k) = \pm 1$. This must happen for $P_k < C$. Then the fundamental unit $\varepsilon = P_k + \omega f Q_k$ is found. [In the exceptional case $d = 5$, $f = 1$, the fundamental unit will be $\omega = (1 + \sqrt{5})/2$.] We now illustrate this by two examples.

Example 1. In order to find the fundamental units of the order $\{1, 3\sqrt{6}\}$ of the field $R(\sqrt{6})$, we must find the continued fraction expansion of $-3\omega' = 3\sqrt{6}$:

$$\sqrt{54} = 7 + (\sqrt{54} - 7),$$

$$\frac{1}{\sqrt{54} - 7} = 2 + \frac{\sqrt{54} - 3}{5},$$

$$\frac{5}{\sqrt{54} - 3} = 1 + \frac{\sqrt{54} - 6}{9},$$

$$\frac{9}{\sqrt{54} - 6} = 6 + \frac{\sqrt{54} - 6}{2},$$

$$\frac{2}{\sqrt{54} - 6} = 1 + \frac{\sqrt{54} - 3}{9}.$$

We fill in the following table simultaneously:

k	0	1	2	3	4	5
q_k	7	2	1	6	1	2
P_k	7	15	22	147	169	485
Q_k	1	2	3	20	23	66
$P_k^2 - 54Q_k^2$	-5	9	-2	9	-5	1

Hence $485 + 66 \cdot 3\sqrt{6} = 485 + 198\sqrt{6}$ is a fundamental unit of the order $\{1, 3\sqrt{6}\}$.

Example 2. Computing a fundamental unit for the field $R(\sqrt{41})$ we have

$$\frac{\sqrt{41} - 1}{2} = 2 + \frac{\sqrt{41} - 5}{2},$$

$$\frac{2}{\sqrt{41} - 5} = 1 + \frac{\sqrt{41} - 3}{8},$$

$$\frac{8}{\sqrt{41} - 3} = 2 + \frac{\sqrt{41} - 5}{4},$$

$$\frac{4}{\sqrt{41} - 5} = 2 + \frac{\sqrt{41} - 3}{4},$$

$$\frac{4}{\sqrt{41} - 3} = 1 + \frac{\sqrt{41} - 5}{8}.$$

k	0	1	2	3	4
q_k	2	1	2	2	1
P_k	2	3	8	19	27
Q_k	1	1	3	7	10
$P_k^2 + P_kQ_k - 10Q_k^2$	-4	2	-2	4	-1

As the fundamental unit for the maximal order of the field $R(\sqrt{41})$ we thus may take

$$27 + 10\,\frac{\sqrt{41}+1}{2} = 32 + 5\sqrt{41}.$$

7.4. *Modules*

We turn to the study of full modules in quadratic fields. Since any module $\{\alpha, \beta\}$ is similar to the module $\{1, \beta/\alpha\}$ without loss of generality, we may study only modules of the form $\{1, \gamma\}$.

Any irrational number γ of $R(\sqrt{d})$ is the root of some polynomial of the form $at^2 + bt + c$ with rational integer coefficients. If we require $(a, b, c) = 1$ and $a > 0$, then the polynomial $at^2 + bt + c$ is uniquely determined. We denote it by $\varphi_\gamma(t)$. If γ' is the conjugate of γ, then we have $\varphi_{\gamma'}(t) = \varphi_\gamma(t)$, and if $\varphi_{\gamma_1}(t) = \varphi_\gamma(t)$, then either $\gamma_1 = \gamma$ or $\gamma_1 = \gamma'$.

Lemma 1. If γ is an irrational number of $R(\sqrt{d})$ with $\varphi_\gamma(t) = at^2 + bt + c$, then the coefficient ring of the module $M = \{1, \gamma\}$ is the order $\{1, a\gamma\}$ with discriminant $D = b^2 - 4ac$.

Proof. Consider the number $\alpha = x + y\gamma$ with rational x and y. Since the inclusion $\alpha M \subset M$ is equivalent to the assertions that $\alpha 1 = x + y\gamma \in M$ and

$$\alpha \cdot \gamma = -\frac{cy}{a} + \left(x - \frac{by}{a}\right)\gamma \in M,$$

then α belongs to the coefficient ring $\mathfrak{O}$ if and only if the rational numbers

$$x,\ y,\ \frac{cy}{a},\ \frac{by}{a}$$

are all integers. Since $(a, b, c) = 1$, this will occur only when x and y are integers and y is divisible by a. This shows that $\mathfrak{O} = \{1, a\gamma\}$. To finish the proof of the lemma we compute the discriminant of the order $\mathfrak{O}$:

$$D = \begin{vmatrix} \operatorname{Sp} 1 & \operatorname{Sp} a\gamma \\ \operatorname{Sp} a\gamma & \operatorname{Sp} a^2\gamma^2 \end{vmatrix} = \begin{vmatrix} 2 & -b \\ -b & b^2 - 2ac \end{vmatrix} = b^2 - 4ac.$$

Corollary. Under the same notations as above, the norm of the module $\{1, \gamma\}$ is equal to $1/a$.

Indeed, the matrix of transition from the basis $\{1, a\gamma\}$ to the basis $\{1, \gamma\}$ is

$$\begin{pmatrix} 1 & 0 \\ 0 & \dfrac{1}{a} \end{pmatrix}$$

Lemma 2. In order that the modules $\{1, \gamma_1\}$ and $\{1, \gamma\}$ be similar, it is necessary and sufficient that the numbers γ and γ_1 be connected by a relation of the form

$$\gamma_1 = \frac{k\gamma + l}{m\gamma + n}, \tag{7.5}$$

where k, l, m, n are rational integers such that

$$\begin{vmatrix} k & l \\ m & n \end{vmatrix} = \pm 1. \tag{7.6}$$

Proof. Since different bases of the same module are connected by unimodular transformations (see Section 2.1), then from the equation $\{\alpha, \alpha\gamma_1\} = \{1, \gamma\}$ it follows that

$$\alpha\gamma_1 = k\gamma + l,$$

$$\alpha = m\gamma + n,$$

where the rational integers k, l, m, n satisfy (7.6). Dividing the first equation by the second, we obtain (7.5). Conversely, let γ_1 and γ be connected by relation (7.5). Then

$$\{1, \gamma_1\} = \frac{1}{m\gamma + n}\{m\gamma + n, k\gamma + l\} = \frac{1}{m\gamma + n}\{1, \gamma\}$$

[$\{m\gamma + n, k\gamma + l\} = \{1, \gamma\}$ in view of (7.6)]. The proof of the lemma is complete.

Consider the set of all modules in the field $R(\sqrt{d})$ which belong to some fixed order $\mathfrak{O}$ (that is, for which $\mathfrak{O}$ is the coefficient ring). By Theorem 3 of Section 6, all such modules are divided into finitely many equivalence classes of similar modules. We now introduce the operation of multiplication of classes and show that under this operation the set of classes belonging to a given order becomes an Abelian group. If $M = \{\alpha, \beta\}$ and $M_1 = \{\alpha_1, \beta_1\}$, then MM_1 denotes the module $\{\alpha\alpha_1, \alpha\beta_1, \beta\alpha_1, \beta\beta_1\}$ (see Problem 7 of Section 2). It is clear that if $\lambda \neq 0$ and $\mu \neq 0$, then

$$(\lambda M)(\mu M_1) = \lambda\mu(MM_1). \tag{7.7}$$

If M is any module, we denote by $[M]$ the class of similar modules which contains M. From (7.7) it follows that the class $[MM_1]$ depends only on the classes $[M]$ and $[M_1]$. The class $[MM_1]$ is called the *product of the classes* $[M]$ *and* $[M_1]$. Hence to multiply two classes we choose arbitrary representatives of each class and multiply them. The class which contains this product will be the product of the classes.

If M is any module, we denote by M' the module consisting of all numbers α', where α is any number of M. If M is a full module, then M' is also a full module. It is easily checked that if $\mathfrak{D}$ is any order, then the conjugate module $\mathfrak{D}'$ coincides with $\mathfrak{D}$. From this it easily follows that conjugate modules have the same coefficient rings.

We shall prove the formula

$$MM' = N(M)\mathfrak{D}, \tag{7.8}$$

where $\mathfrak{D}$ denotes the coefficient ring of M and $N(M)$ the norm of M.

We first assume that the module M has the form $\{1, \gamma\}$. In this case, using the notation of Lemma 1,

$$\begin{aligned} MM' &= \{1, \gamma\}\{1, \gamma'\} = \{1, \gamma, \gamma', \gamma\gamma'\} \\ &= \left\{1, \gamma, -\gamma - \frac{b}{a}, -\frac{c}{a}\right\} \\ &= \left\{1, \gamma, -\frac{b}{a}, -\frac{c}{a}\right\} = \frac{1}{a}\{a, b, c, a\gamma\}. \end{aligned}$$

Since a, b, and c are relatively prime, every rational integer is a linear combination of a, b, c with integer coefficients and hence

$$MM' = \frac{1}{a}\{1, a\gamma\} = \frac{1}{a}\mathfrak{D} = N(M)\mathfrak{D}$$

(by the corollary of Lemma 1). If M is now an arbitrary module, it can be represented in the form $M = \alpha M_1$, where M_1 has the form $\{1, \gamma\}$. By Theorem 2 of Section 6 we have

$$\begin{aligned} MM' = \alpha\alpha' M_1 M_1' &= N(\alpha)N(M_1)\mathfrak{D} \\ &= |N(\alpha)|N(M_1)\mathfrak{D} = N(M)\mathfrak{D}, \end{aligned}$$

and formula (7.8) is proved in general.

Now let M and M_1 be two modules belonging to the same order $\mathfrak{O}$. If $\bar{\mathfrak{O}}$ is the coefficient ring of the product MM_1, then by formula (7.8),

$$MM_1(MM_1)' = N(MM_1)\bar{\mathfrak{O}}.$$

On the other hand, since multiplication of modules is clearly commutative and associative, by use of $MM' = N(M)\mathfrak{O}$ and $M_1M_1' = N(M_1)\mathfrak{O}$, we obtain

$$MM' = N(M)\mathfrak{O} \qquad and \qquad M_1M_1' = N(M_1)\mathfrak{O},$$

$$MM_1(MM_1)' = N(M)N(M_1)\mathfrak{O}.$$

Comparing this equation with the previous one and recalling that two distinct orders cannot be similar, we obtain that $\mathfrak{O} = \bar{\mathfrak{O}}$. Incidentally, since the equality $a\mathfrak{O} = b\mathfrak{O}$ for positive rational a and b is possible only when $a = b$, we obtain

$$N(MM_1) = N(M)N(M_1).$$

Thus if the modules M and M_1 belong to the order $\mathfrak{O}$, then their product MM_1 also belongs to $\mathfrak{O}$. Since for any module M with coefficient ring $\mathfrak{O}$ we have both $M\mathfrak{O} = M$ and $M[(1/N(M))M'] = \mathfrak{O}$, then we obtain the following result.

Theorem 2. The set of all modules of a quadratic field which belong to a fixed order becomes an Abelian group under the operation of multiplication of modules.

From this theorem and Theorem 3 of Section 6, we easily obtain

Theorem 3. The set of classes of similar modules in a quadratic field with given coefficient ring forms a finite Abelian group.

Note that Theorems 2 and 3 hold only for quadratic fields and cease to be true for modules which belong to nonmaximal orders in arbitrary algebraic number fields (see Problem 18 of Section 2).

7.5. *The Correspondence between Modules and Forms*

As shown in Section 1.3, each basis α, β of the full module $M \subset R(\sqrt{d})$ corresponds uniquely to the binary quadratic form $N(\alpha x + \beta y)$ with rational coefficients. Since for different bases of M the corresponding forms are equivalent, the module M corresponds to a class of equivalent forms. If we replace M by the similar module γM, then each corresponding form is multiplied by the constant factor $N(\gamma)$. Hence, considering forms only up to a constant multiple, we may say that any class of similar modules corresponds to a class of equivalent forms. But this correspondence is not one-to-one.

Indeed, conjugate modules M and M' are, in general, not similar, but their corresponding forms coincide.

An analogous phenomenon clearly also holds for decomposable forms of any degree. In general, there is no natural way to rectify this lack of correspondence between modules and forms. But for quadratic fields we shall see that it is possible to establish a one-to-one correspondence by slightly changing the definitions of equivalence of forms and similarity of modules.

Definition. The binary quadratic form $f(x, y) = Ax^2 + Bxy + Cy^2$ with rational integer coefficients is called *primitive* if the greatest common divisor of the coefficients is 1. The integer $B^2 - 4AC$ is called the *discriminant* of the primitive form f.

The discriminant of a primitive form hence differs from its determinant $AC - (B^2/4)$ by a factor of -4.

It is easily seen that any form equivalent to a primitive form is also primitive. Under a linear change of variables with matrix C the determinant of a quadratic form is multiplied by $(\det C)^2$, and hence does not change if $\det C = \pm 1$. Hence equivalent primitive forms have the same discriminant.

Definition. Two primitive forms are called *properly equivalent* if one can be obtained from the other by a linear change of variables with determinant $+1$.

The collection of all primitive binary quadratic forms is broken up into classes of properly equivalent forms. For the rest of this section, when we speak of equivalent forms, we shall always mean properly equivalent forms. It will frequently happen that two forms which are nonproperly equivalent (that is, carried into each other by linear substitutions with determinant -1) will also be properly equivalent.

We now give a new definition for similarity of modules.

Definition. Two modules M and M_1 in a quadratic field are called *strictly similar* if $M_1 = \alpha M$ for some α with positive norm.

Since in imaginary quadratic fields the norm of any nonzero α is positive, in such fields the concept of strict similarity does not differ from the usual concept. The situation will be the same in real quadratic fields when the coefficient ring $\mathfrak{O}$ of the module M contains a unit ε with $N(\varepsilon) = -1$. Indeed, if $M_1 = \alpha M$ and $N(\alpha) < 0$, then, since $\varepsilon M = M$, we have $M_1 = (\alpha\varepsilon)M$, with $N(\alpha\varepsilon) > 0$. Conversely, suppose that the two concepts of similarity coincide. Then if $M_1 = \alpha M$, $N(\alpha) < 0$, there exists a number β for which $N(\beta) > 0$ and $M_1 = \beta M$. Setting $\varepsilon = \alpha\beta^{-1}$, we have $\varepsilon M = M$, and this means that ε is a unit in the coefficient ring $\mathfrak{O}$ with $N(\varepsilon) = -1$.

Hence the concept of strict similarity differs from the usual concept of similarity precisely for those modules in a real quadratic field whose coefficient rings contain only units with norm $+1$. It is clear that in this case every class of modules, similar in the usual sense, breaks up into two classes of strictly similar modules.

We now describe a correspondence between classes of modules and classes of forms.

In each module M of the field $R(\sqrt{d})$ we shall only consider those bases α, β for which the determinant

$$\Delta = \begin{vmatrix} \alpha & \beta \\ \alpha' & \beta' \end{vmatrix} \tag{7.9}$$

satisfies

$$\begin{aligned} \Delta > 0 \qquad & \text{for } d > 0, \\ \frac{1}{i}\Delta > 0 \qquad & \text{for } d < 0. \end{aligned} \tag{7.10}$$

As previously, α' and β' here denote the conjugates of α and β in $R(\sqrt{d})$. [A basis in M which satisfies (7.10) can always be found; if any basis α_1, α_2 does not work, interchange α_1 and α_2.]

We set each basis α, β of the module M which satisfies (7.10) in correspondence with the form

$$\begin{aligned} f(x, y) &= Ax^2 + Bxy + Cy^2 \\ &= \frac{N(\alpha x + \beta y)}{N(M)} = \frac{(\alpha x + \beta y)(\alpha' x + \beta' y)}{N(M)} \end{aligned} \tag{7.11}$$

[$N(M)$ is the norm of the module M]. If for the number $\gamma = -\beta/\alpha$ we consider $\varphi_\gamma(t) = at^2 + bt + c$ (see Section 4), then we shall clearly have

$$N(\alpha x + \beta y) = \frac{N(\alpha)}{a}(ax^2 + bxy + cy^2).$$

On the other hand, by the Corollary of Lemma 1 and by Theorem 2 of Section 6 the module $M = \alpha\{1, \gamma\}$ has norm $|N(\alpha)|/a$. Hence the coefficients A, B, C differ from a, b, c at most in sign. The form (7.11) is primitive and its discriminant $B^2 - 4AC$ coincides with the discriminant $b^2 - 4ac$ of the coefficient ring of the module M (Lemma 1). Thus we have the mapping

$$\{\alpha, \beta\} \to f(x, y), \tag{7.12}$$

which associates to each basis α, β of the field $R(\sqrt{d})$ which satisfies condition (7.10) the primitive form $f(x, y)$ (if the field is real, the coefficient A may be negative). It is clear that if the field is imaginary quadratic, then the form (7.11) will always be positive-definite, so negative-definite forms are not included in the correspondence (7.12).

Theorem 4. Let $\mathfrak{M}$ be the set of all classes of strictly similar modules similar in the narrow sense of the quadratic field $R(\sqrt{d})$. When $d > 0$, let $\mathfrak{F}$ be the set of all classes of properly equivalent binary quadratic forms which split into linear factors in $R(\sqrt{d})$. When $d < 0$, we consider only positive-definite forms. Then the mapping (7.12) establishes a one-to-one correspondence between $\mathfrak{M}$ and $\mathfrak{F}$. If some class of modules has a coefficient ring with discriminant D, then the corresponding forms also have discriminant D.

Let α, β and α_1, β_1 be two bases of the field $R(\sqrt{d})$ for which the determinant (7.9) satisfies (7.10) and let these bases correspond to the forms f and f_1. To prove Theorem 4 we must show that the forms f and f_1 are properly equivalent if and only if the modules $\{\alpha, \beta\}$ and $\{\alpha_1, \beta_1\}$ are strictly similar. Further, we must show that for any irreducible primitive form $g(x, y)$ [which splits up into linear factors in $R(\sqrt{d})$, and is positive-definite in the case $d < 0$] there is a basis α, β satisfying (7.10) for which the form (7.11) coincides with $g(x, y)$. We leave the simple details of this verification to the reader.

In Section 7.4 we defined the product of two classes of similar modules. In precisely the same way one can define the product of two classes of strictly similar modules. Since the mapping $\mathfrak{M} \to \mathfrak{F}$ is one-to-one the multiplication of classes of modules induces a multiplication on the set of classes of forms. The operation of multiplication in $\mathfrak{F}$ is called *composition of classes of forms* (a term introduced by Gauss, who first studied this operation). Since the set of all classes of modules with some fixed coefficient ring is a group, the set of classes of primitive forms with fixed discriminant D (only positive-definite forms for $D < 0$) also forms a group.

7.6. *The Representation of Numbers by Binary Forms and Similarity of Modules*

In this section we show that the problem of finding representations of integers by binary quadratic forms can be reduced to the problem of similarity of modules in a quadratic field.

Let $f(x, y)$ be a primitive binary quadratic form with discriminant $D \neq 0$, which splits into linear factors in the field $R(\sqrt{d})$ and let m be a natural number. In the case $D < 0$ we further assume that f is positive-definite. Our problem is to find all integral solutions of the equation

$$f(x, y) = m. \tag{7.13}$$

(We only consider positive values for m, since in the case $m < 0$, $D > 0$, we can replace f by the form $-f$.) By Theorem 4 we can represent f in the form

$$f(x, y) = \frac{N(\alpha x + \beta y)}{N(M)}, \tag{7.14}$$

where the basis α, β of the module M satisfies (7.10). The mapping $(x, y) \to \xi = \alpha x + \beta y$ establishes a one-to-one correspondence between solutions to (7.13) and numbers $\xi \in M$ with norm $N(\xi) = mN(M)$. Two solutions of (13) are called *associate* if the corresponding numbers of M are associate. It is easily verified that the concept of associate solutions does not depend on the choice of the representation (7.14). We denote the coefficient ring of the module M by $\mathfrak{O}$ and denote the class of strictly similar modules which contains M by C. By Theorem 4 C is uniquely determined by f.

Assume that we have a number $\xi \in M$ with norm $mN(M)$. Consider the module $A = \xi M^{-1}$. Since $AM = \xi M^{-1}M = \xi\mathfrak{O} \subset M$, the module A is contained in $\mathfrak{O}$. Its norm is $N(\xi)N(M)^{-1} = m$. Also it is clear that A is contained in the class C^{-1}, the inverse of the class of M.

Conversely, assume that in the class C^{-1} there is a module A which is contained in the ring $\mathfrak{O}$ and has norm m. Then for some ξ with positive norm we have $A = \xi M^{-1}$, so that $\xi \in MA \subset M$ and $N(\xi) = m$. If A_1 is any other module of the class C^{-1} which is contained in $\mathfrak{O}$ and has norm m, and if $A_1 = \xi_1 M^{-1}$ with $N(\xi_1) > 0$, then $A_1 = \xi_1\xi^{-1}A$. Hence A coincides with A_1 if and only if ξ_1 is associate with ξ.

We have thus proved the following theorem.

Theorem 5. Let the form $f(x, y)$ correspond to the class C of strictly similar modules with coefficient ring $\mathfrak{O}$. The set of classes of associate solutions of (7.13) is in one-to-one correspondence with the set of modules A which are in the class C^{-1}, are contained in the coefficient ring $\mathfrak{O}$, and have norm m. The solutions (x, y) which correspond to the module A are given by the numbers ξ for which $A = \xi M^{-1}$, $N(\xi) > 0$, where M is a module of the class C.

For any natural number m we can easily find the set of all modules A with coefficient ring $\mathfrak{O}$ which are contained in $\mathfrak{O}$ and have norm m. Let A be such a module, and let k be the smallest natural number contained in A. The module A can then be written in the form

$$A = \{k, k\gamma\} = k\{1, \gamma\}.$$

The number γ is determined except for sign and addition or subtraction of integers. We may therefore choose γ so that

$$\begin{aligned} &\operatorname{Im} \gamma > 0 \quad \text{for } d < 0, \\ &\operatorname{Irr} \gamma > 0 \quad \text{for } d > 0 \end{aligned} \tag{7.15}$$

(irr γ denotes the irrational part of the number γ), and also so that the rational part of γ is contained in the interval $(-\frac{1}{2}, \frac{1}{2}]$. In the notation of Lemma 1 we may write

$$\gamma = \frac{-b+\sqrt{D}}{2a}, \tag{7.16}$$

where

$$-a \leqslant b < a \tag{7.17}$$

by our condition on the rational part of γ. Since $\mathfrak{D} = \{1, a\gamma\}$ (see the proof of Lemma 1) and $A \subset \mathfrak{D}$, we easily obtain that k is divisible by a; that is, $k = as$ with integral s. Since $m = N(A) = k^2(1/a)$ (corollary of Lemma 1), then

$$m = as^2. \tag{7.18}$$

We shall show that the representation of the module A in the form

$$A = as\{1, \gamma\}, \tag{7.19}$$

where a, s, and γ satisfy (7.18), (7.15) and (7.17), is unique. Indeed, if $as\{1, \gamma\} = a_1 s_1\{1, \gamma_1\}$, where a_1, s_1, and γ_1 satisfy the same requirements, then $as = a_1 s_1$ and hence $\{1, \gamma\} = \{1, \gamma_1\}$. By the corollary of Lemma 1 we thus have $a = a_1$ and hence also $s = s_1$. Further, since γ in $\{1, \gamma\}$ satisfies (7.15) and (7.17), it is uniquely determined; that is, $\gamma = \gamma_1$.

Conversely, for given m choose a and s so that (7.18) holds. If b and c satisfy the conditions

$$b^2 - 4ac = D, \qquad (a, b, c) = 1, \qquad -a \leqslant b < a, \tag{7.20}$$

then, with γ given by (7.16), the module $A = as\{1, \gamma\}$ will be contained in its coefficient ring $\mathfrak{D} = \{1, a\gamma\}$ and its norm will be $a^2s^2(1/a) = m$.

Thus to obtain the module A we need to find all four numbers $s > 0$, $a > 0$, b, c satisfying conditions (7.18) and (7.20).

If we can devise an algorithm for solving the question of strict similarity of modules of the field $R(\sqrt{d})$, then, after listing all modules $A \subset \mathfrak{D}$ with norm m, we can determine those which are similar to the module M^{-1}. By Theorem 5 this will yield all solutions of (7.13).

The following assertion easily follows from Theorem 5.

Theorem 6. Let m be a natural number. Then m is represented by some primitive binary quadratic form with discriminant D if and only if there is a module A with norm m contained in the order $\mathfrak{D}$ with discriminant D, with $\mathfrak{D}$ the coefficient ring of A. This is in turn equivalent to the existence of integers $s > 0$, $a > 0$, b, c satisfying the conditions: $m = as^2$, $b^2 - 4ac = D$, $(a, b, c) = 1$, $-a \leqslant b < a$.

In case D is the discriminant of a maximal order $\mathfrak{O}$, the second assertion of Theorem 6 is simplified. Namely, we have

Theorem 7. Let D be the discriminant of a quadratic field (that is, the discriminant of a maximal order). In order that the natural number $m = as^2$, where a is square-free, be represented by some primitive binary form with discriminant D, it is necessary and sufficient that the congruence

$$x^2 \equiv D \pmod{4a} \tag{7.21}$$

be solvable.

The proof of Theorem 7 is left to the reader.

7.7. *Similarity of Modules in Imaginary Quadratic Fields*

In the case of an imaginary quadratic field $R(\sqrt{d})$, $d < 0$, there is a particularly simple method for solving the problem of similarity of modules.

The geometric representation of a number $\alpha \in R(\sqrt{d})$ by a point in the space $\mathfrak{R}^2$ (see Section 3.1) coincides with the usual representation of complex numbers in the complex plane. The points of a full module $M \subset R(\sqrt{d})$ correspond to the points (or vectors) of some full lattice in $\mathfrak{R}^2$. The lattice which corresponds to the module M will also be denoted by M. The effect of multiplying all points of the lattice M by the complex number $\xi \neq 0$ is to rotate the lattice M by an angle $\arg \xi$ and to expand it by a factor $|\xi|$, so similar lattices M and ξM are also similar in the sense of elementary geometry. All subsequent results will be based on this simple fact.

The question of similarity of lattices in the plane is solved by constructing a special basis for each lattice, called a *reduced basis*. A reduced basis α, β consists of a shortest nonzero vector α and a shortest vector β which is not collinear with it (and satisfies some further conditions). We now show that in any lattice M such a pair of vectors α and β always forms a basis. For if this were not the case, then M would contain a vector $\xi = u\alpha + v\beta$ with the real numbers u and v not both integers. Adding to this vector a certain integral linear combination of α and β, we may clearly assume that $|u| \leqslant \frac{1}{2}$ and $|v| \leqslant \frac{1}{2}$. If $v \neq 0$, we must have $|\xi| \geqslant |\beta|$, which contradicts the inequality

$$|\xi| < |u\alpha| + |v\beta| \leqslant \tfrac{1}{2}|\alpha| + \tfrac{1}{2}|\beta| \leqslant |\beta|.$$

If $v = 0$, then $|\xi| = |u\alpha| \leqslant \frac{1}{2}|\alpha| < |\alpha|$, which violates the choice of α. Our assertion is proved.

If α is any shortest vector of M and β is any shortest vector among those not collinear with α, then the length of the projection of β on α does not exceed $\frac{1}{2}|\alpha|$. For among the vectors of the form $\beta + n\alpha$ (n an integer) there clearly is one for which the length of the projection is $\leqslant \frac{1}{2}|\alpha|$. On the other hand, the vector of the form $\beta + n\alpha$ with shortest length is also the one with shortest projection.

We now consider the set of all nonzero vectors in M with shortest length, and let w denote the number of such vectors. Since if α is in this set $-\alpha$ also is, the number w is even. Further, the angle between two shortest vectors α and α' cannot be less than $\pi/3$, since otherwise the vector $\alpha - \alpha'$ would be of shorter length. Hence $w \leqslant 6$ and we have the following possible cases: $w = 2$, $w = 4$, $w = 6$.

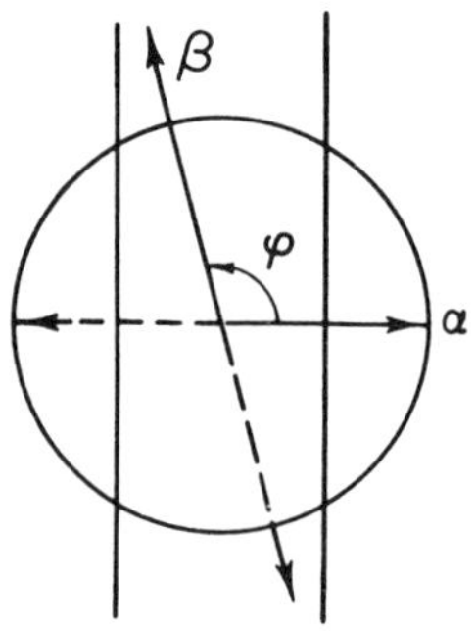

FIG. 1

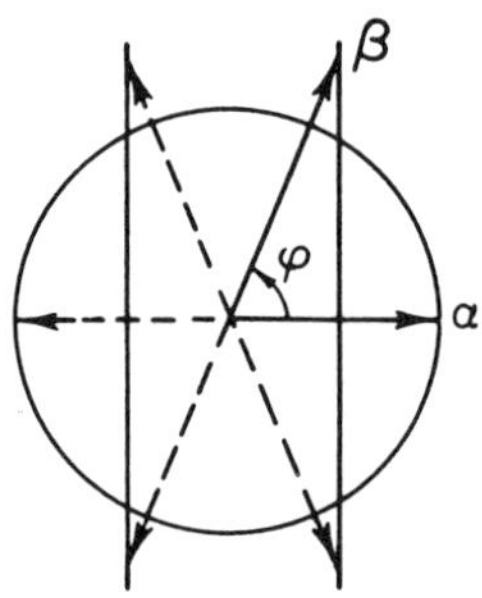

FIG. 2

We now construct a reduced basis for the lattice M. If $w = 2$ we take as α either of the two shortest vectors. There may be two or four vectors in the set of vectors, noncollinear with α, of shortest length (see Figures 1 and 2). As β we choose that one for which the angle φ between α and β in the positive direction (counterclockwise) is smallest. If $w = 4$ or $w = 6$ we take as reduced

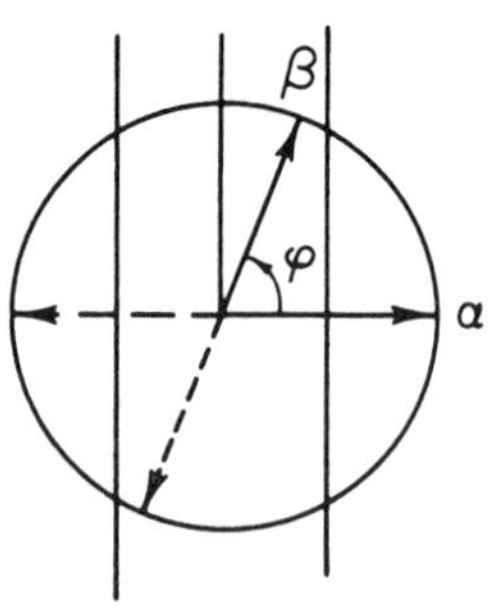

FIG. 3

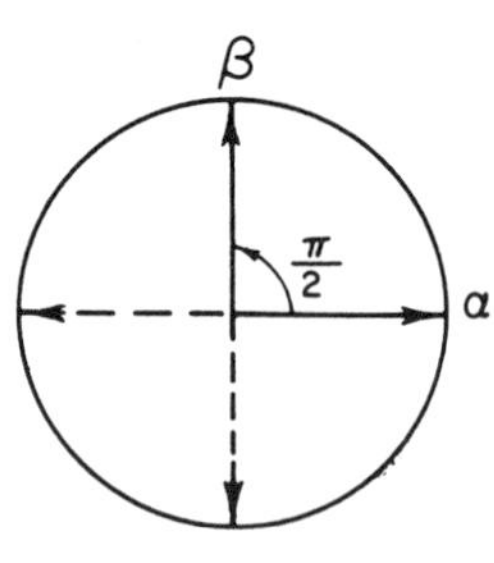

FIG. 4

basis a pair of vectors of shortest length such that the angle between α and β in the positive direction is as small as possible.

It is easily seen that the reduced basis of a lattice is uniquely defined up to a rotation which takes the lattice into itself. In the cases $w = 2$, and $w = 4$ and $\pi/3 < \varphi < \pi/2$ (see Figure 3), there are two reduced bases, which are obtained from each other by a rotation of angle π. For $w = 4$, $\varphi = \pi/2$ (Figure 4) we are dealing with a square lattice with four reduced bases, which can be obtained from one another by rotations through angle $\pi/2$. Finally in the case $w = 6$, we have six reduced bases, and they are transformed into each other by a rotation through angle $\pi/3$ (Figure 5; the circle is divided into six equal parts, since the angle between shortest vectors cannot be less than $\pi/3$). Using the concept of a reduced basis, we can easily solve the question of similarity of lattices in the plane.

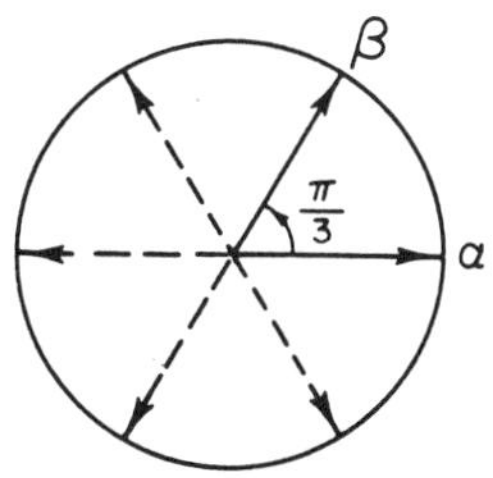

FIG. 5

Theorem 8. The lattices M and M_1 in $\mathfrak{R}^2$ are similar if and only if their reduced bases are similar (that is, are transformed into each other by a rotation and an expansion).

Proof. Let α, β and α_1, β_1 be reduced bases of the lattices M and M_1. If $\xi M = M_1$, then $\xi\alpha$, $\xi\beta$ clearly is a reduced basis for M_1. As we have seen, this basis can be obtained from the basis α_1, β_1 by a rotation. Therefore there is a number η (which is a root of unity of degree 1, 2, 3, 4, or 6) such that $\eta\xi\alpha = \alpha_1$, $\eta\xi\beta = \beta_1$. Hence the basis α_1, β_1 is obtained from the basis α, β by rotation through the angle $\arg(\eta\xi)$ and expansion by a factor $|\eta\xi|$, so that they are similar. The converse is clear.

We now turn to the description of the set of classes of similar modules of an imaginary quadratic field. Let M be any module in $R(\sqrt{d})$, $d < 0$, and let α, β be any reduced bases for M. We pass to the similar module $(1/\alpha)M = \{1, \gamma\}$, where $\gamma = \beta/\alpha$. The basis $\{1, \gamma\}$ here is also reduced. From the definition of a reduced basis it easily follows that γ satisfies

$$\operatorname{Im}\gamma > 0, \tag{7.22}$$

$$-\tfrac{1}{2} < \operatorname{Re}\gamma \leqslant \tfrac{1}{2}, \tag{7.23}$$

$$\begin{aligned} |\gamma| > 1 \qquad & \text{if} \quad -\tfrac{1}{2} < \operatorname{Re}\gamma < 0, \\ |\gamma| \geqslant 1 \qquad & \text{if} \quad 0 \leqslant \operatorname{Re}\gamma \leqslant \tfrac{1}{2}. \end{aligned} \tag{7.24}$$

Definition. The number γ of an imaginary quadratic field is called *reduced* if it satisfies conditions (7.22), (7.23), and (7.24). The module $\{1, \gamma\}$ is called reduced if γ is reduced.

Geometrically, γ is reduced if it lies in the region Γ described in Figure 6 (the indicated part of the boundary, including the point i, is included in Γ; the rest is not).

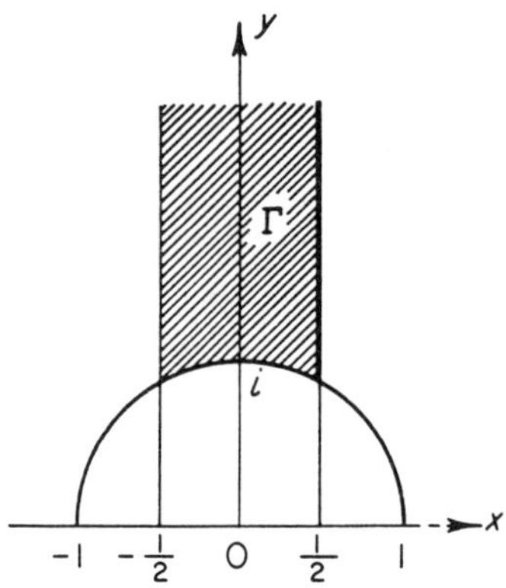

FIG. 6

Theorem 9. Each class of similar modules of the imaginary quadratic number field $R(\sqrt{d})$, $d < 0$, contains one and only one reduced module.

Proof. We have already shown that each class contains a reduced module. We need only show now that distinct reduct modules cannot be similar. We first show that if $\gamma = x + yi$ is reduced, then the numbers $1, \gamma$ form a reduced basis for the lattice $\{1, \gamma\}$. We must show that γ is the smallest of the vectors of the lattice $\{1, \gamma\}$ which do not lie on the real line, that is, that $|k + l\gamma| \geqslant |\gamma|$ for any integers k and $l \neq 0$. Since $|x| \leqslant \tfrac{1}{2}$,

$$|k \pm \gamma|^2 = (k \pm x)^2 + y^2 \geqslant x^2 + y^2 = |\gamma|^2.$$

If $|l| \geqslant 2$, then

$$|k + l\gamma|^2 \geqslant l^2 y^2 > 2y^2 > x^2 + y^2 = |\gamma|^2,$$

which proves our assertion. Now let γ and γ_1 be two reduced numbers. If the modules $\{1, \gamma\}$ and $\{1, \gamma_1\}$ are similar, then by Theorem 8 the bases $\{1, \gamma\}$

and $\{1, \gamma_1\}$ are similar. But this is possible if and only if $\gamma = \gamma_1$. Theorem 9 is completely proved.

To solve the question of the similarity of modules in an imaginary quadratic field we must have an algorithm for finding the reduced module similar to a given module. Such an algorithm is formulated in Problem 24. Two given modules M_1 and M_2 are similar if and only if their reduced modules coincide.

Remark. In the proof of Theorem 9 we never actually used the fact that the module under consideration was contained in some imaginary quadratic field. The assertion of the theorem hence is true for any lattice in the plane: any lattice in the complex plane is similar to one and only one lattice of the form $\{1, \gamma\}$, where γ is some number of the domain Γ, which is described in Figure 6. By Lemma 2, which is applicable to arbitrary lattices in the plane without any changes, two lattices $\{1, \gamma\}$ and $\{1, \lambda\}$ are similar if and only if λ and γ are connected by

$$\lambda = \frac{k\gamma + l}{m\gamma + n}, \qquad kn - ml = \pm 1,$$

with rational integers k, l, m, n. Two such nonreal complex numbers are called *modularly equivalent*. Hence we have shown that every nonreal complex number is modularly equivalent to one and only one number of the region Γ. The region Γ itself is called the *modular domain*. Its points are in one-to-one correspondence with the set of classes of similar lattices in the plane. The question of similarity of planar lattices is connected with many important questions in the theory of elliptic functions. A field of elliptic functions is given by its period lattice, and two such fields are isomorphic if and only if their corresponding period lattices are similar (see, for example, C. Chevalley, "Introduction to the Theory of Algebraic Functions of One Variable," 1951). Hence the points of the modular domain Γ are in one-to-one correspondence with isomorphism classes of fields of elliptic functions.

Consider now the classes of similar modules which belong to some fixed order $\mathfrak{O}$ with discriminant $D < 0$. Let the module $\{1, \gamma\}$, $\gamma \in \Gamma$, belong to the order $\mathfrak{O}$. If we use the notations of Lemma 1 and write γ in the form

$$\gamma = \frac{-b + i\sqrt{|D|}}{2a},$$

then conditions (7.23) and (7.24) yield

$$\begin{gathered} -a \leqslant b < a, \\ c \geqslant a \quad \text{for} \quad b \leqslant 0, \\ c > a \quad \text{for} \quad b > 0. \end{gathered} \tag{7.25}$$

Hence to find a full system of reduced modules of an imaginary quadratic field which belong to the order with discriminant D, we need only find all triples of integers $a > 0$, b, c which satisfy (7.25) and also the condition

$$D = b^2 - 4ac, \qquad (a, b, c) = 1. \tag{7.26}$$

By Theorem 3 of Section 6 the number of such triples is finite, a fact that can be directly verified from the inequalities

$$|D| = 4ac - b^2 \geqslant 4a^2 - a^2 = 3a^2,$$

$$|b| \leqslant a < \sqrt{\frac{|D|}{3}},$$

for given D, so that there are only finitely many possibilities for a and b and hence also for c.

Example 1. We shall find the number of classes of modules which belong to the maximal order of the field $R(\sqrt{-47})$. Since here $D = -47$, then $|b| \leqslant a < \sqrt{47/3}$. Since for odd D the number b is also odd, we have the following possibilities: $b = \pm 1$, $b = \pm 3$. In the second case we would have to have $b^2 - D = 56 = 4ac$, $ac = 14$, $3 \leqslant a \leqslant c$, which is impossible. If $b = \pm 1$, then $b^2 - D = 48 = 4ac$, so that

$$a = 1, \quad c = 12; \qquad a = 2, \quad c = 6; \qquad a = 3, \quad c = 4.$$

Since the case $b = a = 1$ must be excluded, the maximal order of the field $R(\sqrt{-47})$ has five classes of similar modules. Each class contains a reduced module $\{1, \gamma\}$, where γ is one of the numbers

$$\frac{1 + i\sqrt{47}}{2}, \quad \frac{\pm 1 + i\sqrt{47}}{4}, \quad \frac{\pm 1 + i\sqrt{47}}{6}.$$

In the preceding section we remarked that the existence of an algorithm for determining the similarity of modules in quadratic fields allows us to solve equations of the form (7.13).

Example 2. We shall find all numbers in the module $M = \{13, 1 + 5i\}$ with norm 650. In this case the coefficient ring is the order $\mathfrak{O} = \{1, 5i\}$ with discriminant $D = -100$. Since $N(M) = 13$, we must first enumerate the modules $A \subset \mathfrak{O}$, which belong to the order $\mathfrak{O}$ and have norm $m = 650/13 = 50$. From conditions (7.18) and (7.20) we have the following possibilities:

(1) $s = 5, \quad a = 2, \quad b = -2, \quad c = 13;$

(2) $s = 1, \quad a = 50, \quad b = 10, \quad c = 1;$

(3) $s = 1, \quad a = 50, \quad b = -10, \quad c = 1;$

(4) $s = 1, \quad a = 50, \quad b = -50, \quad c = 13.$

For each of these four cases we form the module A of the type (7.19) and find the reduced module similar to it:

$$10\left\{1, \frac{1+5i}{2}\right\},$$

$$50\left\{1, \frac{-1+i}{10}\right\} = (-5+5i)\{1, 5i\},$$

$$50\left\{1, \frac{1+i}{10}\right\} = (5+5i)\{1, 5i\},$$

$$50\left\{1, \frac{5+i}{10}\right\} = 10i\left\{1, \frac{1+5i}{2}\right\}.$$

We also compute the reduced module for M^{-1}:

$$M^{-1} = \left\{1, \frac{1-5i}{13}\right\} = \frac{1-5i}{13}\left\{1, \frac{1+5i}{2}\right\}.$$

We eliminate the module A in cases (2) and (3), since it is not similar to M^{-1}. In cases (1) and (4) the equality $A = \xi M^{-1}$ holds for $\xi = 5 + 25i$ and $\xi = -25 + 5i$. Since $\mathfrak{O}$ contains only two units, ± 1, the module M has four numbers with norm 650: $\pm(5 + 25i)$ and $\pm(-25 + 5i)$.

We have thus also established that the equation $13x^2 + 2xy + 2y^2 = 50$ has four integral solutions:

$$x = 0, \quad y = 5; \qquad x = 0, \quad y = -5;$$

$$x = 2, \quad y = -1; \qquad x = -2, \quad y = 1.$$

Example 3. Which natural numbers are represented by the form $x^2 + y^2$?

The discriminant of the form is $D = -4$. Let $\mathfrak{O} = \{1, i\}$ be the order with discriminant -4, which is contained in the field $R(\sqrt{-1})$. Since conditions (7.25) and (7.26) are satisfied only by $a = c = 1$, $b = 0$, only one reduced module belongs to the order $\mathfrak{O}$. This means that all modules which belong to the order $\mathfrak{O}$ are similar, and hence every binary form with discriminant -4 is equivalent to the form $x^2 + y^2$. But equivalent forms represent the same numbers, and hence by Theorem 6 the form $x^2 + y^2$ represents the number m if and only if there is a module $A \subset \mathfrak{O}$ which belongs to the order $\mathfrak{O}$ and has norm m. If such a module exists, then for some s, a, b, c we have

$$m = as^2, \quad D = -4 = b^2 - 4ac, \quad (a, b, c) = 1.$$

Here the number b must be even, $b = 2z$, where z satisfies

$$z^2 \equiv -1 \pmod{a}. \tag{7.27}$$

Conversely, if (7.27) holds for some $a = m/s^2$; that is, if $z^2 = -1 + ac$, then, as is easily seen, $(a, 2z, c) = 1$, and hence there is a module $A \subset \mathfrak{D}$, belonging to the order $\mathfrak{D}$, and with norm m; that is, m is represented by the form $x^2 + y^2$.

It is well known that the congruence (7.27) is solvable if and only if a is not divisible by 4 and not divisible by any prime number of the form $4k + 3$. Since a must contain all prime factors which occur in m with even exponent, we obtain that m is represented by the form $x^2 + y^2$ if and only if prime numbers of the form $4k + 3$ occur in it only with even exponent.

PROBLEMS

1. Find fundamental units for the fields $R)\sqrt{19})$ and $R(\sqrt{37})$.

2. Show that if $d \equiv 1 \pmod 8$ (positive and square-free), then a fundamental unit for the order $\{1, \sqrt{d}\}$ is also a fundamental unit for the maximal order of the field $R(\sqrt{d})$.

3. Show that if the discriminant of some order $\mathfrak{D}$ in a quadratic field is divisible by at least one prime of the form $4n + 3$, then any unit of $\mathfrak{D}$ has norm $+1$.

4. Let the rational integer $m > 1$ not be a perfect square. Show that in the continued fraction expansion of $\sqrt{m}$ the sequence of entries has the form

$$q_0, q_1, \ldots, q_s, 2q_0, q_1, \ldots, q_s, 2q_0, q_1, \ldots$$

(here $q_{i+1} = q_{s-i}$, $i = 0, \ldots, s-1$).

5. Under the same assumptions, show that if P_s/Q_s is the convergent (s again denoting the period), then $P_s + Q_s\sqrt{m}$ is a fundamental unit of the order $\{1, \sqrt{m}\}$ (in the field $R(\sqrt{m})$).

6. Let the modules M_1 and M_2 of a quadratic field have for coefficient rings the orders $\mathfrak{D}_{f1}$ and $\mathfrak{D}_{f2}$ (using the notation of Section 7.2). Show that the product M_1M_2 belongs to the order $\mathfrak{D}_f$, where f is the greatest common divisor of f_1 and f_2.

7. For any natural number f let $\mathfrak{K}_f$ denote the group of modules in a given quadratic field which belong to the order $\mathfrak{D}_f$ (see Section 7.4). Show that if d is a divisor of f, then the mapping $M \to M\mathfrak{D}_d$ $(M \in \mathfrak{K}_f)$ is a homomorphism from $\mathfrak{K}_f$ to the group $\mathfrak{K}_d$.

8. Let ξ be a number of the maximal order $\tilde{\mathfrak{D}} = \{1, \omega\}$ of a quadratic field which is relatively prime to f. Show that the coefficient ring of the module $M = \{f, f\omega, \xi\}$ is $\mathfrak{D}_f$ and that $M\tilde{\mathfrak{D}} = \tilde{\mathfrak{D}}$. Further, show the converse, that is, that any module M which belongs to the order $\mathfrak{D}_f$ and satisfies the property $M\tilde{\mathfrak{D}} = \tilde{\mathfrak{D}}$ is of the form $M = \{f, f\omega, \xi\}$ for some $\xi \in \tilde{\mathfrak{D}}$ which is relatively prime to f.

9. Let ξ_1 and ξ_2 be two numbers of $\tilde{\mathfrak{D}}$ which are relatively prime to f. Show that $\{f, f\omega, \xi_1\} = \{f, f\omega, \xi_2\}$ if and only if $s\xi_1 \equiv \xi_2 \pmod f$ for some rational integer s.

10. Show that if M_1 and M_2 are any two full modules of a quadratic field (not necessarily belonging to the same order), then

$$N(M_1M_2) = N(M_1)N(M_2).$$

11. Let h denote the number of classes of similar modules belonging to the maximal order $\tilde{\mathfrak{O}}$ of a quadratic field, and let h_f denote the number of classes of similar modules belonging to the order $\mathfrak{O}_f$ (of index f). Show that

$$h_f = h\frac{\Phi(f)}{e_f\varphi(f)},$$

where $\Phi(f)$ is the number of residue classes in $\tilde{\mathfrak{O}}$ modulo f which consist of numbers relatively prime to f (Φ is analogous to the Euler φ-function), and e_f is the index of the group of units of the order $\mathfrak{O}_f$ in the group of units of the maximal order $\tilde{\mathfrak{O}}$.

12. A number γ of a real quadratic field is called reduced if it satisfies $0 < \gamma < 1$ and its conjugate satisfies $\gamma' < -1$. If γ is reduced the module $\{1, \gamma\}$ is also called *reduced.* Using the notation of Lemma 1, show that the number γ is reduced if and only if

$$0 < b < \sqrt{D}, \qquad -b + \sqrt{D} < 2a < b + \sqrt{D}.$$

Deduce that the number of reduced modules which belong to a fixed order of a real quadratic field is finite.

13. Let γ be an irrational number of a real quadratic field such that $0 < \gamma < 1$. Set

$$\gamma_1 = -(\operatorname{sgn}\gamma')\frac{1}{\gamma} - n,$$

where the rational integer n is chosen so that $0 < \gamma_1 < 1$. Show that after a finite number of transformations $\{1, \gamma\} \to \{1, \gamma_1\}$, the module $\{1, \gamma\}$ is transformed into a similar module which is reduced. Hence every class of similar modules (in the usual sense) of a real quadratic field contains a reduced module.

14. Let γ be a reduced number of a real quadratic field. Since $\operatorname{sgn}\gamma' = -1$, the mapping $\gamma \to \gamma_1$ of the preceding problem takes the form

$$\gamma_1 = \frac{1}{\gamma} - n, \qquad n = \left[\frac{1}{\gamma}\right].$$

Show that the number γ_1 is also reduced. It is called the *right neighbor* of the number γ, and γ is called a *left neighbor* of γ_1. Show that any reduced number has one and only one left neighbor γ.

15. Starting from a reduced number γ_0 of a real quadratic field, we construct the sequence of reduced numbers $\gamma_0, \gamma_1, \gamma_2, \ldots$, in which each number is the right neighbor of the preceding one. Show that there exists a natural number m such that $\gamma_0 = \gamma_m$, that is, that the sequence is periodic. If m is the smallest possible such integer, then the numbers $\gamma_0, \gamma_1, \ldots, \gamma_{m-1}$ are distinct. Such a finite sequence of reduced numbers is called a *period.* Show that two reduced modules $\{1, \gamma\}$ and $\{1, \gamma^*\}$ are similar (in the usual sense) if and only if the reduced numbers γ and γ^* belong to the same period.

16. Find the number of classes of similar modules belonging to the maximal order of the field $R(\sqrt{10})$.

17. Show that all integral solutions of the equation

$$17x^2 + 32xy + 14y^2 = 9$$

are given by

$$\pm(15 + 6\sqrt{2})(3 + 2\sqrt{2})^n = \pm[17x_n + (16 + 3\sqrt{2})y_n]$$

(for all integers n).

18. Which of the modules

$$\{1, \sqrt{15}\}, \quad \{2, 1+\sqrt{15}\}, \quad \{3, \sqrt{15}\}, \quad \{35, 20+\sqrt{15}\}$$

of the field $R(\sqrt{15})$ are similar to one another?

19. Find a full system of representatives for the classes of strictly equivalent primitive forms with discriminant 252.

20. What is the number of classes of properly equivalent primitive forms with discriminant 360?

21. Which prime numbers are represented by the forms $x^2 + 5y^2$ and $2x^2 + 2xy + 3y^2$?

22. Find the integral solutions of the equations:

$$5x^2 + 2xy + 2y^2 = 26, \tag{1}$$

$$5x^2 - 2y^2 = 3, \tag{2}$$

$$80x^2 - y^2 = 16. \tag{3}$$

23. Show that the equations

$$13x^2 + 34xy + 22y^2 = 23, \tag{1}$$

$$5x^2 + 16xy + 13y^2 = 23 \tag{2}$$

have no integral solutions.

24. Let γ be a number of an imaginary quadratic field which satisfies $\operatorname{Im} \gamma > 0$, $-\frac{1}{2} < \operatorname{Re} \gamma \leqslant \frac{1}{2}$, but which is not reduced. Set $\gamma_1 = (1/\gamma) + n$, where the rational integer n is chosen so that $-\frac{1}{2} < \operatorname{Re} \gamma_1 \leqslant \frac{1}{2}$. If γ_1 is not reduced, determine $\gamma_2 = -(1/\gamma_1) + n_1$ analogously, etc. Show that after a finite number of steps the module $\{1, \gamma\}$ is transformed into a similar module $\{1, \gamma_s\}$ which is reduced.

25. Determine the coefficient rings of the modules

$$\{11, 6+2i\sqrt{2}\}, \quad \{2, 1+i\sqrt{2}\}, \quad \{4, i\sqrt{2}\}, \quad \{2, i\sqrt{2}\}.$$

Which of these modules are similar?

26. Show that all modules which belong to the maximal order of the field $R(\sqrt{-43})$ are similar.

CHAPTER 3

The Theory of Divisibility

In Chapters 1 and 2 we saw how the solution of number-theoretic problems led to the consideration of broader questions in the theory of algebraic numbers: Thus to find the integral representations of a rational number by a full decomposable form, we had to study the theory of units in orders of algebraic number fields.

Many problems of number theory lead to another important question in the arithmetic of algebraic number fields, the question of decomposition of algebraic numbers into prime factors.

In this chapter we shall construct a general theory of the decomposition of algebraic numbers into prime factors and will apply this theory to several problems in number theory. The results which we shall need from the theory of rings are given in the Supplement, Section 5. These results, along with the theory of finite extensions of fields which has already been used in Chapter 2, form the algebraic tools for this chapter.

The problems of factorization are very closely connected with Fermat's (last) theorem. Historically, it was precisely the problem of Fermat's theorem which led Kummer to his fundamental work on the arithmetic of algebraic numbers.

Therefore we shall start with an exposition of the first results of Kummer on Fermat's theorem as an introduction to the general theory of decomposition of algebraic numbers into prime factors.

1. Some Special Cases of Fermat's Theorem

1.1. *The Connection between Fermat's Theorem and Decomposition into Factors*

The proposition, stated by Fermat, is that the equation

$$x^n + y^n = z^n$$

has no nonzero solutions in rational integers, x, y, z when $n > 2$.

It is clear that if Fermat's theorem is proved for some exponent n, then it is also automatically proved for all exponents which are multiples of n. Since any integer $n > 2$ is either divisible by 4 or by some odd prime, we may limit our consideration to the cases where $n = 4$ or n is an odd prime. For $n = 4$, an elementary proof was given by Euler. We thus consider only

$$x^l + y^l = z^l, \tag{1.1}$$

where the exponent l is an odd prime number. We may clearly assume that the numbers x, y, z in (1.1) are relatively prime.

For those values of l for which a proof of Fermat's theorem has been found, the proof is usually divided into two parts: first, it is shown that (1.1) has no solution in integers x, y, z, which are not divisible by l, and second, that (1.1) has no solution in integers x, y, z precisely one of which is divisible by l. These two are called the first and second cases of Fermat's theorem. From the extant proofs of various cases of Fermat's theorem we can deduce that the principal difficulties in the first and second cases of Fermat's theorem are roughly the same, although the techniques used in the first case are more simple. Here we consider only the first case of Fermat's theorem.

The connection between Fermat's theorem and the problem of the decomposition of algebraic numbers into prime factors is given in the following simple observation. If ζ denotes a primitive lth root of 1, then (1.1) may be written

$$\prod_{k=0}^{l-1} (x + \zeta^k y) = z^l. \tag{1.2}$$

If a product of pairwise relatively prime rational integers is an lth power, then each of the factors is an lth power. The factors on the left side of (1.2) belong to the algebraic field $R(\zeta)$ of degree $l - 1$ over R. (It is easily seen that the polynomial $t^{l-1} + t^{l-2} + \cdots + t + 1$, l a prime, is irreducible over the field of rational numbers; see Problem 6 or Theorem 1 of Section 2, Chapter 5.) Consider in the field $R(\zeta)$ the order $\mathfrak{O} = \{1, \zeta, \ldots, \zeta^{l-2}\}$ [by Theorem 1 of Section 5, Chapter 5, $\mathfrak{O}$ is the maximal order of the field $R(\zeta)$]. Assume that in the ring $\mathfrak{O}$ factorization into primes is unique. Then for any $\alpha \in \mathfrak{O}$, $\alpha \neq 0$, we have a factorization

$$\alpha = \varepsilon {\pi_1}^{a_1} \cdots {\pi_r}^{a_r},$$

where ε is a unit of the ring $\mathfrak{O}$, the prime numbers $\pi_1, \ldots, \pi_r$ are pairwise-nonassociate, and the exponents $a_1, \ldots, a_r$ are uniquely determined (see Section 2.2). Then every prime π which occurs in the factorization of z^l occurs with an exponent divisible by l. But we shall show below that when we are dealing with the first case of Fermat's theorem, the numbers $x + \zeta^k y$ ($k = 0, 1, \ldots, l-1$) are pairwise relatively prime. Hence if we represent $x + \zeta^k y$ as a product of prime factors, each prime will occur with an exponent divisible by l. This means that, up to a unit factor, $x + \zeta^k y$ is an lth power and, in particular,

$$x + \zeta y = \varepsilon \alpha^l, \tag{1.3}$$

where ε is a unit of the ring $\mathfrak{O}$ and $\alpha \in \mathfrak{O}$.

Since l is odd, we may also write (1.1) in the form

$$x^l + (-z)^l = (-y)^l,$$

and analogously we obtain

$$x - \zeta z = \varepsilon_1 \alpha_1{}^l. \tag{1.3'}$$

Equations (1.3) and (1.3′) lead, in a fairly easy manner, to a contradiction. When this is done, we shall prove that (1.1) has no solution in integers x, y, z not divisible by l (under the hypothesis made on the ring $\mathfrak{O}$).

After this introduction we now establish some auxiliary facts concerning the ring $\mathfrak{O}$.

1.2. *The Ring* $Z[\zeta]$.

Lemma 1. In the ring $\mathfrak{O} = Z[\zeta]$ the number $1 - \zeta$ is prime, and l has the factorization

$$l = \varepsilon^*(1 - \zeta)^{l-1} \tag{1.4}$$

where ε^* is a unit in $\mathfrak{O}$.

Proof. In the decomposition

$$t^{l-1} + t^{l-2} + \cdots + t + 1 = (t - \zeta)(t - \zeta^2) \cdots (t - \zeta^{l-1}),$$

set t equal to 1. Then

$$l = (1 - \zeta)(1 - \zeta^2) \cdots (1 - \zeta^{l-1}). \tag{1.5}$$

If $\alpha = r(\zeta)$ is any number of the field $R(\zeta)$ [here $r(t)$ is a polynomial with rational coefficients], then the numbers

$$\sigma_k(\alpha) = r(\zeta^k) \qquad (1 \leqslant k \leqslant l - 1) \tag{1.6}$$

are the images of α under the isomorphisms of the field $R(\zeta)$ into the field of all complex numbers. In the terminology of Section 2.3 of the Supplement, the numbers (1.6) are the conjugates of α, and thus $N(\alpha) = \prod_{k=1}^{l-1} r(\zeta^k)$. In particular, for $s \not\equiv 0 \pmod{l}$, we have

$$N(1 - \zeta^s) = \prod_{k=1}^{l-1} (1 - \zeta^{ks}) = \prod_{k=1}^{l-1} (1 - \zeta^k) = l.$$

From this it follows that $1 - \zeta, 1 - \zeta^2, \dots, 1 - \zeta^{l-1}$ are primes in the ring $\mathfrak{D}$. Indeed, if $1 - \zeta^s = \alpha\beta$, then $N(\alpha)N(\beta) = l$, and then either $N(\alpha) = 1$ or $N(\beta) = 1$; that is, one of the factors is a unit (Theorem 4 of Section 2, Chapter 2). Taking norms in the equation

$$(1 - \zeta^s) = (1 - \zeta)(1 + \zeta + \cdots + \zeta^{s-1}) = (1 - \zeta)\varepsilon_s, \tag{1.7}$$

we obtain $N(\varepsilon_s) = 1$, and thus ε_s is a unit in $\mathfrak{D}$. Hence the numbers $1 - \zeta^s$, for $s \not\equiv 0 \pmod{l}$, are associate with $1 - \zeta$. The decomposition (1.4) now follows from (1.5) and (1.7).

Lemma 2. If the rational integer a is divisible by $1 - \zeta$ (in the ring $\mathfrak{D}$), then it is also divisible by l.

Proof. Let $a = (1 - \zeta)\alpha$, where $\alpha \in \mathfrak{D}$. Taking the norm of both sides, we obtain $a^{l-1} = lN(\alpha)$, where $N(\alpha)$ is a rational integer. Since l is prime, then a is divisible by l.

Lemma 3. The only roots of 1 contained in the field $R(\zeta)$ are those whose degree divides $2l$.

Proof. Any root of 1 in $R(\zeta)$ clearly lies in the maximal order. By Theorem 2 of Section 3, Chapter 2, the set of all roots of 1 in $R(\zeta)$ forms a finite cyclic group. Let m denote the order of this group and let η be any primitive mth root of 1. Since $-\zeta$ belongs to $R(\zeta)$ and is a root of degree $2l$ of 1, m is divisible by $2l$. In Section 2 of Chapter 5 (corollary of Theorem 1), it is proved that the degree of the field $R(\eta)$ over R equals $\varphi(m)$, where $\varphi(m)$ is Euler's function. Set

$$m = l^r m_0, \quad (m_0, l) = 1 \qquad (r \geqslant 1,\, m_0 \geqslant 2).$$

Since $R(\eta)$ is contained in $R(\zeta)$, and the latter field has degree $l - 1$, then

$$\varphi(m) = l^{r-1}(l - 1)\varphi(m_0) \leqslant l - 1.$$

From this inequality it follows that $r = 1$ and $\varphi(m_0) = 1$. Since $\varphi(m_0) = 1$ for $m_0 \geqslant 2$ only when $m_0 = 2$, $m = 2l$, and Lemma 3 is proved.

Lemma 4 (Kummer's Lemma). Any unit of the ring $\mathfrak{D}$ is a product of a power of ζ with a real unit.

Proof. Let

$$\varepsilon = a_0 + a_1\zeta + \cdots + a_{l-2}\zeta^{l-2} = r(\zeta) \qquad (a_i \in Z)$$

be any unit of $\mathfrak{O}$. It is clear that the complex conjugate $\bar{\varepsilon} = r(\zeta^{-1}) = r(\zeta^{l-1})$ is also a unit of $\mathfrak{O}$. Consider the unit $\mu = \varepsilon/\bar{\varepsilon} \in \mathfrak{O}$. By (1.6) any conjugate of μ has the form

$$\sigma_k(\mu) = \frac{r(\zeta^k)}{r(\zeta^{(l-1)k})} = \frac{r(\zeta^k)}{r(\zeta^{-k})}.$$

Since $r(\zeta^k)$ and $r(\zeta^{-k})$ are complex conjugate, then $|\sigma_k(\mu)| = 1$ $(k = 1, \ldots, l-1)$. By Theorem 2 of Section 3, Chapter 2, μ is a root of 1, and then by Lemma 3,

$$\mu = \pm\zeta^a.$$

We shall now show that the plus sign always occurs on the right. For otherwise we would have

$$\varepsilon = -\zeta^a\bar{\varepsilon}.$$

Consider this equation as a congruence in the ring $\mathfrak{O}$ modulo $\lambda = 1 - \zeta$. Since $\zeta \equiv 1 \pmod{\lambda}$, all powers of ζ are congruent to 1 modulo λ, and we have

$$\varepsilon \equiv \bar{\varepsilon} \equiv a_0 + a_1 + \cdots + a_{l-2} = M \pmod{\lambda},$$

which means that $M \equiv -M \pmod{\lambda}$, or $2M \equiv 0 \pmod{\lambda}$. By Lemma 2

$$2M \equiv 0 \pmod{l}, \qquad M \equiv 0 \pmod{l}, \qquad M \equiv 0 \pmod{\lambda},$$

so that

$$\varepsilon \equiv 0 \pmod{\lambda},$$

which contradicts the fact that ε is a unit of the ring $\mathfrak{O}$. Thus

$$\varepsilon = \zeta^a\bar{\varepsilon}.$$

We now take an integer s so that $2s \equiv a \pmod{l}$. Then $\zeta^a = \zeta^{2s}$ and the equation $\varepsilon = \zeta^{2s}\bar{\varepsilon}$ can be written in the form

$$\frac{\varepsilon}{\zeta^s} = \zeta^s\bar{\varepsilon} = \frac{\bar{\varepsilon}}{\zeta^{-s}} = \overline{\left(\frac{\varepsilon}{\zeta^s}\right)}.$$

This shows that the unit $\eta = \varepsilon/\zeta^s$ is real. Hence we have represented ε as the product of ζ^s and the real unit η, and the lemma is proved.

Lemma 5. Let x, y, m, n be rational integers, $m \not\equiv n \pmod{l}$. Then $x + \zeta^m y$ and $x + \zeta^n y$ are relatively prime if and only if x and y are relatively prime and $x + y$ is not divisible by l.

Proof. If x and y have a common divisor $d > 1$, then $x + \zeta^m y$ and $x + \zeta^n y$ are both divisible by d. If $x + y$ is divisible by l, then $x + \zeta^m y$ and $x + \zeta^n y$ have a common divisor $1 - \zeta$ (which is not a unit). Indeed,

$$x + \zeta^m y = x + y + (\zeta^m - 1)y$$
$$= (x + y) - (1 - \zeta)\varepsilon_m y \equiv 0 \pmod{1 - \zeta}.$$

Thus the necessity of both conditions is proved. To prove their sufficiency we shall show that there exist numbers ξ_0 and η_0 in $\mathfrak{O}$ such that

$$(x + \zeta^m y)\xi_0 + (x + \zeta^n y)\eta_0 = 1.$$

Consider the set A of all numbers of the form

$$(x + \zeta^m y)\xi + (x + \zeta^n y)\eta,$$

where ξ and η independently run through all numbers of $\mathfrak{O}$. It is clear that if α and β belong to A, then any linear combination $\alpha\xi' + \beta\eta'$ with coefficients $\xi', \eta' \in \mathfrak{O}$ also belongs to A. We need to show that 1 belongs to A. From

$$(x + \zeta^m y) - (x + \zeta^n y) = \zeta^m(1 - \zeta^{n-m})y = \zeta^m \varepsilon_{n-m}(1 - \zeta)y,$$

$$(x + \zeta^m y)\zeta^n - (x + \zeta^n y)\zeta^m = -\zeta^m(1 - \zeta^{n-m})x = -\zeta^m \varepsilon_{n-m}(1 - \zeta)x,$$

we conclude that $(1 - \zeta)y$ and $(1 - \zeta)x$ belong to A (since $\zeta^m \varepsilon_{n-m}$ is a unit in the ring $\mathfrak{O}$). Since x and y are relatively prime, there exist rational integers a and b such that $ax + by = 1$, and therefore

$$(1 - \zeta)xa + (1 - \zeta)yb = 1 - \zeta \in A.$$

Further,

$$x + y = (x + \zeta^m y) + (1 - \zeta^m)y = (x + \zeta^m y) + (1 - \zeta)\varepsilon_m y,$$

and thus $x + y \in A$. Since l is divisible by $1 - \zeta$, then $l \in A$. But we are also assuming that $x + y$ and l are relatively prime. Hence for some rational integers u and v we have $(x + y)u + lv = 1$, so that $1 \in A$. Lemma 5 is proved.

1.3. *Fermat's Theorem in the Case of Unique Factorization*

Theorem 1. Let l be a prime integer and let ζ be a primitive lth root of 1. If decomposition into prime factors is unique in the order $\mathfrak{O} = Z[\zeta] = \{1, \zeta, \ldots, \zeta^{l-2}\}$ of the field $R(\zeta)$, then the equation

$$x^l + y^l = z^l$$

has no solution in integers x, y, z not divisible by l.

Proof. The prime 3 will play a special role in our proof, so we consider the case $l = 3$ separately. We shall show that not only the equation $x^3 + y^3 = z^3$, but also the congruence

$$x^3 + y^3 \equiv z^3 \pmod 9$$

has no solution in integers not divisible by 3. For assume that there is such a solution of this congruence. But from the congruence $x^3 + y^3 \equiv z^3 \pmod 3$ it easily follows (from the little Fermat theorem) that $x + y \equiv z \pmod 3$; that is, $z = x + y + 3u$, and hence

$$x^3 + y^3 \equiv (x + y + 3u)^3 \equiv x^3 + y^3 + 3x^2y + 3xy^2 \pmod 9,$$

and

$$0 \equiv x^2y + xy^2 = xy(x + y) \equiv xyz \pmod 3.$$

Thus one of the numbers x, y, z is divisible by 3, and our assertion is proved.

Now let $l \geqslant 5$. We prove the theorem by contradiction, assuming that for some rational integers, x, y, z, pairwise relatively prime and not divisible by l, we have $x^l + y^l = z^l$, which we also write in the form (1.2). Since $x + y \equiv x^l + y^l = z^l \not\equiv 0 \pmod l$, and x and y are relatively prime, then by Lemma 5, all the numbers $x + \zeta^k y$ $(k = 0, 1, \ldots, l - 1)$ are pairwise relatively prime. Then, as has already been shown in Section 1.1, from the unique factorization of the numbers appearing in (1.2) it follows that

$$x + \zeta y = \varepsilon\alpha^l, \tag{1.3}$$

$$x - \zeta z = \varepsilon_1 {\alpha_1}^l, \tag{1.3'}$$

where ε and ε_1 are units in the ring $\mathfrak{O}$. We have already remarked that (1.3) and (1.3′) lead to a contradiction. We now show that this contradiction even arises from the corresponding congruences modulo l in the ring $\mathfrak{O}$.

Let $\alpha = a_0 + a_1\zeta + \cdots + a_{l-2}\zeta^{l-2}$ with $a_0, \ldots, a_{l-2}$ rational integers. Then

$$\alpha^l \equiv {a_0}^l + {a_1}^l\zeta^l + \cdots + a_{l-2}^l\zeta^{l(l-2)} \equiv M \pmod l,$$

where $M = a_0 + a_1 + \cdots + a_{l-2}$. By Kummer's lemma the unit ε can be represented in the form $\varepsilon = \zeta^s\eta$, where η is a real unit. Hence from (1.3) we obtain the congruence

$$x + \zeta y \equiv \zeta^s\eta M = \zeta^s\xi \pmod l$$

with the real number $\xi \in \mathfrak{O}$. We may also write this congruence in the form

$$\zeta^{-s}(x + \zeta y) \equiv \xi \pmod l. \tag{1.8}$$

We now note that for any $\alpha \in \mathfrak{O}$ the complex conjugate $\bar{\alpha}$ also belongs to $\mathfrak{O}$. If we have the congruence $\alpha \equiv \beta \pmod l$, then $\alpha - \beta = l\gamma$, so that $\bar{\alpha} - \bar{\beta} = l\bar{\gamma}$

and hence $\bar{\alpha} \equiv \bar{\beta} \pmod{l}$. Passing now from the congruence (1.8) to its complex conjugate, we obtain

$$\zeta^s(x + \zeta^{-1}y) \equiv \bar{\xi} \pmod{l}. \tag{1.9}$$

But $\bar{\xi} = \xi$, and therefore from (1.8) and (1.9) it follows that

$$\zeta^{-s}(x + \zeta y) \equiv \zeta^s(x + \zeta^{-1}y) \pmod{l},$$

or

$$x\zeta^s + y\zeta^{s-1} - x\zeta^{-s} - y\zeta^{1-s} \equiv 0 \pmod{l}. \tag{1.10}$$

It is clear that a number of $\mathfrak{O}$, represented in the canonical form $a_0 + a_1\zeta + \cdots + a_{l-2}\zeta^{l-2}$, is divisible by l if and only if all coefficients $a_0, \ldots, a_{l-2}$ are divisible by l. If the exponents

$$s, \qquad s-1, \qquad -s, \qquad 1-s \tag{1.11}$$

are pairwise-noncongruent modulo l and also noncongruent to $l-1$, then the number on the left side of the congruence (1.10) is in canonical form and hence all its coefficients are divisible by l. Thus in this case $x \equiv 0 \pmod{l}$ and $y \equiv 0 \pmod{l}$, which is impossible, since x and y are relatively prime (and also not divisible by l).

Consider the case when the left side of (1.10) is not in canonical form, that is, when one of the integers (1.11) is congruent to $l-1$ modulo l, or two of them are congruent modulo l. One of the exponents (1.11) will be congruent to $l-1$ modulo l only in the following cases:

s	$s-1$	$-s$	$1-s$
$l-1$	$l-2$	1	2
0	$l-1$	0	1
1	0	$l-1$	0
2	1	$l-2$	$l-1$

We see that in each of these cases only one of the exponents is congruent to $l-1$ (since $l \geqslant 5$). To write the left side of (1.10) in canonical form, we must use the equation

$$\zeta^{l-1} = -1 - \zeta - \cdots - \zeta^{l-2}.$$

Substituting this expression in the left side of (1.10), we replace the term with exponent $l-1$ by a sum of the monomials $1, \zeta, \ldots, \zeta^{l-2}$ each with coefficient $\pm x$ or $\pm y$. Since the number of these terms is equal to $l-1 \geqslant 4$ (since $l \geqslant 5$), after we combine terms in which the exponent of ζ is the same, there will be at least one term in which the coefficient is $\pm x$ or $\pm y$. But this again would imply that $x \equiv 0 \pmod{l}$ or $y \equiv 0 \pmod{l}$, which is impossible, since we have assumed that x and y are not divisible by l.

We now need only consider the case when two of the exponents in (1.11) are congruent modulo l. The congruences $s \equiv s - 1 \pmod{l}$ and $-s \equiv 1 - s \pmod{l}$ are clearly impossible. If $s \equiv -s \pmod{l}$ or $s - 1 \equiv 1 - s \pmod{l}$, then we have $s \equiv 0 \pmod{l}$ or $s \equiv 1 \pmod{l}$, and we have again the cases considered above where $s - 1 \equiv l - 1 \pmod{l}$ or $-s \equiv l - 1 \pmod{l}$. In the remaining (equivalent) possibilities $s \equiv 1 - s \pmod{l}$ and $s - 1 \equiv -s \pmod{l}$, we have $s \equiv (l + 1)/2 \pmod{l}$. In this case the congruence (1.10) takes the form

$$(x - y)\zeta^{(l+1)/2} + (y - x)\zeta^{(l-1)/2} \equiv 0 \pmod{l}.$$

Since the left side of this congruence is in the canonical form [the exponents $(l + 1)/2$ and $(l - 1)/2$ are neither congruent to each other nor to $l - 1$], it follows that

$$x \equiv y \pmod{l}.$$

Analogously, we deduce from (1.3′) that

$$x \equiv -z \pmod{l}.$$

Then from the congruences $x + y \equiv x^l + y^l = z^l \equiv z \pmod{l}$ it follows that $2x \equiv -x \pmod{l}$ or $3x \equiv 0 \pmod{l}$. Since $l \neq 3$, $x \equiv 0 \pmod{l}$ and we again have a contradiction. This completes the proof of Theorem 1.

By using more subtle arguments involving the integers of the field $R(\zeta)$, Kummer showed that if the prime l satisfies the conditions of Theorem 1, then the second case of Fermat's theorem also holds for the prime l.

We shall generalize Theorem 1 to a wider class of exponents in Section 7.3. For this wider class of exponents we shall prove the second case of Fermat's theorem in Section 7.1 of Chapter 5.

We make some remarks about Theorem 1.

Remark 1. The main part of the proof of the theorem is the verification of the impossibility of certain congruences modulo l. Of course it does not follow from this that the congruence $x^l + y^l \equiv z^l \pmod{l}$ is impossible, since this congruence is equivalent to $x + y \equiv z \pmod{l}$, which always has solutions in integers not divisible by l. Moreover, it can be shown that, for example, when $l = 7$, the equation $x^l + y^l = z^l$, when considered as a congruence, has, for any modulus, solutions not divisible by 7.

Thus the proof of the unsolvability of (1.1) is achieved first by using unique factorization in the ring $Z[\zeta]$ to obtain Equations (1.3) and (1.3′), and then by applying the theory of congruences to these latter equations.

Remark 2. It is clear that the methods which we have applied in this section to the solution of Fermat's theorem can also be applied to analogous problems, by using other algebraic number fields instead of the field $R(\zeta)$ (Problem 2).

Remark 3. If we wish to apply the theorem to some particular prime l, we discover that this cannot be done, since we have no means for determining whether factorization into primes is unique for the field $R(\zeta)$.

Hence we come to the following two basic problems of number theory:

(1) In which algebraic number fields K is decomposition into prime factors unique?

(2) What are the arithmetic properties of those fields K in which decomposition into prime factors is not unique?

PROBLEMS

1. Show that the congruence $x^5 + y^5 \equiv z^5 \pmod{5^2}$ has no solution in rational integers x, y, z not divisible by 5.

Let ω be a primitive cube root of 1. Assume it known that decomposition into prime factors is unique in the field $R(\omega)$. Show that the equation $x^3 + y^3 = 5z^3$ has no solution in rational integers x, y, z not divisible by 3.

3. Let l be a prime number, ζ a primitive lth root of 1, x and y rational integers, and d the greatest common divisor of x and y. If $x + y \not\equiv 0 \pmod{l}$ set $\delta = d$, and if $x + y \equiv 0 \pmod{l}$ set $\delta = d(1 - \zeta)$. Show that δ is a common divisor of the numbers $x + \zeta^m y$ and $x + \zeta^n y$ which is divisible by all other common divisors of these numbers.

4. Show that in the order $\{1, \zeta, \dots, \zeta^{l-2}$ of the field $R(\zeta)$ a product $\alpha\beta$ is divisible by $1 - \zeta$ if and only if α or β is divisible by $1 - \zeta$.

5. Using the concept of congruence of integral polynomials (Section 1.1 of Chapter 1), show that

$$t^{l-1} + \cdots + t + 1 \equiv (t-1)^{l-1} \pmod{l}.$$

6. Show that the polynomial $t^{l-1} + \cdots + t + 1$ is irreducible over the field of rational numbers by considering congruence of integral polynomials modulo l^2.

2. Decomposition into Factors

2.1. *Prime Factors*

In Section 1 we saw how a problem of number theory can reduce to a question of decomposition into prime factors in some order of an algebraic number field. We shall see other such examples later. We now consider the general problem of decomposition into prime factors.

In order to speak of decomposition into primes, we must be dealing with a fixed ring $\mathfrak{O}$, the elements of which we are decomposing into factors. We formulate our problem in the general case where $\mathfrak{O}$ is any commutative ring without divisors of zero and possessing a unit element. In the future these conditions will be assumed without special mention.

Definition. An element π of the ring $\mathfrak{O}$, nonzero and not a unit, is called *prime* if it cannot be decomposed into factors $\pi = \alpha\beta$, neither of which is a unit in $\mathfrak{O}$.

Thus an element is prime if it is divisible only by units and associates.

In some rings there are no prime elements and hence not every element of a ring can be represented as a product of primes. For example, let $\mathfrak{O}$ be the ring of *all* algebraic integers. Any $\alpha \neq 0$, which is not a unit, has the factorization $\alpha = \sqrt{\alpha} \cdot \sqrt{\alpha}$, in which both factors lie in $\mathfrak{O}$ and are nonunits. Thus every nonunit of $\mathfrak{O}$ has nontrivial factorizations and there are no prime elements in $\mathfrak{O}$.

For examples of rings in which decomposition into prime factors is always possible, consider orders in algebraic number fields (it is these rings which will interest us most). We shall call prime elements in orders *prime numbers.*

Theorem 1. In any order $\mathfrak{O}$ of an algebraic number field K, every nonzero element which is not a unit can be represented as a product of prime numbers.

Proof. By Theorem 4 of Section 2 of Chapter 2, the units of $\mathfrak{O}$ are characterized by having norm ± 1. We prove the theorem by induction on the absolute value $|N(\alpha)|$ of the number $\alpha \in \mathfrak{O}$. If the number α is itself prime, there is nothing to prove. Otherwise $\alpha = \beta\gamma$, where β and γ are numbers of $\mathfrak{O}$ which are not units, so that

$$1 < |N(\beta)| < |N(\alpha)|, \qquad 1 < |N(\gamma)| < |N(\alpha)|.$$

By the induction assumption, β and γ are products of prime numbers of the ring $\mathfrak{O}$. But then since $\alpha = \beta\gamma$, the number α is also a product of prime numbers of the ring $\mathfrak{O}$. Hence Theorem 1 is proved.

2.2. *Uniqueness of Factorization*

We assume now that in the ring $\mathfrak{O}$ decomposition into prime factors is possible, and we turn to the question of the uniqueness of such factorizations.

Definition. We shall say that decomposition into prime factors in the ring $\mathfrak{O}$ is *unique* if for any two decompositions

$$\alpha = \pi_1 \cdots \pi_r, \quad \alpha = \pi_1{}' \cdots \pi_s{}'$$

the number of factors is always the same ($r = s$) and for suitable indexing of the factors, the prime elements π_i and π_i' are associate ($i = 1, \ldots, r$).

In the decomposition $\alpha = \pi_1 \cdots \pi_r$, associate prime elements can be made equal by multiplying by a suitable unit. We may then group equal factors into powers and obtain a factorization

$$\alpha = \varepsilon \pi_1^{k_1} \cdots \pi_m^{k_m},$$

in which the prime elements $\pi_1, \ldots, \pi_m$ are pairwise-nonassociate and ε is a unit of the ring $\mathfrak{O}$. In case factorization is unique, the prime elements $\pi_1, \ldots, \pi_m$ are determined up to associates and the exponents $k_1, \ldots, k_m$ are uniquely determined.

The classical example of a ring with unique factorization is the ring of rational integers. It is far from true that decomposition into prime factors is unique in all rings. Thus the result of Problem 1 shows that among orders in algebraic number fields, unique factorization can occur only for maximal orders.

Unique factorization for the ring Z of rational integers follows from the theorem on division with remainder, which asserts that for any a and $b \neq 0$ of Z there exist integers q and r, such that $a = bq + r$ and $r < |b|$. If in a ring $\mathfrak{O}$ there is an analog of division with remainder, then we can prove uniqueness of factorization in $\mathfrak{O}$ just as in Z.

Definition. We say that the ring $\mathfrak{O}$ has the division with remainder property if there is a function $\|\alpha\|$ on nonzero elements $\alpha \in \mathfrak{O}$ which takes nonnegative integral values and is such that the following conditions hold:

(1) If $\alpha \neq 0$ is divisible by β, then $\|\alpha\| \geqslant \|\beta\|$.

(2) For any elements α and $\beta \neq 0$ of $\mathfrak{O}$, there exist γ and ρ such that $\alpha = \beta\gamma + \rho$, where either $\rho = 0$, or $\|\rho\| < \|\beta\|$. The ring $\mathfrak{O}$ itself is then called *Euclidean.*

Consider the proof of unique factorization in the ring of rational integers. It uses, in addition to the general properties of rings, only the theorem on division with remainder. Therefore, by translating this proof, we obtain the following result.

Theorem 2. In every Euclidean ring, factorization into primes is unique.

Consider as an example the maximal order $\mathfrak{O}$ of the quadratic field $R(\sqrt{-1})$. We shall show that $\mathfrak{O}$ has the division with remainder property with $\|\alpha\| = N(\alpha)$. Let α and $\beta \neq 0$ be arbitrary numbers of $\mathfrak{O}$. Then

$$\frac{\alpha}{\beta} = u + v\sqrt{-1},$$

where u and v are rational numbers and we choose rational integers x and y so that

$$|u - x| \leqslant \tfrac{1}{2}, \qquad |v - y| \leqslant \tfrac{1}{2}.$$

If we now set $\gamma = x + y\sqrt{-1}$, $\rho = \alpha - \beta\gamma$, then from

$$N\left(\frac{\alpha}{\beta} - \gamma\right) = (u - x)^2 + (v - y)^2 \leqslant \frac{1}{4} + \frac{1}{4} < 1$$

we obtain

$$N(\rho) = N\left(\frac{\alpha}{\beta} - \gamma\right)N(\beta) < N(\beta),$$

and this proves our assertion.

From Theorem 2 we conclude that in the maximal order of the field $R(\sqrt{-1})$ factorization into primes is unique.

In the same fashion we can prove uniqueness of factorization for certain other rings (see Problems 3, 4, and 7). It must be noted that there are rings in which factorization is unique which are not Euclidean. An example is the maximal order of the field $R(\sqrt{-19})$. It follows from Problem 6 that this ring does not have the division with remainder property. But it will follow from Problem 11 of Section 7 that factorization is unique in this ring.

Consider the maximal order of the real quadratic field $R(\sqrt{d})$. We obtain a division algorithm with remainder by using the absolute value of the norm only when d is one of the following sixteen numbers:

2, 3, 5, 6, 7, 11, 13, 17, 19, 21, 29, 33, 37, 41, 57, 73.

2.3. *Examples of Nonunique Factorization*

It is not difficult to construct examples in which the maximal order of an algebraic number field will not have unique factorization. Consider, for example, the field $R(\sqrt{-5})$. As was shown in Section 7.2 of Chapter 2, the numbers of the maximal order of this field are of the form $\alpha = x + y\sqrt{-5}$, where x and y are rational integers, and then $N(\alpha) = x^2 + 5y^2$. In the ring $\mathfrak{O}$ the number 21 has the factorizations

$$(1) \quad 21 = 3 \cdot 7,$$

$$(2) \quad 21 = (1 + 2\sqrt{-5})(1 - 2\sqrt{-5}).$$

We claim that all terms on the right in (1) and (2) are prime. Suppose, for example, that $3 = \alpha\beta$, where α and β are nonunits. Then since $9 = N(\alpha\beta) = N(\alpha)N(\beta)$, we must have $N(\alpha) = 3$. But this is impossible since the equation $x^2 + 5y^2 = 3$ has no integral solutions. In precisely the same manner we could prove that the numbers 7, $1 + 2\sqrt{-5}$, $1 - 2\sqrt{-5}$ are also prime. Since the quantities

$$\frac{1 \pm 2\sqrt{-5}}{3}, \quad \frac{1 \pm 2\sqrt{-5}}{7}$$

are not contained in the ring $\mathfrak{O}$, the numbers 3 and 7 are not associated with $1 + 2\sqrt{-5}$ and $1 - 2\sqrt{-5}$. Hence we see that the ring $\mathfrak{O}$ contains numbers which allow essentially different decompositions into prime factors.

The example of nonunique factorization in the maximal order of the field $R(\sqrt{-5})$ is not an unusual exception. Many such examples are easily found (see Problems 10 and 11).

It might be thought that the phenomenon of nonunique factorization which we have discovered in algebraic number fields would make it impossible to construct a complete theory of the arithmetic of these fields, and that this would dash our hopes for deeper applications to the problems of number theory. But this not the case. In the middle of the last century, Kummer showed that, although the arithmetic of algebraic numbers was radically different from the arithmetic of the rational numbers, it could be developed in great depth, allowing strong applications to number-theoretic problems.

The basic idea of Kummer is that if the maximal order $\mathfrak{O}$ does not possess unique factorization, then the nonzero elements of $\mathfrak{O}$ can be mapped into some new set, in which multiplication is defined and in which factorization into primes is unique. If α is any nonzero element of $\mathfrak{O}$, then its image (α) under this mapping will factor uniquely into a product of primes, but these primes will lie not in the ring but in the new set. Unique factorization, in the sense of Kummer, is restored by virtue of the fact that some prime numbers (perhaps even all of them) are mapped onto nonprime elements of the new set, and therefore their images factor in a nontrivial fashion. Thus, in the example of the maximal order of the field $R(\sqrt{-5})$, there must exist objects $\mathfrak{p}_1$, $\mathfrak{p}_2$, $\mathfrak{p}_3$, $\mathfrak{p}_4$ such that

$$3 = \mathfrak{p}_1\mathfrak{p}_2, \quad 7 = \mathfrak{p}_3\mathfrak{p}_4, \quad 1 + 2\sqrt{-5} = \mathfrak{p}_1\mathfrak{p}_3, \quad 1 - 2\sqrt{-5} = \mathfrak{p}_2\mathfrak{p}_4$$

(in these equations we do not distinguish between numbers and the new objects which correspond to them). The decompositions (1) and (2) now reduce to the decompositions

$$21 = \mathfrak{p}_1\mathfrak{p}_2 \cdot \mathfrak{p}_3\mathfrak{p}_4 = \mathfrak{p}_1\mathfrak{p}_3 \cdot \mathfrak{p}_2\mathfrak{p}_4,$$

which differ only in the order of the factors.

Kummer himself called these new objects *ideal numbers*. Now they are called *divisors*. In Section 3 we give a systematic exposition of the theory of divisors.

PROBLEMS

1. Show that if in the order $\mathfrak{O}$ of the algebraic number field K decomposition into prime factors is unique, then $\mathfrak{O}$ is the maximal order of the field K. And show, in general, that if $\mathfrak{O}$ is a ring in which factorization into primes is unique, then $\mathfrak{O}$ is integrally closed in its field of fractions.

2. Show that if an element $\alpha \neq 0$ of a Euclidean ring is divisible by β, and α and β are not associate, then $\|\alpha\| > \|\beta\|$.

3. Let $\mathfrak{M}$ be a lattice in the complex plane, the points of which represent the numbers of the maximal order $\mathfrak{O}$ of an imaginary quadratic field. Show that we obtain an algorithm for division with remainder in $\mathfrak{O}$ by using the norm $N(\alpha)$ if and only if the translates of the unit disc (without boundary) by all vectors of the lattice $\mathfrak{M}$ completely cover the plane.

4. Show that in the maximal order of the imaginary quadratic field $R(\sqrt{d})$, an algorithm for division with remainder is obtained by using the norm if and only if d is one of the values $-1, -2, -3, -7, -11$.

5. Let $d < 0$ be square-free and not equal to $-1, -2, -3, -7, -11$. Show that the norm of any integer of $R(\sqrt{d})$, except 0 and $\pm$ 1, is greater than 3.

6. Show that, except for the five fields indicated in Problem 4, the maximal order of an imaginary quadratic number field is never a Euclidean ring.

(*Hint:* Carry out the proof by contradiction. Assume that there is a function $\|\alpha\|$ on the elements of the maximal order $\mathfrak{O}$ which satisfies the conditions given in Section 2.2. Among the numbers of $\mathfrak{O}$ which are not units, choose γ so that $\|\gamma\|$ is as small as possible. Then any $\alpha \in \mathfrak{O}$ will be congruent modulo γ to one of the numbers $0, 1, -1$.)

7. Show that there exists an algorithm for division with remainder in the maximal order of the field $R(\sqrt{2})$.

8. Show that in the maximal order of the field $R(\sqrt{-1})$ every odd rational prime p of the form $4k + 3$ remains prime, while every odd rational prime p of the form $4k + 1$ factors into $p = \pi\pi'$, where π and π' are nonassociate primes. Find the decomposition of the number 2 into prime factors.

9. Let $\mathfrak{O}$ be a ring with unique factorization. Show that for any two numbers α and β of $\mathfrak{O}$ (not both equal to zero), there is a common divisor δ which is divisible by all common divisors of α and β (δ is called the *greatest common divisor* of α and β).

10. Show that in the maximal order of the field $R(\sqrt{-6})$ the following are essentially different prime factorizations:

$$55 = 5 \cdot 11 = (7 + \sqrt{-6})(7 - \sqrt{-6}),$$

$$6 = 2 \cdot 3 = -(\sqrt{-6})^2.$$

11. Show that in the maximal order of the field $R(\sqrt{-23})$ the following are essentially different prime factorizations:

$$6 = 2 \cdot 3 = \frac{1 + \sqrt{-23}}{2} \frac{1 - \sqrt{-23}}{2},$$

$$27 = 3 \cdot 3 \cdot 3 = (2 + \sqrt{-23})(2 - \sqrt{-23}).$$

Find all possible factorizations of the number 8 in this ring.

3. Divisors

3.1. *An Axiomatic Description of Divisors*

We consider an arbitrary commutative ring $\mathfrak{O}$ (with unit element and without divisors of zero), and we shall try to clarify the idea mentioned in Section 2.3 of mapping the nonzero elements of the ring $\mathfrak{O}$ into some new domain, in which decomposition into prime factors is unique. Our theory must clearly consist of two parts: the construction of some set $\mathscr{D}$ of new objects in which decomposition into prime factors in unique, and the determination of the mapping of the nonzero elements of the ring $\mathfrak{O}$ into the set $\mathscr{D}$. We start with the first part. In order to be able to speak of decomposition into prime factors in $\mathscr{D}$, we must have an operation of multiplication defined in $\mathscr{D}$; that is, we must associate to each pair of elements of $\mathscr{D}$ a third element, their product. We shall require that this operation be associative and commutative. A set with such an operation is called a *commutative semigroup*. We shall further require that the set $\mathscr{D}$ contain a unit element, that is, an element $\mathfrak{e}$ such that $\mathfrak{e}\mathfrak{a} = \mathfrak{a}$ for all $\mathfrak{a} \in \mathscr{D}$.

In a commutative semigroup $\mathscr{D}$ with unit $\mathfrak{e}$ we may speak of divisibility of elements; an element $\mathfrak{a} \in \mathscr{D}$ is divisible by $\mathfrak{b} \in \mathscr{D}$ if there exists a $\mathfrak{c} \in \mathscr{D}$ such that $\mathfrak{a} = \mathfrak{b}\mathfrak{c}$ (we also say that $\mathfrak{b}$ divides $\mathfrak{a}$).

An element $\mathfrak{p} \in \mathscr{D}$, distinct from $\mathfrak{e}$, is called prime if it is divisible only by itself and by the unit $\mathfrak{e}$. We further say that the semigroup $\mathscr{D}$ has unique factorization into prime elements if every element $\mathfrak{a} \in \mathscr{D}$ can be represented as a product of prime elements

$$\mathfrak{a} = \mathfrak{p}_1 \cdots \mathfrak{p}_r \qquad (r \geqslant 0),$$

and this decomposition is unique up to the order of the factors (for $r = 0$ this product is set equal to $\mathfrak{e}$). Thus uniqueness of factorization implies that $\mathfrak{e}$ is the only invertible element (divisor of $\mathfrak{e}$) in the semigroup $\mathscr{D}$. It is clear that a semigroup with unique factorization is completely determined by its set of prime elements (essentially by the cardinality of this set). As a simple example of a semigroup with unique factorization we may take the set of all natural numbers under the operation of multiplication.

In a semigroup with unique factorization, any two elements have a greatest common divisor (a common divisor which is divisible by all common divisors of the two elements), and also a least common multiple. Two elements of $\mathscr{D}$ are called *relatively prime* if their greatest common divisor is equal to $\mathfrak{e}$. We note some elementary properties of divisibility in $\mathscr{D}$: If a product $\mathfrak{a}\mathfrak{b}$ is divisible by $\mathfrak{c}$ and $\mathfrak{a}$ is relatively prime to $\mathfrak{c}$, then $\mathfrak{b}$ is divisible by $\mathfrak{c}$; if $\mathfrak{c}$ is divisible by the relatively prime elements $\mathfrak{a}$ and $\mathfrak{b}$, then $\mathfrak{c}$ is divisible by their product $\mathfrak{a}\mathfrak{b}$;

if a product $\mathfrak{ab}$ is divisible by a prime element $\mathfrak{p}$, then at least one of the factors is divisible by $\mathfrak{p}$.

We now pass to the second part of the theory, the conditions which must be satisfied by the mapping from the ring $\mathfrak{O}$ to the semigroup $\mathscr{D}$.

Let $\mathfrak{O}^*$ denote the set of all nonzero elements of the ring $\mathfrak{O}$. Since we have assumed that $\mathfrak{O}$ does not have divisors of zero, the set $\mathfrak{O}^*$ is a semigroup under the operation of multiplication.

Suppose that we have a mapping of the semigroup $\mathfrak{O}^*$ into the semigroup $\mathscr{D}$ which has unique factorization. We denote the image of an element $\alpha \in \mathfrak{O}^*$ by (α). It is clear that we can use the semigroup $\mathscr{D}$ to study the multiplicative structure of the ring $\mathfrak{O}$ only if under the mapping $\alpha \to (\alpha)$, the product of two elements in $\mathfrak{O}^*$ is mapped onto the product of their images in $\mathscr{D}$, that is, only if $(\alpha\beta) = (\alpha)(\beta)$ for all α and β in $\mathfrak{O}^*$. Hence we must assume that the mapping $\alpha \to (\alpha)$ is a homomorphism of the semigroup $\mathfrak{O}^*$ into the semigroup $\mathscr{D}$. If α is divisible by β in the ring $\mathfrak{O}$, it will then follow that (α) is divisible by (β) in the semigroup $\mathscr{D}$. In order that divisibility in $\mathfrak{O}$ should closely correspond to divisibility in $\mathscr{D}$, we shall also demand the converse: If (α) is divisible by (β) in $\mathscr{D}$, then α is divisible by β in $\mathfrak{O}$.

We shall also say that the element $\alpha \neq 0$ of $\mathfrak{O}$ is divisible by the element $\mathfrak{a} \in \mathscr{D}$, and shall write $\mathfrak{a}|\alpha$, if (α) is divisible by $\mathfrak{a}$ in the semigroup $\mathscr{D}$. We shall suppose 0 to be divisible by all elements of $\mathscr{D}$.

If $\alpha \in \mathfrak{O}^*$, the set of all elements of $\mathfrak{O}$ which are divisible by α is closed under addition and subtraction. It is natural to assume that this property is preserved for divisors $\mathfrak{a}$ of the semigroup $\mathscr{D}$.

Our last requirement is that $\mathscr{D}$ not contain any "unnecessary" elements. By this we shall mean that distinct elements of $\mathscr{D}$ must not divide precisely the same elements of $\mathfrak{O}^*$.

We thus give the following definition.

Definition. By a theory of divisors for the ring $\mathfrak{O}$ we shall mean the giving of some semigroup $\mathscr{D}$ with unique factorization, along with a homomorphism $\alpha \to (\alpha)$ of the semigroup $\mathfrak{O}^*$ into $\mathscr{D}$, satisfying the following conditions:

(1) An element $\alpha \in \mathfrak{O}^*$ is divisible by $\beta \in \mathfrak{O}^*$ in the ring $\mathfrak{O}$ if and only if (α) is divisible by (β) in the semigroup $\mathscr{D}$.

(2) If α and β of $\mathfrak{O}$ are divisible by $\mathfrak{a} \in \mathscr{D}$, then $\alpha \pm \beta$ are also divisible by $\mathfrak{a}$.

(3) If $\mathfrak{a}$ and $\mathfrak{b}$ are two elements of $\mathscr{D}$ and the set of all elements $\alpha \in \mathfrak{O}$ which are divisible by $\mathfrak{a}$ coincides with the set of all elements $\beta \in \mathfrak{O}$ which are divisible by $\mathfrak{b}$, then $\mathfrak{a} = \mathfrak{b}$.

The elements of the semigroup $\mathscr{D}$ are called *divisors* of the ring $\mathfrak{O}$, and divisors of the form (α), $\alpha \in \mathfrak{O}^*$, are called *principal divisors*. The unit element $\mathfrak{e}$ of the semigroup is called the *unit divisor*.

Condition (1) in the definition of a theory of divisors clearly implies the following assertion: The equality $(\alpha) = (\beta)$ holds if and only if α and β are associate in the ring $\mathfrak{O}$. In particular, units ε of the ring $\mathfrak{O}$ are characterized by $(\varepsilon) = \mathfrak{e}$.

We shall denote a theory of divisors for the ring $\mathfrak{O}$ by $\mathfrak{O}^* \to \mathscr{D}$.

Our definition of a theory of divisors only fixed what we shall mean by such a theory. It does not at all guarantee the existence or uniqueness of the homomorphism $\mathfrak{O} \to \mathscr{D}$.

In the next section we consider the question of the uniqueness of a theory of divisors, assuming that one exists, and in Section 3.3 we indicate an important necessary (but not sufficient) condition for existence.

The existence of a theory of divisors for maximal orders in algebraic number fields will be proved in Section 5 (Theorem 3 implies that such a theory does not exist for nonmaximal orders).

3.2. *Uniqueness*

Theorem 1. If a ring $\mathfrak{O}$ has a theory of divisors, then it has only one. More precisely, if we have two homomorphisms $\mathfrak{O}^* \to \mathscr{D}$ and $\mathfrak{O}^* \to \mathscr{D}'$, satisfying all requirements of the definition, then there is an isomorphism $\mathscr{D} \approx \mathscr{D}'$ under which the principal divisors in $\mathscr{D}$ and $\mathscr{D}'$ which correspond to a given element $\alpha \in \mathfrak{O}^*$ are identified.

Proof. Let $\mathfrak{O}^* \to \mathscr{D}$ and $\mathfrak{O}^* \to \mathscr{D}'$ be two theories of divisors for the ring $\mathfrak{O}$. Let $\mathfrak{p} \in \mathscr{D}$ and $\mathfrak{p}' \in \mathscr{D}'$ be prime divisors. Denote by $\bar{\mathfrak{p}}$ and $\bar{\mathfrak{p}}'$ the sets of elements of the ring which are divisible by $\mathfrak{p}$ and by $\mathfrak{p}'$ (with respect to the theory $\mathfrak{O}^* \to \mathscr{D}$ for $\mathfrak{p}$, and with respect to $\mathfrak{O}^* \to \mathscr{D}'$ for $\mathfrak{p}'$). We now show that for any prime divisor $\mathfrak{p}' \in \mathscr{D}'$ there is a prime divisor $\mathfrak{p} \in \mathscr{D}$ such that $\bar{\mathfrak{p}} \subset \bar{\mathfrak{p}}'$. Assume that this is not the case, that is, that $\bar{\mathfrak{p}} \not\subset \bar{\mathfrak{p}}'$ for all prime divisors $\mathfrak{p} \in \mathscr{D}$. From condition (3) it easily follows that any divisor must divide a nonzero element of the ring $\mathfrak{O}$. Choose in $\mathfrak{O}$ an element $\beta \neq 0$ which is divisible by $\mathfrak{p}'$, and decompose the divisor $(\beta) \in \mathscr{D}$ into prime factors:

$$(\beta) = \mathfrak{p}_1^{k_1} \cdots \mathfrak{p}_r^{k_r}$$

($\mathfrak{p}_1, \ldots, \mathfrak{p}_r$ are prime divisors of the semigroup $\mathscr{D}$). Since we have assumed that $\bar{\mathfrak{p}}_i \not\subset \bar{\mathfrak{p}}'$, then for each $i = 1, \ldots, r$ there is an element $\gamma_i \in \mathfrak{O}$ which is divisible by $\mathfrak{p}_i$ but not divisible by $\mathfrak{p}'$. The product $\gamma = \gamma_1^{k_1} \cdots \gamma_r^{k_r}$ is divisible by $\mathfrak{p}_1^{k_1} \cdots \mathfrak{p}_r^{k_r}$, and this means, by condition (1), that γ is divisible by β in the ring $\mathfrak{O}$. But then γ must be divisible by $\mathfrak{p}'$. Thus we have a contradiction, since the product $\gamma_1^{k_1} \cdots \gamma_r^{k_r}$ cannot be divisible by $\mathfrak{p}'$, since $\mathfrak{p}'$ is prime and does not divide any of the γ_i.

Hence for any prime divisor $\mathfrak{p}' \in \mathscr{D}'$ there is a prime divisor $\mathfrak{p} \in \mathscr{D}$ such that

$\bar{\mathfrak{p}} \subset \bar{\mathfrak{p}}'$. By symmetry, there is a prime divisor $\mathfrak{q}' \in \mathscr{D}'$ for which $\bar{\mathfrak{q}}' \subset \bar{\mathfrak{p}}$. We shall show that $\mathfrak{q}' = \mathfrak{p}'$, and hence $\bar{\mathfrak{q}}' = \bar{\mathfrak{p}} = \bar{\mathfrak{p}}'$. Indeed, by condition (3) there is an element ξ in $\mathfrak{O}$ which is divisible by $\mathfrak{q}'$ and not divisible by $\mathfrak{q}'\mathfrak{p}'$. If we assume that $\mathfrak{q}' \neq \mathfrak{p}'$, then the element ξ will not be divisible by $\mathfrak{p}'$, which is impossible since $\bar{\mathfrak{q}}' \subset \bar{\mathfrak{p}}'$.

Since (for given $\mathfrak{p}' \in \mathscr{D}'$) there is one and only one prime divisor $\mathfrak{p} \in \mathscr{D}$ such that $\bar{\mathfrak{p}} = \bar{\mathfrak{p}}'$ [condition (3)], we obtain a one-to-one correspondence $\mathfrak{p} \leftrightarrow \mathfrak{p}'$ between prime divisors of $\mathscr{D}$ and prime divisors of $\mathscr{D}'$. This correspondence can clearly be extended (in a unique manner) to an isomorphism $\mathscr{D} \approx \mathscr{D}'$. Namely, if $\mathfrak{p}_1 \leftrightarrow \mathfrak{p}_1'$, ..., $\mathfrak{p}_r \leftrightarrow \mathfrak{p}_r'$, then

$$\mathfrak{p}_1^{k_1} \cdots \mathfrak{p}_r^{k_r} \leftrightarrow \mathfrak{p}_1'^{k_1} \cdots \mathfrak{p}_r'^{k_r}.$$

We now need only show that under this isomorphism, the divisors $(\alpha) \in \mathscr{D}$ and $(\alpha)' \in \mathscr{D}'$ (for given $\alpha \in \mathfrak{O}^*$) correspond to one another. Let $\mathfrak{p} \in \mathscr{D}$ and $\mathfrak{p}' \in \mathscr{D}'$ be corresponding prime divisors, and assume that they occur in the factorizations of (α) and $(\alpha)'$ with exponents k and l, respectively. From condition (3) it follows that there is an element $\pi \in \mathfrak{O}$ which is divisible by $\mathfrak{p}$ and not divisible by $\mathfrak{p}^2$. Since $\bar{\mathfrak{p}} = \bar{\mathfrak{p}}'$, the element π is also divisible by $\mathfrak{p}'$. The principal divisor (π) hence has the form $(\pi) = \mathfrak{p}\mathfrak{b}$, where $\mathfrak{b}$ is not divisible by $\mathfrak{p}$. Now choose in $\mathfrak{O}$ an element ω which is divisible by $\mathfrak{b}^k$ and not divisible by $\mathfrak{b}^k\mathfrak{p}$. Since $\mathfrak{p}$ does not divide $\mathfrak{b}^k$, then ω is not divisible by $\mathfrak{p}$ or by $\mathfrak{p}'$. Consider the product $\alpha\omega$. Since α is divisible by $\mathfrak{p}^k$, and ω is divisible by $\mathfrak{b}^k$, then $\alpha\omega$ is divisible by $\mathfrak{p}^k\mathfrak{b}^k = (\pi^k)$, and by condition (1) we have $\alpha\omega = \pi^k\eta$, $\eta \in \mathfrak{O}$. But $\mathfrak{p}'|\pi$, and hence $\alpha\omega$ is divisible by $\mathfrak{p}'^k$, and since $\mathfrak{p}' \nmid \omega$, then $\mathfrak{p}'^k|\alpha$. This means that in the factorization of the divisor $(\alpha)' \in \mathscr{D}'$, the prime divisor $\mathfrak{p}'$ occurs with exponent not less than k; that is, $l \geqslant k$. But by symmetry also $k \geqslant l$, and thus $k = l$.

We have thus shown that if $(\alpha) = \mathfrak{p}_1^{k_1} \cdots \mathfrak{p}_r^{k_r}$ and $\mathfrak{p}_1 \leftrightarrow \mathfrak{p}_1', \ldots, \mathfrak{p}_r \leftrightarrow \mathfrak{p}_r'$, then $(\alpha)' = \mathfrak{p}_1'^{k_1} \cdots \mathfrak{p}_r'^{k_r}$, and this means that under the above isomorphism $\mathscr{D} \approx \mathscr{D}'$, the principal divisors $(\alpha) \in \mathscr{D}$ and $(\alpha)' \in \mathscr{D}'$ correspond to each other.

If the ring $\mathfrak{O}$ has unique factorization, then we can easily construct a theory of divisors $\mathfrak{O}^* \to \mathscr{D}$, and in this theory all divisors will be principal. Indeed, break up the set of all nonzero elements of $\mathfrak{O}$ into classes of associate elements, and consider the set $\mathscr{D}$ of all such classes. For $\alpha \in \mathfrak{O}^*$, denote by (α) the class of elements associate with α. It is easily seen that under the operation of multiplication $(\alpha)(\beta) = (\alpha\beta)$, the set $\mathscr{D}$ becomes a semigroup with unique factorization, and that the mapping $\alpha \to (\alpha)$, $\alpha \in \mathfrak{O}^*$, defines a theory of divisors for the ring $\mathfrak{O}$. [The prime divisors in this theory are just the divisors of the form (π), where π is a prime element of $\mathfrak{O}$.] By Theorem 1 any theory of divisors for this ring coincides with the one just constructed.

Assume now the converse, that we have for some ring $\mathfrak{O}$ a theory of divisors $\mathfrak{O}^* \to \mathscr{D}$, in which all divisors of $\mathscr{D}$ are principal. We now show that an element

$\pi \neq 0$ of the ring $\mathfrak{O}$ will be prime if and only if the corresponding divisor (π) is prime. Indeed, if $(\pi) = \mathfrak{p}$ is a prime divisor and γ divides π in the ring $\mathfrak{O}$, then the divisor (γ) must divide $\mathfrak{p}$ (in the semigroup $\mathscr{D}$) and then, since $\mathfrak{p}$ is prime, either (γ) is equal to $\mathfrak{p}$ or to the unit divisor $\mathfrak{e}$. In the first case γ is associate with π, and in the second case γ is a unit in $\mathfrak{O}$, and this means that π is a prime element of the ring $\mathfrak{O}$. Now let (α) be neither prime nor the unit divisor. Then (α) is divisible by some prime divisor $\mathfrak{p} = (\pi)$, and α is divisible by the prime element π and is not associate with it. Hence α cannot be prime.

We have shown that if every divisor is principal, then the element π is prime if and only if the divisor (π) is prime.

Let α be any element of $\mathfrak{O}^*$. If we have the factorization

$$(\alpha) = \mathfrak{p}_1 \cdots \mathfrak{p}_r \tag{3.1}$$

in $\mathscr{D}$ (the prime divisors $\mathfrak{p}_i$ are not necessarily distinct), and if $\mathfrak{p}_1 = (\pi_1), \ldots, \mathfrak{p}_r = (\pi_r)$, then in the ring $\mathfrak{O}$ we have the factorization

$$\alpha = \varepsilon\pi_1 \cdots \pi_r, \tag{3.2}$$

where ε is a unit of the ring $\mathfrak{O}$. Since any factorization of the form (3.2) induces a factorization of the form (3.1), we must have unique factorization in the ring $\mathfrak{O}$.

We have obtained the following result.

Theorem 2. In order that the ring $\mathfrak{O}$ have unique factorization, it is necessary and sufficient that $\mathfrak{O}$ have a theory of divisors $\mathfrak{O}^* \to \mathscr{D}$ in which every divisor of $\mathscr{D}$ is principal.

3.3. *Divisors and Integrally Closed Rings*

We have already noted that not every ring has a theory of divisors. The existence of a homomorphism $\alpha \to (\alpha)$ which satisfies the requirements of a theory of divisors imposes strong restrictions on a ring. One such restriction is given in the following theorem.

Theorem 3. If the ring $\mathfrak{O}$ has a theory of divisors, then $\mathfrak{O}$ is integrally closed in its quotient field K.

Proof. Assume that the element ξ of K satisfies an equation

$$\xi^n + a_1\xi^{n-1} + \cdots + a_{n-1}\xi + a_n = 0 \qquad (a_1, \ldots, a_n \in \mathfrak{O}),$$

but does not belong to $\mathfrak{O}$. We represent it in the form $\xi = \alpha/\beta$, where $\alpha \in \mathfrak{O}$ and $\beta \in \mathfrak{O}$, and decompose the principal divisors (α) and (β) into prime factors. Since α is not divisible by β in the ring $\mathfrak{O}$ (we have assumed that $\xi \notin \mathfrak{O}$), (α) is not divisible by (β) [by condition (1)]. This means that some prime divisor $\mathfrak{p}$ occurs in (β) with greater exponent than in (α). Let $\mathfrak{p}$ occur in (α) with exponent $k \geqslant 0$. Since (β) is divisible by $\mathfrak{p}^{k+1}$, we deduce by condition (2) that the right side of

$$\alpha^n = -a_1\beta\alpha^{n-1} - \cdots - a_n\beta^n$$

is divisible by $\mathfrak{p}^{kn+1}$. But $\mathfrak{p}$ occurs in $(\alpha^n) = (\alpha)^n$ with exponent kn, and thus α^n is not divisible by $\mathfrak{p}^{kn+1}$. This contradiction shows that $\xi \in \mathfrak{O}$, and Theorem 3 is proved.

Another necessary condition for the existence of a theory of divisors is given in Problem 1.

Since the only orders in algebraic number fields which are integrally closed are the maximal orders, only maximal orders can possibly have a theory of divisors.

3.4. *The Theory of Divisors and Valuations*

We now turn to the question of the practical construction of theories of divisors. We first assume that a theory of divisors $\mathfrak{O}^* \to \mathscr{D}$ exists for the ring $\mathfrak{O}$, and then proceed to clarify how this theory could be constructed.

Taking an arbitrary prime divisor $\mathfrak{p}$, we can construct with it a function $v_\mathfrak{p}(\alpha)$, which is similar to the p-adic valuation with respect to a prime p which was constructed in Chapter 1. Namely, for any $\alpha \neq 0$ of $\mathfrak{O}$, by $v_\mathfrak{p}(\alpha)$ we denote the power to which $\mathfrak{p}$ enters in the factorization of the principal divisor (α) into prime factors. Clearly, $v_\mathfrak{p}(\alpha)$ is characterized by

$$\mathfrak{p}^{v_\mathfrak{p}(\alpha)} | \alpha \qquad \text{and} \qquad \mathfrak{p}^{v_\mathfrak{p}(\alpha)+1} \nmid \alpha.$$

Since zero is divisible by arbitrarily large powers of $\mathfrak{p}$, it is natural to set $v_\mathfrak{p}(0) = \infty$.

It easily follows from the definition that

$$v_\mathfrak{p}(\alpha\beta) = v_\mathfrak{p}(\alpha) + v_\mathfrak{p}(\beta), \tag{3.3}$$

$$v_\mathfrak{p}(\alpha + \beta) \geqslant \min(v_\mathfrak{p}(\alpha), v_\mathfrak{p}(\beta)) \tag{3.4}$$

[for the proof of (3.4) we must use condition (2)].

The function $v_\mathfrak{p}(\alpha)$ can be extended to the quotient field K of the ring $\mathfrak{O}$ in such a way that (3.3) and (3.4) still hold. For any $\xi = \alpha/\beta \in K$ $(\alpha, \beta \in \mathfrak{O})$ set

$$v_\mathfrak{p}(\xi) = v_\mathfrak{p}(\alpha) - v_\mathfrak{p}(\beta).$$

The value of $v_\mathfrak{p}(\xi)$ clearly does not depend on the choice of the representation of ξ in the form $\xi = \alpha/\beta$. It is now easily verified that (3.3) and (3.4) still hold for the extended function $v_\mathfrak{p}$.

We shall now see what values the function $v_\mathfrak{p}(\alpha)$ takes as α ranges through K.

Since the divisors $\mathfrak{p}$ and $\mathfrak{p}^2$ are distinct, by condition (3) there is an element $\gamma \in \mathfrak{O}$ which is divisible by $\mathfrak{p}$ but not by $\mathfrak{p}^2$. For this element we have $v_\mathfrak{p}(\gamma) = 1$. But then $v_\mathfrak{p}(\gamma^k) = k$ for any integer k. Hence the function $v_\mathfrak{p}(\alpha)$ takes on all rational integral values.

Definition. Let K be any field. A function $v(\alpha)$, defined for $\alpha \in K$, is called a *valuation* of the field K, if it satisfies the following conditions:

(1) $v(\alpha)$ takes on all rational integral values as α ranges through the nonzero elements of K; $v(0) = \infty$;

(2) $v(\alpha\beta) = v(\alpha) + v(\beta)$;

(3) $v(\alpha + \beta) \geqslant \min(v(\alpha), v(\beta))$.

We can now say that every prime divisor $\mathfrak{p}$ of the ring $\mathfrak{O}$ determines a valuation $v_\mathfrak{p}(\alpha)$ of the quotient field K. It is easily seen that distinct prime divisors determine different valuations. For if $\mathfrak{p}$ and $\mathfrak{q}$ are distinct prime divisors, then by condition (3) the ring $\mathfrak{O}$ contains an element γ divisible by $\mathfrak{p}$ and not divisible by $\mathfrak{q}$. But then $v_\mathfrak{p}(\gamma) \geqslant 1$ and $v_\mathfrak{q}(\gamma) = 0$, and hence $v_\mathfrak{p} \neq v_\mathfrak{q}$.

All valuations of the field K of the form $v_\mathfrak{p}$ clearly satisfy

$$v_\mathfrak{p}(\alpha) \geqslant 0 \qquad \text{for all} \qquad \alpha \in \mathfrak{O}. \tag{3.5}$$

In terms of valuations we can give a simple expression for the factorization of the principal divisor (α), $\alpha \in \mathfrak{O}^*$. The prime divisors which enter into this decomposition are characterized by $v_{\mathfrak{p}_i}(\alpha) > 0$. Then we have

$$(\alpha) = \prod_i \mathfrak{p}_i^{v_{\mathfrak{p}_i}(\alpha)}, \tag{3.6}$$

where $\mathfrak{p}_i$ runs through all prime divisors for which $v_{\mathfrak{p}_i}(\alpha) > 0$.

We thus see that the semigroup $\mathscr{D}$ of divisors and the homomorphism $\mathfrak{O} \to \mathscr{D}$ are completely determined by the set of all valuations $v_\mathfrak{p}$ of the field K which correspond to prime divisors $\mathfrak{p}$. For the set of all divisors and the operation of multiplication are determined as soon as the set of all prime divisors is known (each divisor is a product of prime divisors with nonnegative exponents, and when divisors are multiplied the corresponding exponents are added). But the prime divisors are in one-to-one correspondence with the valuations $v_\mathfrak{p}$. Finally, the homomorphism $\mathfrak{O}^* \to \mathscr{D}$ is determined from (3.6).

This means that the concept of a valuation can be used as a foundation for the construction of a theory of divisors. We shall proceed to develop this idea.

We must first answer the following important question: How can we characterize the set $\mathfrak{R}$ of valuations of the field K which must be taken to construct a theory of divisors for the ring $\mathfrak{O}$?

The product (3.6) can contain only a finite number of factors. Hence, for any fixed $\alpha \in \mathfrak{O}^*$, the condition $v_\mathfrak{p}(\alpha) = 0$ must hold for almost all valuations of the set $\mathfrak{R}$ (by "almost all" is meant "for all but a finite number").

From (3.5) we see that for all $\nu \in \mathfrak{R}$ we must have $\nu(\alpha) \geqslant 0$ if $\alpha \in \mathfrak{O}$. Conversely, assume that for some $\xi \neq 0$ of K we have $\nu(\xi) \geqslant 0$ for all $\nu \in \mathfrak{R}$. If we represent ξ in the form $\xi = \alpha/\beta$ ($\alpha, \beta \in \mathfrak{O}$), then we have $\nu(\alpha) \geqslant \nu(\beta)$ for all $\nu \in \mathfrak{R}$. But this means that the principal divisor (α) is divisible by the principal divisor (β). Condition (1) now implies that α is divisible by β in the ring $\mathfrak{O}$; that is, $\xi \in \mathfrak{O}$. We hence have a second necessary condition: The set of valuations $\mathfrak{R}$ must be such that $\nu(\alpha) \geqslant 0$ for all $\nu \in \mathfrak{R}$ if and only if α is an element of the ring $\mathfrak{O}$.

We now give another necessary condition for $\mathfrak{R}$. Take any finite set of valuations $\nu_1, \ldots, \nu_m$ of $\mathfrak{R}$ which correspond to the prime divisors $\mathfrak{p}_1, \ldots, \mathfrak{p}_m$. If $k_1, \ldots, k_m$ are fixed nonnegative integers, we consider the divisor $\mathfrak{a} = \mathfrak{p}_1^{k_1} \cdots \mathfrak{p}_m^{k_m}$. From condition (3) it follows that the ring $\mathfrak{O}$ contains an element α_i which is divisible by $\mathfrak{a}_i = \mathfrak{a}\mathfrak{p}_1 \cdots \mathfrak{p}_{i-1}\, \mathfrak{p}_{i+1} \cdots \mathfrak{p}_m$ and not divisible by $\mathfrak{a}_i\mathfrak{p}_i$ ($1 \leqslant i \leqslant m$). Consider the sum

$$\alpha = \alpha_1 + \cdots + \alpha_m .$$

Using condition (2), we easily find that α is divisible by $\mathfrak{p}_i^{k_i}$ and not divisible by $\mathfrak{p}_i^{k_i+1}$. Hence the set $\mathfrak{R}$ must satisfy the following condition: For any valuations $\nu_1, \ldots, \nu_m$ of $\mathfrak{R}$ and for any nonnegative integers $k_1, \ldots, k_m$ there exists an element α in the ring $\mathfrak{O}$ for which $\nu_i(\alpha) = k_i$ ($1 \leqslant i \leqslant m$).

The necessary conditions which we have found on $\mathfrak{R}$ will now be shown sufficient to construct a theory of divisors for the ring $\mathfrak{O}$. To prove this, take a semigroup $\mathscr{D}$ with unique factorization, with the prime elements in one-to-one correspondence with the valuations of the set $\mathfrak{R}$. The valuation $\nu \in \mathfrak{R}$, which corresponds to the prime element $\mathfrak{p} \in \mathscr{D}$, will again be denoted by $\nu_\mathfrak{p}$. By the first and second conditions, for any $\alpha \in \mathfrak{O}^*$ the product (3.6) will make sense [the exponents $\nu_{\mathfrak{p}_i}(\alpha)$ are nonnegative and almost all of them are zero]. Since $\nu(\alpha\beta) = \nu(\alpha) + \nu(\beta)$ the mapping $\alpha \to (\alpha)$ will be a homomorphism from $\mathfrak{O}^*$ to $\mathscr{D}$. From the second condition it easily follows that α is divisible by β in the ring $\mathfrak{O}$ if and only if $\nu(\alpha) \geqslant \nu(\beta)$ for all $\nu \in \mathfrak{R}$. Hence condition (1) is satisfied. Condition (2) is implied by the inequality $\nu(\alpha \pm \beta) \geqslant \min(\nu(\alpha), \nu(\beta))$. If $\mathfrak{a}$ and $\mathfrak{b}$ are two different elements of $\mathscr{D}$, then some prime element $\mathfrak{p}$ occurs in their factorizations with different exponents, say, k and l. Let $k < l$. By the third of the above conditions there is in $\mathfrak{O}$ an element α which is divisible by $\mathfrak{a}$ and for which $\nu_p(\alpha) = k$. But then α is not divisible by $\mathfrak{b}$. This shows that condition (3) is also satisfied. Hence the homomorphism $\mathfrak{O}^* \to \mathscr{D}$ gives us a theory of divisors for the ring $\mathfrak{O}$.

We formulate the results which we have obtained.

Theorem 4. Let $\mathfrak{O}$ be a ring with quotient field K, and let $\mathfrak{R}$ be a set of valuations of K. In order that the valuations of $\mathfrak{R}$ induce a theory of divisors on $\mathfrak{O}$ it is necessary and sufficient that the following conditions hold:

(1) For any $\alpha \neq 0$ of $\mathfrak{D}$, $v(\alpha) = 0$ for almost all valuations $v \in \mathfrak{R}$.

(2) An element α of K belongs to $\mathfrak{D}$ if and only if $v(\alpha) \geqslant 0$ for all $v \in \mathfrak{R}$.

(3) For any finite set of distinct valuations $v_1, \ldots, v_m$ of $\mathfrak{R}$ and for any set of nonnegative integers $k_1, \ldots, k_m$, there is an element $\alpha \in \mathfrak{D}$ for which

$$v_1(\alpha) = k_1, \ldots, v_m(\alpha) = k_m.$$

Hence the construction of a theory of divisors for the ring $\mathfrak{D}$ is reduced to the construction of the corresponding set $\mathfrak{R}$ of valuations of its quotient field K.

We shall not enter here into the determination of those integrally closed rings for which a theory of divisors can be constructed (see, for example, the book "Modern Algebra" by van der Waerden, Section 105, Ungar, New York, 1950). In Section 4 we show that if $\mathfrak{o}$ is a ring with quotient field k and $\mathfrak{D}$ is the integral closure of $\mathfrak{o}$ in some finite extension field K of k, then if $\mathfrak{o}$ has a theory of divisors, so does $\mathfrak{D}$. Since the ring Z has a theory of divisors (being a ring with unique factorization), then we will have proved that there is a theory of divisors for the maximal order of any algebraic number field.

The set of valuations $\mathfrak{R}$ of the field K which must be taken to construct a theory of divisors depends essentially on the ring $\mathfrak{D}$, and, in general, this set will not consist of all valuations of the field K (Problem 6). It can even happen (Problem 7) that condition (1) of Theorem 4 will not hold for the set of all valuations of the field K. We now show, however, that in the case of the ring Z of rational integers, we must take all valuations of the field R of rational numbers (we shall see in the future that this is also true for maximal orders of algebraic number fields).

To each prime number $p \in Z$ (that is, the prime divisor of the ring Z) there corresponds the valuation v_p of the field R, the value of which is given for the nonzero rational number

$$x = p^m \frac{a}{b} \tag{3.7}$$

(a and b integers not divisible by p) by

$$v_p(x) = m. \tag{3.8}$$

This valuation v_p is called the p-adic valuation of the field R [it is clear that the valuation (3.8) coincides with the p-adic valuation of the field R of p-adic numbers; see Section 3.2 of Chapter 1].

Theorem 5. Every valuation of the field of rational numbers is of the form v_p for some prime p.

Proof. Let ν be any valuation of the field R. Since

$$\nu(1 + \cdots + 1) \geqslant \min(\nu(1), \ldots, \nu(1)) = 0,$$

then $\nu(n) \geqslant 0$ for all natural numbers n. If $\nu(p) = 0$ for all primes p, then we would also have $\nu(a) = 0$ for all $a \neq 0$ of R, which is impossible by condition (1) of the definition of a valuation. Hence for some prime p we must have $\nu(p) = e > 0$. Suppose that for the prime $q \neq p$ we also had $\nu(q) > 0$. Then from the equation $pu + qv = 1$ (u and v rational integers) we obtain

$$0 = \nu(pu + qv) \geqslant \min(\nu(pu), \nu(qv)) \geqslant \min(\nu(p), \nu(q)) > 0.$$

This contradiction shows that $\nu(q) = 0$ for all primes q except p. Hence $\nu(a) = 0$ for all integers a not divisible by p. For the rational number (3.7) we thus have

$$\nu(x) = m\nu(p) + \nu(a) - \nu(b) = me = e\nu_p(x).$$

Since the valuation ν must take on all integral values, $e = 1$ and hence $\nu = \nu_p$. Theorem 5 is proved.

Note that Theorem 5 could easily have been deduced from Theorem 3 of Section 4, Chapter 1, the second part of the proof of which we have essentially repeated above.

We conclude this section by considering another special case.

Assume that for some ring $\mathfrak{O}$ we have a theory of divisors $\mathfrak{O}^* \to \mathcal{D}$ with only finitely many prime divisors $\mathfrak{p}_1, \ldots, \mathfrak{p}_m$. Denote by $\nu_1, \ldots, \nu_m$ the corresponding valuations of the quotient field K. By condition (3) of Theorem 4 for any divisor $\mathfrak{a} = \mathfrak{p}_1^{k_1} \cdots \mathfrak{p}_m^{k_m}$ ($k_i \geqslant 0$), there is an element $\alpha \in \mathfrak{O}$ for which $\nu_1(\alpha) = k_1, \ldots, \nu_m(\alpha) = k_m$. But this means that the divisor $\mathfrak{a}$ coincides with the principal divisor (α). Thus all divisors of $\mathcal{D}$ are principal, and the ring $\mathfrak{O}$ has unique factorization (Theorem 2). If $\mathfrak{p}_1 = (\pi_1), \ldots, \mathfrak{p}_m = (\pi_m)$, then the elements $\pi_1, \ldots, \pi_m$ constitute a complete set of pairwise-nonassociate prime elements of the ring $\mathfrak{O}$ and every element $\alpha \in \mathfrak{O}^*$ has a unique representation in the form

$$\alpha = \varepsilon\pi_1^{k_1} \cdots \pi_m^{k_m},$$

where ε is a unit of the ring $\mathfrak{O}$. The prime elements $\pi_1, \ldots, \pi_m$ are clearly characterized by

$$\nu_i(\pi_i) = 1, \quad \nu_j(\pi_i) = 0 \qquad \text{for } j \neq i.$$

We have obtained the following result.

Theorem 6. If for some ring $\mathfrak{O}$ we have a theory of divisors with only a finite number of prime divisors, then $\mathfrak{O}$ has unique factorization into primes.

PROBLEMS

1. If the ring $\mathfrak{O}$ has a theory of divisors, show that every element of $\mathfrak{O}$ has only a finite number of (nonassociate) factors.

2. Show that in any theory of divisors any divisor is the greatest common divisor of two principal divisors.

3. Let $K = k(x)$ be the field of rational functions over a field k and let φ be some irreducible polynomial in $k[x]$. Every nonzero rational function u of K can be written in the form $u = \varphi^k(f/g)$, where f and g are polynomials in $k[x]$ which are not divisible by φ. Show that the function ν_φ, given by $\nu_\varphi(u) = k$, is a valuation of the field K.

4. If f and g are nonzero polynomials of $k[x]$ of degrees n and m, and $u = f/g \in k(x)$, set $\nu^*(u) = m - n$. Show that the function ν^* is a valuation of the field $K = k(x)$.

5. Let ν be a valuation of the field $k(x)$ such that $\nu(a) = 0$ for all nonzero a in k. Show that ν is either of the form ν_φ (for some irreducible polynomial $\varphi \in k[x]$) or else $\nu = \nu^*$ (see Problems 3 and 4).

6. If we set $\mathfrak{O} = k[x]$, determine the set $\mathfrak{R}$ of valuations of the field $K = k(x)$ which satisfies the conditions of Theorem 4. Further, determine the set $\mathfrak{R}$ for the ring $\mathfrak{O}' = k[1/x]$.

7. Let $K = k(x, y)$ be a field of rational functions in two variables over the field k. For any natural number n set $x_n = x/y^n$. A nonzero rational function $u = u(x, y) \in K$ can be represented in the form

$$u = u(x_n y^n, y) = y^k \frac{f(x_n, y)}{g(x_n, y)},$$

where the polynomials f and g are not divisible by y. If we set $\nu_n(u) = k$, show that the function ν_n is a valuation of the field K. Further, show that the valuations ν_n $(n \geqslant 1)$ are all distinct, and that for all of them $\nu_n(x) > 0$.

8. Formulate and prove an "Eisenstein irreducibility criterion" for polynomials over any ring $\mathfrak{D}$ with a theory of divisors.

9. Show that if a ring $\mathfrak{O}$ has a theory of divisors, then its quotient field K has algebraic extensions of all degrees.

10. Let f be a nonzero polynomial in the ring $\mathfrak{O} = k[x, y]$ of polynomials in two variables over the field k. Denote by $(\tilde{\nu} f)$ the smallest degree of a monomial which appears in f with nonzero coefficient. Show that the function $\tilde{\nu}$ can be extended to a valuation of the field of rational functions $k(x, y)$. Denote by $\mathfrak{R}$ the set of all valuations of the field $k(x, y)$ which correspond to irreducible polynomials of the ring $\mathfrak{O}$, and let $\mathfrak{R}_1$ be obtained from $\mathfrak{R}$ by adjoining $\tilde{\nu}$. Which of the conditions of Theorem 4 are not fulfilled for the ring $\mathfrak{O}$ and the set $\mathfrak{R}_1$ of valuations?

4. Valuations

Theorem 4 of Section 3 reduces the problem of constructing a theory of divisors for an integrally closed ring $\mathfrak{O}$ to the determination of a set of valuations of the quotient field K which satisfy the conditions of the theorem. We turn to a systematic study of valuations.

4.1. *Simple Properties of Valuations*

From the definition of a valuation of a field K (Section 3.4) we immediately obtain

$$v(\pm 1) = 0$$

$$v(-\alpha) = v(\alpha)$$

$$v\left(\frac{\alpha}{\beta}\right) = v(\alpha) - v(\beta), \qquad (\beta \neq 0),$$

$$v(\alpha^n) = nv(\alpha) \qquad n \in Z,$$

$$v(\alpha_1 + \cdots + \alpha_n) \geqslant \min(v(\alpha_1), \ldots, v(\alpha_n)).$$

Now assume that $v(\alpha) \neq v(\beta)$. If $v(\alpha) > v(\beta)$, then $v(\alpha + \beta) \geqslant v(\beta)$. On the other hand, since $\beta = (\alpha + \beta) - \alpha$, $v(\beta) \geqslant \min(v(\alpha + \beta), v(\alpha))$, so that $v(\beta) \geqslant v(\alpha + \beta)$. Hence

$$v(\alpha + \beta) = \min(v(\alpha), v(\beta)) \qquad \text{if } v(\alpha) \neq v(\beta). \tag{4.1}$$

By induction we obtain

$$v(\alpha_1 + \cdots + \alpha_n) = \min(v(\alpha_1), \ldots, v(\alpha_n)),$$

provided that the minimum value of $v(\alpha_1), \ldots, v(\alpha_n)$ occurs only once.

Definition. Let v be a valuation of the field K. The subring $\mathfrak{O}_v$ of the field K consisting of all elements $\alpha \in K$ for which $v(\alpha) \geqslant 0$ is called the *ring of the valuation v*. The elements of $\mathfrak{O}_v$ are called *integral* with respect to the valuation v.

It is clear that all three conditions of Theorem 4 of Section 3 are fulfilled for the ring $\mathfrak{O}_v$ and the set $\mathfrak{R}$ consisting of the single valuation v. Hence the ring $\mathfrak{O}_v$ has a theory of divisors with a single prime divisor. From Theorems 3 and 6 of Section 3 we obtain the following results.

Theorem 1. The ring $\mathfrak{O}_v$ of the valuation v of the field K is integrally closed in K.

Theorem 2. The ring $\mathfrak{O}_v$ has (up to associates) a single prime element π, and any element $\alpha \neq 0$ of $\mathfrak{O}_v$ has a unique (for fixed π) representation in the form $\alpha \doteq \varepsilon\pi^m$, where ε is a unit in $\mathfrak{O}_v$ $(m \geqslant 0)$.

The prime element π is clearly characterized by $v(\pi) = 1$.

In the ring $\mathfrak{O}_v$, as in any ring, we can consider congruences with respect to the elements of $\mathfrak{O}_v$ (see the Supplement, Section 4.1). Since congruences modulo associate elements are equivalent, the ring of residue classes modulo the prime element π does not depend on the choice of π but is completely determined by the ring $\mathfrak{O}_v$. We denote this ring of residue classes by Σ_v and will now show that it is a field. For if $\alpha \in \mathfrak{O}_v$ and $\alpha \not\equiv 0 \pmod{\pi}$, then $v(\alpha) = 0$ and this means that α is a unit in $\mathfrak{O}_v$. Then α has an inverse ξ and $\alpha\xi \equiv 1 \pmod{\pi}$, since $\alpha\xi = 1$.

The field Σ_v is called the *residue class field* of the valuation v.

4.2. *Independence of Valuations*

Let the ring $\mathfrak{O}$ have a theory of divisors $\mathfrak{O}^* \to \mathscr{D}$, and let $\mathfrak{p}_1, \ldots, \mathfrak{p}_m$ be distinct prime divisors of $\mathscr{D}$. By Theorem 4 of Section 3 there correspond to these prime divisors valuations $v_1, \ldots, v_m$ of the quotient field K, and these valuations are independent in the sense that there exist elements in K on which they take on any given set of values $k_1, \ldots, k_m$. For if we set $k_i' = \max(0, k_i)$, $i = 1, \ldots, m$, and $k_i'' = \min(0, k_i)$, then by condition (3) of Theorem 4 of Section 3 we can find elements α and β in $\mathfrak{O}$ for which $v_i(\alpha) = k_i'$ and $v_i(\beta) = -k_i''$, and then for the quotient $\xi = \alpha/\beta$ we will have $v_i(\xi) = k_i$ $(1 \leqslant i \leqslant m)$.

We now show that this property of independence does not depend on the fact that the valuations v_i corresponded to prime divisors in some theory of divisors, but is true for any finite set of valuations.

Theorem 3. If $v_1, \ldots, v_m$ are distinct valuations of the field K, then for any rational integers $k_1, \ldots, k_m$ there exists an element $\xi \in K$ for which

$$v_1(\xi) = k_1, \ldots, v_m(\xi) = k_m.$$

Let $\mathfrak{O}_1, \ldots, \mathfrak{O}_m$ denote the rings of the valuations $v_1, \ldots, v_m$ and set $\mathfrak{O} = \bigcap_{i=1}^m \mathfrak{O}_i$. Conditions (1) and (2) of Theorem 4 of Section 3 are clearly fulfilled for the ring $\mathfrak{O}$ and the set $\mathfrak{R}$ consisting of the valuation s$v_1, \ldots, v_m$. From the formulation of Theorem 3 we see that condition (3) also holds, and hence the ring $\mathfrak{O}$ has a theory of divisors with a finite number of prime divisors. Thus Theorem 3 implies that for any finite set of valuations $v_1, \ldots, v_m$ of the field K we have a theory of divisors for the ring $\mathfrak{O} = \bigcap_{i=1}^m \mathfrak{O}$. From Theorem 6 of Section 3 we then derive the following result.

Corollary. If $v_1, \ldots, v_m$ are distinct valuations of the field K with rings $\mathfrak{O}_1, \ldots, \mathfrak{O}_m$, then the intersection $\mathfrak{O} = \bigcap_{i=0}^m \mathfrak{O}_i$ is a ring with unique factorization. Further, each nonzero element of $\mathfrak{O}$ has a unique representation in the form $\alpha = \varepsilon\pi_1^{k_1} \cdots \pi_m^{k_m}$, where ε is a unit in $\mathfrak{O}$, and $\pi_1, \ldots, \pi_m$ are fixed prime elements of $\mathfrak{O}$ characterized by

$$v_i(\pi_i) = 1, \quad v_j(\pi_i) = 0 \qquad (j \neq i).$$

Proof of Theorem 3. For $m = 1$, the assertion of the theorem is contained in the definition of a valuation. Assume that $m \geqslant 2$ and that the case of $m - 1$ valuations has already been proved. We show that then there do not exist rational integers $c_1, \ldots, c_m$, not all zero, for which

$$c_1 v_1(\xi) + \cdots + c_m v_m(\xi) = 0 \tag{4.2}$$

for all nonzero $\xi \in K$. Assume the contrary, that is, that (4.2) does hold. Among the coefficients at least two must be nonzero and have the same sign [otherwise there would be only two nonzero coefficients, say, c_1 and c_2, with $c_1 > 0$ and $c_2 < 0$, and then from $c_1 v_1(\xi) + c_2 v_2(\xi) = 0$ we would obtain $v_1(\xi) = e v_2(\xi)$ with positive e, and this is possible only for $e = 1$ and $v_1 = v_2$]. Changing the numeration, if necessary, we can write (4.2) in the form

$$v_1(\xi) = a_2 v_2(\xi) + \cdots + a_m v_m(\xi), \tag{4.3}$$

where at least one of the rational coefficients a_i is negative. By the induction hypothesis there exist elements β and β' in the field K such that

$$v_i(\beta) = 0, \quad v_i(\beta') = 1 \quad \text{if} \quad a_i \geqslant 0,$$

$$v_i(\beta) = 1, \quad v_i(\beta') = 0 \quad \text{if} \quad a_i < 0,$$

for all $i = 2, \ldots, m$. Then

$$v_1(\beta) < 0, \quad v_1(\beta') \geqslant 0. \tag{4.4}$$

Consider the sum $\beta + \beta'$. Since one of the numbers $v_i(\beta)$ and $v_i(\beta')$ $(i = 2, \ldots, m)$ equals 0 and the other equals 1, then $v_i(\beta + \beta') = \min(v_i(\beta), v_i(\beta')) = 0$. From the relation (4.3) we therefore obtain $v_1(\beta + \beta') = 0$. On the other hand, from (4.4) we obtain

$$v_i(\beta + \beta') = \min(v_i(\beta), v_1(\beta')) < 0.$$

This contradiction proves that (4.2) is impossible.

Let $\mathfrak{O}$ now denote the intersection of the rings of the valuations $v_2, \ldots, v_m$, and let E denote the group of units of this ring. Let $\pi_2, \ldots, \pi_m$ denote prime elements of $\mathfrak{O}$, numbered so that $v_i(\pi_i) = 1$ $(i = 2, \ldots, m)$ (recall that for the case of $m - 1$ valuations, Theorem 3, and hence also its corollary, are assumed). We now show that the valuation v_1 cannot be identically zero on the group E. Any element $\xi \in K^*$ can be written in the form

$$\xi = \varepsilon \pi_2^{k_2} \cdots \pi_m^{k_m}, \tag{4.5}$$

where $\varepsilon \in E$, $k_i = v_i(\xi)$ $(2 \leqslant i \leqslant m)$. If $v_1(\varepsilon) = 0$ for all $\varepsilon \in E$, then from (4.5) we would obtain

$$v_1(\xi) = k_2 v_1(\pi_2) + \cdots + k_m v_1(\pi_m),$$

which can also be written in the form

$$v_1(\xi) = a_2 v_2(\xi) + \cdots + a_m v_m(\xi),$$

where the rational integers $a_i = v_i(\pi_i)$ do not depend on ξ, and this contradicts the fact that a relation of the form (4.2) is impossible. Hence the group E contains elements on which the valuation v_1 is nonzero.

Choose an element γ in the group E on which the valuation v_1 takes its smallest positive value l. It is clear that all values of v_1 on E are divisible by l. We shall show that $l = 1$. If all the values $a_2 = v_1(\pi_2), \ldots, a_m = v_1(\pi_m)$ were divisible by l, then we would deduce from (4.5) that all values $v_1(\xi)$ of the valuation v_1 are divisible by l, which is possible only for $l = 1$. Consider the case where not all a_i are divisible by l, say, a_2 is not divisible by l. Consider the element

$$\alpha = \pi_2(\pi_3 \cdots \pi_m)^l \gamma^s,$$

where s is an integer chosen so that the number

$$a_2 + l(a_3 + \cdots + a_m) + sl = l_1$$

satisfies the inequality $0 < l_1 < l$. It is clear that $v_1(\alpha) = l_1$ and $v_i(\alpha) > 0$ for $i = 2, \ldots, m$. Set

$$\varepsilon = \gamma + \alpha.$$

Since $v_i(\varepsilon) = \min(v_i(\gamma), v_i(\alpha)) = 0$ for all $i = 2, \ldots, m$, then $\varepsilon \in E$. But at the same time,

$$v_1(\varepsilon) = \min(l, l_1) = l_1,$$

which contradicts the choice of γ. This shows that the case when not all a_i are divisible by l is impossible, and hence $l = 1$.

We may now assume that the prime elements π_i $(2 \leqslant i \leqslant m)$ of the ring $\mathfrak{O}$ are chosen so that $v_i(\pi_i) = a_i = 0$. For each π_i can be replaced by $\pi_i' = \pi_i \gamma^{-a_i}$, for which $v_1(\pi_i') = a_i - a_i v_i(\gamma) = 0$.

Setting $\pi_1 = \gamma$, we have obtained a system of elements $\pi_1, \pi_2, \ldots, \pi_m$, for which $v_i(\pi_i) = 1$ and $v_j(\pi_i) = 0$ for $j \neq i$. If now $k_1, \ldots, k_m$ are any integers, for the element $\xi = \pi_1^{k_1} \cdots \pi_m^{k_m}$ we have

$$v_1(\xi) = k_1, \ldots, v_m(\xi) = k_m.$$

Theorem 3 is proved.

From Theorem 3 we easily deduce the following stronger result.

Theorem 4 (Approximation Theorem). If $v_1, \ldots, v_m$ are distinct valuations of the field K, then for any elements $\xi_1, \ldots, \xi_m$ of K and any integer N, there exists an element $\xi \in K$ for which

$$v_1(\xi - \xi_1) \geqslant N, \ldots, v_m(\xi - \xi_m) \geqslant N.$$

Proof. Choose in K elements $\alpha_1, \ldots, \alpha_m$ such that $v_i(\alpha_i) = -1$, $v_j(\alpha_i) = 1$ $(j \neq i)$ and set

$$\xi = \frac{\alpha_1{}^k}{1 + \alpha_1{}^k}\, \xi_1 + \cdots + \frac{\alpha_m{}^k}{1 + \alpha_m{}^k}\, \xi_m .$$

Since $v_j(\alpha_i{}^k) \neq 0 = v_j(1)$ for all natural numbers k, then by (4.1) the value of $v_j(1 + \alpha_i{}^k)$ equals 0 for $i \neq j$ and equals $-k$ for $i = j$, so that

$$v_j\left(\frac{\alpha_i^k}{1 + \alpha_i^k}\right) = k \qquad \text{for} \qquad i \neq j \quad \text{and} \quad v_j\left(\frac{-1}{1 + \alpha_j^k}\right) = k.$$

Hence

$$v_j(\xi - \xi_j) \geqslant \min_i \, (k + v_j(\xi_i)).$$

It is clear that ξ will satisfy the theorem provided

$$k \geqslant N - \min_{i,j} v_j(\xi_i).$$

4.3. *Extension of Valuations*

Let k be a field and K a finite extension of k. If v is some valuation of the field K, then by restricting v to the field k we obtain a function which clearly satisfies conditions (2) and (3) in the definition of a valuation (Section 3.4). The first condition may not be satisfied; that is, the values of v on the elements of k may not exhaust the group Z. But v cannot be identically zero on k. If this were the case, then the field k would be contained in the ring of the valuation v, and since that ring is integrally closed (Theorem 1), then K would also be contained in it, and this is impossible. Thus $v(a)$, $a \in k^*$, takes on both negative and positive values [if $v(a) < 0$, then $v(a^{-1}) > 0$].

Let p denote any element of k on which v takes its least positive value $v(p) = e$. Then for any $a \in k^*$ the value $v(a) = m$ is divisible by e. For if $m = es + r$, $0 \leqslant r < e$, then $v(ap^{-s}) = m - se = r$, so by the minimality of e we have $r = 0$. Now setting

$$v_0(a) = \frac{v(a)}{e}, \qquad (a \in k^*), \quad v_0(0) = \infty, \tag{4.6}$$

we obtain on k a function v_0, which takes all integral values and which consequently is a valuation of the field k.

Definition. Let K be a finite extension of the field k. If the valuation v_0 of the field k is related to the valuation v of the field K by (4.6), then we say that v_0 is induced on k by the valuation v, and v is an extension of v_0 to the field K. The uniquely determined integer e which appears in (4.6) is called the *ramification index* of v with respect to v_0 (or with respect to the subfield k).

Note that in this definition when $e > 1$, the term *extension of a valuation* does not correspond to the usual concept of the extension of functions to larger domains of definition.

It follows from the above that every valuation v of K is induced by a unique valuation v_0 of k. The converse assertion is also valid, that is, for any valuation v_0 of k there exists an extension to K (which is, in general, not unique). The proof of this fact is fairly difficult, and we shall give it in the next section. First, we consider some properties of extensions of a given v_0, assuming that such extensions exist.

Let $k \subset K \subset K'$ be a tower of finite extensions, and let v_0, v, v' be valuations of the fields k, K, K'. It is clear that if v is an extension of v_0 with ramification index e, and v' an extension of v with ramification index e', then v' is an extension of v_0 to the field K', and the ramification index of v' with respect to v_0 is equal to ee'. It is also easy to see that if v and v_0 are induced by the valuation v', then v is an extension of v_0.

Lemma 1. If K is a finite extension of the field k of degree n, then any valuation v_0 of the field k has at most n extensions to the field K.

Proof. Let $v_1, \dots, v_m$ be distinct extensions of v_0 to the field K. By Theorem 3 we can find elements $\xi_1, \dots, \xi_m$ for which $v_i(\xi_i) = 0$ and $v_j(\xi_i) = 1$ for $j \neq i$. We shall show that these elements are linearly independent over k. Consider the linear combination

$$\gamma = a_1\xi_1 + \cdots + a_m\xi_m$$

with coefficients a_j of k, not all zero. Set $k = \min(v_0(a_1), \dots, v_0(a_m))$, and let i_0 be such that $v_0(a_{i_0}) = k$. Denoting by e the ramification index of v_{i_0} with respect to k, we have

$$v_{i_0}(a_{i_0}\xi_{i_0}) = ev_0(a_{i_0}) + v_{i_0}(\xi_{i_0}) = ek,$$

$$v_{i_0}(a_j\xi_j) = ev_0(a_j) + v_{i_0}(\xi_j) \geqslant ek + 1 \qquad (j \neq i_0),$$

and therefore

$$v_{i_0}(\gamma) = \min(v_{i_0}(a_1\xi_1), \dots, v_{i_0}(a_m\xi_m)) = ek,$$

so that $\gamma \neq 0$, which proves our assertion. From the linear independence of the elements $\xi_1, \dots, \xi_m$ over the field k, it follows that $m \leqslant (K:k)$, and this means that the number of extensions v_i is not greater than n. Lemma 1 is proved.

Assume now that $v_1, \ldots, v_m$ are all extensions of a fixed valuation v_0 of a field k to a finite extension K. Let $\mathfrak{o}$ denote the ring of the valuation v_0 and $\mathfrak{O}$ its integral closure in the field K, and let $\mathfrak{O}_1, \ldots, \mathfrak{O}_m$ denote the rings of the valuations $v_1, \ldots, v_m$. Since $\mathfrak{o} \subset \mathfrak{O}_i$, and the ring $\mathfrak{O}_i$ is integrally closed in K, then $\mathfrak{O} \subset \mathfrak{O}_i$ for $i = 1, \ldots, m$, and hence

$$\mathfrak{O} \subset \bigcap_{i=1}^{m} \mathfrak{O}_i .$$

Later we shall see that equality actually occurs here. If this is so, then by the corollary of Theorem 3, $\mathfrak{O}$ has unique factorization with a finite number of nonassociate prime elements. Since the nonassociate prime elements $\pi_1, \ldots, \pi_m$ of the ring $\mathfrak{O}$ are in one-to-one correspondence with the valuations $v_1, \ldots, v_m$, we obtain a method for constructing valuations of K which are extensions of the valuation v_0.

So assume that we know that the ring $\mathfrak{O}$, the integral closure of the ring of the valuation v_0 in the field K, has unique factorization with a finite number of nonassociate prime elements. From Theorem 6 of Section 3, this assumption is equivalent to the existence in $\mathfrak{O}$ of a theory of divisors with a finite set of prime divisors $\mathfrak{p}_1, \ldots, \mathfrak{p}_m$. We shall show that then the valuation v_0 has precisely m extensions to the field K, these being the valuations $v_1, \ldots, v_m$ of the field K which correspond to the prime divisors $\mathfrak{p}_1, \ldots, \mathfrak{p}_m$.

Let p be any prime element of the ring $\mathfrak{o}$ of the valuation v_0 [that is, any element of k such that $v_0(p) = 1$], and let $\pi_1, \ldots, \pi_m$ be a complete set of nonassociate prime elements of the ring $\mathfrak{O}$ [numbered so that $v_i(\pi_i) = 1$]. Since $\mathfrak{o} \subset \mathfrak{O}$, the element p has a factorization

$$p = \varepsilon \pi_1{}^{e_1} \cdots \pi_m{}^{e_m} \tag{4.7}$$

in the ring $\mathfrak{O}$, with nonnegative exponents e_i (ε a unit in $\mathfrak{O}$). Now if a is any element of k^* and $v_0(a) = s$, that is, $a = p^s u$, where u is a unit in $\mathfrak{o}$, then in $\mathfrak{O}$ we have

$$v_i(a) = e_i s = e_i v_0(a). \tag{4.8}$$

If $e_i = 0$, then v_i would be identically zero on k^*, and we saw at the start of this section that this is impossible. Hence $e_i > 0$. Formula (4.8) now implies that each of the valuations v_i $(i = 1, \ldots, m)$ is an extension of v_0 to the field K. We also obtain that e_i, the ramification index of v_i with respect to v_0, is given by (4.7)

Assume now that v is an extension of the valuation v_0 to the field K. Since $\mathfrak{o}$ is contained in the ring of the valuation v, so is its integral closure $\mathfrak{O}$, that is, $v(a) \geqslant 0$ for all $\alpha \in \mathfrak{O}$, and this means that $v(\varepsilon) = 0$ for all units $\varepsilon \in \mathfrak{O}$. If v were distinct from $v_1, \ldots, v_m$, then by Theorem 3 there would be a unit ε of the ring $\mathfrak{O}$ such that $v(\varepsilon) \neq 0$. Hence v must be one of the v_i.

Hence every extension of the valuation v_0 to the field K is one of the valuations $v_1, \ldots, v_m$. By condition (2) of Theorem 4 of Section 3 we also obtain that the integral closure $\mathfrak{O}$ of the ring $\mathfrak{o}$ in the field K consists of all elements $\alpha \in K$ for which $v_i(\alpha) \geqslant 0$ for all extensions v_i. If we again denote the ring of the valuation v_i by $\mathfrak{O}_i$, we can state this last result as

$$\mathfrak{O} = \bigcap_{i=1}^{m} \mathfrak{O}_i. \tag{4.9}$$

We have shown that to guarantee the existence of extensions of the valuation v_0 to the field K and to give a complete description of all extensions, it suffices to verify that the ring $\mathfrak{O}$ has unique factorization (with a finite number of nonassociate prime elements).

4.4 *Existence of Extensions*

Let, as before, k be a field with a valuation v_0, $\mathfrak{o}$ the ring of the valuation v_0, and p a prime element of the ring $\mathfrak{o}$. We denote the residue class field of the valuation v_0 by Σ_0. For each element $a \in \mathfrak{o}$ we denote the corresponding residue class modulo p by $\bar{a}$. We then have $\bar{a} = \bar{b}$ in the field Σ_0 if and only if $a \equiv b \pmod{p}$ in the ring $\mathfrak{o}$.

Now let K be a finite extension of k and let $\mathfrak{O}$ be the integral closure of $\mathfrak{o}$ in K.

Lemma 2. If the number of elements of the residue class field Σ_0 of the valuation v_0 is not less than the degree of the extension K/k (in particular, if the field Σ_0 is infinite), then the ring $\mathfrak{O}$ is Euclidean and hence has unique factorization. The ring $\mathfrak{O}$ then contains only a finite number of pairwise-nonassociate prime elements.

Proof. We define $\alpha \in K^*$ a function $\|\alpha\|$, by setting

$$\|\alpha\| = 2^{v_0(N_{K/k}\alpha)}.$$

It is clear that this function satisfies $\|\alpha\beta\| = \|\alpha\| \cdot \|\beta\|$ $(\alpha, \beta \in K^*)$. If $\alpha \in \mathfrak{O}^*$, then $\|\alpha\|$ is clearly a natural number. We must show that for any pair of elements α and $\beta \neq 0$ of $\mathfrak{O}$ there exist $\xi, \rho \in \mathfrak{O}$, such that

$$\alpha = \beta\xi + \rho, \tag{4.10}$$

where ρ is either zero or else $\|\rho\| < \|\beta\|$.

If α is divisible by β in the ring $\mathfrak{O}$, that is, $\alpha = \beta\gamma$, where $\gamma \in \mathfrak{O}$, then (4.10) holds with $\xi = \gamma$ and $\rho = 0$. Assume that α is not divisible by β, that is, that the element $\gamma = \alpha\beta^{-1}$ does not belong to $\mathfrak{O}$. Let $f(t) = t^n + c_1 t^{n-1} + \cdots + c_n$ $(c_i \in k)$ be the characteristic polynomial of the element γ with respect to the extension K/k. Since $\gamma \notin \mathfrak{O}$, not all the coefficients c_i belong to $\mathfrak{o}$. If $\min_{1 \leqslant i \leqslant n} v_0(c_i) = -r < 0$, then all coefficients of the polynomial $\varphi(t) = p^r f(t)$ will belong to the ring $\mathfrak{o}$, and at least one of them will be a unit in $\mathfrak{o}$. We now replace all coefficients of $\varphi(t)$ by the corresponding residue classes modulo p. Since the leading coefficient of $\varphi(t)$ is p^r, which is divisible by p, we obtain a polynomial $\bar{\varphi}(t)$ of the ring $\Sigma_0[t]$ of degree $\leqslant n - 1$, where not all coefficients are zero. Since we have assumed that the field Σ_0 contains at least n elements, there is an element $a \in \mathfrak{o}$ for which the residue class $\bar{a}$ is not a root of $\bar{\varphi}(t)$. This means that $\varphi(a) \not\equiv 0 \pmod{p}$; that is, $\varphi(a)$ is a unit in the ring $\mathfrak{o}$. We now compute $\|\gamma - a\|$. The characteristic polynomial of $\gamma - a$ equals $f(t + a)$, and therefore

$$N_{K/k}(\gamma - a) = (-1)^n f(a) = (-1)^n \varphi(a) p^{-r},$$

so that

$$\|\gamma - a\| = 2^{-r} < 1, \quad \|\alpha - a\beta\| < \|\beta\|.$$

Hence (4.10) is satisfied if we set $\xi = a$, $\rho = \alpha - a\beta$.

We have shown that $\mathfrak{O}$ is a Euclidean ring, and thus by Theorem 2 of Section 2 it has unique factorization.

Let π be any prime element of the ring $\mathfrak{O}$. Since for any $\alpha \in \mathfrak{O}^*$, its norm $N_{K/k}(\alpha)$ is always divisible by α, then $N_{K/k}(\pi) = p^f u$ is divisible by π (u is a unit of $\mathfrak{o}$, $f \geqslant 1$). But since π is prime and factorization into primes is unique, the element p is also divisible by π. Hence, if p has the factorization

$$p = \varepsilon \pi_1^{e_1} \cdots \pi_m^{e_m}$$

in the ring $\mathfrak{O}$ (ε a unit in $\mathfrak{O}$), then the prime elements $\pi_1, \ldots, \pi_m$ form a complete set of nonassociate primes of $\mathfrak{O}$.

The proof of Lemma 2 is complete.

We now turn to the proof of the basic results of this section.

Theorem 5. Any valuation v_0 of the field k can be extended to any finite extension K of k.

Theorem 6. Let $\mathfrak{o}$ be the ring of the valuation v_0, and let $\mathfrak{O}$ be its integral closure in the field K. If $v_1, \ldots, v_m$ are all extensions of the valuation v_0 to the field K, and $\mathfrak{O}_1, \ldots, \mathfrak{O}_m$ are their rings, then

$$\mathfrak{O} = \bigcap_{i=1}^{m} \mathfrak{O}_i.$$

Theorem 7. Under the same notations, the ring $\mathfrak{O}$ has unique factorization, and the set of valuations of K which correspond to prime elements of $\mathfrak{O}$ is precisely the set of all extensions $v_1, \ldots, v_m$ of the valuation v_0 to the field K.

If the prime elements $\pi_1, \ldots, \pi_m$ of the ring $\mathfrak{O}$ are ordered so that $v_i(\pi_i) = 1$, and if the prime element p of the ring $\mathfrak{o}$ has the factorization

$$p = \varepsilon\pi_1^{e_1} \cdots \pi_m^{e_m} \qquad (\varepsilon \text{ a unit in } \mathfrak{O}),$$

then e_i is the ramification index of the valuation v_i with respect to v_0.

Proof. If we assume that Theorems 5 and 6 have already been proved, then by the corollary of Theorem 3 the ring has unique factorization (with a finite number of nonassociate prime elements), and hence all results obtained in the second half of Section 4.3 are valid. But these results are precisely the contents of Theorem 7.

Theorems 5 and 6 will be proved by induction on the degree n of the extension K/k. For $n = 1$, there is nothing to prove. Let $n > 1$ and assume that Theorems 5 and 6 have already been proved for all extensions of degree $<n$ for any ground field k.

If the residue class field Σ_0 of the valuation v_0 contains at least n elements, then by Lemma 2 the ring $\mathfrak{O}$ has unique factorization, and hence the theorems are valid by what was proved in Section 4.3 [see (4.9)].

Hence we need only consider the case when the number q of elements of the residue class field Σ_0 is finite and less than n. We reduce this case to ones already considered by extending the ground field k to a field k', so that, first, the degree $(k' : k) = n - 1$ (by the induction hypothesis there is then an extension of the valuation v_0 to a valuation v_0' of the field k'), and, second, the residue class field Σ' of the valuation v_0' already contains not less than n elements. If we then denote by K' the smallest field containing both k' and K, the conditions of Lemma 2 will be fulfilled for the extension K'/k' and the valuation v_0'. We carry out this plan as follows.

We know (Supplement, Section 3) that over any finite field there are irreducible polynomials of all degrees. Let $\bar{\varphi}(t)$ be an irreducible polynomial of degree $n - 1$ with coefficients in the field Σ_0, and leading coefficient equal to 1. Each of its coefficients is a residue class of the ring $\mathfrak{o}$ modulo p. Replacing each class by one of its elements (and taking the leading coefficient to be 1), we obtain a polynomial $\varphi(t)$ of the ring $\mathfrak{o}[t]$, which is irreducible over the field k. Indeed, if $\varphi(t)$ were reducible over the field k, then it could be factored as a product of polynomials with coefficients in $\mathfrak{o}$, and after passing to the residue class field we would obtain a factorization for $\bar{\varphi}(t)$, which contradicts the choice $\bar{\varphi}(t)$. We now construct the extension field $K' = K(\theta)$, where θ is a root of the polynomial $\varphi(t)$. The degree of the extension K'/K does not exceed $n - 1$ [the polynomial $\varphi(t)$ may be reducible over the field K]. In K' we consider the subfield $k' = k(\theta)$. Since $\varphi(t)$ is irreducible over k, we have $(k' : k) = n - 1$. Let v_0' be any valuation of the field k' which is an extension of the valuation v_0 (the existence of v_0' is guaranteed by the induction hypothesis). Let $\mathfrak{o}'$, p', and

Σ' denote the ring of the valuation v_0', a prime element of the ring $\mathfrak{o}'$, and the residue class field of $\mathfrak{o}'$ modulo p'. Two elements a and b of $\mathfrak{o}$ are congruent modulo p' (in the ring $\mathfrak{o}'$) if and only if they are congruent modulo p in the ring $\mathfrak{o}$. Hence those residue classes of $\mathfrak{o}'$ modulo p' which contain an element of $\mathfrak{o}$ form a subfield of Σ' isomorphic to Σ_0. Having in mind this uniquely defined isomorphism $\Sigma_0 \to \Sigma'$, we shall assume that $\Sigma_0 \subset \Sigma'$. Since the element θ is the root of a polynomial with coefficients in $\mathfrak{o}$ and with leading coefficient 1, then $\theta \in \mathfrak{o}'$ (since $\mathfrak{o}'$ is integrally closed). Let $\bar{\theta}$ denote its residue class in Σ'. The equation $\varphi(\theta) = 0$ can be reduced modulo p' and gives us $\bar{\varphi}(\bar{\theta}) = 0$. Since $\bar{\varphi}$ was chosen to be irreducible over the field Σ_0, the powers $\bar{1}, \bar{\theta}, \ldots, \bar{\theta}^{n-2}$ are linearly independent over Σ_0. This means that the field Σ' contains q^{n-1} elements (q is the number of elements of the field Σ_0). But now

$$(K' : k') = \frac{(K' : K)(K : k)}{(k' : k)} \leqslant \frac{(n-1)n}{n-1} = n.$$

Since $q \geqslant 2$ and $n \geqslant 2$, we have

$$q^{n-1} \geqslant n.$$

Since the number of elements in the residue class field Σ' of the valuation v_0' is not less than $(K' : k')$, v_0' extends to a valuation v' of the field K'. Since v' is an extension of v_0 to K', the valuation v, induced by v' on the field K, is also an extension of the valuation v_0 (see Section 4.3). Theorem 5 is proved.

To complete the proof of Theorem 6 we first show that v_0' is the only extension of v_0 to a valuation of the field k'. Assume that v_0'' is another extension of the valuation v_0 to the field k'. By Theorem 3 the field k' contains an element γ such that $v_0'(\gamma) = 0$, $v_0''(\gamma) > 0$. Since the powers $1, \theta, \ldots, \theta^{n-2}$ form a basis for k' over k, the element γ can be represented in the form

$$\gamma = p^k(c_0 + c_1\theta + \cdots + c_{n-2}\theta^{n-2}) = p^k\alpha,$$

where all coefficients c_i belong to $\mathfrak{o}$ and at least one of them is a unit in $\mathfrak{o}$. We saw above that $\theta \in \mathfrak{o}'$ and the classes $1, \theta, \ldots, \theta^{n-2}$ of Σ' are linearly independent over Σ_0. Hence the residue class

$$\bar{\alpha} = \bar{c}_0 + \bar{c}_1\bar{\theta} + \cdots + \bar{c}_{n-2}\bar{\theta}^{n-2}$$

is nonzero (since at least one of the coefficients c_i is nonzero). This means that α is not divisible by p' (in the ring $\mathfrak{o}'$); that is, $v_0'(\alpha) = 0$. Analogously we obtain $v_0''(\alpha) = 0$. Comparing the conditions $v_0'(\gamma) = 0$ and $v_0'(\alpha) = 0$ with $\gamma = p^k\alpha$, we see that $k = 0$, and hence $v_0''(\gamma) = v_0''(\alpha) = 0$. But this contradicts the choice of γ. Thus the valuation v_0 has only one extension to the field k'.

Since Theorem 6 is assumed valid by induction for the extension k'/k, the ring $\mathfrak{o}'$ of the valuation v_0' coincides with the integral closure of the ring $\mathfrak{o}$ in the field k'. Let $\mathfrak{O}'$ denote the integral closure of the ring $\mathfrak{o}$ in the field K'.

Since $\mathfrak{o}' \subset \mathfrak{O}'$ and the ring $\mathfrak{O}'$ is integrally closed in K' (Supplement, Section 4.3), then $\mathfrak{O}'$ is also the integral closure of the ring $\mathfrak{o}'$ in the field K'. Let $\nu_1', \ldots, \nu_r'$ be all extensions of the valuation ν_0' to the field K', and let $\mathfrak{O}_1', \ldots, \mathfrak{O}_r'$ be their rings. Since Theorem 6 holds for the extension K'/k' (by Lemma 2), then

$$\mathfrak{O}' = \bigcap_{j=1}^{r} \mathfrak{O}_j'. \tag{4.11}$$

The set of valuations ν_j' is also the set of all extensions of the valuation ν_0 to the field K'. Equation (4.11) thus can be considered a proof of Theorem 6 for the extension K'/k and the valuation ν_0.

Let $\nu_1, \ldots, \nu_m$ denote all valuations of the field K which are induced by one of the valuations ν_j', and let $\mathfrak{O}_1, \ldots, \mathfrak{O}_m$ denote their rings. If ν_j' is an extension of ν_i, then clearly $\mathfrak{O}_j' \cap K = \mathfrak{O}_i$. Noting that the intersection $\mathfrak{O}' \cap K$ coincides with the integral closure $\mathfrak{O}$ of the ring $\mathfrak{o}$ in the field K, we have

$$\mathfrak{O} = \mathfrak{O}' \cap K = \bigcap_{j=1}^{r} (\mathfrak{O}_j' \cap K) = \bigcap_{i=1}^{m} \mathfrak{O}_i. \tag{4.12}$$

Suppose now that there is an extension ν of the valuation ν_0 to K different from $\nu_1, \ldots, \nu_m$. Then by Theorem 3 there would be an element α in K for which $\nu_1(\alpha) \geqslant 0, \ldots, \nu_m(\alpha) \geqslant 0$ (and hence $\alpha \in \mathfrak{O}$) and $\nu(\alpha) < 0$. But this would contradict the fact that $\mathfrak{O}$ must be contained in the ring $\mathfrak{O}_\nu$ of the valuation ν. Thus $\nu_1, \ldots, \nu_m$ are the only extensions of the valuation ν_0 to the field K. Formula (4.12) coincides with the assertion of Theorem 6.

PROBLEMS

1. Show that an algebraically closed field has no valuations.

2. Let $K = k(x)$ be a field of rational functions over the field k and let ν be the valuation of K corresponding to the polynomial $x - a$ ($a \in k$). Show that the residue class field Σ_ν of the valuation ν is isomorphic to k. Show further that two elements $f(x)$ and $g(x)$ of the ring lie in the same residue class if and only if $f(a) = g(a)$.

3. Let $K = k(x)$ be a field of rational functions over the field k of real numbers, and let ν be the valuation of K corresponding to the irreducible polynomial $x^2 + 1$. Find the residue classfield Σ_ν of the valuation ν.

4. Let $\mathfrak{O}_1$ and $\mathfrak{O}_2$ be the rings of the valuations ν_1 and ν_2 of some field K. Show that if $\mathfrak{O}_1 \subset \mathfrak{O}_2$, then $\nu_1 = \nu_2$.

5. Find the integral closure of the ring of 3-integral rational numbers in the field $R(\sqrt{-5})$ and determine all extensions of the 3-adic valuation ν_3 to this field.

6. For all prime numbers p find all extensions of the p-adic valuation ν_p to the field $R(\sqrt{-1})$ and determine the corresponding ramification indices.

7. Let K/k be a normal extension and ν_0 a valuation of the field k. Show that if ν is any extension of ν_0 to the field K, then all extensions have the form

$$\nu'(\alpha) = \nu(\sigma(\alpha)) \qquad (\alpha \in K),$$

where σ runs through all automorphisms of K/k.

8. Let k be a field of characteristic p. Let K/k be a purely inseparable extension. Show that a valuation ν_0 of the field k has only one extension to the field K. [The extension K/k is called purely inseparable if every element of K is a root of degree p^s ($s \geqslant 0$) of some element of k.]

9. Let $k = k_0(x, y)$ be a field of rational functions in x and y over some field k_0. In the field of formal power series $k_0\{t\}$ (see Problem 7 of Section 4, Chapter 1, or Section 1.5 of Chapter 4) choose a series $\xi(t) = \sum_{n=0}^{\infty} c_n t^n$ ($c_n \in k_0$) which is transcendental over the field of rational functions $k_0(t)$ [the existence of such series follows from the fact that the field $k_0\{t\}$ has higher cardinality than the field $k_0(t)$, and hence higher cardinality than the set of elements of $k_0\{t\}$ which are algebraic over $k_0(t)$]. For a nonzero polynomial $f = f(x, y) \in k_0[x, y]$, the series $f(t, \xi(t))$ will be nonzero by choice of ξ. If t^n is the smallest power of t which occurs in this series with nonzero coefficient, set $\nu_0(f) = n$. Show that the function ν_0 (after suitable extension) is a valuation of the field k and that the residue class field of the valuation is isomorphic to the field k_0.

5. Theories of Divisors for Finite Extensions

5.1. *Existence*

Theorem 1. Let the ring $\mathfrak{o}$ with quotient field k have a theory of divisors $\mathfrak{o}^* \to \mathscr{D}$ which is determined by the set $\mathfrak{R}_0$ of valuations. If K is a finite extension of the field k, then the set $\mathfrak{R}$ of all valuations of K which are extensions of valuations of $\mathfrak{R}_0$ determines a theory of divisors for the integral closure $\mathfrak{O}$ of the ring $\mathfrak{o}$ in the field K.

Proof. By Theorem 4 of Section 3 we need only verify that the set $\mathfrak{R}$ satisfies all conditions of that theorem. We first verify the second condition. For any valuation $\nu \in \mathfrak{R}$ and any $a \in \mathfrak{o}$ we clearly have $\nu(a) \geqslant 0$. This means that $\mathfrak{o}$ is contained in the ring of the valuation ν. But then by Theorem 1 of Section 4 the integral closure of the ring $\mathfrak{o}$ in the field K is also contained in the ring of the valuation ν. In other words, $\nu(\alpha) \geqslant 0$ for all $\alpha \in \mathfrak{O}$. Conversely, let $\alpha \in K$ be an element such that $\nu(\alpha) \geqslant 0$ for all $\nu \in \mathfrak{R}$. Let $t^r + a_1 t^{r-1} + \cdots + a_r$ denote the minimal polynomial of α with respect to k. Let ν_0 be any valuation of k belonging to the set $\mathfrak{R}_0$, and let $\nu_1, \ldots, \nu_m$ be its extensions to the field K. Since $\nu_1(\alpha) \geqslant 0, \ldots, \nu_m(\alpha) \geqslant 0$, then by Theorem 6 of Section 4 the element α lies in the integral closure in K of the ring of the valuation ν_0. But in this case all the coefficients $a_1, \ldots, a_r$ must lie in the ring of the valuation ν_0 (see the Supplement, Section 4.3); that is, $\nu_0(a_1) \geqslant 0, \ldots, \nu_0(a_r) \geqslant 0$. Since this holds for all $\nu_0 \in \mathfrak{R}_0$, the coefficients $a_1, \ldots, a_r$ belong to $\mathfrak{o}$, and hence $\alpha \in \mathfrak{O}$.

We now turn to the first condition. Let $\alpha \in \mathfrak{O}$, $\alpha \neq 0$, and let a_r be determined as above. Then for all but a finite number of valuations v_0 of $\mathfrak{R}_0$ we have $v_0(a_r) = 0$. Hence for all but a finite number of valuations v of $\mathfrak{R}$ we have $v(\alpha^{-1}) = a_r^{-1}(\alpha^{r-1} + \cdots + a_{r-1})) \geqslant 0$, and this means that $v(\alpha) = 0$. Hence $v(\alpha) = 0$ for almost all $v \in \mathfrak{R}$.

Only the third condition now remains to be verified. Let $v_1, \ldots, v_m$ be distinct valuations of $\mathfrak{R}$ and $k_1, \ldots, k_m$ be nonnegative integers. Let $v_{01}, \ldots, v_{0m}$ be the corresponding valuations of $\mathfrak{R}_0$ (the v_{0i} are not necessarily all distinct). Expand the original set of valuations to the set $v_1, \ldots, v_m, v_{m+1}, \ldots, v_s$, consisting of all extensions of the valuations v_{0i} to the field K. By Theorem 3 of Section 4 there is an element γ in the field K for which

$$v_1(\gamma) = k_1, \ldots, v_m(\gamma) = k_m, \quad v_{m+1}(\gamma) = 0, \ldots, v_s(\gamma) = 0.$$

If this element γ belongs to the ring $\mathfrak{O}$, just set $\alpha = \gamma$. Assume that γ does not belong to $\mathfrak{O}$. In this case denote by $v_1', \ldots, v_r'$ the valuations of $\mathfrak{R}$ which take negative values on γ:

$$v_1'(\gamma) = -l_1, \ldots, v_r'(\gamma) = -l_r,$$

and by $v_{01}', \ldots, v_{0r}'$ the corresponding valuations of $\mathfrak{R}_0$ (various v_{0j}' also may be equal). Since each of the valuations v_{0j}' is different from each of the valuations v_{0i}, in $\mathfrak{o}$ there is an element a such that

$$v_{0i}(a) = 0 \quad (1 \leqslant i \leqslant m), \qquad v_{0j}'(a) = l \quad (1 \leqslant j \leqslant r),$$

where l is taken equal to max $(l_1, \ldots, l_r)$. Set $\alpha = \gamma a$. Since

$$v_j'(\alpha) = v_j'(\gamma) + v_j'(a) \geqslant -l_j + v_{0j}'(a) = -l_j + l \geqslant 0,$$

then $\alpha \in \mathfrak{O}$. Thus in any case we have in the ring $\mathfrak{O}$ an element α such that $v_1(\alpha) = k_1, \ldots, v_m(\alpha) = k_m$, so that condition (3) of Theorem 4 of Section 3 also holds for the set $\mathfrak{R}$ of valuations. The proof of Theorem 1 is complete.

We apply Theorem 1 to the case of algebraic number fields.

The maximal order $\mathfrak{O}$ of the algebraic number field K is, as we have seen, the integral closure in K of the ring Z of rational integers. Since Z has a theory of divisors (since it has unique factorization), then by Theorem 1 $\mathfrak{O}$ also has a theory of divisors. By Theorem 5 of Section 3 the theory of divisors for Z is induced by the set of all valuations of the field R of rational numbers, and since every valuation of the field K is an extension of some valuation of the field R we find that the theory of divisors for the ring $\mathfrak{O}$ is induced by the set of all valuations of the field K. We hence have the following theorem.

Theorem 2. If $\mathfrak{O}$ is the maximal order of the algebraic number field K, there exists a theory of divisors $\mathfrak{O}^* \to \mathfrak{D}$, and this theory is induced by the set of all valuations of the field K.

5.2. *Norms of Divisors*

Let $\mathfrak{o}$ be a ring with theory of divisors $\mathfrak{o}^* \to \mathfrak{D}_0$ and with quotient field k; let K be a finite extension of k, with $\mathfrak{O}$ the integral closure of $\mathfrak{o}$ in K, and let $\mathfrak{O}^* \to \mathfrak{D}$ be a theory of divisors for the ring $\mathfrak{O}$. In this paragraph we establish a connection between the semigroups of divisors $\mathfrak{D}_0$ and $\mathfrak{D}$.

Since $\mathfrak{o} \subset \mathfrak{O}$, elements of $\mathfrak{o}^*$ correspond to principal divisors both in $\mathfrak{D}_0$ and in $\mathfrak{D}$. To distinguish these principal divisors, if $a \in \mathfrak{o}^*$, we denote the corresponding principal divisor in $\mathfrak{D}_0$ by $(a)_k$, and for any $\alpha \in \mathfrak{O}^*$ we denote the corresponding principal divisor by $(\alpha)_K$.

We have an inclusion isomorphism of semigroups $\mathfrak{o}^* \to \mathfrak{O}^*$. Since any unit of the ring $\mathfrak{O}$ which is contained in $\mathfrak{o}$ is a unit of the ring $\mathfrak{o}$, this inclusion induces an isomorphism $(a)_k \to (a)_K$, for $a \in \mathfrak{o}^*$, of the semigroup of principal divisors of the ring $\mathfrak{O}$. We now show that this isomorphism can be extended to an isomorphism $\mathfrak{D}_0 \to \mathfrak{D}$ (which will not be onto).

Theorem 3. There is an isomorphism of the semigroup $\mathfrak{D}_0$ into the semigroup $\mathfrak{D}$, which on principal divisors coincides with the isomorphism $(a)_k \to (a)_K$, $a \in \mathfrak{o}^*$.

The isomorphism $\mathfrak{D}_0 \to \mathfrak{D}$ is clearly characterized by the commutativity of the diagram

$$\begin{array}{ccc} \mathfrak{o}^* & \to & \mathfrak{O}^* \\ \downarrow & & \downarrow \\ \mathfrak{D}_0 & \to & \mathfrak{D} \end{array}$$

that is, by the fact that the two composite homomorphisms $\mathfrak{o}^* \to \mathfrak{O}^* \to \mathfrak{D}$ and $\mathfrak{o}^* \to \mathfrak{D}_0 \to \mathfrak{D}$ coincide (the vertical homomorphisms denote the homomorphisms of the multiplicative semigroups of the rings onto the semigroups of principal divisors).

Let $\mathfrak{p}$ be any prime divisor of the ring $\mathfrak{o}$, $v_\mathfrak{p}$ the corresponding valuation of the field k, and $v_{\mathfrak{P}_1}, \ldots, v_{\mathfrak{P}_m}$ all extensions of $v_\mathfrak{p}$ to the field K ($\mathfrak{P}_1, \ldots, \mathfrak{P}_m$ are the corresponding prime divisors of the ring $\mathfrak{O}$). Let $e_1 \ldots, e_m$ denote the respective ramification indices of the valuations $v_{\mathfrak{P}_1}, \ldots, v_{\mathfrak{P}_m}$ with respect to $v_\mathfrak{p}$. Since $v_{\mathfrak{P}_i}(a) = e_i v_\mathfrak{p}(a)$ for all $a \in \mathfrak{o}^*$, then the factor $\mathfrak{p}^{v_\mathfrak{p}(a)}$ of the principal divisor $(a)_k \in \mathfrak{D}_0$ will become $(\mathfrak{P}_1^{e_1} \cdots \mathfrak{P}_m^{e_m})^{v_\mathfrak{p}(a)}$ in the principal divisor $(a)_K \in \mathfrak{D}$. This means that the isomorphism from $\mathfrak{D}_0$ to $\mathfrak{D}$ defined by the mapping

$$\mathfrak{p} \to \mathfrak{P}_1^{e_1} \cdots \mathfrak{P}_m^{e_m} \tag{5.1}$$

(for all $\mathfrak{p}$) satisfies the requirements of Theorem 3.

It is easily seen that the isomorphism $\mathfrak{D}_0 \to \mathfrak{D}$, satisfying the requirements of Theorem 3, is unique (Problem 5).

By means of the isomorphism $\mathscr{D}_0 \to \mathscr{D}$ we can identify the semigroup $\mathscr{D}_0$ with its image in $\mathscr{D}$. But prime divisors in $\mathscr{D}_0$ will not, in general, remain prime in $\mathscr{D}$. For by (5.1) each prime $\mathfrak{p}$ has the decomposition

$$\mathfrak{p} = \mathfrak{P}_1^{e_1} \cdots \mathfrak{P}_m^{e_m} \tag{5.2}$$

in the semigroup $\mathscr{D}$.

Using the embedding $\mathscr{D}_0 \to \mathscr{D}$ we may speak of the divisibility of divisors of $\mathfrak{o}$ by divisors of $\mathfrak{O}$. From (5.2) we see that a prime divisor $\mathfrak{p}$ of the ring $\mathfrak{o}$ is divisible by a prime divisor $\mathfrak{P}$ of the ring $\mathfrak{O}$ if and only if the valuation $v_{\mathfrak{P}}$ is an extension of the valuation $v_{\mathfrak{p}}$. It is further clear that relatively prime divisors in $\mathscr{D}_0$ remain relatively prime in $\mathscr{D}$.

Definition. Let $\mathfrak{P}|\mathfrak{p}$. The ramification index $e = e_{\mathfrak{P}}$ of the valuation $v_{\mathfrak{P}}$ with respect to the valuation $v_{\mathfrak{p}}$ is also called the *ramification index* of the prime $\mathfrak{P}$ relative to $\mathfrak{p}$ (or relative to k).

The ramification index is thus the largest natural number e such that $\mathfrak{P}^e|\mathfrak{p}$.

If α is any element of $\mathfrak{O}$, its norm $N(\alpha) = N_{K/k}(\alpha)$ lies in $\mathfrak{o}$. The mapping $\alpha \to N(\alpha)$, $\alpha \in \mathfrak{O}^*$, is a homomorphism from the multiplicative semigroup $\mathfrak{O}^*$ to the semigroup $\mathfrak{o}$. Since the norm of any unit of the ring $\mathfrak{O}^*$ is a unit in $\mathfrak{o}$, this homomorphism induces a homomorphism $(\alpha)_K \to (N(\alpha))_k$ of the semigroup of principal divisors of the ring $\mathfrak{O}$ to the semigroup of principal divisors of the ring $\mathfrak{o}$. We shall show that this homomorphism can be extended to a homomorphism of the entire semigroup $\mathscr{D}$ to $\mathscr{D}_0$.

Theorem 4. There is a homomorphism from the semigroup of divisors $\mathscr{D}$ to $\mathscr{D}_0$, $N: \mathscr{D} \to \mathscr{D}_0$, such that

$$N((\alpha)_K) = (N_{K/k}(\alpha))_k \tag{5.3}$$

for any $\alpha \in \mathfrak{O}^*$.

We can express (5.3) by saying that the diagram

$$\begin{array}{ccc} \mathfrak{O}^* & \xrightarrow{N} & \mathfrak{o}^* \\ \downarrow & & \downarrow \\ \mathscr{D} & \xrightarrow{N} & \mathscr{D}_0 \end{array}$$

is commutative.

For a fixed prime divisor $\mathfrak{p} \in \mathscr{D}_0$ we denote the ring of the valuation $v_{\mathfrak{p}}$ by $\mathfrak{o}_{\mathfrak{p}}$ and its integral closure in the field K by $\mathfrak{O}_{\mathfrak{p}}$. By Theorem 7 of Section 4 the prime divisors $\mathfrak{P}_1, \ldots, \mathfrak{P}_m$ of the ring $\mathfrak{O}$ which divide $\mathfrak{p}$ correspond uniquely to a complete set of pairwise-nonassociate prime elements $\pi_1, \ldots, \pi_m$ of the ring $\mathfrak{O}_\mathfrak{p}$. The correspondence $\mathfrak{P}_i \leftrightarrow \pi_i$ has the property that for any nonzero element $\alpha \in K$, if

$$\alpha = \varepsilon \pi_1^{k_1} \cdots \pi_m^{k_m}, \tag{5.4}$$

where ε is a unit of the ring $\mathfrak{O}_\mathfrak{p}$, then

$$k_i = \nu_{\mathfrak{P}_i}(\alpha). \tag{5.5}$$

Let $\mathfrak{P}$ be one of the prime divisors $\mathfrak{P}_i$ which divides $\mathfrak{p}$, and let π be the corresponding prime element of the ring $\mathfrak{O}_\mathfrak{p}$. Set

$$d_\mathfrak{P} = \nu_\mathfrak{p}(N_{K/k}(\pi)). \tag{5.6}$$

It is clear that $d_\mathfrak{P}$ does not depend on the choice of π. Taking the norm in (4) and comparing with (5) and (6) we obtain the relation

$$\nu_\mathfrak{p}(N_{K/k}(\alpha)) = \sum_{\mathfrak{P}|\mathfrak{p}} d_\mathfrak{P} \nu_\mathfrak{P}(\alpha) \tag{5.7}$$

($\mathfrak{P}$ runs through all prime divisors of the ring $\mathfrak{O}$ which divide $\mathfrak{p}$).

We can now construct the desired homomorphism $N: \mathscr{D} \to \mathscr{D}_0$.

A divisor $\mathfrak{A} = \mathfrak{P}_1^{A_1} \cdots \mathfrak{P}_r^{A_r}$ of the semigroup $\mathscr{D}$ can be conveniently written as an infinite product

$$\mathfrak{A} = \prod_\mathfrak{P} \mathfrak{P}^{A(\mathfrak{P})}$$

over all prime divisors $\mathfrak{P}$ of $\mathscr{D}$, in which, however, only a finite number of the exponents $A(\mathfrak{P})$ are nonzero. [$A(\mathfrak{P})$ equals A_i, if $\mathfrak{P} = \mathfrak{P}_i$, and equals zero if the divisor $\mathfrak{P}$ is not one of $\mathfrak{P}_1, \ldots, \mathfrak{P}_r$.] We can write divisors of the ring $\mathfrak{o}$ in an analogous fashion.

If $(\alpha)_K$ is the principal divisor corresponding to a nonzero element α of $\mathfrak{O}$, then we have the representation

$$(\alpha)_K = \prod_\mathfrak{P} \mathfrak{P}^{\nu_\mathfrak{P}(\alpha)}. \tag{5.8}$$

From (5.7) we see that if

$$(N(\alpha))_k = \prod_\mathfrak{p} \mathfrak{p}^{c(\mathfrak{p})}, \tag{5.9}$$

then $c(\mathfrak{p})$ must satisfy

$$c(\mathfrak{p}) = \sum_{\mathfrak{P}|\mathfrak{p}} d_\mathfrak{P} \nu_\mathfrak{P}(\alpha). \tag{5.10}$$

This suggests the following definition.

Definition. Let $\mathfrak{A} = \prod_\mathfrak{P} \mathfrak{P}^{A(\mathfrak{P})}$ be a divisor of the ring $\mathfrak{O}$. For any prime divisor $\mathfrak{p}$ of the ring $\mathfrak{o}$, set

$$a(\mathfrak{p}) = \sum_{\mathfrak{P}|\mathfrak{p}} d_\mathfrak{P} A(\mathfrak{P}).$$

The divisor $\prod_{\mathfrak{p}} \mathfrak{p}^{a(\mathfrak{p})}$ of the ring $\mathfrak{o}$ is called the *norm of the divisor* $\mathfrak{A}$ with respect to the extension K/k and is denoted by $N_{K/k}(\mathfrak{A})$, or simply by $N(\mathfrak{A})$.

Since $A(\mathfrak{P})$ equals zero for almost all $\mathfrak{P}$ (that is, all but a finite number of $\mathfrak{P}$), then $a(\mathfrak{p})$ also equals zero for almost all $\mathfrak{p}$, and hence the expression $\prod_{\mathfrak{p}} \mathfrak{p}^{a(\mathfrak{p})}$ actually is a divisor of the ring $\mathfrak{o}$.

From the definition it is clear that

$$N(\mathfrak{A}\mathfrak{B}) = N(\mathfrak{A})N(\mathfrak{B})$$

for any two divisors $\mathfrak{A}$ and $\mathfrak{B}$ of $\mathfrak{O}$. The mapping $\mathfrak{A} \to N(\mathfrak{A})$ is thus a homomorphism of the semigroup $\mathfrak{O}$ to the semigroup $\mathscr{D}_0$.

In the case of a prime divisor $\mathfrak{A} = \mathfrak{P}$ we clearly have

$$N(\mathfrak{P}) = \mathfrak{p}^{d_{\mathfrak{P}}(\mathfrak{P}|\mathfrak{p})}. \tag{5.11}$$

In view of (5.10), the norm of the divisor (5.8) equals the divisor (5.9) and hence we have proved the existence of a homomorphism $N: \mathscr{D} \to \mathscr{D}_0$, which satisfies the requirements of Theorem 3.

As in the case of the isomorphism $\mathscr{D}_0 \to \mathscr{D}$, it can be shown (Problem 4) that the homomorphism $N: \mathscr{D} \to \mathscr{D}_0$ is uniquely determined by condition (5.3).

One of the central problems of the theory of divisors is to determine a rule for the decomposition of the prime divisor $\mathfrak{p}$ of the ring $\mathfrak{o}$ into prime divisors of the integral closure $\mathfrak{O}$ of $\mathfrak{o}$ in some finite extension field. In the general case this problem is still unsolved (however, see the end of Section 8.2). Each decomposition (5.2) is characterized by the number m of prime divisors and by the various ramification indices $e_i = e_{\mathfrak{P}_i}$. The natural numbers e_i, however, cannot be taken arbitrarily (for a given extension K/k). For they are related to the numbers $d_{\mathfrak{P}}$ [see (5.6)] by the formula

$$\sum_{\mathfrak{P}|\mathfrak{p}} d_{\mathfrak{P}} e_{\mathfrak{P}} = n = (K:k), \tag{5.12}$$

for the proof of which it suffices to apply formula (5.7) to the case of a prime element p of the ring $\mathfrak{o}_{\mathfrak{p}}$ [recall that $v_{\mathfrak{P}_i}(p) = e_i$].

5.3. *The Degree of Inertia*

The definition of the homomorphism $N: \mathscr{D} \to \mathscr{D}_0$ depended on the numbers $d_{\mathfrak{P}}$, which were defined in a rather formal manner in (5.6). We now clarify the deep arithmetical significance of these numbers.

Let $\mathfrak{P}|\mathfrak{p}$. Let $\mathfrak{o}_{\mathfrak{p}}$ and $\mathfrak{O}_{\mathfrak{P}}$ denote the rings of the valuations $v_{\mathfrak{p}}$ and $v_{\mathfrak{P}}$, and p and π the prime elements in these rings. Since for elements a and b of $\mathfrak{o}_{\mathfrak{p}}$ the congruences $a \equiv b \pmod{p}$ in the ring $\mathfrak{o}_{\mathfrak{p}}$) and $a \equiv b \pmod{\pi}$ in the ring $\mathfrak{O}_{\mathfrak{P}}$ are equivalent, each residue class in $\mathfrak{o}_{\mathfrak{p}}$ modulo p is contained in a single residue class modulo π in $\mathfrak{O}_{\mathfrak{P}}$. This determines an isomorphic embedding of the residue class field $\Sigma_{\mathfrak{p}} = \mathfrak{o}_{\mathfrak{p}}/(p)$ of the valuation $v_{\mathfrak{p}}$ into the residue class field $\Sigma_{\mathfrak{P}} = \mathfrak{O}_{\mathfrak{P}}/(\pi)$ of the valuation $v_{\mathfrak{P}}$. Using this isomorphism we assume that $\Sigma_{\mathfrak{p}} \subset \Sigma_{\mathfrak{P}}$. For any $\xi \in \mathfrak{O}_{\mathfrak{P}}$ we denote the residue class modulo π which contains ξ by $\bar{\xi}$. The subfield $\Sigma_{\mathfrak{p}}$ of the field $\Sigma_{\mathfrak{P}}$ then consists of those residue classes of the form $\bar{a}$, where $a \in \mathfrak{o}_{\mathfrak{p}}$.

Let the residue classes $\bar{\omega}_1, \ldots, \bar{\omega}_m$ of $\Sigma_{\mathfrak{P}}$ ($\omega_i \in \mathfrak{O}_{\mathfrak{P}}$) be linearly independent over the field $\Sigma_{\mathfrak{p}}$. We now show that then the representatives $\omega_1, \ldots, \omega_m$ of these classes are linearly independent over the field k. Assume that this is not so, that is, that for some coefficients $a_i \in k$, not all zero, we have

$$a_1\omega_1 + \cdots + a_m\omega_m = 0.$$

Multiplying this relation by a suitable power of p, we may assume that all a_i lie in the ring $\mathfrak{o}_{\mathfrak{p}}$ and that at least one of them is not divisible by p. Passing now to the residue class field $\Sigma_{\mathfrak{P}}$, we obtain

$$\bar{a}_1\bar{\omega}_1 + \cdots + \bar{a}_m\bar{\omega}_m = \bar{0},$$

in which not all coefficients $\bar{a}_i \in \Sigma_{\mathfrak{p}}$ are zero. This contradiction proves our assertion.

From the linear independence of $\omega_1, \ldots, \omega_n$ over the field k it follows that $m \leqslant n = (K:k)$. Thus the residue class field $\Sigma_{\mathfrak{P}}$ is a finite extension of the field $\Sigma_{\mathfrak{p}}$ for which

$$(\Sigma_{\mathfrak{P}}:\Sigma_{\mathfrak{p}}) \leqslant (K:k).$$

Definition. Let the prime divisor $\mathfrak{P}$ of the ring $\mathfrak{O}$ divide the prime divisor $\mathfrak{p}$ of the ring $\mathfrak{o}$. The degree $f = f_{\mathfrak{P}} = (\Sigma_{\mathfrak{P}}:\Sigma_{\mathfrak{p}})$ of the residue class field of the valuation $v_{\mathfrak{P}}$ over the residue class field of the valuation $v_{\mathfrak{p}}$ is called the *degree of inertia* of the prime divisor $\mathfrak{P}$ relative to $\mathfrak{p}$ (or relative to k).

As in Section 5.2 we denote by $\mathfrak{O}_{\mathfrak{p}}$ the integral closure of the ring $\mathfrak{o}_{\mathfrak{p}}$ in the field K. In analogy with the definition of a fundamental basis of an algebraic number field, we make the following definition.

Definition. A basis $\omega_1, \ldots, \omega_n$ of the extension K/k is called a *fundamental basis* for the ring $\mathfrak{O}_{\mathfrak{p}}$ relative to $\mathfrak{o}_{\mathfrak{p}}$ if all its elements lie in $\mathfrak{O}_{\mathfrak{p}}$ and every element $\alpha \in \mathfrak{O}_{\mathfrak{p}}$ is represented by a linear combination

$$\alpha = a_1\omega_1 + \cdots + a_n\omega_n \tag{5.13}$$

with coefficients a_i in $\mathfrak{o}_{\mathfrak{p}}$.

We shall see below that in the case of a separable extension K/k, a fundamental basis for the ring $\mathfrak{O}_{\mathfrak{p}}$ (for any $\mathfrak{p}$) always exists. On the other hand, by Problems 11 and 12, for nonseparable extensions K/k it may occur that the ring $\mathfrak{O}_{\mathfrak{p}}$ has no fundamental basis relative to $\mathfrak{o}_{\mathfrak{p}}$.

The value of the concept of a fundamental basis is indicated in the following theorem.

Theorem 5. Let $\mathfrak{P}$ be a prime divisor of the ring $\mathfrak{O}$ which divides $\mathfrak{p}$, and let π be a prime element of the ring $\mathfrak{O}_\mathfrak{p}$ which corresponds to it. If the ring $\mathfrak{O}_\mathfrak{p}$ has a fundamental basis relative to $\mathfrak{o}_\mathfrak{p}$, then

$$f_\mathfrak{P} = d_\mathfrak{P} = v_\mathfrak{p}(N_{K/k}(\pi)).$$

Proof. The prime element $\pi \in \mathfrak{O}_\mathfrak{p}$ is clearly also a prime element of the ring $\mathfrak{O}_\mathfrak{P}$. We shall show that each residue class ξ of the ring $\mathfrak{O}_\mathfrak{P}$ modulo π contains a representative in $\mathfrak{O}_\mathfrak{p}$; that is, for any $\xi \in \mathfrak{O}_\mathfrak{P}$ there is an element $\alpha \in \mathfrak{O}_\mathfrak{p}$ such that

$$\xi \equiv \alpha \pmod{\pi}.$$

Let $\mathfrak{P} = \mathfrak{P}_1, \mathfrak{P}_2, \ldots, \mathfrak{P}_m$ be all prime divisors of the ring $\mathfrak{O}$ which divide $\mathfrak{p}$. By Theorem 6 of Section 4, $\gamma \in \mathfrak{O}_\mathfrak{p}$ if and only if $v_{\mathfrak{P}_i}(\gamma) \geqslant 0$ for all $i = 1, \ldots, m$. Hence the element α must satisfy

$$v_\mathfrak{P}(\xi - \alpha) \geqslant 1,$$

$$v_{\mathfrak{P}_i}(\alpha) \geqslant 0 \qquad (i = 2, \ldots, m),$$

and the proof of its existence is given by Theorem 4 of Section 4.

Now let $\omega_1, \ldots, \omega_n$ be a fundamental basis of the ring $\mathfrak{O}_\mathfrak{p}$ relative to $\mathfrak{o}_\mathfrak{p}$. By the above, every element of $\Sigma_\mathfrak{P}$ can be represented in the form $\bar{a}_1\bar{\omega}_1 + \cdots + \bar{a}_n\bar{\omega}_n$, where $a_i \in \mathfrak{o}_\mathfrak{p}$, and hence $\bar{a}_i \in \Sigma_\mathfrak{p}$. This means that the residue classes $\bar{\omega}_1, \ldots, \bar{\omega}_n$ generate $\Sigma_\mathfrak{P}$ as a vector space over $\Sigma_\mathfrak{p}$. If $f = (\Sigma_\mathfrak{P} : \Sigma_\mathfrak{p}) = f_\mathfrak{P}$, then we can choose from among them f elements which are linearly independent over $\Sigma_\mathfrak{p}$. Let these be $\bar{\omega}_1, \ldots, \bar{\omega}_f$. It is clear that the congruence

$$a_1\omega_1 + \cdots + a_f\omega_f \equiv 0 \pmod{\pi},$$

with $a_i \in \mathfrak{o}_\mathfrak{p}$, holds in the ring $\mathfrak{O}_\mathfrak{p}$ if and only if $a_i \equiv 0 \pmod{p}$, p being a prime element of the ring $\mathfrak{o}_\mathfrak{p}$.

Since each of the residue classes $\bar{\omega}_j \in \Sigma_\mathfrak{P}$ for $j = f + 1, \ldots, n$ can be expressed in terms of $\bar{\omega}_1, \ldots, \bar{\omega}_f$, then

$$\omega_j \equiv \sum_{s=j}^{f} b_{js}\omega_s \pmod{\pi} \qquad (j = f + 1, \ldots, n)$$

for some b_{js} from $\mathfrak{o}_\mathfrak{p}$. Set

$$\theta_i = \omega_i \qquad \text{for } i = 1, \ldots, f,$$

$$\theta_j = -\sum_{s=1}^{f} b_{js}\omega_s + \omega_j \qquad \text{for } j = f + 1, \ldots, n.$$

It is clear that $\theta_1, \ldots, \theta_n$ also form a fundamental basis of $\mathfrak{O}_\mathfrak{p}$ relative to $\mathfrak{o}_\mathfrak{p}$ (since all ω_s can be expressed in terms of θ_s with coefficients in $\mathfrak{o}_\mathfrak{p}$). Each element $\theta_{f+1}, \ldots, \theta_n$ is divisible in the ring $\mathfrak{O}_\mathfrak{p}$ by π, and therefore the congruence

$$a_1\theta_1 + \cdots + a_n\theta_n \equiv 0 \pmod{\pi}$$

holds if and only if

$$a_1 \equiv \cdots \equiv a_f \equiv 0 \pmod{p}.$$

Consider the set $\mathfrak{M}$ of all elements of the ring $\mathfrak{O}_\mathfrak{p}$ which are divisible by π. By what was just proved, the set $\mathfrak{M}$ consists of all linear combinations of the elements

$$p\theta_1, \ldots, p\theta_f, \quad \theta_{f+1}, \ldots, \theta_n \tag{5.14}$$

with coefficients in $\mathfrak{o}_\mathfrak{p}$. On the other hand, it is clear that $\mathfrak{M}$ also coincides with the set of all linear combinations of the elements

$$\pi\theta_1, \ldots, \pi\theta_n \tag{5.15}$$

with coefficients in $\mathfrak{o}_\mathfrak{p}$. Let C denote the transition matrix from the basis (5.14) to the basis (5.15). Since every element $\pi\theta_j$ can be expressed in terms of the basis (5.14) with coefficients from $\mathfrak{o}_\mathfrak{p}$, then $\det C$ is an element of $\mathfrak{o}_\mathfrak{p}$. By symmetry this also holds for $\det C^{-1}$. Hence $\det C$ is a unit in the ring $\mathfrak{o}_\mathfrak{p}$; that is, $\nu_\mathfrak{p}(\det C) = 0$. If we multiply the first f columns of the matrix C by p, then we clearly obtain a matrix $A = (a_{ij})$, for which

$$\pi\theta_i = \sum_{j=1}^{n} a_{ij}\theta_j.$$

Therefore,

$$N_{K/k}(\pi) = \det A = p^f \det C,$$

so that

$$\nu_\mathfrak{p}(N_{K/k}(\pi)) = f,$$

and Theorem 5 is proved.

Theorem 6. If the extension K/k is separable, then $\mathfrak{O}_\mathfrak{p}$ always has a fundamental basis relative to $\mathfrak{o}_\mathfrak{p}$.

Before starting the proof of this theorem we note that it is analogous to the proof of Theorem 6 of Section 2, Chapter 2.

Since every element of K, after multiplication by a suitable power of a prime element of the ring $\mathfrak{o}_\mathfrak{p}$, becomes integral with respect to $\mathfrak{o}_\mathfrak{p}$, the extension K/k

has a basis, $\alpha_1, \dots, \alpha_n$, all elements of which lie in $\mathfrak{O}_\mathfrak{p}$. Consider the dual basis $\alpha_1^*, \dots, \alpha_n^*$ (see the Supplement, Section 2.3; here we have already assumed that K/k is separable). If $\alpha \in \mathfrak{O}_\mathfrak{p}$ and

$$\alpha = c_1\alpha_1^* + \cdots + c_n\alpha_n^*, \tag{5.16}$$

where $c_i \in k$, then $c_i = \operatorname{Sp}(\alpha\alpha_i)$, and this means that $c_i \in \mathfrak{o}_\mathfrak{p}$ (since $\alpha\alpha_i \in \mathfrak{O}_\mathfrak{p}$). For each $s = 1, \dots, n$ we consider in the ring $\mathfrak{O}_\mathfrak{p}$ those elements which when expressed in terms of the basis $\alpha_1^*, \dots, \alpha_n^*$ have the form

$$c_s\alpha_s^* + \cdots + c_n\alpha_n^* \qquad (c_i \in \mathfrak{o}_\mathfrak{p}), \tag{5.17}$$

and we choose among these an element

$$\omega_s = c_{ss}\alpha_s^* + \cdots + c_{sn}\alpha_n^* \qquad (c_{sj} \in \mathfrak{o}_\mathfrak{p})$$

such that $v_\mathfrak{p}(c_s) \geqslant v_\mathfrak{p}(c_{ss})$ for all coefficients c_s of elements of the form (5.17) of $\mathfrak{O}_\mathfrak{p}$. It is clear that $c_{ss} \neq 0$ for all s, so that the elements $\omega_1, \dots, \omega_n$ of $\mathfrak{O}_\mathfrak{p}$ are linearly independent over k. Let α be any element of $\mathfrak{O}_\mathfrak{p}$. If we represent it in the form (5.16), then $c_1 = c_{11}a_1$, where $a_1 \in \mathfrak{o}_\mathfrak{p}$, by choice of ω_1. For the difference $\alpha - a_1\omega_1$ we have the expansion

$$\alpha - a_1\omega_1 = c_2'\alpha_2^* + \cdots + c_n'\alpha_n^* \qquad (c_i' \in \mathfrak{o}_\mathfrak{p}),$$

and here $c_2' = c_{22}a_2$, where $a_2 \in \mathfrak{o}_\mathfrak{p}$ by choice of ω_2. Continuing this process n times we finally arrive at the expansion (5.13), in which all coefficients a_i belong to $\mathfrak{o}_\mathfrak{p}$. The basis ω_j is hence a fundamental basis relative to $\mathfrak{o}_\mathfrak{p}$, and Theorem 6 is proved.

From Theorems 5 and 6 and formula (5.12) we easily obtain the following assertion.

Theorem 7. If the extension K/k is separable and $\mathfrak{p}$ is a fixed prime divisor of the ring $\mathfrak{o}$, then the ramification indices $e_\mathfrak{P}$ and degrees of inertia $f_\mathfrak{P}$ of the prime divisors $\mathfrak{P}$ of the ring $\mathfrak{O}$ which divide $\mathfrak{p}$ are connected by the relation

$$\sum_{\mathfrak{P}|\mathfrak{p}} e_\mathfrak{P} f_\mathfrak{P} = n = (K:k).$$

Hence for separable extensions K/k formula (5.7) can be written in the form

$$v_\mathfrak{p}(N_{K/k}(\alpha)) = \sum_{\mathfrak{P}|\mathfrak{p}} f_\mathfrak{P} v_\mathfrak{P}(\alpha). \tag{5.18}$$

Remark. For nonseparable extensions, Theorem 7 is no longer necessarily valid. However, the inequality $\sum_{\mathfrak{P}|\mathfrak{p}} e_\mathfrak{P} f_\mathfrak{P} \leqslant n$ always holds (see Problem 13). It can further be shown that, in general, $f_\mathfrak{P} \leqslant d_\mathfrak{P}$.

5.4. *Finiteness of the Number of Ramified Prime Divisors*

Definition. The prime divisor $\mathfrak{p}$ of the ring $\mathfrak{o}$ is called *ramified* in the ring $\mathfrak{O}$ if it is divisible by the square of some prime divisor of the ring $\mathfrak{O}$, and is called *unramified* otherwise.

Hence $\mathfrak{p}$ is unramified if and only if all e_i in (5.2) are equal to 1.

Under the assumption that the extension K/k is separable, we shall obtain an important condition for $\mathfrak{p}$ to be unramified.

Assume that the ring $\mathfrak{O}_\mathfrak{p}$ contains some primitive element θ for the extension K/k, such that the discriminant $D(f)$ of its minimum polynomial $f(t)$ is a unit in $\mathfrak{o}_\mathfrak{p}$. We now show that in this case the powers $1, \theta, \ldots, \theta^{n-1}$, where $n = (K:k)$ form a fundamental basis for the ring $\mathfrak{O}_\mathfrak{p}$ over $\mathfrak{o}_\mathfrak{p}$. Let $\omega_1, \ldots, \omega_n$ be any fundamantal basis for $\mathfrak{O}_\mathfrak{p}$, and let C be the matrix of transition from the basis ω_i to the basis θ_j. Then

$$D(f) = D(1, \theta, \ldots, \theta^{n-1}) = (\det C)^2 D(\omega_1, \ldots, \omega_n)$$

[see the Supplement, formula (2.12)]. Since $D(f)$ is a unit in $\mathfrak{o}_\mathfrak{p}$, and both terms on the right belong to $\mathfrak{o}_\mathfrak{p}$, then $\det C$ is a unit in $\mathfrak{o}_\mathfrak{p}$, and hence $1, \theta, \ldots, \theta^{n-1}$ is also a fundamental basis.

Let p be a prime element of the ring $\mathfrak{o}_\mathfrak{p}$ and $\Sigma_\mathfrak{p}$ the residue class field of the valuation $v_\mathfrak{p}$. For any polynomial $g(t)$ with coefficients in $\mathfrak{o}_\mathfrak{p}$ we denote by $\bar{g}(t)$ the polynomial obtained by replacing all coefficients of $g(t)$ by their residue classes modulo p. Since the discriminant $D(\bar{f}) \in \Sigma_\mathfrak{p}$ of the polynomial $\bar{f}(t) \in \Sigma_\mathfrak{p}[t]$ is equal to the residue class modulo p of the discriminant $D(f) \in \mathfrak{o}_\mathfrak{p}$, then by our assumption the discriminant $D(f)$ is nonzero. Hence, all factors in the decomposition

$$\bar{f}(t) = \bar{\varphi}_1(t) \cdots \bar{\varphi}_m(t) \tag{5.19}$$

into irreducible factors in the ring $\Sigma_\mathfrak{p}[t]$ are distinct (here $\bar{\varphi}_i$ is some polynomial of $\mathfrak{o}_\mathfrak{p}[t]$). If we denote the degree of $\bar{\varphi}_i$ by d_i, then we clearly have

$$d_1 + \cdots + d_m = n = (K:k). \tag{5.20}$$

Theorem 8. If the discriminant of the minimum polynomial $f(t)$ of a primitive element $\theta \in \mathfrak{O}_\mathfrak{p}$ is a unit in $\mathfrak{o}_\mathfrak{p}$, then the prime divisor $\mathfrak{p}$ is unramified in $\mathfrak{O}$ and the prime divisors $\mathfrak{P}_i$ of the decomposition

$$\mathfrak{p} = \mathfrak{P}_1 \cdots \mathfrak{P}_m$$

can be put in one-to-one correspondence with the irreducible polynomials $\bar{\varphi}_i \in \Sigma_\mathfrak{p}[t]$ of the decomposition (5.19) in such a way that the degree of inertia f_i of the prime divisor $\mathfrak{P}_i$ coincides with the degree d_i of the corresponding polynomial $\bar{\varphi}_i(t)$.

Proof. Let $g(t)$ be any polynomial of $\mathfrak{o}_\mathfrak{p}[t]$. We shall show that if the polynomials $\bar{g}$ and $\bar{\varphi}_i$ are relatively prime in the ring $\Sigma_\mathfrak{p}[t]$, then the elements $g(\theta)$ and $\varphi_i(\theta)$ are relatively prime in the ring $\mathfrak{O}_\mathfrak{p}$. For then there exist polynomials $u(t)$, $v(t)$, and $l(t)$ in the ring $\mathfrak{o}_\mathfrak{p}[t]$ such that

$$g(t)u(t) + \varphi_i(t)v(t) = 1 + pl(t).$$

If $g(\theta)$ and $\varphi_i(\theta)$ were divisible in the ring $\mathfrak{O}_\mathfrak{p}$ by some prime element π, then since $\pi|p$ (Theorem 7 of Section 4), from the preceding equation (for $t = \theta$), it would follow that $\pi|1$. This contradiction proves our assertion.

Since the irreducible polynomials $\bar{\varphi}_i$ are distinct, then the elements $\varphi_1(\theta), \ldots, \varphi_m(\theta)$ are pairwise relatively prime.

Assume that $\varphi_i(\theta)$ is a unit in $\mathfrak{O}_\mathfrak{p}$, that is, that $\varphi_i(\theta)\xi = 1$, $\xi \in \mathfrak{O}_\mathfrak{p}$. Since $1, \theta, \ldots, \theta^{n-1}$ form a fundamental basis for $\mathfrak{O}_\mathfrak{p}$ over $\mathfrak{o}_\mathfrak{p}$, then $\xi = h(\theta)$, where $h(t) \in \mathfrak{o}_\mathfrak{p}[t]$. The equation $\varphi_i(\theta)h(\theta) = 1$ implies that $\varphi_i(t)h(t) = 1 + f(t)q(t)$, where $q(t) \in \mathfrak{o}_\mathfrak{p}[t]$ [since the leading coefficient of $f(t)$ equals 1]. Passing to the residue class field $\Sigma_\mathfrak{p}$ we obtain $\bar{\varphi}_i\bar{h} = 1 + \bar{\varphi}_1 \cdots \bar{\varphi}_m q$, and we again have a contradiction. Hence none of the elements $\varphi_1(\theta), \ldots, \varphi_m(\theta)$ are units in $\mathfrak{O}_\mathfrak{p}$.

For each i choose in $\mathfrak{O}_\mathfrak{p}$ a prime element $\pi_i|\varphi_i(\theta)$. Since we have proved that the $\varphi_i(\theta)$ are pairwise relatively prime, the prime elements $\pi_1, \ldots, \pi_m$ are pairwise-nonassociate. Let $\mathfrak{P}_1, \ldots, \mathfrak{P}_m$ denote the corresponding prime divisors of the ring $\mathfrak{O}$, and $f_1, \ldots, f_m$ denote the degrees of inertia of these divisors. In the residue class field $\Sigma_{\mathfrak{P}_i}$ of the valuation $\nu_{\mathfrak{P}_i}$, the residue classes $1, \bar{\theta}, \ldots, \bar{\theta}^{d_i-1}$ are linearly independent over $\Sigma_\mathfrak{p}$ (d_i is the degree of $\bar{\varphi}_i$). For if there is a polynomial $g(t) \in \mathfrak{o}_\mathfrak{p}[t]$ of degree $< d_i$ for which $\bar{g}(\bar{\theta}) = 0$, then the element $g(\theta)$ is divisible by π_i in the ring $\mathfrak{O}_\mathfrak{p}$ and hence $g(\theta)$ and $\varphi_i(\theta)$ are not relatively prime. But we saw at the beginning of the proof that then $\bar{g}(t)$ must be divisible by $\bar{\varphi}_i(t)$, and this can happen only if all coefficients of $\bar{g}(t)$ are zero.

We have shown that

$$d_i \leqslant f_i \qquad (i = 1, \ldots, m).$$

Comparing this inequality with (5.20) and considering Theorem 7, we see that $\mathfrak{P}_1, \ldots, \mathfrak{P}_m$ are the only prime divisors which divide $\mathfrak{p}$, that their ramification indices e_i all equal 1, and that $d_i = f_i$. This proves Theorem 8. Finally, we note that since $\varphi_i(\theta)$ is divisible by π_i but not divisible by any other prime element π_j, then π_i can be determined as the greatest common divisor in the ring $\mathfrak{O}_\mathfrak{p}$ of the elements $\varphi_i(\theta)$ and p.

Corollary. If the extension K/k is separable, then there are only finitely many prime divisors $\mathfrak{p}$ of the ring $\mathfrak{o}$ which are ramified in $\mathfrak{O}$.

Let θ be any primitive element of the extension K/k which is contained in $\mathfrak{O}$. The discriminant $D = D(1, \theta, \ldots, \theta^{n-1})$ is an element of $\mathfrak{o}^*$. If $\mathfrak{p} \nmid D$, then by the theorem $\mathfrak{p}$ is not ramified in $\mathfrak{O}$. Thus only those prime divisors of the ring $\mathfrak{o}$ which divide D can be ramified in $\mathfrak{O}$.

PROBLEMS

1. Let $\mathfrak{o}$ be a ring with a theory of divisors, k its quotient field, and $k \subset K \subset K'$ a tower of finite extensions. Let $\mathfrak{O}$ and $\mathfrak{O}'$ denote the integral closure of the ring $\mathfrak{o}$ in the fields K and K'. If $\mathfrak{P}'$ is any prime divisor of the ring $\mathfrak{O}'$, denote the prime divisor of the ring $\mathfrak{O}$ which is divisible by $\mathfrak{P}'$ by $\mathfrak{P}$, and the prime divisor of the ring $\mathfrak{o}$ which is divisible by $\mathfrak{P}'$ by $\mathfrak{p}$. Show that the degree of inertia of $\mathfrak{P}'$ relative to k equals the product of the degree of inertia of $\mathfrak{P}'$ relative to K and the degree of inertia of $\mathfrak{P}$ relative to k. Formulate and prove an analogous assertion for the index of ramification.

2. Let the ring $\mathfrak{o}$ with quotient field k have a theory of divisors with only a finite number of prime divisors, and let the prime divisor $\mathfrak{p}$ correspond to the prime element p of the ring $\mathfrak{o}$. Show that the residue class ring $\mathfrak{o}/(p)$ is isomorphic to the residue class field $\Sigma_{\mathfrak{p}}$ of the valuation $\nu_{\mathfrak{p}}$.

3. Let $\nu_{\mathfrak{p}}$ be a valuation of the field k, $\mathfrak{o}_{\mathfrak{p}}$ its ring, K/k a finite separable extension, $\mathfrak{O}_{\mathfrak{p}}$ the integral closure of the ring $\mathfrak{o}$ in the field K, and $\omega_1, \ldots, \omega_n$ a basis for the field K over k, all elements of which lie in the ring $\mathfrak{O}_{\mathfrak{p}}$. Show that if the discriminant $D(\omega_1, \ldots, \omega_n)$ is a unit in the ring $\mathfrak{o}_{\mathfrak{p}}$, then $\omega_1, \ldots, \omega_n$ is a fundamental basis for the ring $\mathfrak{O}_{\mathfrak{p}}$ over $\mathfrak{o}_{\mathfrak{p}}$.

4. Show that the homomorphism $N: \mathscr{D} \to \mathscr{D}_0$, satisfying the conditions of Theorem 4, is unique.

5. Show that the isomorphic embedding $\mathscr{D}_0 \to \mathscr{D}$, satisfying the conditions of Theorem 3, is unique.

6. Let $\mathfrak{a}$ be a divisor of the ring $\mathfrak{o}$. Considering it as a divisor of the ring $\mathfrak{O}$ (using the embedding $\mathscr{D} \to \mathscr{D}_0$) show that

$$N_{K/k}(\mathfrak{a}) = \mathfrak{a}^n \qquad (n = (K:k)).$$

7. Let K/k be a separable extension of degree n. Show that if the divisor $\mathfrak{a}$ of the ring $\mathfrak{o}$ becomes a principal divisor of the ring $\mathfrak{O}$, then $\mathfrak{a}^n$ is a principal divisor of $\mathfrak{o}$.

8. Let K/k be separable. Show that the norm $N_{K/k}(\mathfrak{a})$ of a divisor $\mathfrak{a}$ of the ring $\mathfrak{O}$ is the greatest common divisor of the principal divisors $(N_{K/k}(\alpha))_k$, where α runs through all elements of $\mathfrak{O}$ divisible by $\mathfrak{a}$.

9. The polynomial $f(t) = t^n + a_1 t^{n-1} + \cdots + a_n$ with coefficients in the ring $\mathfrak{o}$ is called an *Eisenstein polynomial* relative to the prime divisor $\mathfrak{p}$, if $a_1, \ldots, a_n$ are all divisible by $\mathfrak{p}$, and a_n, while divisible by $\mathfrak{p}$, is not divisible by $\mathfrak{p}^2$. If the ring $\mathfrak{O}$ contains a primitive element θ for the extension K/k of degree n, with the minimum polynomial of θ an Eisenstein polynomial relative to $\mathfrak{p}$, show that $\mathfrak{p}$ is divisible by only one prime divisor $\mathfrak{P}$ of the ring $\mathfrak{O}$ and that

$$\mathfrak{p} = \mathfrak{P}^n$$

(the degree of inertia of $\mathfrak{P}$ relative to $\mathfrak{p}$ hence equals 1).

10. Under the same hypotheses show that the basis $1, \theta, \ldots, \theta^{n-1}$ is a fundamental basis of the ring $\mathfrak{O}_{\mathfrak{p}}$ relative to $\mathfrak{o}_{\mathfrak{p}}$.

11. Let k_0 be any field of characteristic p and $k = k_0(x, y)$ a field of rational functions in x and y over the field k_0. Consider the valuation ν_0 of k, which was defined in Problem 9 of Section 4, where for the series $\xi(t) \in k_0\{t\}$ [transcendental over $k_0(t)$] we take a series of the form

$$\xi(t) = \eta(t)^p = \left(\sum_{n=0}^{\infty} a_n t^n\right)^p = \sum_{n=0}^{\infty} a_n{}^p t^{np} \qquad (a_n \in k_0).$$

By Problem 8 of Section 4 there is a unique extension of the valuation ν to the purely inseparable extension $K = k(\sqrt[p]{y})$ of degree p over k. Show that the ramification index of ν relative to ν_0 equals 1, and that the residue class field of the valuation ν coincides with the residue class field of the valuation ν_0 (under the inclusion isomorphism). It now follows from Theorem 5 and (5.12) that the ring $\mathfrak{O}$ of the valuation ν, which is the integral closure in K of the ring $\mathfrak{o}$ of the valuation ν_0, does not have a fundamental basis relative to $\mathfrak{o}$.

12. Under the same assumptions as in the preceding problem, give a direct proof (without involving Theorem 5) that $\mathfrak{O}$ has no fundamental basis over $\mathfrak{o}$.

13. Let $\mathfrak{o}$ be a ring with a theory of divisors, k its quotient field, K/k a finite extension of degree n, $\mathfrak{O}$ the integral closure of $\mathfrak{o}$ in K, $\mathfrak{p}$ a prime divisor of the ring $\mathfrak{o}$, $\mathfrak{P}_1, \dots, \mathfrak{P}_m$ the prime divisors of the ring $\mathfrak{O}$ which divide $\mathfrak{p}$, $e_1, \dots, e_m$ their ramification indices, and $f_1, \dots, f_m$ their degrees of inertia relative to $\mathfrak{p}$. For any $s = 1, \dots, m$ we denote by $\bar{\alpha}^{\mathfrak{P}_s}$ the residue class in the field $\Sigma_{\mathfrak{P}_s}$ which contains $\alpha \in \mathfrak{O}_{\mathfrak{P}_s}$. Choose elements $\omega_{si} \in \mathfrak{O}_\mathfrak{p}$ $(1 \leqslant i \leqslant f_s)$ so that the classes $\bar{\omega}_{si}{}^{\mathfrak{P}_s}$ form a basis for $\Sigma_{\mathfrak{P}_s}/\Sigma_\mathfrak{p}$ and also so that $\nu_{\mathfrak{P}_j}(\omega_{si}) \geqslant e_j$ for $j \neq s$, $1 \leqslant j \leqslant m$. Prime elements of the ring $\mathfrak{O}_\mathfrak{p}$, corresponding to the divisors $\mathfrak{P}_1, \dots, \mathfrak{P}_m$ are denoted by $\pi_1, \dots, \pi_m$. Show that the system of elements

$$\omega_{si}\pi_{sj} \qquad (s = 1, \dots, m;\ i = 1, \dots, f_s;\ j = 0, 1, \dots, e_s - 1) \qquad (*)$$

is linearly independent over k.

Hint: Consider linear combinations

$$\alpha = \sum c_{sij}\omega_{si}\pi_s{}^j$$

with coefficients from $\mathfrak{o}_\mathfrak{p}$, at least one of which is a unit in $\mathfrak{o}_\mathfrak{p}$. Let $\nu_\mathfrak{p}(c_{s_0 i_0 j_0}) = 0$, where j_0 is chosen so that $\nu_\mathfrak{p}(c_{s_0 ij}) > 0$ for all $j \leqslant j_0$ and all i. Then

$$\nu_{\mathfrak{P}_{s_0}}(\alpha) = j_0.$$

14. Show that if the extension K/k is separable, then the system (*) forms a fundamental basis for $\mathfrak{O}_\mathfrak{p}$ over $\mathfrak{o}_\mathfrak{p}$.

15. Show that if the extension K/k is separable, then for any $\alpha \in \mathfrak{O}_\mathfrak{p}$ we have

$$\overline{\mathrm{Sp}_{K/k}(\alpha)^\mathfrak{p}} = \sum_{s=1}^{m} e_s \mathrm{Sp}_{\Sigma_{\mathfrak{P}_s}/\Sigma_\mathfrak{p}}(\bar{\alpha}^{\mathfrak{P}_s}).$$

16. Let $f(t)$ be the characteristic polynomial of the element $\alpha \in \mathfrak{O}_\mathfrak{p}$ relative to K/k. Taking the corresponding residue classes in $\Sigma_\mathfrak{p}$, we obtain a polynomial $\bar{f}(t) \in \Sigma_\mathfrak{p}[t]$. For $s = 1, \dots, m$ let $\varphi_s(t)$ denote the characteristic polynomial of the element $\bar{\alpha}^{\mathfrak{P}_s} \in \Sigma_{\mathfrak{P}_s}$ relative to the extension $\Sigma_{\mathfrak{P}_s}/\Sigma_\mathfrak{p}$. Generalizing the preceding problem (for separable K/k), show that

$$\bar{f}(t) = \varphi_1(t)^{e_1} \cdots \varphi_m(t)^{e_m}.$$

17. Let K/k be separable. For each $\mathfrak{p}$ choose in the ring $\mathfrak{o}$ a fundamental basis $\alpha_1, \dots, \alpha_n$ over $\mathfrak{o}_\mathfrak{p}$. Set

$$d_\mathfrak{p} = \nu_\mathfrak{p}(D(\alpha_1, \dots, \alpha_n)).$$

Show that the integer $d_{\mathfrak{p}} \geqslant 0$ is almost always zero. The integral divisor

$$\mathfrak{d}_{K/k} = \prod_{\mathfrak{p}} \mathfrak{p}^{d_{\mathfrak{p}}}$$

of the ring $\mathfrak{o}$ is called the *discriminant* of the extension K/k (relative to the ring $\mathfrak{o}$).

18. Show that the prime divisor $\mathfrak{p}$ of the ring $\mathfrak{o}$ does not occur in the discriminant $\mathfrak{d}_{K/k}$ (that is, $d_{\mathfrak{p}} = 0$) if and only if $\mathfrak{p}$ is unramified in $\mathfrak{O}$ and the extensions $\Sigma_{\mathfrak{P}_s}/\Sigma_{\mathfrak{p}}$ $(s = 1, \ldots, m)$ are all separable.

19. Let the ring $\mathfrak{O}$ have a fundamental basis $\omega_1, \ldots, \omega_n$ over $\mathfrak{o}$. Show that the discriminant $\mathfrak{d}_{K/k}$ coincides with the principal divisor $(D(\omega_1, \ldots, \omega_n))$.

6. Dedekind Rings

6.1. *Congruences Modulo Divisors*

We consider a ring $\mathfrak{O}$ with quotient field K for which there exists a theory of divisors $\mathfrak{O}^* \to \mathscr{D}$.

Definition. We say that the elements α and β of the ring $\mathfrak{O}$ are *congruent modulo* the divisor $\mathfrak{a} \in \mathscr{D}$, and write

$$\alpha \equiv \beta \pmod{\mathfrak{a}},$$

if the difference $\alpha - \beta$ is divisible by $\mathfrak{a}$.

In the case of a principal divisor (μ) the congruence $\alpha \equiv \beta \pmod{(\mu)}$ is clearly equivalent to the congruence $\alpha \equiv \beta \pmod{\mu}$ in the sense of the definition of Section 4.1 of the Supplement.

We indicate some elementary properties of congruences which easily follow from the definition.

(1) Congruences modulo $\mathfrak{a}$ can be added and multiplied termwise.

(2) If a congruence holds modulo $\mathfrak{a}$, then it also holds modulo $\mathfrak{b}$ for any divisor $\mathfrak{b}$ dividing $\mathfrak{a}$.

(3) If a congruence holds modulo $\mathfrak{a}$ and modulo $\mathfrak{b}$, then it also holds modulo their least common multiple.

(4) If an element $\alpha \in \mathfrak{O}$ is relatively prime to $\mathfrak{a}$ [that is, if the divisors (α) and $\mathfrak{a}$ are relatively prime], then from the congruence $\alpha\beta \equiv 0 \pmod{\mathfrak{a}}$ it follows that $\beta \equiv 0 \pmod{\mathfrak{a}}$.

(5) If α divides both sides of a congruence modulo $\mathfrak{a}$, and α is relatively prime to $\mathfrak{a}$, then we may cancel α from the congruence.

(6) If $\mathfrak{p}$ is a prime divisor and $\alpha\beta \equiv 0 \pmod{\mathfrak{p}}$ then either $\alpha \equiv 0 \pmod{\mathfrak{p}}$ or $\beta \equiv 0 \pmod{\mathfrak{p}}$.

It follows from property (1) that the residue classes of the ring $\mathfrak{O}$ modulo a given divisor $\mathfrak{a}$ can be added and multiplied. It is easily verified that under these operations the set of residue classes becomes a ring. It is called the *ring of residue classes* modulo the divisor $\mathfrak{a}$ and is denoted by $\mathfrak{O}/\mathfrak{a}$.

Property (6) can then be interpreted as saying that for a prime divisor $\mathfrak{p}$ the ring $\mathfrak{O}/\mathfrak{p}$ has no divisors of zero.

Assume now that $\mathfrak{O}$ is the maximal order of an algebraic number field K. The divisors of the ring $\mathfrak{O}$ we call in this case the *divisors* of the field K.

Since every divisor $\mathfrak{a}$ of the field K divides some nonzero number $\alpha \in \mathfrak{O}$, and the number α in its turn divides some natural number a [for example, $|N(\alpha)|$ is divisible by α], then for each divisor $\mathfrak{a}$ there is a natural number a which is divisible by $\mathfrak{a}$. By property (2) numbers in distinct residue classes modulo $\mathfrak{a}$ remain in distinct classes modulo a. Recalling now that in the order $\mathfrak{O}$ the number of residue classes modulo a is finite (actually equal to a^n, where n is the degree of the field K; see the proof of Theorem 5 of Section 2, Chapter 2), we obtain the following theorem.

Theorem 1. For any divisor $\mathfrak{a}$ of the algebraic number field K, the residue class ring $\mathfrak{O}/\mathfrak{a}$ is finite.

Let $\mathfrak{p}$ be any prime divisor of the field K. The corresponding valuation $\nu_\mathfrak{p}$ induces on R the p-adic valuation ν_p for some prime p. Since $\nu_p(p) = 1$, then $\nu_\mathfrak{p}(p) > 0$; that is, $p \equiv 0 \pmod{\mathfrak{p}}$. If the prime number q is different from p, then $\nu_p(q) = 0$, and therefore $\nu_\mathfrak{p}(q) = 0$; that is, $q \not\equiv 0 \pmod{\mathfrak{p}}$.

The residue class ring $\mathfrak{O}/\mathfrak{p}$, being finite and without divisors of zero, is a finite field (Supplement, Section 3). Since for any $\alpha \in \mathfrak{O}$ we have $p\alpha \equiv 0 \pmod{\mathfrak{p}}$, then the characteristic of this field is p. Hence we have

Theorem 2. Any prime divisor $\mathfrak{p}$ of an algebraic number field divides one and only one rational prime p. The residue class ring $\mathfrak{O}/\mathfrak{p}$ is a finite field of characteristic p.

A theory of divisors for an algebraic number field hence has the property that the residue class ring modulo a prime divisor is a field. In general, this is not the case. For example, in the ring of polynomials $k[x, y]$ in two variables over a field k the residue class ring of the prime divisor (x) is isomorphic to the ring of polynomials $k[y]$ and hence is not a field.

The residue class ring $\mathfrak{O}/\mathfrak{p}$ is a field if and only if the congruence $\alpha\xi \equiv 1 \pmod{\mathfrak{p}}$ is always solvable when $\alpha \not\equiv 0 \pmod{\mathfrak{p}}$. Hence only under this assumption can we expect to construct a completely adequate theory of congruences in the ring $\mathfrak{O}$.

6.2. *Congruences in Dedekind Rings*

Definition. A ring $\mathfrak{O}$ is called a Dedekind ring if it has a theory of divisors $\mathfrak{O}^* \to \mathscr{D}$ and for every prime divisor $\mathfrak{p} \in \mathscr{D}$ the residue class ring $\mathfrak{O}/\mathfrak{p}$ is a field.

Examples of Dedekind rings, other than the maximal orders of algebraic number fields, can be obtained by taking the integral closure of the polynomial ring $k[x]$ in a single variable in a finite extension of the field of rational functions $f(x)$ (Problems 1 and 2). The valuation ring $\mathfrak{O}_\nu$ of any valuation ν is also a Dedekind ring (see Section 4.1), as is any ring which has a theory of divisors with only a finite number of prime divisors (Problem 3).

Lemma 1. If $\mathfrak{O}$ is a Dedekind ring and $\alpha \in \mathfrak{O}$ is not divisible by the prime divisor $\mathfrak{p}$, then the congruence $\alpha\xi \equiv 1 \pmod{\mathfrak{p}^m}$ is solvable in $\mathfrak{O}$ for any natural number m.

Proof. For $m = 1$ the congruence is solvable by the definition of a Dedekind ring. The lemma will be proved by induction on m. Suppose that for some $\xi_0 \in \mathfrak{O}$ we have $\alpha\xi_0 \equiv 1 \pmod{\mathfrak{p}^m}$. Choose an element ω in the ring $\mathfrak{O}$ for which $\nu_\mathfrak{p}(\omega) = m$. The principal divisor (ω) has the form $(\omega) = \mathfrak{p}^m\mathfrak{a}$, where $\mathfrak{a}$ is not divisible by $\mathfrak{p}$. Choose an element $\gamma \in \mathfrak{O}$ for which $\nu_\mathfrak{p}(\gamma) = 0$ and $\gamma \equiv 0 \pmod{\mathfrak{a}}$. The product $\gamma(\alpha\xi_0 - 1)$ will be divisible by $\mathfrak{p}^m\mathfrak{a} = (\omega)$, and hence $\gamma(\alpha\xi_0 - 1) = \omega\mu$ with $\mu \in \mathfrak{O}$. We now try to solve the congruence $\alpha\xi \equiv 1 \pmod{\mathfrak{p}^{m+1}}$, taking as ξ an element in the form $\xi = \xi_0 + \omega\lambda$, where λ is to be chosen suitably in $\mathfrak{O}$. Since

$$\gamma(\alpha\xi - 1) = \gamma(\alpha\xi_0 - 1) + \gamma\alpha\omega\lambda = \omega(\mu + \gamma\alpha\lambda)$$

and $\omega \equiv 0 \pmod{\mathfrak{p}^m}$, then we shall achieve our goal if λ satisfies the congruence $\lambda\alpha\gamma \equiv -\mu \pmod{\mathfrak{p}}$. But since $\alpha\gamma$ is not divisible by $\mathfrak{p}$, this congruence is solvable. Hence there is an element $\xi \in \mathfrak{O}$ for which $\gamma(\alpha\xi - 1) \equiv 0 \pmod{\mathfrak{p}^{m+1}}$ and since $\nu_\mathfrak{p}(\gamma) = 0$, dividing by γ, we obtain $\alpha\xi - 1 \equiv 0 \pmod{\mathfrak{p}^{m+1}}$. Lemma 1 is proved.

Theorem 3. If $\mathfrak{p}_1, \ldots, \mathfrak{p}_m$ are distinct prime divisors of the Dedekind ring $\mathfrak{O}$, and $\beta_1, \ldots, \beta_m$ are any elements of $\mathfrak{O}$, then there is an element ξ in $\mathfrak{O}$ which satisfies

$$\xi \equiv \beta_1 \pmod{{\mathfrak{p}_1}^{k_1}},$$
$$\cdot \quad \cdot \quad \cdot \quad \cdot \quad \cdot \quad \cdot$$
$$\xi \equiv \beta_m \pmod{{\mathfrak{p}_m}^{k_m}}$$

($k_1, \ldots, k_m$ are any natural numbers).

Proof. For each divisor

$$\mathfrak{a}_i = {\mathfrak{p}_1}^{k_1} \cdots \mathfrak{p}_{i-1}^{k_{i-1}}\mathfrak{p}_{i+1}^{k_{i+1}} \cdots {\mathfrak{p}_m}^{k_m} \qquad (i = 1, \ldots, m)$$

we can find an element $\alpha_i \in \mathfrak{O}$ which is divisible by $\mathfrak{a}_i$ but not divisible by $\mathfrak{p}_i$. Lemma 1 guarantees that we can solve the congruence $\alpha_i\xi_i \equiv \beta_i \pmod{\mathfrak{p}_i^{k_i}}$ in $\xi_i \in \mathfrak{O}$. It is easily seen that the element

$$\xi = \alpha_1\xi_1 + \cdots + \alpha_m\xi_m$$

satisfies the requirements of the theorem.

Theorem 4. If $\alpha \neq 0$ and β are elements of a Dedekind ring $\mathfrak{O}$, then the congruence

$$\alpha\xi \equiv \beta \pmod{\mathfrak{a}} \tag{6.1}$$

is solvable if and only if β is divisible by the greatest common divisor of the divisors (α) and $\mathfrak{a}$.

Proof. We first assume that the divisors (α) and $\mathfrak{a}$ are relatively prime, and will show that in this case the congruence (6.1) is solvable for any β. Let $\mathfrak{a} = \mathfrak{p}_1^{k1} \cdots \mathfrak{p}_m^{km} = \mathfrak{p}_i^{k_i}\mathfrak{a}_i$, where $\mathfrak{p}_1, \ldots, \mathfrak{p}_m$ are distinct prime divisors. By Lemma 1 for each $i = 1, \ldots, m$ in the ring $\mathfrak{O}$ there is an element ξ_i' such that $\alpha\xi_i' \equiv \beta \pmod{\mathfrak{p}_i^{k_i}}$. By Theorem 3 we can find for each i an element ξ_i for which $\xi_i \equiv \xi_i' \pmod{\mathfrak{p}_i^{k_i}}$ and $\xi_i \equiv 0 \pmod{\mathfrak{a}_i}$. It is now clear that the sum $\xi_1 + \cdots + \xi_m = \xi$ will satisfy the congruences $\alpha\xi \equiv \beta \pmod{\mathfrak{p}_i^{k_i}}$ for $i = 1, \ldots, m$, and hence will also satisfy (6.1).

We now prove the theorem in general. Let $\mathfrak{d} = \mathfrak{p}_1^{l_1} \cdots \mathfrak{p}_m^{l_m}$ be the greatest common divisor of the divisors (α) and $\mathfrak{a}$. If (6.1) holds modulo $\mathfrak{a}$, then it also holds modulo $\mathfrak{d}$, and since $\alpha \equiv 0 \pmod{\mathfrak{d}}$, then we must also have $\beta = 0 \pmod{\mathfrak{d}}$. This proves the necessity of this condition.

Assume now that β is divisible by $\mathfrak{d}$. By Theorem 3 of Section 4, there is an element $\mu \in K$ for which

$$v_{\mathfrak{p}_i}(\mu) = -\, l_i \qquad (i = 1, \ldots, m). \tag{6.2}$$

We shall show that we can choose μ so that also

$$v_{\mathfrak{q}}(\mu) \geqslant 0 \tag{6.3}$$

for all prime divisors $\mathfrak{q}$, distinct from $\mathfrak{p}_1, \ldots, \mathfrak{p}_m$. Suppose that μ does not satisfy condition (6.3), and let $\mathfrak{q}_1, \ldots, \mathfrak{q}_s$ be the prime divisors, different from $\mathfrak{p}_1, \ldots, \mathfrak{p}_m$, for which $v_{\mathfrak{q}_j}(\mu) = -r_j < 0$. Choose in $\mathfrak{O}$ an element γ such that $v_{\mathfrak{q}_j}(\gamma) = r_j$ $(1 \leqslant j \leqslant s)$ and $v_{\mathfrak{p}_i}(\gamma) = 0$ $(1 \leqslant i \leqslant m)$. It is clear that the element $\mu' = \mu\gamma$ satisfies both conditions (6.2) and (6.3), and our assertion is proved. Let the divisor $\mathfrak{b}$ be determined by $\mathfrak{a} = d\mathfrak{b}$. If μ satisfies conditions (6.2) and (6.3), then the element $\alpha\mu$ belongs to $\mathfrak{O}$ and is relatively prime to $\mathfrak{b}$. Since we have assumed β divisible by d, then $\beta\mu$ also belongs to $\mathfrak{O}$. We have already proved that then there exists an element ξ in the ring $\mathfrak{O}$ such that $\alpha\mu\xi \equiv \beta\mu$ (mod $\mathfrak{b}$). For $i = 1, \ldots, m$ we have

$$v_{\mathfrak{p}_i}(\alpha\xi - \beta) = v_{\mathfrak{p}_i}(\alpha\mu\xi - \beta\mu) + l_i \geqslant k_i - l_i + l_i = k_i,$$

and this means that ξ satisfies (6.1).

6.3. *Divisors and Ideals*

In this section we show that there is a one-to-one correspondence between divisors and nonzero ideals in a Dedekind ring.

For each divisor $\mathfrak{a}$ denote by $\bar{\mathfrak{a}}$ the set of all elements of the ring $\mathfrak{O}$ which are divisible by $\mathfrak{a}$. It is clear that $\bar{\mathfrak{a}}$ is a nonzero ideal of the ring $\mathfrak{O}$.

Theorem 5. In a Dedekind ring $\mathfrak{O}$ the mapping $\mathfrak{a} \to \bar{\mathfrak{a}}$ $(\mathfrak{a} \in \mathscr{D})$ is an isomorphism of the semigroup of divisors $\mathscr{D}$ onto the semigroup of all nonzero ideals of the ring $\mathfrak{O}$.

We first verify the following lemma.

Lemma 2. If $\alpha_1, \ldots, \alpha_s$ are any nonzero elements of the Dedekind ring $\mathfrak{O}$ and $\mathfrak{d}$ is the greatest common divisor of the principal divisors $(\alpha_1), \ldots, (\alpha_s)$, then any element $\alpha \in \mathfrak{O}$ which is divisible by $\mathfrak{d}$ can be written in the form

$$\alpha = \xi_1\alpha_1 + \cdots + \xi_s\alpha_s \qquad (\xi_i \in \mathfrak{O}).$$

The proof of the lemma is by induction on s. For $s = 1$ the lemma is obvious. Let $s \geqslant 2$. Let $\mathfrak{d}_1$ denote the greatest common divisor of the divisors $(\alpha_1), \ldots, (\alpha_{s-1})$. Then $\mathfrak{d}$ is the greatest common divisor of the divisors $\mathfrak{d}_1$ and (α_s). Let α be divisible by $\mathfrak{d}$. By Theorem 4 the congruence $\alpha_s\xi \equiv \alpha$ (mod $\mathfrak{d}_1$) has a solution $\xi \in \mathfrak{O}$. By the induction hypothesis there are elements $\xi_1, \ldots, \xi_{s-1}$ in the ring $\mathfrak{O}$ such that $\alpha - \xi\alpha_s = \xi_1\alpha_1 + \cdots + \xi_{s-1}\alpha_{s-1}$. Lemma 2 is proved.

Proof of Theorem 5. By condition (3) of the definition of a theory of divisors the mapping $\mathfrak{a} \to \bar{\mathfrak{a}}$ takes distinct divisors to distinct ideals.

Let A be any nonzero ideal of the ring $\mathfrak{O}$. For each prime divisor $\mathfrak{p}$ set

$$a(\mathfrak{p}) = \min_{\alpha \in A} v_{\mathfrak{p}}(\alpha).$$

It is clear that $a(\mathfrak{p})$ will be nonzero for only a finite number of prime divisors $\mathfrak{p}$. Hence the product $\mathfrak{a} = \prod_{\mathfrak{p}} \mathfrak{p}^{a(\mathfrak{p})}$, in which $\mathfrak{p}$ runs through all prime divisors for which $a(\mathfrak{p}) \neq 0$, is a divisor. We shall show that $\bar{\mathfrak{a}} = A$. Let α be any element of $\bar{\mathfrak{a}}$. It is clear that we can find a finite set of elements $\alpha_1, \ldots, \alpha_s$ in A such that $a(\mathfrak{p}) = \min(v_{\mathfrak{p}}(\alpha_1), \ldots, v_{\mathfrak{p}}(\alpha_s))$ for all $\mathfrak{p}$. This means that the divisor $\mathfrak{a}$ is the greatest common divisor of the principal divisors $(\alpha_1), \ldots, (\alpha_s)$. By Lemma 2 the element α can be represented in the form $\alpha = \xi_1\alpha_1 + \cdots + \xi_s\alpha_s$ with coefficients $\xi_i \in \mathfrak{O}$. It follows that $\alpha \in A$, and hence that $\bar{\mathfrak{a}} \subset A$. Since it is clear that $A \subset \bar{\mathfrak{a}}$, we obtain $A = \bar{\mathfrak{a}}$. We have thus proved that $\mathfrak{a} \to \bar{\mathfrak{a}}$ is a one-to-one mapping of the set of all divisors of the ring $\mathfrak{O}$ onto the set of all nonzero ideals of $\mathfrak{O}$.

We shall now show that this mapping is an isomorphism, that is, that for any two divisors $\mathfrak{a}$ and $\mathfrak{b}$ we have

$$\bar{\mathfrak{a}}\bar{\mathfrak{b}} = \overline{\mathfrak{a}\mathfrak{b}}. \tag{6.4}$$

Denote the product $\bar{\mathfrak{a}}\bar{\mathfrak{b}}$ by C. Since C is a nonzero ideal of $\mathfrak{O}$, there is a divisor $\mathfrak{c}$ such that $\bar{\mathfrak{c}} = C$. We must prove that $\mathfrak{c} = \mathfrak{a}\mathfrak{b}$. Let the prime divisor $\mathfrak{p}$ enter into the divisors $\mathfrak{a}$ and $\mathfrak{b}$ with exponents a and b. Then

$$\min_{\gamma \in C} v_{\mathfrak{p}}(\gamma) = \min_{\alpha \in \bar{\mathfrak{a}}, \beta \in \bar{\mathfrak{b}}} v_{\mathfrak{p}}(\alpha\beta) = \min_{\alpha \in \bar{\mathfrak{a}}} v_{\mathfrak{p}}(\alpha) + \min_{\beta \in \bar{\mathfrak{b}}} v_{\mathfrak{p}}(\beta) = a + b.$$

Since this is true for all prime divisors $\mathfrak{p}$, then $\mathfrak{c} = \mathfrak{a}\mathfrak{b}$ and (6.4) is proved.

From the fact that the mapping $\mathfrak{a} \to \bar{\mathfrak{a}}$ is an isomorphism, it follows, in particular, that the set of all nonzero ideals of the Dedekind ring $\mathfrak{O}$ form a semigroup with unique factorization under the operation of multiplication. To construct a theory of divisors in Dedekind rings (in particular, in the maximal order of an algebraic number field), we could take the semigroup of nonzero ideals for the semigroup $\mathscr{D}$. The image of the element α under the homomorphism $\mathfrak{O}^* \to \mathscr{D}$ would then be the principal ideal (α) generated by this element. This construction of a theory of divisors is due to Dedekind.

6.4. *Fractional Divisors*

If we construct a theory of divisors $\mathfrak{O}^* \to \mathscr{D}$ for the ring $\mathfrak{O}$, then we obtain some information on the structure of the semigroup $\mathfrak{O}^*$. It is natural to try an analogous procedure with the multiplicative group K^* of the quotient field K. To do this we need to extend the concept of a divisor.

Following an established tradition, we shall reserve the term "divisor" for this broader concept, and will call divisors in the earlier sense "integral divisors."

Definition. Let $\mathfrak{O}$ be a ring with a theory of divisors, with quotient field K, and let $\mathfrak{p}_1, \ldots, \mathfrak{p}_m$ be a finite system of prime divisors. An expression

$$\mathfrak{a} = \mathfrak{p}_1^{k_1} \cdots \mathfrak{p}_m^{k_m} \tag{6.5}$$

with integer exponents $k_1, \ldots, k_m$ (not necessarily positive) is called a divisor of the field K. If all the exponents k_i are nonnegative, then the divisor is called *integral* (or a divisor of the ring $\mathfrak{O}$). Otherwise it is called *fractional.*

It is sometimes convenient to write a divisor (6.5) as a formal infinite product

$$\mathfrak{a} = \prod_{\mathfrak{p}} \mathfrak{p}^{a(\mathfrak{p})}, \tag{6.6}$$

over all prime divisors $\mathfrak{p}$, in which almost all exponents $a(\mathfrak{p})$ are zero.

Multiplication of divisors is determined by the formula

$$\left(\prod_{\mathfrak{p}} \mathfrak{p}^{a(\mathfrak{p})}\right)\left(\prod_{\mathfrak{p}} \mathfrak{p}^{b(\mathfrak{p})}\right) = \prod_{\mathfrak{p}} \mathfrak{p}^{a(\mathfrak{p})+b(\mathfrak{p})}.$$

For integral divisors this definition coincides with the definition of multiplication in the semigroup $\mathscr{D}$. It is easily seen that under this operation the set of all divisors of the field K is an Abelian group, which we shall denote by $\hat{\mathscr{D}}$. The unit element of this group is the divisor $\mathfrak{e}$ for which all exponents $a(\mathfrak{p})$ in (6.6) are zero.

Since every nonzero element $\xi \in K$ is the quotient of two elements of $\mathfrak{O}$, it follows from condition (1) of Theorem 4 of Section 3 that for all but a finite number of the valuations $\nu_{\mathfrak{p}}$, which correspond to the prime divisors $\mathfrak{p}$, we have $\nu_{\mathfrak{p}}(\xi) = 0$. We denote this finite set by $\nu_{\mathfrak{p}_1}, \ldots, \nu_{\mathfrak{p}_m}$. The divisor

$$\prod_{i=1}^{m} \mathfrak{p}_i^{\nu_{\mathfrak{p}_i}(\xi)} = \prod_{\mathfrak{p}} \mathfrak{p}^{\nu_{\mathfrak{p}}(\xi)}$$

is called the *principal divisor* corresponding to the element $\xi \in K^*$, and is denoted by (ξ). When applied to elements of the ring $\mathfrak{O}$, the new concept of a principal divisor coincides with the previous one (Section 3.4). By condition (2) of Theorem 4 of Section 3 the principal divisor (ξ) will be integral if and only if ξ belongs to $\mathfrak{O}$.

From the definition of a valuation (Section 3.4) it easily follows that the mapping $\xi \to (\xi)$, $\xi \in K^*$, is a homomorphism $K^* \to \hat{\mathscr{D}}$ of the multiplicative group of the field K to the group of divisors $\hat{\mathscr{D}}$. By Theorem 2 of Section 3 this homomorphism maps onto the entire group $\hat{\mathscr{D}}$ (that is, is an epimorphism) if and only if $\mathfrak{O}$ has unique factorization. The kernel of this map clearly is the group of units of the ring $\mathfrak{O}$, and this means that for elements ξ, η of K^* we have $(\xi) = (\eta)$ if and only if $\xi = \eta\varepsilon$, where ε is a unit of the ring $\mathfrak{O}$.

We now define a concept of divisibility for arbitrary divisors. Let $\mathfrak{a} = \prod_{\mathfrak{p}} \mathfrak{p}^{a(\mathfrak{p})}$ and $\mathfrak{b} = \prod_{\mathfrak{p}} \mathfrak{p}^{b(\mathfrak{p})}$ be two divisors (not necessarily integral). We say that $\mathfrak{a}$ is divisible by $\mathfrak{b}$ if there is an integral divisor $\mathfrak{c}$ such that $\mathfrak{a} = \mathfrak{b}\mathfrak{c}$. In other words, $\mathfrak{a}$ is divisible by $\mathfrak{b}$ if and only if $a(\mathfrak{p}) \geqslant b(\mathfrak{p})$ for all $\mathfrak{p}$.

For any $\mathfrak{a}$ and $\mathfrak{b}$ set $d(\mathfrak{p}) = \min(a(\mathfrak{p}), b(\mathfrak{p}))$. Since the rational integer $d(\mathfrak{p})$ is equal to zero for almost all $\mathfrak{p}$, then the product $\mathfrak{d} = \prod_{\mathfrak{p}} \mathfrak{p}^{d(\mathfrak{p})}$ is a divisor. The divisor $\mathfrak{d}$ is called the *greatest common divisor* of the divisors $\mathfrak{a}$ and $\mathfrak{b}$ ($\mathfrak{a}$ and $\mathfrak{b}$ are both divisible by $\mathfrak{d}$ and $\mathfrak{d}$ is divisible by every common divisor of

$\mathfrak{a}$ and $\mathfrak{b}$). The least common multiple of the divisors $\mathfrak{a}$ and $\mathfrak{b}$ is defined analogously.

The element $\alpha \in K$ is called *divisible by the divisor* $\mathfrak{a} = \prod_{\mathfrak{p}} \mathfrak{p}^{a(\mathfrak{p})}$, if $\alpha = 0$ or the principal divisor (α) is divisible by $\mathfrak{a}$. In terms of valuations this is characterized by $v_{\mathfrak{p}}(\alpha) \geqslant a(\mathfrak{p})$ for all $\mathfrak{p}$.

The correspondence of the preceding section between integral divisors and ideals of a Dedekind ring can be extended to fractional divisors, providing the proper generalization of the concept of ideal is used.

As in Section 6.3, we denote the set of all elements of the field K which are divisible by the divisor $\mathfrak{a}$ by $\bar{\mathfrak{a}}$ (these elements may now be nonintegral). From condition (3) in the definition of a valuation (Section 3.4) it follows that if α and β are divisible by $\mathfrak{a}$, then $\alpha \pm \beta$ are also divisible by $\mathfrak{a}$. This means that the set $\bar{\mathfrak{a}}$ is a group under addition. Further, if $\alpha \in \bar{\mathfrak{a}}$ and $\xi \in \mathfrak{O}$, then the product $\xi\alpha$ also belongs to $\bar{\mathfrak{a}}$. We now verify the following formula:

$$\overline{(\gamma)\mathfrak{a}} = \gamma\bar{\mathfrak{a}} \qquad (\gamma \in K^*,\ \mathfrak{a} \in \mathfrak{O}). \tag{6.7}$$

For the element ξ is divisible by $(\gamma)\mathfrak{a}$ if and only if any of the following hold: $v_{\mathfrak{p}}(\xi) \geqslant v_{\mathfrak{p}}(\gamma) + a(\mathfrak{p})$ for all $\mathfrak{p}$; $v_{\mathfrak{p}}(\xi/\gamma) \geqslant a(\mathfrak{p})$ for all $\mathfrak{p}$; $\xi/\gamma \in \mathfrak{a}$; $\xi \in \gamma\mathfrak{a}$ [here $a(\mathfrak{p})$ denotes the power to which $\mathfrak{p}$ appears in the divisor $\mathfrak{a}$]. It is clear that for any divisor we can find an element $\gamma \in \mathfrak{O}^*$ such that the divisor $(\gamma)\mathfrak{a}$ is integral. Formula (6.7) shows that for such a γ we will have $\gamma\bar{\mathfrak{a}} \subset \mathfrak{O}$.

Definition. Let $\mathfrak{O}$ be a Dedekind ring with quotient field K. A subset $A \subset K$, containing at least one nonzero element, is called an *ideal* of the field K (relative to $\mathfrak{O}$), if it satisfies:

(1) A is a group under the operation of addition.

(2) For any $\alpha \in A$ and any $\xi \in \mathfrak{O}$, the product $\xi\alpha$ lies in A.

(3) There is a nonzero element γ of the field K such that $\gamma A \subseteq \mathfrak{O}$.

The ideal A is called *integral* if it is contained in $\mathfrak{O}$ and otherwise is called *fractional.*

An integral ideal in K is clearly just a nonzero ideal in $\mathfrak{O}$.

If A and B are two ideals of the field K, then by their product AB we mean the set of all elements $\gamma \in K$ which can be represented in the form

$$\gamma = \alpha_1\beta_1 + \cdots + \alpha_m\beta_m \quad (m \geqslant 1) \qquad \alpha_i \in A,\ \beta_i \in B \quad (1 \leqslant i \leqslant m).$$

It is clear that the product of two ideals of a field K is again an ideal of the field K. (When the ideals are integral, the definition of product coincides with the usual notion of the product of two ideals in a commutative ring.)

We have already verified that for any divisor $\mathfrak{a}$ of K, the set $\bar{\mathfrak{a}}$ is an ideal in K. Assume that for two divisors $\mathfrak{a}$ and $\mathfrak{b}$ we have $\bar{\mathfrak{a}} = \bar{\mathfrak{b}}$. Choose a nonzero element γ so that the divisors $(\gamma)\mathfrak{a}$ and $(\gamma)\mathfrak{b}$ are both integral. From formula (6.7) we have $\overline{(\gamma)\mathfrak{a}} = \overline{(\gamma)\mathfrak{b}}$, so that $(\gamma)\mathfrak{a} = (\gamma)\mathfrak{b}$ and hence $\mathfrak{a} = \mathfrak{b}$. Hence the mapping $\mathfrak{a} \to \bar{\mathfrak{a}}$ is a monomorphism. Now let A be any ideal of the field K. If the element $\gamma \neq 0$ is chosen so that $\gamma A \subset \mathfrak{O}$, then γA will be a nonzero ideal of the ring $\mathfrak{O}$, and hence by Theorem 5 there exists an integral divisor $\mathfrak{c}$ such that $\bar{\mathfrak{c}} = \gamma A$. Set $\mathfrak{a} = \mathfrak{c}(\gamma)^{-1}$. Then $\gamma A = \overline{(\gamma)\mathfrak{a}} = \gamma\bar{\mathfrak{a}}$, so that $A = \bar{\mathfrak{a}}$. Thus each ideal of the field K is the image of some divisor under the mapping $\mathfrak{a} \to \bar{\mathfrak{a}}$. If $\mathfrak{a}$ and $\mathfrak{b}$ are two divisors, then, taking elements $\gamma \neq 0$ and $\gamma' \neq 0$ so that $(\gamma)\mathfrak{a}$ and $(\gamma')\mathfrak{b}$ are integral divisors, we have [from Theorem 5 and formula (6.7)]

$$\gamma\gamma'\overline{\mathfrak{a}\mathfrak{b}} = \overline{(\gamma)\mathfrak{a}\cdot(\gamma')\mathfrak{b}} = \overline{(\gamma)\mathfrak{a}}\cdot\overline{(\gamma')\mathfrak{b}} = \gamma\bar{\mathfrak{a}}\cdot\gamma'\bar{\mathfrak{b}} = \gamma\gamma'\bar{\mathfrak{a}}\bar{\mathfrak{b}},$$

so that $\overline{\mathfrak{a}\mathfrak{b}} = \bar{\mathfrak{a}}\bar{\mathfrak{b}}$. The mapping $\mathfrak{a} \to \bar{\mathfrak{a}}$ hence is an isomorphism. It follows that the set of all ideals of the field K is a group under multiplication. The unit element in this group is the ring $\mathfrak{O} = \bar{\mathfrak{e}}$. The inverse of the ideal $\bar{\mathfrak{a}}$ will be the ideal $\overline{\mathfrak{a}^{-1}}$.

We formulate this generalization of Theorem 5.

Theorem 6. Let $\mathfrak{O}$ be a Dedekind ring with quotient field K. For every divisor $\mathfrak{a}$, denote by $\bar{\mathfrak{a}}$ the set of all elements of K which are divisible by $\mathfrak{a}$. The mapping $\mathfrak{a} \to \bar{\mathfrak{a}}$ is an isomorphism of the group of all divisors of the field K onto the group of all ideals of the field K. This mapping takes integral divisors to integral ideals and conversely.

PROBLEMS

1. Show that the ring $k[x]$ of polynomials in one variable over a field k is Dedekind.

2. Let $\mathfrak{o}$ be a Dedekind ring with quotient field k. Show that the integral closure $\mathfrak{O}$ of the ring $\mathfrak{o}$ in any finite extension of the field k is also Dedekind.

3. Show that any ring which has a theory of divisors with a finite number of prime divisors is Dedekind.

4. Show that a system of congruences

$$\xi \equiv \alpha_1 \pmod{\mathfrak{a}_1},$$
$$\cdots\cdots\cdots$$
$$\xi \equiv \alpha_m \pmod{\mathfrak{a}_m}$$

in a Dedekind ring is solvable if and only if $\alpha_i \equiv \alpha_j \pmod{\mathfrak{d}_{ij}}$, $i \neq j$, where $\mathfrak{d}_{ij}$ is the greatest common divisor of the divisors $\mathfrak{a}_i$ and $\mathfrak{a}_j$.

5. Let $\mathfrak{O}$ be a Dedekind ring and $\mathfrak{a}$ a divisor of $\mathfrak{O}$. Show that the set of those residue classes in $\mathfrak{O}/\mathfrak{a}$ which consist of elements relatively prime to $\mathfrak{a}$ is a group under the operation of multiplication.

6. Let $f(x)$ be a polynomial of degree m with coefficients in the Dedekind ring $\mathfrak{O}$, with not all coefficients divisible by a prime divisor $\mathfrak{p}$. Show that the congruence $f(x) \equiv 0 \pmod{\mathfrak{p}}$ has at most m solutions (noncongruent modulo $\mathfrak{p}$), in $\mathfrak{O}$.

7. Let $\mathfrak{O}$ be a Dedekind ring, $\mathfrak{p}$ a prime divisor of $\mathfrak{O}$, and $f(x)$ a polynomial with coefficients in $\mathfrak{O}$. If for some element $\alpha \in \mathfrak{O}$ we have

$$f(\alpha) \equiv 0 \pmod{\mathfrak{p}}, \qquad f'(\alpha) \not\equiv 0 \pmod{\mathfrak{p}},$$

show that for every $m \geqslant 2$ there exists an element ξ in the ring $\mathfrak{O}$ such that

$$f(\xi) \equiv 0 \pmod{\mathfrak{p}^m}, \qquad \xi \equiv \alpha \pmod{\mathfrak{p}}.$$

8. Show that in a Dedekind ring every ideal is either principal or is generated by two elements.

9. Let $\mathfrak{O}$ be a Dedekind ring with quotient field K. Show that under the isomorphism $\mathfrak{a} \to \bar{\mathfrak{a}}$ of the group of divisors of the field K onto the group of ideals of the field K, the greatest common divisor of divisors corresponds to the sum of the corresponding ideals, and the least common multiple of divisors corresponds to the intersection of the corresponding ideals. (By the sum $A + B$ of the ideals A and B we mean the set of all sums $\alpha + \beta$, where $\alpha \in A$ and $\beta \in B$.)

10. The ring $\mathfrak{O} = k[x, y]$ of polynomials in two variables over the field k has unique factorization and hence has a theory of divisors. Show that the ideal $A = (x, y)$ of the ring $\mathfrak{O}$ which is generated by the elements x and y does not correspond to any divisor.

11. Show that if $\mathfrak{O}$ is a ring with a theory of divisors $\mathfrak{O}^* \to \mathscr{D}$ in which every nonzero ideal of $\mathfrak{O}$ is of the form $\bar{\mathfrak{a}}$ (where $\mathfrak{a} \in \mathscr{D}$), then $\mathfrak{O}$ is Dedekind.

12. Let $\mathfrak{O}$ be a ring in which the nonzero ideals form a semigroup with unique factorization under multiplication. Show that $\mathfrak{O}$ is Dedekind.

13. Let $\mathfrak{O}$ be a Dedekind ring with quotient field K. If A and B are ideals of the field K (relative to $\mathfrak{O}$), we say that A is divisible by B if there is an integral ideal C such that $A = BC$. Show that A is divisible by B if and only if $A \subset B$.

14. Let $\mathfrak{O}$ be any ring with a theory of divisors and let $\mathfrak{p}$ be a prime divisor. Show that the set $\bar{\mathfrak{p}}$ of all elements $\alpha \in \mathfrak{O}$ which are divisible by $\mathfrak{p}$ is a minimal prime ideal of the ring $\mathfrak{O}$. (An ideal P in a ring $\mathfrak{O}$ is called *prime* if the quotient ring $\mathfrak{O}/P$ has no divisors of zero, that is, if the product of any two elements of $\mathfrak{O}$, which do not lie in P, does not lie in P. The prime ideal P is called *minimal* if it does not contain any other prime ideal except the zero ideal.)

15. If $\mathfrak{O}$ is a ring with a theory of divisors, show that any nonzero prime ideal P of $\mathfrak{O}$ contains a prime ideal of the form $\bar{\mathfrak{p}}$, where $\mathfrak{p}$ is some prime divisor of the ring $\mathfrak{O}$.

7. Divisors in Algebraic Number Fields

7.1. *The Absolute Norm of a Divisor*

By Theorem 2 of Section 5 the maximal order $\mathfrak{O}$ of any algebraic number field K is a ring with a theory of divisors. Further, we saw in Section 6.1 that the residue class ring $\mathfrak{O}/\mathfrak{p}$ modulo a prime divisor $\mathfrak{p}$ is a finite field, and hence that the ring $\mathfrak{O}$ is Dedekind.

Consider the algebraic number field K as an extension of the field of rational numbers R (of finite degree). Since the divisors of the ring Z are in one-to-one correspondence with the natural numbers, we can assume that the group of all divisors (integral and fractional) of the field R coincides with the multiplicative group of positive rational numbers. In Section 5.2 we defined the concept of the norm of a divisor of the ring $\mathfrak{O}$ relative to a given extension K/k. If $\mathfrak{a}$ is a divisor of the order $\mathfrak{O}$ of the algebraic number field K, then we call the norm $N(\mathfrak{a}) = N_{K/R}(\mathfrak{a})$ the absolute norm of $\mathfrak{a}$. We extend the concept of absolute norm to fractional divisors by setting

$$N\left(\frac{\mathfrak{m}}{\mathfrak{n}}\right) = \frac{N(\mathfrak{m})}{N(\mathfrak{n})},$$

where $\mathfrak{m}$ and $\mathfrak{n}$ are integral divisors. The mapping $\mathfrak{a} \rightarrow N(\mathfrak{a})$ will then be a homomorphism from the group of all divisors of the field K to the multiplicative group of positive rational numbers.

The absolute norm of a principal divisor (ξ), $\xi \in K^*$, equals the absolute value of the norm of the number ξ:

$$N((\xi)) = |N(\xi)|. \tag{7.1}$$

Indeed, if ξ is integral this is just (5.3). If $\xi = \alpha/\beta$ with integral α and β, then

$$N((\xi)) = \frac{N((\alpha))}{N((\beta))} = \frac{|N(\alpha)|}{|N(\beta)|} = |N(\xi)|.$$

The degree of inertia f of a prime divisor $\mathfrak{p}$ of the field K relative to R is called the *absolute degree of inertia* of $\mathfrak{p}$ (or simply the *degree* of $\mathfrak{p}$). The ramification index e of the divisor $\mathfrak{p}$ relative to R is called the *absolute ramification index* of $\mathfrak{p}$.

If $\mathfrak{p}$ divides the rational prime p and if $\mathfrak{p}$ has degree f, then by (5.11),

$$N(\mathfrak{p}) = p^f. \tag{7.2}$$

Let $\mathfrak{p}_1, \ldots, \mathfrak{p}_m$ be all prime divisors of the field K which divide p, and let $e_1, \ldots, e_m$ be their ramification indices. Then in the field K we have the decomposition

$$p = \mathfrak{p}_1^{e_1} \ldots \mathfrak{p}_m^{e_m}.$$

By Theorem 7 of Section 5 the ramification indices e_i and degrees f_i of the divisors $\mathfrak{p}_i$ are connected by the relation

$$f_1 e_1 + \cdots + f_m e_m = n = (K:R). \tag{7.3}$$

Theorem 1. The absolute norm of an integral divisor $\mathfrak{a}$ of the algebraic number field K is equal to the number of residue classes in the maximal order $\mathfrak{O}$ modulo $\mathfrak{a}$.

Proof. We first prove the theorem for a prime divisor $\mathfrak{p}$. Let p be the rational prime which is divisible by $\mathfrak{p}$. The degree of inertia f of the divisor $\mathfrak{p}$ (by the definition of Section 5.3) equals the degree of the residue class field $\Sigma_{\mathfrak{p}}$ of the valuation $v_{\mathfrak{p}}$ over the residue class field Σ_p of the valuation v_p. Since Σ_p clearly consists of p elements, $\Sigma_{\mathfrak{p}}$ is a finite field with p^f elements. Hence it suffices to show that the fields $\Sigma_{\mathfrak{p}}$ and $\mathfrak{O}/\mathfrak{p}$ are isomorphic, that is, that the inclusion isomorphism $\mathfrak{O}/\mathfrak{p} \to \Sigma_{\mathfrak{p}}$ maps the field $\mathfrak{O}/\mathfrak{p}$ onto the entire field $\Sigma_{\mathfrak{p}}$. To do this it suffices to show that for any $\xi \in K$ for which $v_{\mathfrak{p}}(\xi) \geqslant 0$, there exists an element $\alpha \in \mathfrak{O}$ such that $v_{\mathfrak{p}}(\xi - \alpha) \geqslant 1$. We denote by $\mathfrak{q}_1, \ldots, \mathfrak{q}_s$ all those prime divisors of the field K for which $v_{\mathfrak{q}_i}(\xi) = -k_i < 0$. By Theorem 3 of Section 6 there is an element γ in the order $\mathfrak{O}$ such that

$$\gamma \equiv 1 \pmod{\mathfrak{p}},$$

$$\gamma \equiv 0 \pmod{\mathfrak{q}_i^{k_i}}, \qquad (i = 1, \ldots, s).$$

It is clear that $\alpha = \gamma\xi \in \mathfrak{O}$ and $v_{\mathfrak{p}}(\alpha - \xi) \geqslant 1$. Hence Theorem 1 is proved in the case of prime divisors.

To prove the theorem in general it suffices to prove that if it holds for integral divisors $\mathfrak{a}$ and $\mathfrak{b}$, then it also holds for their product $\mathfrak{ab}$. By condition (3) of Theorem 4 of Section 3 there is an element $\gamma \neq 0$ in the maximal order $\mathfrak{O}$ such that $\mathfrak{a}|\gamma$ and the divisor $(\gamma)\mathfrak{a}^{-1}$ is relatively prime to $\mathfrak{b}$. Let $\alpha_1, \ldots, \alpha_r$ $[r = N\mathfrak{a})]$ be a complete set of residue-class representatives of $\mathfrak{O}$ modulo the divisor $\mathfrak{a}$, and $\beta_1, \ldots, \beta_s$ $[s = N(\mathfrak{b})]$ be a complete set for $\mathfrak{b}$. We shall show that then the rs numbers

$$\alpha_i + \beta_j\gamma \tag{7.4}$$

form a complete system of residue-class representatives modulo $\mathfrak{ab}$. Let α be any number of $\mathfrak{O}$. For some i $(1 \leqslant i \leqslant r)$

$$\alpha \equiv \alpha_i \pmod{\mathfrak{a}}.$$

Consider the congruence

$$\gamma\xi \equiv \alpha - \alpha_i \pmod{\mathfrak{ab}}. \tag{7.5}$$

Since by choice of γ the greatest common divisor of the divisors (γ) and $\mathfrak{ab}$ equals $\mathfrak{a}$, and $\alpha - \alpha_i$ is divisible by $\mathfrak{a}$, then by Theorem 4 of Section 6 this congruence has a solution $\xi \in \mathfrak{O}$. If $\xi \equiv \beta_j \pmod{\mathfrak{b}}$ for some j $(1 \leqslant j \leqslant s)$, then $\gamma\xi \equiv \gamma\beta_j \pmod{\mathfrak{ab}}$. Along with (7.5) this shows that

$$\alpha \equiv \alpha_i + \gamma\beta_j \pmod{\mathfrak{ab}}.$$

We have proved that every residue class modulo $\mathfrak{ab}$ has a representative of the form (7.4). We now must show that the numbers (7.4) are pairwise-non-congruent modulo $\mathfrak{ab}$. Let

$$\alpha_i + \gamma\beta_j \equiv \alpha_k + \gamma\beta_l \pmod{\mathfrak{ab}}.$$

Since this congruence also holds modulo $\mathfrak{a}$, and $\gamma \equiv 0 \pmod{\mathfrak{a}}$, we obtain $\alpha_i \equiv \alpha_k \pmod{\mathfrak{a}}$, and this means that $i = k$, and we obtain

$$\gamma(\beta_j - \beta_l) \equiv 0 \pmod{\mathfrak{ab}}. \tag{7.6}$$

Let the prime divisor $\mathfrak{p}$ occur in the divisors $\mathfrak{a}$ and $\mathfrak{b}$ with exponents a and $b > 0$. Since $v_{\mathfrak{p}}(\gamma) = a$, it follows from (7.6) that $v_{\mathfrak{p}}(\beta_j - \beta_l) \geqslant b$. Since this is true for all prime divisors $\mathfrak{p}$ which occur in $\mathfrak{b}$ with positive exponent, then $\beta_j \equiv \beta_l \pmod{\mathfrak{b}}$, so that $j = l$.

Hence the numbers (7.4) for a complete set of residue representatives modulo $\mathfrak{ab}$. The number of residue classes of the ring $\mathfrak{O}$ modulo $\mathfrak{ab}$ hence equals $rs = N(\mathfrak{a})N(\mathfrak{b}) = N(\mathfrak{ab})$.

Theorem 1 is proved.

If $\mathfrak{a}$ is any divisor of the field K (integral or fractional) as in Section 6.3, we denote by $\bar{\mathfrak{a}}$ the ideal of the field K, consisting of all $\alpha \in K$ which are divisible by $\mathfrak{a}$. Let the number γ be chosen so that $\gamma\bar{\mathfrak{a}} \subset \mathfrak{O}$. By the corollary of Theorem 2 of Section 2, Chapter 2, the set $\gamma\bar{\mathfrak{a}}$ is a module of the field K (a submodule of the ring $\mathfrak{O}$). But then the ideal $\bar{\mathfrak{a}}$ also is a module of the field K. If $\alpha \in \bar{\mathfrak{a}}$, $\alpha \neq 0$, and $\omega_1, \ldots, \omega_n$ is a basis for the ring $\mathfrak{O}$, then all the products $\alpha\omega_1, \ldots, \alpha\omega_n$ lie in $\bar{\mathfrak{a}}$, and hence $\bar{\mathfrak{a}}$ contains $n = (K : R)$ linearly independent (over R) numbers of the field K. Hence, for any divisor $\mathfrak{a}$, the ideal $\bar{\mathfrak{a}}$ is a full module of the field K. Its coefficient ring will clearly be the maximal order $\mathfrak{O}$. Conversely, if A is a full module of the field K, whose coefficient ring is the maximal order $\mathfrak{O}$, then A fulfills all the conditions of being an ideal (Section 6.4). Thus the set of all ideals $\bar{\mathfrak{a}}$ coincides with the set of all full modules of the field K which belong to the maximal order $\mathfrak{O}$.

In Section 6.1 of Chapter 2 we introduced the concept of the norm of a full module of an algebraic number field. We can therefore speak of the norm of the ideal $\bar{\mathfrak{a}}$. We shall show that the norm of any divisor coincides with the norm of its ideal:

$$N(\mathfrak{a}) = N(\bar{\mathfrak{a}}). \tag{7.7}$$

For integral divisors this follows from Theorem 1 of this section and Theorem 1 of Section 6, Chapter 2. If the divisor $\mathfrak{a}$ is fractional, then we can find a $\gamma \in K^*$ such that the divisor $(\gamma^{-1})\mathfrak{a} = \mathfrak{b}$ is integral. Then by Theorem 2 of Section 6, Chapter 2, we have

$$N(\mathfrak{a}) = N(\mathfrak{b})|N(\gamma)| = N(\bar{\mathfrak{b}})|N(\gamma)| = N(\gamma\bar{\mathfrak{b}}) = N(\overline{\gamma\mathfrak{b}}) = N(\bar{\mathfrak{a}}),$$

and (7.7) is proved for all $\mathfrak{a}$.

As a simple application of the concept of norm we give a more precise estimate for the number $\omega(a)$ of nonassociate numbers in the maximal order whose norm has absolute value equal to a (in the proof of Theorem 5 of Section 2, Chapter 2, we showed that $\omega(a) \leqslant a^n$).

Let $\psi(a)$ denote the number of integral divisors with norm a. Since the numbers α and β are associate if and only if the principal divisors (α) and (β) are equal, then from (7.1) we have

$$\omega(a) \leqslant \psi(a).$$

We will find an estimate for (a). Let

$$a = p_1^{k_1} \cdots p_s^{k_s},$$

where the p_i are distinct primes. If $N(\mathfrak{a}) = a$, then $\mathfrak{a} = \mathfrak{a}_1 \ldots \mathfrak{a}_s$, where $\mathfrak{a}_i$ consists of those prime divisors $\mathfrak{p}$ which divide p_i. By formula (7.2) and the multiplicativity of the norm, we have $N(\mathfrak{a}_i) = p_i^{k_i}$, and this means that $\psi(a) = \psi(p_i^{k_1}) \cdots (\mathfrak{p}_s^{k_s})$. It therefore suffices to obtain an estimate for $\psi(p^k)$. Let $\mathfrak{p}_1, \ldots, \mathfrak{p}_m$ be the distinct prime divisors which divide p, and let $f_1, \ldots, f_m$ be their degrees. Since

$$N(\mathfrak{p}_1^{x_1} \cdots \mathfrak{p}_m^{x_m}) = p^{f_1x_1 + \cdots + f_mx_m}$$

the problem reduces to the determination of all solutions of the equation

$$f_1x_1 + \cdots + f_mx_m = k$$

in nonnegative x_i. Since $0 \leqslant x_i \leqslant k$, then the number of solutions cannot exceed $(k+1)^m$. But $m \leqslant n = (K: R)$, and thus

$$\psi(a) \leqslant ((k_1 + 1) \cdots (k_s + 1))^n.$$

The expression in parentheses on the right equals, as is well known, the number $\tau(a)$ of all divisors of a. We have hence obtained the estimate

$$\omega(a) \leqslant \psi(a) \leqslant (\tau(a))^n. \tag{7.8}$$

To compare our estimate (7.8) with the previous estimate $\omega(a) \leqslant a^n$, we note that for any $\varepsilon > 0$, the quantity $\tau(a)/a^\varepsilon$ converges to zero as $a \to \infty$.

7.2. *Divisor Classes*

Definition. Two divisors $\mathfrak{a}$ and $\mathfrak{b}$ of the algebraic number field K are called *equivalent*, and we write $\mathfrak{a} \sim \mathfrak{b}$, if they differ by a factor which is a principal divisor: $\mathfrak{a} = \mathfrak{b}(\alpha)$, $\alpha \in K^*$. The set of all divisors of K which are equivalent to a given divisor $\mathfrak{a}$, is called a *divisor class* and denoted by $[\mathfrak{a}]$.

In the terminology of group theory, the equivalence $\mathfrak{a} \sim \mathfrak{b}$ denotes that the divisors $\mathfrak{a}$ and $\mathfrak{b}$ belong to the same coset of the subgroup of all principal divisors, and the divisor class $[\mathfrak{a}]$ can be defined as the coset of the subgroup of all principal divisors which contains the divisor $\mathfrak{a}$. We clearly have $[\mathfrak{a}] = [\mathfrak{b}]$ if and only if $\mathfrak{a} \sim \mathfrak{b}$.

For any two divisor classes $[\mathfrak{a}]$ and $[\mathfrak{b}]$ set

$$[\mathfrak{a}]\cdot[\mathfrak{b}] = [\mathfrak{ab}].$$

It is easily verified that this definition is independent of the choice of $\mathfrak{a}$ and $\mathfrak{b}$ in these divisor classes, and that under this operation the set of all divisor classes becomes a group, the divisor class group of the field K. The unit element is the class $[\mathfrak{e}]$, consisting of all principal divisors. The inverse of the class $[\mathfrak{a}]$ is the class $[\mathfrak{a}^{-1}]$.

In the terminology of group theory the divisor-class group is the factor group of the group of all divisors by the subgroup of principal divisors.

The divisor-class group, and particularly its order, is an important arithmetic invariant of the algebraic number field K. If the number of divisor classes equals 1, then this means that every divisor is principal, which is equivalent to the maximal order of the field K having unique factorization (Theorem 2 of Section 3). The question of whether the algebraic integers of the number field K have unique factorization is hence a part of the problem of determining the number of divisor classes of this field. We shall now show that this number is always finite.

Theorem 2. The divisor class group of an algebraic number field is a finite group.

Proof. From the definition of equivalence of divisors it easily follows that the divisors $\mathfrak{a}$ and $\mathfrak{b}$ are equivalent if and only if the corresponding ideals $\bar{\mathfrak{a}}$ and $\bar{\mathfrak{b}}$ are similar (in the sense of similarity of modules; Section 1.3 of Chapter 2). The partitioning of divisors into classes of equivalent divisors hence corresponds to the partitioning of ideals of the field K (that is, of full modules whose coefficient ring is the maximal order of the field K) into classes of similar ideals. By Theorem 3 of Section 6, Chapter 2, the number of classes of similar modules with given coefficient ring is finite. Thus the number of classes of similar ideals, and the number of classes of equivalent divisors, are also finite.

Remark 1. Theorem 2 was obtained as a simple corollary of Theorem 3 of Section 6, Chapter 2. The proof of the latter theorem was based on geometric considerations, in particular, on Minkowski's lemma on convex bodies. Hence, the proof of Theorem 2 is also based on Minkowski's lemma.

Remark 2. From the proof of Theorem 3 of Section 6, Chapter 2, we can deduce the following strengthening of Theorem 2. In each divisor class of an algebraic number field K of degree $n = s + 2t$, there exists an integral divisor with norm $(2/\pi)^t\sqrt{|D|}$, where D is the discriminant of the field K (that is, the discriminant of the ring of integers of the field K). Let $[\mathfrak{b}]$ be any divisor class. Then there is an ideal $A = \alpha\overline{\mathfrak{b}^{-1}}$, similar to the ideal $\overline{\mathfrak{b}^{-1}}$, for which $A \supset \mathfrak{O}$ and $(A:\mathfrak{O}) \leqslant (2/\pi)^t\sqrt{|D|}$ (see the proof of Theorem 3 of Section 6, Chapter 2). Since the ideal A contains $\mathfrak{O}$, its inverse will be integral: $A = \overline{\mathfrak{a}^{-1}}$ with integral $\mathfrak{a}$. From $\overline{\mathfrak{a}^{-1}} = \alpha\overline{\mathfrak{b}^{-1}}$ it follows that $\mathfrak{a}(\alpha) = \mathfrak{b}$, that is, that the integral divisor $\mathfrak{a}$ is contained in the class $[\mathfrak{b}]$, and here (Problem 2)

$$N(\mathfrak{a}) = \frac{N(\mathfrak{e})}{N(\mathfrak{a}^{-1})} = (\overline{\mathfrak{a}^{-1}}:\bar{\mathfrak{e}}) = (A:\mathfrak{O}) \leqslant \left(\frac{2}{\pi}\right)^t\sqrt{|D|}.$$

Theorem 3. If the divisor-class group of the field K has order h, then the hth power of any divisor is principal.

Proof. The assertion of the theorem is a simple corollary of elementary group theory. The order of every element of a finite group divides the order of the group. Let $\mathfrak{a}$ be any divisor. Since $[\mathfrak{a}]^h$ is the unit element of the divisor-class group, then $[\mathfrak{a}^h] = [\mathfrak{e}]$, and this means that the divisor $\mathfrak{a}^h$ is principal.

Corollary. If the number h of divisor classes of the field K is not divisible by the prime number l, and if the divisor $\mathfrak{a}^l$ is principal, then $\mathfrak{a}$ is also principal.

Since l and h are relatively prime, we can find rational integers u and v such that $lu + hv = 1$. Since the divisors $\mathfrak{a}^l$ and $\mathfrak{a}^h$ are principal (the first by assumption, and the second by Theorem 3), it follows that $\mathfrak{a}^{lu}$ and $\mathfrak{a}^{hv}$ are also principal. But then so is the product $\mathfrak{a}^{lu+hv} = \mathfrak{a}$.

By Problem 20, every algebraic number field K can be embedded in a larger algebraic number field $\bar{K}$, so that every divisor of the field K will be principal in $\bar{K}$. We cannot, however, assert that every divisor of the field K is principal. Moreover, it has recently been shown (by Golod and Shafarevich) that there exist algebraic number fields, for example, $K = R(\sqrt{-3\cdot5\cdot7\cdot11\cdot13\cdot17\cdot19})$, which are not contained in any extension field with $h = 1$.

The following question is still open: Are there infinitely many algebraic number fields with $h = 1$? Examination of tables show that such fields occur rather frequently (see the tables of h for real quadratic fields and totally real cubic fields).

For certain classes of fields (for example, for quadratic and cyclotomic fields, see Chapter 5) formulas for the number of divisor classes have been found, but in the general case little is known about h and the divisor-class group. Among the few general theorems about the number h is the theorem of Siegel and Brauer, which asserts that for all fields with a fixed degree n, the number h of divisor classes, the regulator R, and the discriminant D are related by the following asymptotic formula:

$$\frac{\ln(hR)}{\ln\sqrt{|D|}} \to 1 \qquad \text{for } |D| \to \infty \tag{*}$$

[R. Brauer, On the zeta functions of algebraic number fields, *Am. J. Math*, **69**, No. 2, 243–250 (1947)]. Since for imaginary quadratic fields the regulator of the field is equal to 1, then it follows from (*) that as $|D| \to \infty$, so does $h \to \infty$. In particular, we may deduce that there are only a finite number of imaginary quadratic fields with $h = 1$. In tables we see nine imaginary quadratic fields with $h = 1$ (their discriminants are -3, -4, -7, -8, -11, -19, -43, -67, -163). It is known that there is at most one more imaginary quadratic field with $h = 1$. It is not known whether or not it exists.

In the general case we can say almost nothing from (*) about the behavior of the number h, since we know very little about the value of the regulator R.

7.3. *Applications to Fermat's Theorem*

The results of the preceding section allow us to prove the validity of Theorem 1 of Section 1 for a much wider class of exponents l.

Theorem 4. Let l be an odd prime and let ζ be a primitive lth root of 1. If the number of divisor classes of the field $R(\zeta)$ is not divisible by l, then the first case of Fermat's theorem holds for the exponent l.

Proof. Assume that, contrary to the theorem, there exist rational integers x, y, and z, not divisible by l, and satisfying the equation

$$x^l + y^l = z^l.$$

We may further assume that x, y, and z are pairwise relatively prime. In the ring of integers of the field $R(\zeta)$ our equation can be written in the form

$$\prod_{k=0}^{l-1} (x + \zeta^k y) = z^l.$$

Since $x + y \equiv x^l + y^l = z^l \equiv z \pmod{l}$ and z is not divisible by l, then $x + y$ is also not divisible by l. Then, as we proved in Lemma 5 of Section 1, for $m \not\equiv n \pmod{l}$ there are numbers ξ_0 and η_0 in the ring $Z[\zeta]$ such that

$$(x + \zeta^n y)\xi_0 + (x + \zeta^m y)\eta_0 = 1.$$

Hence the principal divisors $(x + \zeta^k y)$ $(k = 0, 1, \ldots, l-1)$ are pairwise relatively prime. Since their product is an lth power [of the divisor (z)], then each of them separately must be an lth power. In particular,

$$(x + \zeta y) = \mathfrak{a}^l,$$

where $\mathfrak{a}$ is an integral divisor of the field $R(\zeta)$. Since we have assumed that the number of divisor classes of the field $R(\zeta)$ is not divisible by l, it follows from the corollary to Theorem 3 that the divisor $\mathfrak{a}$ is principal; that is, $\mathfrak{a} = (\alpha)$, where α belongs to the maximal order $\mathfrak{O} = Z[\zeta]$ of the field $R(\zeta)$. From the equality

$$(x + \zeta y) = (\alpha^l)$$

it now follows that

$$x + \zeta y = \varepsilon\alpha^l,$$

where ε is a unit of the ring $\mathfrak{O}$. Analogously we obtain

$$x - \zeta z = \varepsilon_1 {\alpha_1}^l$$

($\alpha_1 \in \mathfrak{O}$, ε_1 a unit in $\mathfrak{O}$). We have reached equations which, as was shown in Section 1.3, lead to a contradiction (in that part of the proof of Theorem 1 of Section 1 unique factorization was not used). Hence Theorem 4 is proved.

Those odd primes l for which the number of divisor classes of the field $R(\zeta)$, $\zeta^l = 1$, is not divisible by l, are called *regular primes*, and all others are called *irregular*. By very beautiful number-theoretic and analytic arguments Kummer obtained a fairly simple criterion (which we shall present in Section 6.4 of Chapter 5), which allows one to check easily whether a given prime l is regular or not. Using this method it can be verified that among the prime numbers <100 only three, 37, 59, and 67, are irregular, and all the rest are regular. To show how much broader the class of exponents for which Theorem 4 holds is than the class for which Theorem 1 of Section 1 holds, we note that among the prime numbers < 100, only for the first seven, 3, 5, 7, 11, 13, 17, 19, do we have unique factorization in the ring $\mathfrak{O} = Z[\zeta]$ where $\zeta^l = 1$.

In his first paper Kummer stated the hypothesis that the number of irregular primes was finite. He later retracted this and hypothesized that regular primes occur twice as frequently as irregular ones. With the aid of electronic computers, it has been shown that of the 550 odd primes $\leqslant$ 4001, there are 334 regular ones and 216 irregular ones. A table of all irregular primes $\leqslant$ 4001 is given at the end of the book. Jensen (see Section 7.2 of Chapter 5) showed that the number of irregular primes is infinite. It is not known whether there are infinitely many regular primes; there are no indications that there are only finitely many.

The first case of Fermat's theorem for the exponent l is also connected with the number h_0 of divisor classes of the field $R(\zeta + \zeta^{-1}) = R[2\cos(2\pi/l)]$. It is easily seen that $R(\zeta + \zeta^{-1})$ consists of all real numbers of the field $R(\zeta)$. Vandiver showed that if the number h_0 of divisor classes of the field $R(\zeta + \zeta^{-1})$ is not divisible by l, then the first case of Fermat's theorem is valid for the exponent l [H. S. Vandiver, Fermat's last theorem and the second factor in the cyclotomic class number, *Bull. Am. Math. Soc.* **40**, No. 2, 118–126 (1934)]. It is not known if there exist prime numbers l for which the number h_0 is divisible by l. It has only been verified that there are none among numbers $\leqslant 4001$.

We note here some other results which bear on the first case of Fermat's theorem. Wieferich showed that the first case of Fermat's theorem is valid for all primes l such that $2^{l-1} \not\equiv 1 \pmod{l^2}$ [A. Wieferich, Zum letzen Fermatschen Theorem, *J. Math.* **136**, 293–302 (1909)]. To indicate the strength of this result we note that only two prime numbers $l \leqslant 200, 183$, namely, 1093 and 3511, satisfy the condition $2^{l-1} \equiv 1 \pmod{l^2}$ [Erna H. Pearson, *Math. Comp.* 17, No. 82, 194–195 (1963)]. However, it is not known whether there are infinitely many such l. Several other authors have shown that the first case of Fermat's theorem holds for all l such that $q^{l-1} \not\equiv 1 \pmod{l^2}$ for some prime number $q \leqslant 43$ (D. Mirimanoff, H. S. Vandiver, G. Frobenius, F. Pollaczek, T. Morishima, J. B. Rosser). This has made it possible to verify the first case of Fermat's theorem for all prime numbers $<253, 747, 889$ [D. H. Lehmer and Emma Lehmer, On the first case of Fermat's last theorem, *Bull. Am. Math. Soc.* **47**, No. 2, 139–142 (1941)].

7.4. *The Question of Effectiveness*

Up to this time we have avoided the question of the practical construction of a theory of divisors for a given algebraic number field K. Since all divisors are determined once we know all prime divisors, and the prime divisors are determined by the valuations of the field K, our question reduces to the effective construction of all extensions to the field K of the valuation ν_p of the field R for any fixed p. In addition to enumerating the prime divisors, it is important to have a finite algorithm for computing the number h of divisor classes of the field K. For only then will the results of the preceding section concerning Fermat's theorem have any real value.

In this section we shall show how to construct all extensions of the valuation ν_p and how to compute the number h, both in a finite number of steps.

Let $\mathfrak{o}_p$ be the ring of the valuation ν_p of the field R (that is, the ring of p-integral rational numbers, see Section 3.2 of Chapter 1) and $\mathfrak{O}_p$ its integral closure in the field K. Every number $\xi \in \mathfrak{O}_p$ is the root of a polynomial $t^k + a_1 t^{k-1} + \cdots + a_k$ with p-integral coefficients a_k. If m is a common denominator for all a_i, then the number $m\xi = \alpha$ will be a root of the polynomial $t^k + ma_1 t^{k-1} + \cdots + m^k a_k$ with coefficients in Z; that is, it will lie in the ring of integers $\mathfrak{O}$ of the field K (the maximal order). The converse assertion also holds: If $\alpha \in \mathfrak{O}$ and if the rational integer m is not divisible by p, then $\alpha/m \in \mathfrak{O}_p$. Thus the ring $\mathfrak{O}_p$ consists of all numbers of the form α/m, where $\alpha \in \mathfrak{O}$ and the rational integer m is not divisible by p. Choose some fundamental basis

$\omega_1, \ldots, \omega_n$ of the field K (that is, a basis for the ring $\mathfrak{O}$ over Z). Then we have shown that the number $\xi \in K$, which has the representation

$$\xi = a_1\omega_1 + \cdots + a_n\omega_n \qquad (a_i \in R),$$

will lie in the ring $\mathfrak{O}_\mathfrak{p}$ if and only if all a_i are p-integers.

By Theorem 7 of Section 4 our first problem (that is, the construction of all extensions of the valuation $\nu_\mathfrak{p}$) reduces to the determination of a complete system of pairwise-nonassociate prime elements $\pi_1, \ldots, \pi_m$ of the ring $\mathfrak{O}_\mathfrak{p}$. Once the elements π_i have been found, then for any $\xi \in \mathfrak{O}_\mathfrak{p}^*$ we can easily obtain the factorization

$$\xi = \eta\pi_1^{k_1} \cdots \pi_m^{k_m}, \tag{7.9}$$

where η is a unit in $\mathfrak{O}_\mathfrak{p}$. To do this we divide successively by each of the π_i until the quotient would not lie in the ring $\mathfrak{O}_\mathfrak{p}$; at some stage we obtain a quotient η which cannot be divided by any of the π_i and hence is a unit in $\mathfrak{O}_\mathfrak{p}$. Since each element of K is the quotient of two elements of $\mathfrak{O}_\mathfrak{p}$ (even of $\mathfrak{O}$), the representation (7.9) can also be found for any $\xi \in K^*$. But this determines the valuations $\nu_1, \ldots, \nu_m$ of K which are extensions of $\nu_\mathfrak{p}$. The ramification indices of these valuations are found in the factorization $p = \varepsilon\pi_1^{e_1} \cdots \pi_m^{e_m}$ (ε a unit in $\mathfrak{O}_\mathfrak{p}$).

Let π be any prime element of the ring $\mathfrak{O}_\mathfrak{p}$. Since rational integers which are not divisible by p are units in $\mathfrak{O}_\mathfrak{p}$, we may assume that $\pi \in \mathfrak{O}$. For any $\alpha \in \mathfrak{O}$ the number $\pi + p^2\alpha = \pi\;[1 + (p^2/\pi)\alpha]$ is associate with π, since the factor $1 + (p^2/\pi)\alpha$ lies in $\mathfrak{O}_\mathfrak{p}$ and is not divisible by any of the prime elements $\pi_1, \ldots, \pi_m$. Thus a complete set of pairwise-nonassociate prime elements in $\mathfrak{O}_\mathfrak{p}$ can be chosen from the system of numbers

$$x_1\omega_1 + \cdots + x_n\omega_n,$$

where $0 \leqslant x_i < p^2$ $(i = 1, \ldots, n)$. Since the set of all such numbers is finite, the set of prime elements can be found, and the valuations $\nu_1, \ldots, \nu_m$ determined, in a finite number of steps.

To find the degrees $f_1, \ldots, f_m$ of the prime divisors $\mathfrak{p}_1, \ldots, \mathfrak{p}_m$, corresponding to the valuations $\nu_1, \ldots, \nu_m$, we can use Theorem 5 of Section 5. By this theorem for each prime element $\pi_i \in \mathfrak{O}$ of the ring $\mathfrak{O}_\mathfrak{p}$ we have

$$N(\pi_i) = p^{f_i}a,$$

where the rational integer a is not divisible by p. Hence the degree f_i of the prime divisor $\mathfrak{p}_i$ is just the exponent with which p occurs in the rational integer $N(\pi_i)$.

We now turn to our second question, the effective computation of h, the number of divisor classes.

In a remark after Theorem 2 it was noted that every divisor class contains a divisor $\mathfrak{a}$ for which

$$N(\mathfrak{a}) \leqslant \left(\frac{2}{\pi}\right)^t \sqrt{|D|} \tag{7.10}$$

(see also Problem 9). Let

$$\mathfrak{a}_1, \ldots, \mathfrak{a}_N \tag{7.11}$$

be all integral divisors of the field K which satisfy (7.10). The number of such divisors is finite, since there are only finitely many divisors with given norm in K [for fixed a, from $N(\mathfrak{p}_1^{k_1} \cdots \mathfrak{p}_r^{k_r}) = a$ we easily deduce bounds on the prime numbers p which are divisible by the $\mathfrak{p}_i$, and on the exponents k_1]. To determine the number of divisor classes we must find in the set (7.11) a maximal set of pairwise-nonequivalent divisors. To make this effective, we must have a practical method for determining whether or not two given divisors are equivalent. Let $\mathfrak{a}$ and $\mathfrak{b}$ be two integral divisors. Choose in K a number $\beta \neq 0$, which is divisible by $\mathfrak{b}$, and consider the divisor $\mathfrak{a}\mathfrak{b}^{-1}(\beta)$. The divisors $\mathfrak{a}$ and $\mathfrak{b}$ are equivalent if and only if the divisor $\mathfrak{a}\mathfrak{b}^{-1}(\beta)$ is principal. Hence we need to be able to determine whether or not a given integral divisor is principal.

Denote the norm of such a divisor by a. In Section 5.4 of Chapter 2 we showed that we could find, in a finite number of steps, a finite set of numbers

$$\alpha_1, \ldots, \alpha_r \tag{7.12}$$

with norm $\pm a$, such that any $\alpha \in \mathfrak{O}$ with norm $\pm a$ is associate with a number of (7.12). If the divisor $\mathfrak{a}$ is principal; that is, if $\mathfrak{a} = (\alpha)$ with $\alpha \in \mathfrak{O}^*$, then $|N(\alpha)| = a$, and hence for some i $(1 \leqslant i \leqslant r)$ we shall have $\mathfrak{a} = (\alpha_i)$. Hence if we have already found the system (7.12), then to determine if the given divisor is principal, we need only compare it with each of the principal divisors $(\alpha_1), \ldots, (\alpha_r)$.

Hence we have shown that the number h can be computed for the field K in a finite number of steps.

The determination of the decomposition of the rational prime p into a product of prime divisors is often more easily done by considering the norms of k-nomials of numbers ($k \geqslant 2$). To describe this method we need some auxiliary results.

Let θ be an integral primitive element of the algebraic number field K of degree n. Then the index of the order $\mathfrak{O}' = \{1, \theta, \ldots, \theta^{n-1}\}$ in the maximal order $\mathfrak{O}$ is called the *index* of the number θ.

Lemma. If the prime divisor $\mathfrak{p}$ does not divide the index k of the number θ, then any number $\alpha \in \mathfrak{O}$ is congruent modulo $\mathfrak{p}$ to some number of the order $\mathfrak{O}' = \{1, \theta, \ldots, \theta^{n-1}\}$.

Since $\mathfrak{p} \nmid k$, then $kx \equiv 1 \pmod{\mathfrak{p}}$ for some integer x. Set $\gamma = kx\alpha$. Since $k\alpha \in \mathfrak{O}'$, then also $\gamma \in \mathfrak{O}'$, and $\alpha \equiv \gamma \pmod{\mathfrak{p}}$.

Corollary. If $\mathfrak{p}$ does not divide the discriminant $D' = D(1, \theta, \ldots, \theta^{n-1})$, then any integer $\alpha \in \mathfrak{O}$ is congruent modulo $\mathfrak{p}$ to some number of the order $\mathfrak{O}' = \{1, \theta, \ldots, \theta^{n-1}\}$.

From the formula $D' = Dk^2$, where D is the discriminant of the field K [Lemma 1 of Section 6, Chapter 2, and (2.12) of the Supplement], it follows that if $\mathfrak{p}$ does not divide D', then also $\mathfrak{p}$ does not divide k.

Assume now that the rational prime p does not divide the index of the integer $\theta \in K$. Let $\mathfrak{p}$ be a prime divisor with degree f which divides p, and let $\bar{\theta}$ be the residue class of θ modulo $\mathfrak{p}$. By the lemma the residue-class field $\mathfrak{O}/\mathfrak{p}$ is generated by the residue class $\bar{\theta}$ containing θ. If $x_1, \ldots, x_f$ independently run through a full system of residues modulo p (in the ring Z), then among the numbers

$$\gamma = x_1 + x_2\theta + \cdots + x_f\theta^{f-1} + \theta^f$$

there will be one and only one which is divisible by $\mathfrak{p}$. Computing the norms $N(\gamma)$, we can easily determine which γ are divisible by some prime divisor which divides p. If, for example, for $f = 1$, we find s numbers γ whose norms are divisible precisely by the first power of p, then we have found s prime divisors of the first degree which divide p. Assume now that all prime divisors of first degree which occur in p have been found (so we have a set of numbers $\beta_1, \ldots, \beta_u$ with norms pa_i, $p \nmid a_i$). Now setting $f = 2$, we isolate those numbers γ whose norm is divisible by p^2. Dividing by the numbers β_i already found, we can eliminate those γ which arose from prime divisors of degree 1, and if after this $N(\gamma) = p^2(b/c)$ $(bc, p) = 1$, then γ is divisible by a prime divisor of degree 2. If by this method we can find all prime divisors of degree 2 which divide p, then we take $f = 3$, and so on. Of course, for large n the computations will generally be large; for $n = 3$ and $n = 4$ we can often achieve our goal quite quickly. Some refinements of this method are given in Problems 25 to 27.

Example 1. We shall factor the numbers 2, 3, 5, and 7 into prime divisors in the field of fifth degree $R(\theta)$, where $\theta^5 = 2$. The discriminant $D(1, \theta, \theta^2, \theta^3, \theta^4)$ equals $2^4 5^5$, and hence only the primes 2 and 5 can divide the index of θ. By Problem 15, the number 2 does not occur in the index. Since $\theta^2 = 2$, then $\mathfrak{p}_2 = (\theta)$ is a prime divisor of the first degree, and we have

$$2 = \mathfrak{p}_2{}^5.$$

From

$$N(\theta) = 2, \quad N(\theta + 1) = 3, \quad N(\theta - 1) = 1 \tag{7.13}$$

it follows that only one prime divisor of first degree divides the number 3, namely, $\mathfrak{p}_3 = (\theta + 1)$, and $\mathfrak{p}_3{}^2 \nmid 3$ by Theorem 8 of Section 5. Further,

$$N(\theta + 2) = 2\cdot 17, \quad N(\theta - 2) = -2\cdot 3\cdot 5. \tag{7.14}$$

The second of these equations shows that the number 5 has a prime divisor $\mathfrak{p}_5$ of degree 1, and since $\theta - 2 = (\theta + 1) - 3$ is divisible by $\mathfrak{p}_3$, we have $(\theta - 2) = \mathfrak{p}_2\mathfrak{p}_3\mathfrak{p}_5$. The number $\theta - 2$ satisfies the equation

$$(\theta - 2)^5 + 10(\theta - 2)^4 + 40(\theta - 2)^3 + 80(\theta - 2)^2 + 80(\theta - 2) + 30 = 0.$$

By Problem 9 of Section 5 we have

$$5 = \mathfrak{p}_5{}^5.$$

The result of Problem 15 also shows that 5 does not divide the index of the number θ, and this means that the ring of all integers of the field $R(\theta)$ coincides with the order $\{1, \theta, \theta^2, \theta^3, \theta^4\}$.

Combining (7.13) and (7.14) with

$$N(\theta + 3) = 5\cdot 7^2, \quad N(\theta - 3) = -241,$$

we see that there are two possibilities. Either the number $\theta + 3$ is divisible by the square of a prime divisor (which divides 7) of first degree, or it is divisible by a prime divisor of second degree (which divides 7). But for the number $\theta - 4 = (\theta + 3) - 7$ we have $N(\theta - 4) = -2\cdot 7\cdot 73$, and hence the first possibility holds. This means that 7 has one (and only one) prime divisor of first degree $\mathfrak{p}_7$, where $\mathfrak{p}_7{}^2 | 7$.

To determine whether 3 and 7 have prime divisors of second degree, we consider trinomials of the form $\theta^2 + \theta x + y$. We have

$$N(\theta^2 + x\theta + y) = 2x^5 + y^5 - 10x^3y + 10xy^2 + 4. \tag{7.15}$$

Substituting for x and y the values 0, 1, -1, we obtain nine numbers, none of which is divisible by 9. This means that no prime divisor of degree 2 divides 3. By formula (7.3) there is now only one possibility for the decomposition of the number 3:

$$3 = \mathfrak{p}_3\mathfrak{p}_3{}',$$

where $\mathfrak{p}_3{}'$ is a prime divisor of fourth degree. Now if x and y take the values 0, ± 1, ± 2, ± 3 in (7.15), then of the 49 numbers which arise only one is divisible by 7^2:

$$N(\theta^2 + 2\theta - 3) = 5\cdot 7^2.$$

But $\theta^2 + 2\theta - 3 = (\theta + 3)(\theta - 1)$, and therefore we have only the square of the divisor $\mathfrak{p}_7$, so that also for 7 we have the factorization

$$7 = \mathfrak{p}_7\mathfrak{p}_7{}',$$

where $\mathfrak{p}_7{}'$ is a prime divisor of fourth degree.

Example 2. Consider the cubic field $R(\theta)$, $\theta^3 - 9\theta - 6 = 0$. Since $D(1, \theta, \theta^2) = 3^5 \cdot 2^3$, by Problem 15 only 2 can divide the index of θ (it can be shown that the order $\{1, \theta, \theta^2\}$ is maximal, but we shall not need this). By Problem 9 of Section 5 we have the decomposition

$$3 = \mathfrak{p}_3{}^3.$$

From

$$N(\theta) = 6, \quad N(\theta + 1) = -4, \quad N(\theta - 1) = 14, \tag{7.16}$$

we conclude that the number 2 has at least two prime divisors of first degree, $\mathfrak{p}_2$ and $\mathfrak{p}_2'$:

$$(\theta) = \mathfrak{p}_2\mathfrak{p}_3, \qquad (\theta - 1) = \mathfrak{p}_2'\mathfrak{p}_7 \tag{7.17}$$

(that there are only two would follow from the maximality of the order $\{1, \theta, \theta^2\}$, for then 2 would not divide the index of the number θ). But from the equation

$$(\theta - 1)^3 + 3(\theta - 1)^2 - 6(\theta - 1) - 14 = 0$$

we see that 2 is divisible by $\mathfrak{p}_2'^2$, and hence

$$2 = \mathfrak{p}_2\mathfrak{p}_2'^2, \qquad (\theta + 1) = \mathfrak{p}_2'^2, \tag{7.18}$$

The norms (7.16) and also

$$N(\theta + 2) = -4, \qquad N(\theta - 2) = 16 \tag{7.19}$$

are not divisible by 5. This means that 5 has no prime divisor of first degree. Since the field is cubic, it follows that the principal divisor 5 is prime. To decompose the number 7, we must also consider the norms

$$N(\theta + 3) = 6, \qquad N(\theta - 3) = 6.$$

Since there is only one norm divisible by 7, then 7 has only one prime divisor of first degree. Since $\mathfrak{p}_2{}^2 \nmid 7$, we must have $7 = \mathfrak{p}_7\mathfrak{p}_7'$, where $\mathfrak{p}_7'$ is a prime divisor of second degree.

In the process of decomposing rational primes into the products of prime divisors by our method of examining the values of the norms of integers, we have also obtained a series of equivalences among divisors. These equivalences allow us to reduce the number of divisors of the system (7.11) from which we must choose a maximal set of pairwise-nonequivalent divisors to determine the number h. Thus, in Example 2, by Problem 9 the system (7.11) consists of integral divisors with norm $\leqslant (3!/3^3)\sqrt{3^5 \cdot 2^3} < 10$, that is, of the divisors

$$1,\ \mathfrak{p}_2,\ \mathfrak{p}_2',\ \mathfrak{p}_3,\ \mathfrak{p}_2{}^2,\ \mathfrak{p}_2'^2,\ \mathfrak{p}_2\mathfrak{p}_2',\ \mathfrak{p}_2\mathfrak{p}_3,\ \mathfrak{p}_2'\mathfrak{p}_3,\ \mathfrak{p}_7,\ \mathfrak{p}_2{}^3,\ \mathfrak{p}_2{}^2\mathfrak{p}_2',\ 2,\ \mathfrak{p}_2'^3,\ \mathfrak{p}_3{}^2. \tag{7.20}$$

It follows from (7.18) that $\mathfrak{p}_2'^2 \sim 1$ and $\mathfrak{p}_2 \sim 1$ (1 being the unit divisor), and then from (7.17) and $(\theta + 3) = \mathfrak{p}_2'\mathfrak{p}_3$ that $\mathfrak{p}_3 \sim 1$, $\mathfrak{p}_2' \sim 1$, $\mathfrak{p}_7 \sim 1$. Hence all divisors of the system (7.20) are principal and we have $h = 1$ for the field $R(\theta)$, $\theta^2 - 9\theta - 6 = 0$.

Sometimes (for small discriminants) the system of divisors (7.11) consists only of the unit divisor. In these cases we obtain $h = 1$ without further computations. For example, for the field $R(\theta)$, $\theta^3 - \theta - 1 = 1$, the discriminant of the basis, $1, \theta, \theta^3$ equals -23, so by Problem 8 of Section 2, Chapter 2, this basis is fundamental and -23 is the discriminant of the field. By Problem 9 there is an integral divisor in each divisor class of the field $R(\theta)$ with norm

$$\leqslant \frac{4}{\pi}\frac{3!}{3^3}\sqrt{23} < 2,$$

and this means that in the field $R(\theta)$ every divisor is principal.

For quadratic fields the number of divisor classes can also be computed by the theory of reduction, considered in Problems 12 to 15 and 24 of Section 7 of Chapter 2.

PROBLEMS

1. Show that in an algebraic number field of degree n the number $\psi(a)$ of integral divisors with given norm a does not exceed the number $\tau_n(a)$ of all solutions to the equation $x_1x_2 \cdots x_n = a$ ($x_1, \ldots, x_n$ independently taking all natural values).

2. Let $\mathfrak{a}$ and $\mathfrak{b}$ be two divisors of an algebraic number field (integral or fractional), with $\bar{\mathfrak{a}}$ and $\bar{\mathfrak{b}}$ the corresponding ideals. Show that if $\mathfrak{a}$ is divisible by $\mathfrak{b}$, then

$$(\bar{\mathfrak{b}} : \bar{\mathfrak{a}}) = (N\mathfrak{a}\mathfrak{b}^{-1}).$$

3. Show that in any two distinct divisor classes there exist relatively prime integral divisors.

4. If $\mathfrak{a}$ is an integral divisor of an algebraic number field, let $\varphi(\mathfrak{a})$ denote the number of residue classes modulo $\mathfrak{a}$ which consist of numbers relatively prime to $\mathfrak{a}$ (this generalizes Euler's function). Show that if the integral divisors $\mathfrak{a}$ and $\mathfrak{b}$ are relatively prime, then

$$\varphi(\mathfrak{a}\mathfrak{b}) = \varphi(\mathfrak{a})\varphi(\mathfrak{b}).$$

5. Prove the formula

$$\varphi(\mathfrak{a}) = N(\mathfrak{a}) \prod_{\mathfrak{p}} \left(1 - \frac{1}{N(\mathfrak{p})}\right),$$

in which $\mathfrak{p}$ runs through all prime divisors which divide the integral divisor $\mathfrak{a}$.

6. Show that for any integer α which is relatively prime to the integral divisor $\mathfrak{a}$, we have

$$\alpha^{\varphi(\mathfrak{a})} \equiv 1 \pmod{\mathfrak{a}}$$

(this generalizes Euler's theorem). Further, show that for any integer α and prime divisor $\mathfrak{p}$ of an algebraic number field,

$$\alpha^{N(\mathfrak{p})} \equiv \alpha \pmod{\mathfrak{p}}$$

(this generalizes the small Fermat theorem).

7. Prove the formula

$$\sum_{\mathfrak{c}} \varphi(\mathfrak{c}) = N(\mathfrak{a}),$$

where the sum is taken over all divisors $\mathfrak{c}$ which divide the integral divisor $\mathfrak{a}$ (including $\mathfrak{e}$ and $\mathfrak{a}$).

8. Let $\xi_1, \dots, \xi_s$ $(s = N(\mathfrak{p}) - 1)$ be a system of residues for the prime divisor $\mathfrak{p}$, not divisible by $\mathfrak{p}$. Show that then

$$\xi_1 \dots \xi_s \equiv -1 \pmod{\mathfrak{p}}$$

(Wilson's theorem).

9. Let K be an algebraic number field of degree $n = s + 2t$ and discriminant D. Use Problem 2 of Section 6, Chapter 2, to show that in every divisor class of K there is an integral divisor $\mathfrak{a}$ with

$$N(\mathfrak{a}) \leqslant \left(\frac{4}{\pi}\right)^t \frac{n!}{n^n} \sqrt{|D|}.$$

10. Show that for the quadratic fields with discriminants 5, 8, 12, 13, -3, -4, -7, -8, -11 the number of divisor classes is 1.

11. Show that the number of divisor classes of the field $R(\sqrt{-19})$ equals 1.

12. Show that the ring of integers of the field $R(\zeta)$, where ζ is a primitive fifth root of 1, has unique factorization.

13. Show that the number of divisor classes in the field $R(\sqrt{-23})$ is equal to 3.

14. Let K_1, K_2, and K_3 be the three cubic fields described in Problem 21 of Section 2, Chapter 2. Show that the number 5 remains a prime divisor in the fields K_1 and K_2, and in the field K_3 it factors as a product of three distinct prime divisors of first degree: $5 = \mathfrak{p}\mathfrak{p}'\mathfrak{p}''$. Further, show that the number 11 factors as a product of three distinct prime divisors in the field K_1, $11 = \mathfrak{q}\mathfrak{q}'\mathfrak{q}''$, and that 11 remains prime in K_2. (It follows that the fields K_1, K_2, and K_3 are distinct.)

15. Let the primitive element $\theta \in K$ be the root of an Eisenstein polynomial relative to the prime number p. Use Problem 9 of Section 5 to show that p does not divide the index of the number θ.

16. Let the prime number p be less than the degree n of the algebraic number field K. If there is an integral primitive element in K whose index is not divisible by p, show that p cannot factor in K as the product of n distinct prime divisors of first degree.

17. Use Problems 18 and 19 of Section 5 to show that a rational prime number is ramified in the algebraic number field K (that is, is divisible by the square of a prime divisor) if and only if it divides the discriminant of the field K.

18. Let $f(x_1, \dots, x_n)$ be a quadratic form whose coefficients are integers in the algebraic number field K, and let δ be its determinant. Let $\mathfrak{p}$ be a prime divisor which does not divide 2 or δ. If α is an integer of K not divisible by $\mathfrak{p}$, set $(\alpha/\mathfrak{p}) = +1$ if the congruence $\xi^2 \equiv \alpha \pmod{\mathfrak{p}}$ is solvable, and $(\alpha/\mathfrak{p}) = -1$ otherwise. If N is the number of solutions of the congruence

$$f(x_1, \dots, x_n) \equiv 0 \pmod{\mathfrak{p}}$$

show that

$$N = N(\mathfrak{p})^{n-1}, \qquad \text{if } n \text{ is odd},$$

$$N = N(\mathfrak{p})^{n-1} + \left(\frac{(1-)^{n/2}\,\delta}{\mathfrak{p}}\right) N(\mathfrak{p})^{(n-2)/2} (N(\mathfrak{p}) - 1) \qquad \text{if } n \text{ is even}.$$

19. Let $\mathfrak{a}$ be a divisor of the algebraic number field K, such that $\mathfrak{a}^m = (a)$ is a principal divisor. Show that the divisor $\mathfrak{a}$ becomes principal in the field $K(\sqrt[m]{a})$.

20. Show that for any algebraic number field K there is a finite extension $\bar{K}/K$ such that every divisor $\mathfrak{a}$ of the field K becomes principal in the field $\bar{K}$.

21. Let K be a cubic field and p a prime which factors as a product of three distinct prime divisors in K: $p = \mathfrak{p}\mathfrak{p}'\mathfrak{p}''$. If α is an integer in K with $\mathrm{Sp}(\alpha) = 0$ and $\mathfrak{p}\mathfrak{p}'|\alpha$, show that $\mathfrak{p}''|\alpha$ and hence $p|\alpha$.

22. Show that the field $R(\theta)$, $\theta^3 = 6$, has only one divisor class. [By Problem 24 of Section 2, Chapter 2, the numbers $1, \theta, \theta^2$ form a fundamental basis for the field $R(\theta)$.]

23. Let K be the cubic field $K = R(\theta)$, $\theta^3 = 6$. Show that there is no number $\alpha \neq 0$ of K of the form $\alpha = x + y\theta$, x and y relatively prime rational integers, for which $N(\alpha) = 10z^3$ (z a rational integer). Deduce that the equation $x^3 + 6y^3 = 10z^3$ (and hence also the equation $3x^3 + 4y^3 + 5z^3 = 0$) has no nontrivial solution in rational integers.

Hint: Assume that the number α exists and show that it must have the form $\alpha = \alpha_0\xi^3$, where ξ is an integer of the field K and α_0 is one of the following six numbers:

$$\lambda\mu,\ \lambda\mu\varepsilon,\ \lambda\mu\varepsilon^2,\ \lambda\nu,\ \lambda\nu\varepsilon,\ \lambda\nu\varepsilon^2.$$

Here $\lambda = 2 - \theta$ $[N(\lambda) = 2]$; $\mu = \theta - 1$ $[N(\mu) = 5]$; $\nu = (\theta^2 + \theta + 1)^2 = 13 + 8\theta + 3\theta^2$ $[N(\nu) = 5 \cdot 5^3]$; and $\varepsilon = 1 - 6\theta + 3\theta^2$ is a fundamental unit of the field K (Problem 4 of Section 5, Chapter 2). For the proof use Problem 21, applied to the number $\alpha\theta$, Problems 17 and 22, and also prime factorizations in the field K of the numbers 2, 3, and 5. Further, setting $\xi = u + v\theta + w\theta^2$, write

$$\alpha = \alpha_0\xi^3 = \Phi + \Psi\theta + \Omega\theta^2,$$

where Φ, Ψ, and Ω are integral cubic forms in the variables u, v, and w. Show that for any of the six values of α_0 the equation $\Omega(u, v, w) = 0$ has only the trivial solution in rational (and in 3-adic) numbers.]

24. Let a and b be natural numbers which are square-free and relatively prime, and let $d = ab^2 > 1$. Show that in the field $R(\sqrt[3]{d})$ the number 3 factors into prime divisors as follows:

$$3 = \mathfrak{p}^3 \quad \text{if} \quad d \not\equiv \pm 1 \pmod 9,$$

$$3 = \mathfrak{p}^2\mathfrak{q}\ (\mathfrak{p} \neq \mathfrak{q}), \quad \text{if} \quad d \equiv \pm 1 \pmod 9).$$

Hint: In the case $d \equiv \pm 1 \pmod 9$ consider the norms $N(\omega - 1)$, $N(\omega)$, $N(\omega + 1)$, where

$$\omega = \frac{1}{3}(1 + \sigma\sqrt[3]{ab^2} + \tau\sqrt[3]{a^2b}),$$

$$\sigma = \pm 1, \quad \tau = \pm 1, \quad \sigma a \equiv \tau b \equiv 1 \pmod 3).$$

25. Let θ be an integral primitive element of the algebraic number field K, with minimum polynomial $\varphi(t)$, and let p be a rational prime which does not divide the index of θ. Suppose that, modulo p, we have the factorization

$$\varphi(t) \equiv \varphi_1(t)^{e_1} \cdots \varphi_m(t)^{e_m} \pmod p,$$

where $\varphi_1, \ldots, \varphi_m$ are distinct irreducible polynomials modulo p with degrees $f_1, \ldots, f_m$. Show that the decomposition of the number p in the field K takes the form

$$p = \mathfrak{p}_1^{e_1} \cdots \mathfrak{p}_m^{e_m},$$

where the distinct prime divisors $\mathfrak{p}_1, \ldots, \mathfrak{p}_m$ have degrees $f_1, \ldots, f_m$ and where $\varphi_i(\theta) \equiv 0 \pmod{\mathfrak{p}_i}$ for $i = 1, \ldots, m$.

Hint: Use the fact that every integer of K is congruent modulo $\mathfrak{p}_i$ to a linear combination (with rational integer coefficients) of the powers θ^s ($s \geqslant 0$).

26. Let θ be an integral primitive element of the field K, and p a prime number which does not divide the index of θ. For any rational integer x, show that the number $\theta + x$ is not divisible by any prime divisor which divides p and is of degree greater than 1. Further, show that $\theta + x$ is not divisible by the product of any two distinct prime divisors which divide p.

27. Under the same assumptions, let $\mathfrak{p}_1, \ldots, \mathfrak{p}_s$ be distinct prime divisors which divide p, with degrees $f_1, \ldots, f_s$, and let $r < f_1 + \cdots + f_s$. If $x_0, \ldots, x_{r-1}$ are any rational integers, show that the number $\theta^r + x_{r-1}\theta^{r-1} + \cdots + x_0$ is not divisible by the product $\mathfrak{p}_1 \cdots \mathfrak{p}_s$.

8. Quadratic Fields

In this section we consider the somewhat simpler theory of divisors for the case of quadratic fields. We start with a description of the prime divisors.

8.1. *Prime Divisors*

Since every prime divisor divides one and only one rational prime, to give a complete description of the set of all prime divisors it suffices to show how each rational prime p factors as a product of prime divisors. For quadratic fields it follows from (7.3) that there are only three possibilities for the numbers m, f_i, e_i:

(1) $m = 2,\quad f_1 = f_2 = 1,\quad e_1 = e_2 = 1;$

(2) $m = 1,\quad f = 2,\quad e = 1;$

(3) $m = 1,\quad f = 1,\quad e = 2.$

Corresponding to these we have the following types of decomposition:

(1) $p = \mathfrak{p}\mathfrak{p}',\quad N(\mathfrak{p}) = N(\mathfrak{p}') = p,\quad \mathfrak{p} \neq \mathfrak{p}';$

(2) $p = \mathfrak{p},\quad N(\mathfrak{p}) = p^2;$

(3) $p = \mathfrak{p}^2,\quad N(\mathfrak{p}) = p.$

Our problem is to discover what determines the type of factorization for any prime p. The answer will be easily derived from Theorem 8 of Section 5.

In Section 7.1 of Chapter 2 we showed that every quadratic field has a unique representation in the form $R(\sqrt{d})$, where d is a square-free rational integer.

First, let p be an odd prime. If p does not divide d, then it also does not divide the discriminant of the polynomial $x^2 - d$, a root of which generates the field. We then deduce from Theorem 8 of Section 5 that p has a decomposition of either the first or the second type, depending on whether the polynomial $x^2 - d$ is reducible modulo p or not. This in turn depends on whether d is a quadratic residue modulo p or not.

If $p|d$, then $d = pd_1$, where d_1 is not divisible by p, since d is square-free. It follows from

$$pd_1 = (\sqrt{d})^2, \qquad (d_1, p) = 1,$$

that all prime divisors which divide p occur with an even exponent in its factorization, which is possible only for the third type of decomposition. Thus for odd p we have the first, second, or third type of decomposition in the following three cases, respectively: (1) $p \nmid d$, $(d/p) = 1$; (2) $p \nmid d$, $(d/p) = -1$; (3) $p|d$. Note that since the discriminant D of the field $R(\sqrt{d})$ is either d or $4d$ (Theorem 1 of Section 7, Chapter 2), then in each of these conditions we could replace d by D.

The case $p = 2$ remains. Assume first that $2 \nmid D$. By Theorem 1 of Section 7, Chapter 2, this means that $D = d \equiv 1 \pmod 4$. It is clear that $R(\sqrt{d}) = R(\omega)$, where $\omega = (-1 + \sqrt{D})/2$. The minimum polynomial of ω is

$$x^2 + x + \frac{1 - D}{4}. \tag{8.1}$$

Since the discriminant of the basis $1, \omega$ is odd, we obtain from Theorem 8 of Section 5 that the number 2 will have either the first or the second type of decomposition, depending on whether the polynomial (8.1) is reducible or not. But the polynomial $x^2 + x + a$ is reducible modulo 2 if and only if $2|a$. Thus for $2 \nmid D$ we obtain the first and the second types of decomposition in the respective cases $D \equiv 1 \pmod 8$ and $D \equiv 5 \pmod 8$.

We now show that if $2|D$, then 2 always has the third type of decomposition. If $2|d$, then $d = 2d'$, $2 \nmid d'$, and from

$$2d' = (\sqrt{d})^2, \quad 2 \nmid d',$$

just as in the case of odd p, we see that 2 has the third type of decomposition. If $2 \nmid d$, then $d \equiv 3 \pmod 4$ (Theorem 1 of Section 7 of Chapter 2) and we have

$$(1 + \sqrt{d})^2 = 2\alpha$$

with the integer $\alpha = (1 + d)/2 + \sqrt{d}$ relatively prime to 2, since its norm

$$N(\alpha) = \frac{(1 + d)^2}{4} - d = \left(\frac{1 - d}{2}\right)^2$$

is not divisible by 2. Thus we again obtain the third type of decomposition for 2.

We have obtained the following theorem.

Theorem 1. In a quadratic field with discriminant D the prime number p has the decomposition

$$p = \mathfrak{p}^2, \quad N(\mathfrak{p}) = p,$$

if and only if p divides D. If p is odd and does not divide D, then

$$p = \mathfrak{p}\mathfrak{p}', \quad \mathfrak{p} \neq \mathfrak{p}', \quad N(\mathfrak{p}) = N(\mathfrak{p}') = p \qquad \text{for } \left(\frac{D}{p}\right) = 1;$$

$$p = \mathfrak{p}, \quad N(\mathfrak{p}) = p^2 \qquad \text{for } \left(\frac{D}{p}\right) = -1.$$

If 2 does not divide D [and hence $D \equiv 1 \pmod 4$], then

$$2 = \mathfrak{p}\mathfrak{p}', \quad \mathfrak{p} \neq \mathfrak{p}', \quad N(\mathfrak{p}) = N(\mathfrak{p}') = 2 \qquad \text{for } D = 1 \pmod 8;$$

$$2 = \mathfrak{p}, \quad N(\mathfrak{p}) = 4 \qquad \text{for } D = 5 \pmod 8.$$

8.2. *Rules of Decomposition*

Theorem 1 tells us that the type of decomposition of the odd prime p is determined by the residue of D (or d) modulo p, and even by the Legendre symbol $(D/p) = (d/p)$ as a function of p. It is natural to ask if the theorem can be reformulated so that the decomposition depends on the residue of p with respect to some modulus (the modulus depending only on the field). To achieve such a formulation, we use the reciprocity law for the Jacobi symbol.

It is well known that the Jacobi symbol (c/b) is defined for odd c and positive odd b, relatively prime to c. The reciprocity law for this symbol states that

$$\left(\frac{c}{b}\right) = (-1)[(b-1)/2] \cdot [(c-1)/2]\left(\frac{b}{|c|}\right)$$

(the proof for $c < 0$ is easily reduced to the positive case).

Let p be any odd prime. If $d = D \equiv 1 \pmod 4$, then

$$\left(\frac{D}{p}\right) = \left(\frac{d}{p}\right) = (-1)[(p-1)/2] \cdot [(d-1)/2]\left(\frac{p}{|d|}\right) = \left(\frac{p}{|D|}\right), \tag{8.2}$$

since $(d-1)/2$ is even. If $d \equiv 3 \pmod 4$, then

$$\left(\frac{D}{p}\right) = \left(\frac{d}{p}\right) = (-1)^{[(p-1)/2] \cdot [(d-1)/2]}\left(\frac{p}{|d|}\right) = (-1)^{(p-1)/2}\left(\frac{p}{|d|}\right), \tag{8.3}$$

since $(d-1)/2$ is odd. Finally, for $d = 2d'$, $2 \nmid d'$, we have

$$\left(\frac{D}{p}\right) = \left(\frac{d}{p}\right) = \left(\frac{2}{p}\right)\left(\frac{d'}{p}\right) = (-1)^{[(p^2-1)/8]+[(p-1)/2][(d'-1)/2]}\left(\frac{p}{|d'|}\right). \tag{8.4}$$

The value of the Jacobi symbol $(p/|d|)$ [or $(p/|d'|)$] depends only on the residue of p modulo $|d|$ (or $|d'|$). If $d \equiv 1 \pmod 4$, so that $d = D$, then (D/p) depends only on the residue of p modulo $|d| = |D|$. If $d \equiv 3 \pmod 4$, so that $D = 4d$, then (D/p) depends not only on the residue of p modulo $|d|$, but also on the number $(-1)^{(p-1)/2}$, that is, on the residue of p modulo 4; hence (D/p) depends on the residue of p modulo $4|d| = |D|$. Finally, if $d = 2d', D = 4d = 8d'$, then (D/p) depends on the residue of p modulo $|d'|$, $(-1)^{(p-1)/2}$ depends on the residue of p modulo 4, and $(-1)^{(p^2-1)/8}$ depends on the residue of p modulo 8. Hence in this case the value of (D/p) depends on the residue of p modulo $8|d'| = |D|$. Thus in all cases the type of decomposition of the prime number p depends only on its residue modulo $|D|$, so that all prime numbers having the same residue have the same decomposition. This conclusion, which is completely nonobvious a priori, is the most important property of the decomposition rules for prime numbers in quadratic fields.

To make this new form of the decomposition rule more clear, we introduce a new function. For all x relatively prime to the discriminant D, we set

$$\chi(x) = \begin{cases} \left(\dfrac{x}{|d|}\right) & \text{for } d \equiv 1 \pmod 4, \\ (-1)^{(x-1)/2}\left(\dfrac{x}{|d|}\right) & \text{for } d \equiv 3 \pmod 4, \\ (-1)^{[(x^2-1)/8]+[(x-1)/2]\cdot[(d'-1)/2]}\left(\dfrac{x}{|d'|}\right) & \text{for } d = 2d' \end{cases} \tag{8.5}$$

[in case $d \equiv 2, 3 \pmod 4$ the expressions $(-1)^{(x-1)/2}$ and $(-1)^{(x^2-1)/8}$ make sense since the discriminant $D = 4d$ is even and so x is odd].

In the arguments above which showed that for odd p the value of (D/p) depends only on the residue of p modulo $|D|$, we never used the fact that p was prime. Hence by the same arguments it follows that $\chi(x)$ depends only on the residue of x modulo $|D|$. Further, it is easily verified that if $(x, D) = 1$ and $(x', D) = 1$, then $\chi(xx') = \chi(x)\chi(x')$. This means that the function χ can be considered as a homomorphism of the multiplicative group of residue classes modulo $|D|$, relatively prime to $|D|$, to the group of order 2 consisting of $+1$ and -1. If we extend such a function by giving it to the value 0 on all numbers not relatively prime to D, it is called a *numerical character*.

Definition. The numerical character χ with modulus $|D|$, where the value of $\chi(x)$ for x relatively prime to D is given by (8.5), is called the character of the field $R(\sqrt{d})$.

Returning to (8.2), (8.3) and (8.4), we see that the decomposition of an odd prime p which does not divide D will be of the first type if $\chi(p) = +1$ and of the second type if $\chi(p) = -1$. This result remains true for $p = 2$. For if $2 \nmid D$, then $D \equiv 1 \pmod 4$ and this means that $\chi(2) = (2/|D|)$, which equals $+1$ for $D \equiv 1 \pmod 8$ and -1 for $D \equiv 5 \pmod 8$.

Hence we have the following new formulation of the rule for decomposition in quadratic fields.

Theorem 2. If χ is the character of the quadratic field $R(\sqrt{d})$, then the decomposition of a prime p in $R(\sqrt{d})$ is given by the conditions:

$$p = \mathfrak{p}\mathfrak{p}', \quad \mathfrak{p} \neq \mathfrak{p}', \quad N(\mathfrak{p}) = N(\mathfrak{p}') = p \qquad \text{if } \chi(p) = 1;$$

$$p = \mathfrak{p}, \quad N(\mathfrak{p}) = p^2, \qquad \text{if } \chi(p) = -1;$$

$$p = \mathfrak{p}^2, \quad N(\mathfrak{p}) = p, \qquad \text{if } \chi(p) = 0.$$

All rational integers are partitioned into three sets, depending on the value of χ. Each of these sets consists of certain residue classes modulo $|D|$. By Theorem 2 the type of decomposition of p depends only on which of the three sets contains p.

A law of decomposition like that in quadratic fields, where the type of decomposition depends only on the residue of the prime p with respect to a certain fixed modulus, also occurs for certain other fields. This is the case, for example, for cyclotomic fields (see Section 2.2. of Chapter 5). But it is far from being the case in general. Since the knowledge of such a law of decomposition allows us to solve many number-theoretic problems (see, for example, the following section and Section 2 of Chapter 5), it would be interesting to know for precisely which fields we have such a simple law of decomposition. The answer to this question leads into class field theory. It can be shown that any such field is a normal extension of the field of rational numbers, the Galois group of which is Abelian. Among such fields lie, of course, quadratic fields, which have a cyclic group of order 2 as Galois group. The simplest examples of non-Abelian fields are cubic fields whose discriminant is not a perfect square. An example is the field $R(\theta)$, where $\theta^3 - \theta - 1 = 0$. Hence for this field there does not exist any integer M such that the type of decomposition into prime divisors of the prime number p depends only on the residue of p modulo M.

Class field theory solves much more general problems than those we have mentioned. It allows one to describe the law of decomposition for prime divisors of an arbitrary algebraic number field k in some extension K/k, provided that the Galois group of this extension is Abelian (we spoke above of the special case when $k = R$). Class field theory has many number-theoretic applications. It allows us to carry over the theorem on quadratic forms with rational coefficients, proved in Chapter 1, to the case of quadratic forms with coefficients from an arbitrary algebraic number field k; to get a better understanding of the theory of genera, which we shall describe in Section 8.4; to prove a theorem on the existence of prime divisors in given classes of divisors; and so on. The following works deal with class field theory:

H. Hasse, Bericht uber neuere Untersuchungen und Probleme aus der Theorie de algebraischen Zahlkorper, *Jahresbericht Deutschen Mathematiker Vereinigung*, Teil I (Klassenkorpertheorie): **35**, 1–55 (1926); Teil Ia (Beweise zu Teil I): **36**, 233–311 (1927), Teil II (Reziprozitatsgesetz): **6**, 1–204 (1930).

C. Chevalley, Sur la theorie du corps de classes dans les corps finis et les corps locaux, *J. Fac. Sci. Univ. Tokyo* **2**, 366–476 (1933).

E. Artin and J. Tate, "Class Field Theory," Printed privately, 1961.

Many arithmetic questions lead to questions within the domain of class field theory concerning the decomposition of prime numbers in fields with non-Abelian Galois group. Very little is known about this subject at present.

8.3. *Representation of Numbers by Binary Quadratic Forms*

In Section 7.5 of Chapter 2 we saw that there is a one-to-one correspondence between classes of properly equivalent primitive binary quadratic forms and classes of strictly similar modules of a quadratic field (in the case $D < 0$ we consider only positive-definite forms). On the other hand, by Theorem 6 of Section 6 there is a one-to-one correspondence between full modules which belong to the maximal order of the field (that is, ideals of the field) and divisors. It is natural to expect that the theory of divisors in a quadratic field will have a definite connection with the theory of those primitive forms whose discriminant coincides with the discriminant of the field.

To extend the correspondence between classes of forms and modules to divisors, we clearly must change the concept of equivalence of divisors slightly.

Definition. Two divisors $\mathfrak{a}$ and $\mathfrak{b}$ of the quadratic field $R(\sqrt{d})$ are called *strictly equivalent* if there exists a number $\alpha \neq 0$ of $R(\sqrt{d})$ such that $N(\alpha) > 0$ and $\mathfrak{a} = \mathfrak{b}(\alpha)$.

In an imaginary quadratic field, where the norm of every number is positive, this definition coincides with the usual one (Section 7.2). The same arguments that were used for modules (Section 7.5 of Chapter 2) show that the new concept of equivalence coincides with the usual one in the real quadratic field $R(\sqrt{d})$ if and only if the norm of a fundamental unit ε of the field equals -1. If $N(\varepsilon) = +1$, then each class of divisors, equivalent in the usual sense, splits into precisely two classes of strictly equivalent divisors. Thus the number $\bar{h}$

of strict divisor classes is also finite, and is related to the number h of divisor classes in the usual sense by

$$\bar{h} = h \qquad \text{for } d < 0;$$

$$\bar{h} = h \qquad \text{for } d > 0, \qquad N(\varepsilon) = -1;$$

$$\bar{h} = 2h \qquad \text{for } d > 0, \qquad N(\varepsilon) = +1.$$

Theorem 4 of Section 7 of Chapter 2, when applied to modules which belong to the maximal order of the field $R(\sqrt{d})$ with discriminant D, can be reformulated as follows: the strict divisor classes of the quadratic field $R(\sqrt{d})$ are in one-to-one correspondence with the classes of properly equivalent primitive binary quadratic forms of discriminant D (which are positive definite if $D < 0$).

We shall try to apply the results of Sections 8.1 and 8.2 to the question of the representation of numbers by binary forms.

By Theorem 6 of Section 7 of Chapter 2 the natural number a is represented by some form of discriminant D if and only if there is an integral divisor of the field $R(\sqrt{d})$ with norm a (the norm of a divisor coincides with the norm of the corresponding module). Now we can use Theorem 2 to characterize all numbers which are norms of divisors. By this theorem the norm $N(\mathfrak{p})$ of a prime divisor $\mathfrak{p}$ equals the prime number p if $\chi(p) = 0$ or if $\chi(p) = 1$, and equals p^2 if $\chi(p) = -1$. Hence the number a is representable in the form $N(\mathfrak{a})$ for some integral divisor $\mathfrak{a} = \prod_{\mathfrak{p}} \mathfrak{p}^{a(\mathfrak{p})}$ of the field $R(\sqrt{d})$ if and only if all prime factors p, for which $\chi(p) = -1$, occur in a with even exponent.

By using the Hilbert symbol (Section 6.3 of Chapter 1) we can put this result in somewhat different form. We compute $(a, D/p)$ for all primes which do not divide D. Let $a = p^k b$, where b is not divisible by p. From the properties of the Hilbert symbol we obtain

$$\left(\frac{a, D}{p}\right) = \left(\frac{b, D}{p}\right)\left(\frac{D}{p}\right)^k = \left(\frac{D}{p}\right)^k = \chi(p)^k \qquad \text{for } p \neq 2,\ p \nmid D;$$

$$\left(\frac{a, D}{2}\right) = (-1)^{[(b-1)/2]\cdot[(D-1)/2]+k[(D^2-1)/8]} = (-1)^{k[(D^2-1)/8]} = \chi(2)^k \qquad \text{for } p = 2,\ 2 \nmid D$$

[for $p = 2$, $2 \nmid D$, we have used the fact that $D \equiv 1 \pmod 4$]. This formula proves the second part of the following theorem.

Theorem 3. A natural number a is represented by some binary form of discriminant D if and only if every prime p, for which $\chi(p) = -1$, occurs with even exponent in the prime factorization of a. This in turn occurs if and only if

$$\left(\frac{a, D}{p}\right) = +1 \qquad \text{for all } p \nmid D.$$

Since the integers a and ab^2 are either both represented or both not represented by forms of discriminant D, we may limit our consideration to square-free numbers a.

If $p \neq 2$, $p \nmid D$, and $p \nmid a$, then we know that $(a, D/p) = +1$. Hence Theorem 3 only imposes a finite number of conditions on the number a, and these conditions only involve the residues modulo $|D|$ of the prime divisors of the square-free number a.

Theorem 3 could have been deduced easily from Theorem 7 of Section 7 of Chapter 2. We gave a proof based on Theorem 2 to point out the connection between the question of the representation of numbers by forms of discriminant D and the question of the decomposition into factors in the corresponding quadratic field.

This result is not all that we might wish to obtain. We would like to have a criterion for the representation of the number a by forms from a given class of properly equivalent forms, and Theorem 3 only gives us a condition for the representability of a by forms from some class. The following question then arises: Can we partition the classes of forms into nonintersecting collections, so that, for any a, all forms which represent the number a (if any exist) are contained in the same collection? Such a partition was found by Gauss. It is connected with rational equivalence of quadratic forms.

Definition. We say that two primitive binary quadratic forms with discriminant D belong to the same *genus* if they are rationally equivalent.

Since integrally equivalent forms are certainly rationally equivalent, all forms of the same class lie in the same genus. Hence each genus is the union of certain classes. It follows that the number of genera of forms (for given discriminant D) is finite.

In Section 7.5 of Chapter 1 we defined the invariant $e_p(f)$ for a nonsingular binary rational form f, where p is a prime number or the symbol ∞. In the case of a primitive form f with discriminant D, the determinant equals $-\frac{1}{4}D$, and therefore

$$e_p(f) = \left(\frac{a, D}{p}\right),$$

where $a \neq 0$ is any number which is rationally represented by the form f.

Let G be any genus of forms. Since all forms of G have the same invariant, we can set

$$e_p(G) = e_p(f),$$

where f is any form of G.

Let a be any nonzero number represented by the form f. By the second assertion of Theorem 3 we have $e_p(f) = (a, D/p) = 1$ for all primes p which do not divide D. Further $e_\infty(f) = 1$, since in the case $D < 0$ we are considering only positive-definite forms. Hence for any genus G of forms with discriminant D we have

$$e_p(G) = 1 \qquad \text{for } p \nmid D \text{ and } p = \infty. \tag{8.6}$$

Hence each genus G is uniquely determined by the invariants $e_p(G)$, where p runs through all prime divisors of the discriminant D.

We can now give conditions for a number to be represented by some form of a fixed genus.

Theorem 4. Let a be a natural number and G be a genus of forms with discriminant D. In order that a be integrally represented by some form of G, it is necessary and sufficient that

$$\left(\frac{a, D}{p}\right) = e_p(G)$$

for all primes p.

Proof. The condition is clearly necessary. If for some a we have $(a, D/p) = e_p(G)$ for all p, then by (8.6) $[(a, D)/p] = 1$ for all $p \nmid D$. Theorem 3 implies that a is represented by some form f of discriminant D, and since $e_p(f) = [(a, D)/p] = e_p(G)$, then f belongs to the genus G. The theorem is proved.

The assertion of Theorem 4 is interesting in that the condition for the representability of a by some form of the genus G only involves the residue of a modulo $|D|$ [assuming that a is represented by some form of discriminant D, that is, that $[(a, D)/p] = 1$ for all $p \nmid D$]. In the case when each genus consists of one and only one class, Theorem 4 gives us an ideal answer to the question of the representation of numbers by binary forms.

In the general case this result cannot be improved, in the following sense. Suppose we take a set S of classes of forms, which is not the union of some genera. Then there does not exist any modulus m such that the representation of a number a by some form of our set S depends only on the residue of a modulo m. In particular, if a genus consists of several classes, then it is not possible to characterize the numbers represented by forms of one of those classes in terms of the residues of the numbers for some modulus. These facts can be proved by class field theory. The proof has the following flavor. The representation of a prime number p by some form from our set of classes S can be interpreted in terms of the type of splitting of this prime into prime divisors in some field L. The field L will have an Abelian Galois group over the rational field if and only if our set S is a union of genera [H. Hasse, Zur Geschlechtertheorie in quadratischen Zahlkorpern, *J. Math. Japan* **3**, No. 1, 45–51 (1951)].

We now investigate the question of the number of genera. Let $p_1, \ldots, p_t$ be all prime divisors of the discriminant D. By (8.6) every genus is uniquely determined by the invariants $e_i = e_{p_i}(G)$. These invariants cannot be arbitrary, since, if $f \in G$ and the number $a \neq 0$ is represented by f, we have

$$e_1 \cdots e_t = \prod_p e_p(G) = \prod_p \left(\frac{a, D}{p}\right) = 1$$

[see (7.17) of Chapter 1; the product is taken over all prime numbers p and the symbol ∞].

We now show that the relation

$$e_1 \cdots e_t = 1 \tag{8.7}$$

between the numbers $e_i = \pm 1$ is not only necessary, but also sufficient, in order that these numbers be the invariants of some genus G.

Denote by k_i the power to which p divides D (k_i equals 1 for all $p_i \neq 2$ and equals 2 or 3 for $p_i = 2$). For $i = 1, \ldots, t$ choose an integer a_i, not divisible by p_i, such that $(a_i, D/p_i) = e_i$, and then determine a by the system of congruences

$$a \equiv a_i \pmod{p_i^{k_i}} \qquad (1 \leqslant i \leqslant t).$$

For any a which satisfies these congruences we have (by the properties of the Hilbert symbol)

$$\left(\frac{a, D}{p_i}\right) = \left(\frac{a_i, D}{p_i}\right) = e_i .$$

Our problem is then to find, among all such values of a, one for which $(a, D/p) = 1$ for all $p \nmid D$. We use here the theorem of Dirichlet on prime numbers in arithmetic progressions (Section 3 of Chapter 5). Since the set of all such values of a is a residue class modulo $|D| = \prod p_i^{k_i}$ consisting of numbers relatively prime to D, we may choose among them a prime value q, by Dirichlet's theorem. We then have

$$\left(\frac{q, D}{p_i}\right) = \left(\frac{a, D}{p_i}\right) = e_i;$$

$$\left(\frac{q, D}{p}\right) = 1 \qquad \text{for } p \nmid D,\ p \neq 2 \quad \text{and} \quad p \neq q;$$

$$\left(\frac{q, D}{2}\right) = (-1)^{[(q-1)/2][(D-1)/2]} = 1 \qquad \text{for } 2 \nmid D.$$

The relation $\prod_p (q, D/p) = 1$ then yields $e_1 \cdots e_t(q, D/q) = 1$, so it follows from (8.7) that the value of the symbol $(q, D/q)$ is also 1.

Thus there exists a natural number a (which is also prime) such that

$$\left(\frac{a, D}{p_i}\right) = e_i\,(1 \leqslant i \leqslant t) \quad \text{and} \quad \left(\frac{a, D}{p}\right) = 1 \qquad \text{for } p \nmid D.$$

By Theorem 3, a is represented by some form f with discriminant D. If this form belongs to the genus G, then

$$e_{p_i}(G) = \left(\frac{a, D}{p_i}\right) = e_i \qquad (1 \leqslant i \leqslant t).$$

This proves our assertion on the existence of a genus with given invariants [satisfying, of course, (8.7)]. Since there are 2^{t-1} possible choices of $e_i = \pm 1$ which satisfy (8.7), the number of genera of forms of discriminant D also equals 2^{t-1}.

Theorem 5. Let $p_1, \ldots, p_t$ be the distinct prime divisors of the discriminant D of the quadratic field $R(\sqrt{d})$. For any choice of the values $e_i = \pm 1$ $(1 \leqslant i \leqslant t)$ such that $e_1 \cdots e_t = 1$, there is a genus G of forms of discriminant D for which $e_{p_i}(G) = e_i$. Hence there are 2^{t-1} genera of forms of discriminant D.

Remark 1. The theory of genera of forms, given in this section when the discriminant coincides with the discriminant of the maximal order of a quadratic field, can also be developed for forms with discriminant Df^2.

Remark 2. If every genus of forms with negative discriminant Df^2 consists of a single class, then there is a simple formula (Problem 18) for the number of representations of an integer relatively prime to f by a fixed form of discriminant Df^2. A table of the known values for the discriminant $Df^2 < 0$ with each genus consisting of a single class is given at the end of the book. It is not known if this table is complete. It has been proved that the number of such discriminants is finite. For the even numbers Df^2 in this table the numbers $-\frac{1}{4}Df^2$ were found by Euler, who called them *convenient numbers.* They were used by Euler to find large prime numbers because of the following property: If a and b are relatively prime numbers whose product ab is a convenient number and if the form $ax^2 + by^2$ represents the number q in essentially only one way (with relatively prime x and y), then the number q is prime (see Problem 19). For example, the difference $3049 - 120y^2$ is a square only when $y = 5$, and this means that the number 3049 is represented by the form $x^2 + 120y^2$ in only one way: $3049 = 7^2 + 120 \cdot 5^2$, and hence is prime. By this method Euler found many primes which were very large for those times. It is clear that for larger convenient numbers, the work involved in proving uniqueness is less.

8.4. *Genera of Divisors*

The results on genera of forms obtained in Section 8.3 allow us to draw some conclusions on the structure of the group of divisor classes (in the strict sense) of a quadratic field. We carry over the concept of genus to divisors.

By Theorem 6 of Section 6 every divisor $\mathfrak{a}$ (integral or fractional) corresponds to a unique ideal $\bar{\mathfrak{a}}$, which consists of all numbers of the field which are divisible by $\mathfrak{a}$. For a quadratic field every basis $\{\alpha, \beta\}$ of the module $\bar{\mathfrak{a}}$, which satisfies condition (7.10) of Chapter 2, corresponds to a primitive form

$$f(x, y) = \frac{N(\alpha x + \beta y)}{N(\mathfrak{a})}. \tag{8.8}$$

If we pass to another basis of the module $\mathfrak{a}$ [which also satisfies (7.10) of Chapter 2], the form f will be taken to a strictly equivalent form. Hence by (8.8) the divisor is associated with a whole class of strictly equivalent forms. This mapping sets up a one-to-one correspondence between classes of divisors in the narrow sense and classes of strictly equivalent forms of discriminant D, which was already remarked at the beginning of Section 8.3.

Definition. Two divisors of a quadratic field are in the same genus if their corresponding classes of forms are contained in the same genus of forms (that is, are rationally equivalent).

Since divisors which are strictly equivalent correspond to the same class of forms, then each genus of divisors is a union of classes (in the strict sense) of divisors.

A genus of divisors, corresponding to the genus of forms G, will also be denoted by G. By the invariants $e_p(G)$ of a genus G of divisors, we mean the analogous invariants for the corresponding genus of forms. We then have the formula

$$e_p(G) = \left(\frac{N(\mathfrak{a}), D}{p}\right), \tag{8.9}$$

where $\mathfrak{a}$ is any divisor of the genus G. For by definition we have $e_p(G) = (a, D/p)$, where a is some nonzero rational number which is represented by the form $f(x, y)$ given by (8.8). But the form $N(\alpha x + \beta y)$ represents all squares of rational numbers, so in particular it represents $N(\mathfrak{a})^2$. Hence, $f(x, y)$ represents $N(\mathfrak{a})$, which proves (8.9).

The genus of divisors G_0, all invariants of which equal 1, is called the *principal genus*. The divisors of the principal genus are characterized by the condition $(N(\mathfrak{a}), D/p) = 1$ for all p. Hence the principal genus is a subgroup (with respect to the operation of multiplication of divisors) of the group of all divisors. Further, any genus G of divisors is a coset of $\mathfrak{a}G_0$ of the subgroup

G_0, where $\mathfrak{a}$ is any divisor of the genus G. The set of all cosets of the subgroup G_0 is a group, the factor group of the group of all divisors modulo the subgroup G_0. Hence we can consider the set of all genera as a group. It is called the *group of genera*. By Theorem 5 the order of the group of genera is 2^{t-1}, where t is the number of distinct prime divisors of the discriminant D.

We now characterize the genera of divisors in terms of divisors, without mentioning forms.

Theorem 6. Two divisors $\mathfrak{a}$ and $\mathfrak{a}_1$ of a quadratic field belong to the same genus if and only if there is an element of positive norm in the field such that

$$N(\mathfrak{a}_1) = N(\mathfrak{a})N(\gamma).$$

Proof. Choose bases $\{\alpha, \beta\}$ and $\{\alpha_1, \beta_1\}$ for the ideals $\bar{\mathfrak{a}}$ and $\bar{\mathfrak{a}}_1$ which satisfy (7.10) of Chapter 2. Then the forms

$$f(x, y) = \frac{N(\alpha x + \beta y)}{N(\mathfrak{a})}, \qquad f_1(x, y) = \frac{N(\alpha_1 x + \beta_1 y)}{N(\mathfrak{a}_1)},$$

correspond to the divisors $\mathfrak{a}$ and $\mathfrak{a}_1$. By Theorem 11 of Section 1 of the Supplement, the forms f and f_1 are rationally equivalent if and only if there is a nonzero rational number which is represented by both of these forms. But this would mean that

$$\frac{N(\xi)}{N(\mathfrak{a})} = \frac{N(\xi_1)}{N(\mathfrak{a}_1)} \qquad (\xi, \xi_1 \neq 0),$$

and the assertion of the theorem follows.

Divisors of the principal genus have the following important characterization.

Theorem 7. The divisor $\mathfrak{a}$ belongs to the principal genus if and only if it is strictly equivalent to the square of some divisor.

Proof. Suppose that the divisor $\mathfrak{a}$ belongs to the principal genus. Since the unit divisor belongs to the principal genus, it follows from Theorem 6 that there is a number γ for which $N(\mathfrak{a}) = N(\gamma)$. Replacing $\mathfrak{a}$ by the equivalent divisor $\mathfrak{a}(\gamma^{-1})$, we may assume that $N(\mathfrak{a}) = 1$. Now write $\mathfrak{a}$ as a product of prime divisors. Here we distinguish between those prime divisors $\mathfrak{p}_i$ for which their exists another prime divisor $\mathfrak{p}_i'$ with the same norm (the first type of decomposition in the terminology of Section 8.1), and all other prime divisors $\mathfrak{q}_j$

$$\mathfrak{a} = \prod_i \mathfrak{p}_i^{a_i} \mathfrak{p}_i'^{b_i} \prod_j \mathfrak{q}_j^{c_j}.$$

Since $N(\mathfrak{p}_i) = N(\mathfrak{p}_i') = p_i$ and $N(\mathfrak{q}_j) = q_j^{r_j}$ (where r_j equals 1 or 2), then we have

$$\prod_i p_i^{a_i+b_i} \prod_j q_j^{r_j c_j} = 1$$

[since $N(\mathfrak{a}) = 1$]. Since the primes p_i and q_j are all distinct, $b_i = -a_i$ and $c_j = 0$, so that

$$\mathfrak{a} = \prod_i \mathfrak{p}_i^{a_i} \mathfrak{p}_i'^{-a_i}.$$

But $\mathfrak{p}_i \mathfrak{p}_i' = p_i$, so that $\mathfrak{p}_i'^{-1} \sim \mathfrak{p}_i$, from which it follows that

$$\mathfrak{a} \sim \left(\prod_i \mathfrak{p}_i^{a_i}\right)^2$$

(here the sign $\sim$ denotes strict equivalence of divisors).

Conversely, if $\mathfrak{a} \sim \mathfrak{b}^2$, that is, $\mathfrak{a} = \mathfrak{b}^2(\alpha)$, with $N(\alpha) > 0$, then $N(\mathfrak{a}) = N(\beta)$, where $\beta = N(\mathfrak{b})\alpha$, and then, by Theorem 6, $\mathfrak{a}$ belongs to the principal genus.

Theorem 7 is proved.

Let $\mathfrak{C}$ denote the group of classes of strictly equivalent divisors. If we map each class $C \in \mathfrak{C}$ to that genus G which contains the class C, we obtain a homomorphism of the group of classes onto the group of genera. Its kernel is the set of all classes which are contained in the principal genus. By Theorem 7 the class C' is contained in the principal genus if and only if it is the square of some class of $\mathfrak{C}$. Hence the kernel of the homomorphism of the group $\mathfrak{C}$ onto the group of genera is the subgroup $\mathfrak{C}^2$ which consists of all squares C^2 of classes $C \in \mathfrak{C}$. Using an elementary theorem on homomorphisms in group theory and the fact that the group of genera has order 2^{t-1}, we arrive at the following result.

Theorem 8. The factor group $\mathfrak{C}/\mathfrak{C}^2$ of the group of classes of strictly equivalent divisors by the subgroup of squares has order 2^{t-1}, where t is the number of distinct prime numbers which divide the discriminant D of the quadratic field.

The value of Theorem 8 lies in the information which it gives on the structure of the group $\mathfrak{C}$. By Theorem 1 of Section 5 of the Supplement, the group $\mathfrak{C}$ can be decomposed into the direct product of cyclic subgroups. From Theorem 8 it easily follows that precisely $t - 1$ of these subgroups have even order. In particular, we obtain the following fact.

Corollary. The number of classes of strictly equivalent divisors of a quadratic field is odd if and only if the discriminant of the field is only divisible by one prime.

Such fields are $R(\sqrt{-1})$, $R(\sqrt{2})$, $R(\sqrt{-2}$, $R(\sqrt{p})$, with p of the form $4n+1$, and $R(\sqrt{-q})$ with q of the form $4n+3$.

This fact is the basis for the little we know about the structure of the divisor class group.

PROBLEMS

1. Let χ be the character of the quadratic field with discriminant D. Show that χ can be expressed in terms of the Hilbert symbol by the formula

$$\chi(a) = \prod_{p \mid D} \left(\frac{a, D}{p}\right) \qquad (a, D) = 1.$$

2. If γ is any integer of a quadratic field which is relatively prime to the discriminant D, show that the congruence

$$x^2 \equiv N(\gamma) \pmod{|D|}$$

is always solvable.

3. Let G be the group of residue classes of rational integers (mod $|D|$) which are relatively prime to D, and let H be the subgroup of those classes which contain the norm of some integer of the quadratic field with discriminant D. Show that the index $(G:H)$ is equal to 2^t, where t is the number of distinct primes which divide D.

4. Under the same notations as Problem 3, let H^* denote the subgroup of G consisting of all residue classes which contain the norm of some integral divisor of the quadratic field with discriminant D. Show that $(G:H^*) = 2$.

5. If γ is any number with positive norm of the quadratic field with discriminant D, show that for all p,

$$\left(\frac{N(\gamma), D}{p}\right) = 1.$$

6. Let $\mathfrak{a}$ and $\mathfrak{b}$ be integral ideals which are relatively prime to D. Show that $\mathfrak{a}$ and $\mathfrak{b}$ belong to the same genus if and only if for some integer γ we have

$$N(\mathfrak{a}) \equiv N(\gamma)N(\mathfrak{b}) \pmod{|D|}.$$

7. If the discriminant of a real quadratic field is divisible by only one prime, show that the norm of a fundamental unit is -1.

8. Show that the automorphism $\sigma : \alpha \to \alpha^\sigma$ of the quadratic field $R(\sqrt{d})$ (not the identity automorphism) induces an automorphism $\mathfrak{a} \to \mathfrak{a}^\sigma$ of the group of divisors for which $(\alpha^\sigma) = (\alpha)^\sigma$ for all $\alpha \neq 0$. What is the behavior of this automorphism on prime divisors?

9. The automorphism σ of the group of divisors (Problem 8) induces an automorphism $\sigma : C \to C^\sigma$ of the group of classes of strictly equivalent divisors. Namely, if $\mathfrak{a} \in C$, then C^σ is that class which contains $\mathfrak{a}^\sigma$. The class C is called *invariant* if $C^\sigma = C$. Show that a class C is invariant if and only if C^2 is the principal class.

10. Show that the subgroup of the group of classes of strictly equivalent divisors which consists of all invariant classes is of order 2^{t-1} (t is the number of distinct primes which divide the discriminant).

11. If β is an element of a quadratic field with $N(\beta) = 1$, show that there exists an α such that

$$N(\alpha) > 0, \quad \beta = \pm \frac{\alpha^\sigma}{\alpha}.$$

12. Show that every invariant class C contains a divisor $\mathfrak{a}$ for which $\mathfrak{a}^\sigma = \mathfrak{a}$.

13. Let $\mathfrak{p}_1, \ldots, \mathfrak{p}_t$ be the distinct prime divisors which divide the discriminant D. Show that each invariant class C contains precisely two divisors of the type

$$\mathfrak{p}_{i_1} \ldots \mathfrak{p}_{i_k}, \quad 1 \leqslant i_1 < \cdots < i_k \leqslant t \qquad (k = 0, 1, \ldots, t).$$

14. The subgroup of all invariant classes contained in the principal genus is a direct product of cyclic subgroups of order 2. Let $\mathfrak{C}$ be the group of classes of strictly equivalent divisors, and recall the definition of the invariants of a finite Abelian group (Supplement, Section 5.1). Show that the number of cyclic components of the above subgroup equals the number of invariants of $\mathfrak{C}$ which are divisible by 4.

15. Show that the number of positive integers r which divide the discriminant D are square-free, and satisfy

$$\left(\frac{r, D}{p}\right) = 1 \qquad \text{for all } p,$$

is a number of the form 2^u. Show further that the number of invariants of the group $\mathfrak{C}$ which are divisible by 4 equals $u - 1$.

16. Let m be a natural number which is relatively prime to the index f of the order $\mathfrak{O}_f$ in the maximal order of the quadratic field $R(\sqrt{d})$. Show that the number of modules in $R(\sqrt{d})$ which have coefficient ring $\mathfrak{O}_f$, are contained in $\mathfrak{O}_f$, and have norm m, equals the number of integral divisors of the field $R(\sqrt{d})$ with norm m.

17. Show that the number of integral divisors of the quadratic field $R(\sqrt{d})$ with norm m equals

$$\sum_{r \mid m} \chi(r),$$

where χ is the character of the field $R(\sqrt{d})$ and r runs through all natural numbers which divide m.

18. Let $g_1(x, y), \ldots, g_s(x, y)$ be a full system of pairwise nonequivalent positive primitive quadratic forms with discriminant $Df^2 < 0$ [D being the discriminant of the maximal order of the field $R(\sqrt{d})$], and let m be a natural number which is relatively prime to f. Show that the number N of all representations of the number m by all the forms $g_1, \ldots, g_s$ is given by

$$N = x \sum_{r \mid m} \chi(r),$$

where

$$x = \begin{cases} 6 & \text{for } D = -3, f = 1; \\ 4 & \text{for } D = -4, f = 1; \\ 2 & \text{for } Df^2 < -4. \end{cases}$$

19. Let $g(x, y)$ be a positive form with discriminant $Df^2 < -4$ and let q be a natural number relatively prime to Df^2. Assume that every genus of forms with discriminant Df^2 consists of a single class. If $g(x, y) = q$ has precisely four solutions in integral, relatively prime x and y, show that q is a prime.

20. Let h_f be the number of classes of similar (in the usual sense) modules of a quadratic field which belong to the order $\mathfrak{D}_f$ (Problem 11 of Section 7, Chapter 2). Show that

$$h_f = h \frac{f}{e_f} \prod_{p \mid f} \left(1 - \frac{\chi(p)}{p}\right),$$

where χ is the character of the quadratic field and p runs through all primes which divide f.

21. Show that a prime number is represented by the form $x^2 + 3y^2$ if and only if it has the form $3n + 1$.

22. Show that the form $x^2 - 5y^2$ represents all prime numbers of the form $10n \pm 1$ and does not represent any primes of the form $10n \pm 3$.

23. Show that the natural number m is represented by the form $x^2 + 2y^2$ with relatively prime x and y if and only if in the representation

$$m = 2^{\alpha} p_1^{\alpha_1} \cdots p_r^{\alpha_r},$$

α is either 0 or 1, and each odd prime p_i is of the form $8n + 1$ or $8n + 3$.

24. Show that there exist quadratic fields (both real and imaginary) with arbitrarily large numbers of divisor classes.

25. Let $p_1, \ldots, p_s$ be the distinct prime numbers which divide the discriminant D of the quadratic field $R(\sqrt{d})$. From

$$\left(\frac{p_i, D}{p_j}\right) = (-1)^{a_{ij}} \qquad (1 \leqslant i, j \leqslant s)$$

we obtain a matrix (a_{ij}) whose elements lie in the field of residue classes modulo 2. Let ρ denote the rank of this matrix [over the field $GF(2)$]. Show that the number of invariants of the group of classes of strictly equivalent divisors, which are divisible by 4, equals $s - \rho - 1$.

26. Let p and q be prime integers, with $p \neq 2$ and $q \not\equiv p \pmod 4$. Show that the number of divisor classes of the field $R(\sqrt{-pq})$ is divisible by 4 if and only if $(q/p) = 1$.

27. Let $p_1, \ldots, p_s$ be distinct prime numbers of the form $4n + 1$, and let $d = p_1 \cdots p_s \equiv 1 \pmod 8$. Show that every genus of divisors of the field $R(\sqrt{-d})$ consists of an even number of classes.

28. Let ε be a fundamental unit of the real quadratic field $R(\sqrt{d})$, whose discriminant is not divisible by any prime number of the form $4n + 3$. Show that if the principal genus of divisors of the field $R(\sqrt{d})$ consists of an odd number of classes of strictly equivalent divisors, then $N(\varepsilon) = -1$.

29. Let p be a prime number of the form $8n + 1$. Show that the number of divisor classes of the field $R(\sqrt{-p})$ is divisible by 4.

CHAPTER 4

Local Methods

In Section 7 of Chapter 1 we gave a proof of the Hasse-Minkowski theorem on the representation of zero by rational quadratic forms. Both in the formulation and in the proof of this theorem we had to embed the field R in the p-adic fields R_p, and in the real field R_∞, that is, in all completions of the field R. A method of solving problems in number theory by use of the embeddings of the ground field in its completions is called a *local method.* Such methods have important number-theoretic consequences, not only when applied to the field of rational numbers, but also when applied to algebraic number fields. Local methods are also instrumental in the study of algebraic function fields.

In this chapter we describe the basic facts that hold for local methods for any ground field, and then make an intense application to prove some deep results on the representation of numbers by nonfull decomposable forms (Section 1.3 of Chapter 2). We refer particularly to the theorem of Thue, which states that the equation $f(x, y) = c$, where $f(x, y)$ is an integral homogeneous irreducible polynomial of degree $\geqslant 3$, has only a finite number of solutions in integers. Thue himself proved this theorem by using the theory of rational approximations to algebraic numbers. A proof based on local methods was given by Skolem. Actually Skolem's proof involves putting a small restriction on the polynomial $f(x, y)$. On the other hand, his proof has the advantage of allowing a general approach to the problem of the representation of numbers by a fairly wide class of nonfull decomposable forms. We return to this point in Section 6.4.

The basic idea in Skolem's method can be clarified in the following simple example. We shall try to establish the finiteness of the number of solutions in integers to the equation

$$x^3 + dy^3 = c, \tag{0.1}$$

where c and d are integers, and d is not a cube. Consider the cubic field $R(\theta)$, where $\theta = \sqrt[3]{d}$. We can now write (0.1) in the form

$$N(x + y\theta) = c. \tag{0.2}$$

Hence the problem reduces to finding all numbers in the nonfull module $\{1, \theta\}$ of the field $R(\theta)$ with given norm. We embed the module $\{1, \theta\}$ in the full module $\{1, \theta, \theta^2\}$, which coincides in this case with its coefficient ring $\mathfrak{O}$. The solutions of (0.2) are hence numbers $\alpha \in \mathfrak{O}$ with norm c, for which in the representation $\alpha = x + y\theta + z\theta^2$ the coefficient z equals 0. But we have already solved the problem of finding all numbers with given norm in a full module (Theorem 1 of Section 5, Chapter 2). In this case we have $s = 1$, $t = 1$ (since the polynomial $x^3 - d$ has one real root and two complex roots). Hence there is a unit $\varepsilon \in \mathfrak{O}$ with norm $+1$, and a finite set of numbers $\mu_1, \dots, \mu_k$ each with norm c, so that every $\alpha \in \mathfrak{O}$ with norm c has a unique representation in the form $\mu_i\varepsilon^u$ for some $i = 1, \dots, k$ and some rational integer u. To prove that (0.1) has only a finite number of solutions, it will suffice to show that only a finite number of the numbers $\mu_i\varepsilon^u$ have the form $x + y\theta$.

Along with the field $R(\theta)$ consider its conjugate fields $R(\theta')$ and $R(\theta'')$ and, for every $\alpha \in R(\theta)$, denote by $\alpha' \in R(\theta')$ and $\alpha'' \in R(\theta'')$ the corresponding conjugates. If we set

$$\mu\varepsilon^u = x + y\theta + z\theta^2,$$

then, taking conjugates, we shall also have

$$\mu'\varepsilon'^u = x + y\theta' + z\theta'^2,$$

$$\mu''\varepsilon''^u = x + y\theta'' + z\theta''^2.$$

From these three equations we can find an expression for z. It will have the form

$$z = \gamma_0\varepsilon^u + \gamma_1\varepsilon'^u + \gamma_2\varepsilon''^u,$$

where $\gamma_0, \gamma_1, \gamma_2$ are certain (nonzero) numbers of the field $K = R(\theta, \theta', \theta'')$.

Solutions of (0.1) hence lead to solutions of the equation

$$\gamma_0\varepsilon^u + \gamma_1\varepsilon'^u + \gamma_2\varepsilon''^u = 0 \tag{0.3}$$

in the rational integer u. Since this equation contains only one unknown, it is natural to expect that it will have only a finite number of solutions. But it is not at all simple to give a proof.

Skolem's method is based on the consideration of the left side of (0.3) as an analytic function $F(u)$ in the p-adic domain. If (0.1) has an infinite number of integral solutions, then $F(u)$ has an infinite number of integral zeros. In Section 3.4 of Chapter 1 we saw that the collection of p-adic integers is a compact set, and hence the function $F(u)$ would vanish at an infinite sequence of points which converge to a limit point (in its domain of definition). In the theory of analytic functions of a complex variable it follows from the uniqueness theorem that such a function is identically zero. The proof of this fact translates word-for-word to the case of p-adic analytic functions. Hence the function $F(u)$ must be identically zero and we obtain a contradiction.

Already in this example the p-adic numbers introduced in Chapter 1 are insufficient. Since the numbers γ_0, γ_1, γ_2, ε, ε', ε'' in (0.3) are algebraic numbers, we must develop a theory, analogous to the theory of p-adic numbers, with the rational field R replaced by an arbitrary algebraic number field k and the prime number p replaced by a prime divisor $\mathfrak{p}$. This leads us to Section 1.

1. Fields Complete with Respect to a Valuation

1.1. *The Completion of a Field with Respect to a Valuation*

In Section 4 of Chapter 1 we showed that to every prime number p, that is, every prime divisor of the rational field R, corresponds the p-adic metric φ_p of the field R, the completion of which is the field R_p of p-adic numbers. The definition of the metric used no properties of the field R other than the existence of the p-adic valuation v_p [see formula (4.1) of Chapter 1]. Therefore, the construction of analogous completions can be carried out for any field k, provided we have a theory of divisors in it. If $\mathfrak{p}$ is any prime divisor of the field k, and $v = v_\mathfrak{p}$ is the corresponding valuation, then by taking any real number ρ with $0 < \rho < 1$, we can define a metric $\varphi = \varphi_\mathfrak{p}$ on k by setting

$$\varphi(x) = \rho^{v(x)} \qquad (x \in k). \tag{1.1}$$

Then, following the construction of Section 4.1 of Chapter 1, we may form the completion $\bar{k} = \bar{k}_\mathfrak{p}$ of the field k with respect to this metric. [The fact that the function (1.1) is a metric is easily verified.] The field $\bar{k}_\mathfrak{p}$ is called the $\mathfrak{p}$-*adic completion* of the field k. The completion $\bar{k} = \bar{k}_\mathfrak{p}$ clearly does not depend on what theory of divisors we have in mind for the field k. It is completely determined by the single valuation $v = v_\mathfrak{p}$. Hence we shall also call it the completion of k with respect to the valuation v. In this section we study such completions and their finite extensions.

Let $\bar{k}$ be the completion of the field k with respect to the valuation v. We now show that the valuation v can be extended in a unique fashion to a valuation

$\bar{v}$ of the field $\bar{k}$. In Section 4.1 of Chapter 1 we saw that the metric φ of the field k [see (1.1)] can be extended to a metric $\bar{\varphi}$ of the field $\bar{k}$, so that if $\alpha \in \bar{k}$ and $\alpha = \lim_{n\to\infty} a_n$, where $a_n \in k$, then $\bar{\varphi}(\alpha) = \lim_{n\to\infty} \varphi(a_n)$. But in this case zero is the only limit point of the set of numbers $\varphi(a)$, $a \in k$, and hence the sequence $\{\varphi(a_n)\}$ must either converge to zero (if $\alpha = 0$), or it must remain constant from some point on (if $\alpha \neq 0$). Hence the sequence $\{v(a_n)\}$ converges to infinity if $\alpha = 0$ and eventually becomes constant if $\alpha \neq 0$. We can therefore set

$$\bar{v}(\alpha) = \lim_{n\to\infty} v(a_n).$$

It is now easily verified that the function $\bar{v}(\alpha)$ (whose value clearly does not depend on the choice of the sequence $\{a_n\}$) is a valuation of the field $\bar{k}$, with $\bar{v}(a) = v(a)$ for all $a \in k$. It is also clear that the metric $\bar{\varphi}$ of the field $\bar{k}$ is related to the valuation $\bar{v}$ by

$$\bar{\varphi}(\alpha) = \rho^{\bar{v}(\alpha)}, \qquad (\alpha \in \bar{k}).$$

It will be convenient later to express convergence in the field k in terms of the valuation $\bar{v}$ instead of the metric $\bar{\varphi}$ (just as was done for the p-adic numbers in Section 3.4 of Chapter 1).

Let $\mathfrak{o}$ be the ring of the valuation v, that is, the ring of all $a \in k$ such that $v(a) \geqslant 0$ (Section 4.1 of Chapter 3). We now show that the closure $\bar{\mathfrak{o}}$ of the ring $\mathfrak{o}$ in the field $\bar{k}$ coincides with the ring of the valuation $\bar{v}$ (if $A \subset \bar{k}$ is any set, by the closure $\bar{A}$ of A we mean the set of all elements of $\bar{k}$ which are limits of sequences of elements of A). If $\alpha \in \bar{\mathfrak{o}}$, then $\alpha = \lim_{n\to\infty} a_n$, where $a_n \in \mathfrak{o}$, so that $\bar{v}(\alpha) = \lim_{n\to\infty} v(a_n) \geqslant 0$. Assume, conversely, that $\bar{v}(\alpha) \geqslant 0$. Since α is the limit of a sequence of elements of k, then for any natural number n we can find an element $\alpha \in k$ such that $\bar{v}(\alpha - a_n) \geqslant n$. Then $\alpha = \lim_{n\to\infty} a_n$, where

$$v(a_n) = \bar{v}(\alpha - (\alpha - a_n)) \geqslant \min(\bar{v}(\alpha), (\bar{v}(\alpha - a_n)) \geqslant 0;$$

that is, $a_n \in \mathfrak{o}$. Our assertion is proved.

By Theorem 2 of Section 4, Chapter 4, the ring $\mathfrak{o}$ has, up to associates, only one prime element π, which is characterized by $v(\pi) = 1$. It thus remains a prime element in the ring $\bar{\mathfrak{o}}$ [since $\bar{v}(\pi) = 1$]. Let Σ_v and $\Sigma_{\bar{v}}$ denote the residue class fields of the valuations v and $\bar{v}$ (see Section 4.1 of Chapter 3). Since elements of $\mathfrak{o}$ are congruent modulo π in $\mathfrak{o}$ if and only if they are congruent modulo π in $\bar{\mathfrak{o}}$, we have an isomorphism from the field Σ_v to the field $\Sigma_{\bar{v}}$. On the other hand, for any $\alpha \in \bar{\mathfrak{o}}$ there is an element $a \in \mathfrak{o}$ such that $\bar{v}(\alpha - a) \geqslant 1$; that is, $\alpha \equiv a \pmod{\pi}$. This means that the mapping $\Sigma_v \to \Sigma_{\bar{v}}$ is an isomorphism onto the entire field $\Sigma_{\bar{v}}$. Because of this isomorphism we shall frequently identify the field $\Sigma_{\bar{v}}$ with Σ_v.

1.2. *Representation of Elements by Series*

In this section we assume that k is complete with respect to the valuation v [that is, that it is complete under the metric (1.1)].

We shall call the ring $\mathfrak{o}$ of the valuation v the ring of integral elements (or integers) of the field k. We denote some fixed prime element of the ring $\mathfrak{o}$ by π.

The residue class field Σ of the valuation v will also be called the *residue class field* of the field k.

Everything that was said about p-adic series in Section 3.4 of Chapter 1 clearly remains true for series in the field k. In particular, Theorem 8 of Section 3, Chapter 1, is valid.

Taking arbitrary integers $\alpha_n (m \leqslant n < \infty)$, we consider the series

$$\sum_{n=m}^{\infty} \alpha_n \pi^n . \tag{1.2}$$

Since $v(\alpha_n \pi^n) = v(\alpha_n) + n \geqslant n$, then $\alpha_n \pi^n \to 0$ as $n \to \infty$; that is, the general term of the series (1.2) converges to zero. Hence the series (1.2) converges and its sum is some element of the field k. It is now natural to ask if every element of k has a representation in the form (1.2), and, if so, if there is a canonical representation such as that obtained for p-adic numbers (Theorem 10 of Section 3, Chapter 1). The answer will be affirmative.

We choose in the ring $\mathfrak{o}$ some complete system of residues modulo π. We assume that $0 \in S$, that is, that the class of elements of the ring $\mathfrak{o}$ which are divisible by π is represented by 0.

Theorem 1. Let k be a complete field under the valuation v, with $\mathfrak{o}$ the ring of integers of k, π a prime element of $\mathfrak{o}$, and S a complete system of residues (containing 0) of the ring $\mathfrak{o}$ modulo π. Then every element $\alpha \in k$ has a unique representation as the sum of a series

$$\alpha = \sum_{i=m}^{\infty} a_i \pi^i, \tag{1.3}$$

where $a_i \in S$ $(m \leqslant i < \infty)$.

Proof. For $\alpha = 0$ we have the representation $0 = \sum_{i=0}^{\infty} 0 \cdot \pi^i$. Let $\alpha \neq 0$. If $v(\alpha) = m$, then $v(\alpha\pi^{-m}) = 0$. The element $\alpha\pi^{-m}$ is congruent modulo π to some nonzero element of S, say, to a_m. Since $\alpha\pi^{-m} - a_m = \pi\xi$, where $\xi \in \mathfrak{o}$, then

$$\alpha = a_m \pi^m + \xi \pi^{m+1}.$$

Assume that for some $n > m$ we have found the representation

$$\alpha = a_m \pi^m + \cdots + a_{n-1} \pi^{n-1} + \eta_n \pi^n,$$

where $a_i \in S$ $(m \leqslant i \leqslant n-1)$, $\eta_n \in \mathfrak{o}$. Choose $a_n \in S$, so that $\eta_n \equiv a_n \pmod{\pi}$. Since $\eta_n = a_n + \eta_{n+1}\pi$, where $\eta_{n+1} \in \mathfrak{o}$, then we have the representation

$$\alpha = a_m\pi^m + \cdots + a_n\pi^n + \eta_{n+1}\pi^{n+1}$$

for α. We continue this process indefinitely. Since $v(\eta_n\pi^n) \geqslant n$, then $\eta_n\pi^n \to 0$ as $n \to \infty$, and hence $\alpha = \sum_{i=m}^{\infty} a_i\pi^i$.

If not all coefficients a_i in (1.3) are zero, then we may assume that $a_m \neq 0$. In this case $v(a_m) = 0$, since all elements of $\mathfrak{o}$ which are not divisible by π are units. Then

$$v\left(\sum_{i=m}^{\infty} a_i\pi^i\right) = v(a_m\pi^m) = m.$$

From this it follows that the representation for $\alpha = 0$ is unique. Assume that for some $\alpha \neq 0$ we have two representations:

$$\alpha = \sum_{i=m}^{\infty} a_i\pi^i = \sum_{i=m'}^{\infty} a_i'\pi^i \qquad (a_i, a_i' \in S).$$

If $a_m \neq 0$ and $a_{m'}' \neq 0$ in these representations, then it follows that $m = m'$. Suppose we have already established that $a_i = a_i'$ for $m \leqslant i < n$ $(n \geqslant m)$. Multiply the equation $\sum_{i=n}^{\infty} a_i\pi^i = \sum_{i=n}^{\infty} a_i'\pi^i$ by π^{-n}. Turning to congruences modulo π we find that $a_n \equiv a_n' \pmod{\pi}$, and since both a_n and a_n' lie in S, then $a_n = a_n'$. This proves Theorem 1.

Note that in the case when $k = R_p$, $\pi = p$, and $S = (0, 1, \ldots, p-1)$, Theorem 1 reduces to Theorem 10 of Section 3 of Chapter 1.

Corollary. Using the notations of Theorem 1, every integral element $\alpha \in k$ has a unique representation in the form

$$\alpha = a_0 + a_1\pi + \cdots + a_n\pi^n + \cdots \qquad (a_n \in S). \tag{1.4}$$

It is easily seen that Theorem 9 of Section 3 of Chapter 1 is valid for series in the field k. Hence convergent series in k can be multiplied by the usual method in analysis. In particular, we can treat series of the form (1.3) as a power series in π. But when we operate with series of the form (1.2) using the rules for power series we must keep in mind that we will obtain series of the form (1.2), in which the coefficients α_n do not belong to the system S. We can translate the obtained series into the form (1.3) by replacing in turn each coefficient α_n by its residue $a_n \in S$, where $\alpha_n = a_n + \pi\gamma_n$, adding at each stage the element $\gamma_n \in \mathfrak{o}$ to the following coefficient.

Remark 1. The representation (1.3) clearly depends on the choice of the system S. Among all such systems there is, in many important cases, a "best" system which has the property of multiplicative closure, or even is a subfield of the field k (see Problems 7 to 11).

Remark 2. These results generalize the analogous facts for p-adic fields (Section 3.4 of Chapter 1). We must warn that Theorem 6 of Section 3, Chapter 1, is no longer valid for arbitrary fields complete with respect to a valuation. It holds only for those fields k where the residue class field Σ has a finite number of elements. The same is the case for Theorems 1 and 2 of Section 5, Chapter 1 (when F is a form with coefficients in the ring $\mathfrak{o}$). But Theorem 3 of Section 5, Chapter 1, carries over word-for-word to the case of an arbitrary field k which is complete with respect to a valuation. In the future we shall use the corollary to this theorem in the following form: If $F(x)$ is a polynomial with integral coefficients in k and there is an integer $\xi \in k$ for which $F(\xi) \equiv 0 \pmod{\pi}$ and $F'(\xi) \not\equiv 0 \pmod{\pi}$, then there is an integral element $\theta \in k$ for which $\xi \equiv \theta \pmod{\pi}$ and $F(\theta) = 0$.

1.3. *Finite Extensions of a Field Complete with Respect to a Valuation*

Let k be complete under the valuation v_0. Then k has algebraic extensions of all degrees (Problem 9 of Section 3, Chapter 3). Let K be an extension of k of degree n. By Theorem 5 of Section 4, Chapter 3, there is a valuation v of K which is an extension of v_0. Our goal is to show that in this case v is unique and that K is complete under v.

Let L be a subspace of K, considered as a vector space over the field k, and let $\omega_1, \dots, \omega_s$ be a basis for L over k. Each element α of L has a unique representation in the form

$$\alpha = a_1\omega_1 + \cdots + a_s\omega_s \qquad (a_i \in k). \tag{1.5}$$

If $v_0(a_i) \geqslant N$ $(i = 1, \dots, s)$, then, by the properties of valuations,

$$v(\alpha) \geqslant \min v(a_i\omega_i) \geqslant eN + \min v(\omega_i),$$

where e denotes the ramification index of v relative to v_0 (see Section 4.3 of Chapter 3). Conversely, we shall show that if the element $\alpha \in L$ is very small under the valuation v (recall that "small" elements are characterized by large values of the valuation), then all the coefficients a_i in (1.5) will be small under the valuation v_0. More precisely, this means that for any N we can find an M such that whenever $v(\alpha) \geqslant M$, then $v_0(\alpha_i) \geqslant N$ $(i = 1, \dots, s)$. For $s = 1$ the assertion is clear. We prove the general case by induction on s. Let $s \geqslant 2$, and assume that the assertion is false, that is, that there exists a number N and elements $\alpha \in L$ so that the value of $v(\alpha)$ can be made arbitrarily large, but for which there is at least one coefficient a_i with $v_0(a_i) < N$. We may clearly assume that this inequality always holds for the first coefficient a_1. Hence for each natural number k we choose an element $\alpha_k \in L$ for which $v(\alpha_k) \geqslant k + eN$ and the coefficient $a_1^{(k)}$ in the decomposition

$$\alpha_k = a_1^{(k)}\omega_1 + \cdots + a_s^{(k)}\omega_s, \qquad (a_i^{(k)} \in k),$$

satisfies $v_0(a_1^{(k)}) < N$. Consider the sequence $\{\beta_k\}$, where

$$\beta_k = \alpha_k a_1^{(k)-1} = \omega_1 + b_2^{(k)}\omega_2 + \cdots + b_s^{(k)}\omega_s. \tag{1.6}$$

Since $v(\beta_k) = v(\alpha_k) - ev_0(a_1^{(k)})$, then

$$v(\beta_k) > k.$$

The differences

$$\beta_{k+1} - \beta_k = \sum_{i=2}^{s} (b_i^{(k+1)} - b_i^{(k)})\omega_i$$

all lie in the subspace of dimension $s - 1$ which is spanned by the elements $\omega_2, \ldots, \omega_s$, and they satisfy

$$v(\beta_{k+1} - \beta_k) \geqslant \min(v(\beta_{k+1}), \quad v(\beta_k)) > k;$$

that is, $v(\beta_{k+1} - \beta_k) \to \infty$ as $k \to \infty$. Then by the induction hypothesis we also have (for $i = 2, \ldots, s$)

$$v(b_i^{(k+1)} - b_i^{(k)}) \to \infty \qquad \text{for } k \to \infty.$$

Hence since the field k is complete (see Theorem 7 of Section 3, Chapter 1) the sequence $\{b_i^{(k)}\}_{k=1}^{\infty}$ converges to some element $b_i \in k$. Passing to the limit in (1.6) as $k \to \infty$ and noting that $\beta_k \to 0$, we obtain

$$\omega_1 + b_2\omega_2 + \cdots + b_s\omega_s = 0,$$

which contradicts the linear independence of the elements $\omega_1, \ldots, \omega_s$ over the field k. This contradiction proves our assertion.

We now take for L the whole field K. If the sequence $\{\alpha_k\}$ of elements of K is a Cauchy sequence, that is, if $v(\alpha_{k+1} - \alpha_k) \to \infty$ as $k \to \infty$, then by what we have just shown, the sequences $\{a_i^{(k)}\}_{k=1}^{\infty}$, which arise from the decompositions

$$\alpha_k = a_1^{(k)}\omega_1 + \cdots + a_n^{(k)}\omega_n \qquad (a_i^{(k)} \in k) \tag{1.7}$$

(here $\omega_1, \ldots, \omega_n$ is a basis for K over k), will converge in the field k. Then the sequence $\{\alpha_k\}$ will also converge. This shows that K is complete with respect to the valuation v. In addition, we see that convergence in K, relative to the valuation v, is completely determined by convergence in k (relative to the valuation v_0).

From this fact it easily follows that there is only one extension of the valuation v_0 to the field K. For suppose that v and v' are two different extensions. By the independence of valuations, there is an element $\alpha \in K$, for which $v(\alpha) > 0$ and $v'(\alpha) = 0$. The sequence $\{\alpha^k\}$ converges to zero relative to the valuation v, but does not converge relative to v' [since $v'(\alpha^{k+1} - \alpha^k) = v'(\alpha - 1)$ does not converge to infinity]. This contradicts the fact that convergence in K is determined by the valuation v_0.

We have hence obtained the following theorem.

Theorem 2. Let k be complete with respect to the valuation v_0, and let K be a finite extension field. The valuation v_0 has a unique extension v to the field K. K is complete with respect to v. If $\omega_1, \ldots, \omega_n$ is any basis for K over k, then a sequence $\{\alpha_k\}$ of elements of K is convergent if and only if each of the sequences $\{a_i^{(k)}\}$ $(1 \leqslant i \leqslant n)$, which are determined by (1.7), converges in k.

1.4. *Integral Elements*

We now study the relationship between the ring $\mathfrak{o}$ of integral elements of the field k, complete under the valuation v_0, and the ring $\mathfrak{O}$ of integral elements of the finite extension K of k. Since the valuation v_0 has only one extension v to the field K, then by Theorem 6 of Section 4, Chapter 3, the ring $\mathfrak{O}$ (the ring of the valuation v) is the integral closure of the ring $\mathfrak{o}$ in the field K. Hence for any $\alpha \in \mathfrak{O}$ the norm $N(\alpha) = N_{K/k}(\alpha)$ belongs to $\mathfrak{o}$, and then the norm $N(\varepsilon)$ of any unit ε of the ring $\mathfrak{O}$ will be a unit in the ring $\mathfrak{o}$. Now let $\alpha \notin \mathfrak{O}$. Since $\alpha^{-1} \in \mathfrak{O}$ and is not a unit, then $N(\alpha^{-1}) = N(\alpha)^{-1}$ belongs to $\mathfrak{o}$ and is not a unit in $\mathfrak{o}$. But in this case $N(\alpha) = (N(\alpha)^{-1})^{-1}$ does not belong to $\mathfrak{o}$. We have proved the following theorem.

Theorem 3. Let α be an element of the finite extension K of the field k, complete with respect to a valuation. Then α is an integral element if and only if $N_{K/k}(\alpha)$ is integral in k.

Corollary. An element $\varepsilon \in K$ is a unit in the ring $\mathfrak{O}$ if and only if its norm $N(\varepsilon)$ is a unit in the ring $\mathfrak{o}$.

The rings $\mathfrak{o}$ and $\mathfrak{O}$ can be considered as rings with a theory of divisors. Let $\mathfrak{p}$ and $\mathfrak{P}$ denote the (only) prime divisors of these rings. The degree of inertia f of the divisor $\mathfrak{P}$ relative to $\mathfrak{p}$, that is, the degree of the extension $(\Sigma : \Sigma_0)$ of the residue class field Σ of the field K over the residue class field Σ_0 of the field k, is also called in this case the *degree of inertia* of the extension K/k. Analogously, the ramification index e of the divisor $\mathfrak{P}$ relative to $\mathfrak{p}$ is called the *ramification index* of the extension K/k. If π_0 and π are prime elements in $\mathfrak{o}$ and $\mathfrak{O}$, then we know that

$$\pi_0 = \pi^e \varepsilon, \tag{1.8}$$

where ε is a unit of $\mathfrak{O}$.

Let S_0 be a complete system of residues in the ring $\mathfrak{o}$ modulo π_0. As before, we assume that $0 \in S_0$. It is easily seen that if the residue classes $\bar{\omega}_1, \ldots, \bar{\omega}_f$ of Σ form a basis over Σ_0, then the set S which consists of the linear combinations

$$a_1\omega_1 + \cdots + a_f\omega_f, \tag{1.9}$$

where $a_1, \ldots, a_f$ independently run through all elements of S_0, is a complete system of residues in the ring $\mathfrak{O}$ modulo π.

Definition. A basis $\theta_1, \ldots, \theta_n$ for the field K over k is called a *fundamental basis* if all θ_i are integral and for any integral $\alpha \in K$ all coefficients a_i in

$$\alpha = a_1\theta_1 + \cdots + a_n\theta_n \qquad (a_i \in k)$$

are integral in k.

Theorem 4. Let k be complete with respect to the valuation v_0 and let K be a finite extension with ramification index e and degree of inertia f. Let Σ_0 and Σ be the residue class fields of k and K, and let π be a prime element of the ring of integers of the field K. If $\bar{\omega}_1, \ldots, \bar{\omega}_f$ are residue classes of the field Σ which form a basis over the field Σ_0, then the system of elements

$$\omega_i \pi^j, \qquad (i = 1, \ldots, f; \quad j = 0, \quad 1 \ldots, e-1), \tag{1.10}$$

is a fundamental basis for the extension K/k.

Proof. We first show that the elements (1.10) are linearly independent over k. Assume, on the contrary, that we have

$$\sum_{i=1}^{f} \sum_{j=0}^{e-1} a_{ij}\omega_i\pi^j = 0,$$

where ai_j are elements of k, not all equal to zero. We may assume that all a_{ij} are integers and that at least one of them in a unit in $\mathfrak{o}$ (if this is not so, multiply by a suitable power of the prime element $\pi_0 \in \mathfrak{o}$). Let j_0 $(0 \leqslant j_0 \leqslant e-1)$ be the smallest index for which there exists an i_0 $(1 \leqslant i_0 \leqslant f)$ with $a_{i_0 j_0}$ a unit in $\mathfrak{o}$. Hence if $j < j_0$, then $v_0(a_{ij}) \geqslant 1$ for all i. Since $\sum_{i=1}^{f} \bar{a}_{ij_0}\bar{\omega}_i \neq \hat{\alpha}, \bar{0}$ the sum $\sum_{i=1}^{f} a_{ij_0}\omega_i$ is not divisible by π, and hence for the element

$$\gamma = \sum_{i=1}^{f} a_{ij_0}\omega_i\pi^{j_0}$$

we have

$$v(\gamma) = j_0 + v\left(\sum_{i=1}^{f} a_{ij_0}\omega_i\right) = j_0.$$

On the other hand,

$$\gamma = -\sum_{i=1}^{f} \sum_{j \neq j_0} a_{ij}\omega_i\pi^j.$$

If $j < j_0$, then

$$v(a_{ij}\omega_i\pi^j) = j + v(a_{ij}) \geqslant ev_0(a_{ij}) \geqslant e > j_0.$$

If $j > j_0$, then

$$v(a_{ij}\omega_i\pi^j) = j + v(a_{ij}) \geqslant j > j_0.$$

Hence

$$v(\gamma) \geqslant \min_{j \neq j_0} v(a_{ij}\omega_i\pi^j) > j_0.$$

This contradiction shows the linear independence of the elements (1.10) over the field k.

Now let α be any element of $\mathfrak{O}$. By the corollary of Theorem 1 we have

$$\alpha \equiv \xi_0 + \xi_1\pi + \cdots + \xi_{e-1}\pi^{e-1} \pmod{\pi^e},$$

where ξ_i are elements of some fixed system S of residues, which we may take to consist of numbers of the form (1.9). Since π_0 and π^e are associate in $\mathfrak{O}$ [see (1.8)], then congruences in $\mathfrak{O}$ modulo π_0 and modulo π^e are equivalent. Hence we have

$$\alpha \equiv \sum_{i=1}^{f} \sum_{j=0}^{e-1} a_{ij}^{(0)}\omega_i\pi^j \pmod{\pi_0}, \qquad (a_{ij}^{(0)} \in S_0),$$

and this means that

$$\alpha = \sum_{i=1}^{f} \sum_{j=0}^{e-1} a_{ij}^{(0)}\omega_i\pi^j + \pi_0\alpha_1, \qquad (\alpha_1 \in \mathfrak{O}).$$

Analogously,

$$\alpha_1 = \sum_{i=1}^{f} \sum_{j=0}^{e-1} a_{ij}^{(1)}\omega_i\pi^j + \pi_0\alpha_2, \qquad (\alpha_2 \in \mathfrak{O}, \qquad a_{ij}^{(1)} \in S_0).$$

Continuing this process indefinitely, we obtain a sequence of equations

$$\alpha_n = \sum_{i=1}^{f} \sum_{j=0}^{e-1} a_{ij}^{(n)}\omega_i\pi^j + \pi_0\alpha_{n+1}, \qquad (\alpha_{n+1} \in \mathfrak{O}, \qquad a_{ij}^{(n)} \in S_0).$$

For fixed i and j we have an infinite sequence $\{a_{ij}^{n}\}$. Consider the series

$$\sum_{n=0}^{\infty} a_{ij}^{(n)}\pi_0{}^n.$$

Since the $a_{ij}^{(n)}$ are integers, this series converges and its sum a_{ij} is an integral element of k; that is, $a_{ij} \in \mathfrak{o}$. We shall show that

$$\alpha = \sum_{i=1}^{f} \sum_{j=0}^{e-1} a_{ij}\omega_i\pi^j. \tag{1.11}$$

By the construction of the elements $\alpha_1, \alpha_2, \ldots$ we have

$$\alpha = \sum_{k=0}^{n-1}\left(\sum_{i=1}^{f} \sum_{j=0}^{e-1} a_{ij}^{(k)}\omega_i\pi^j\right)\pi_0{}^k + \pi_0{}^n\alpha_n,$$

from which it follows that the difference

$$\alpha - \left(\sum_{i=1}^{f} \sum_{j=0}^{e-1} a_{ij}\omega_i\pi^j\right)$$

is divisible by π_0^n (in the ring $\mathfrak{O}$). Since this holds for all n, this difference must be zero and (1.11) is valid.

If β is any element of K, then for some m, the element $\beta\pi_0^m$ will be integral. Representing it in the form (1.11), we see that β is a linear combination of the elements (1.10) with coefficients in k. Hence the system (1.10) is a basis for K over k, and since for integral $\alpha \in K$ all coefficients a_{ij} in (1.11) lie in $\mathfrak{o}$, this basis is fundamental. Theorem 4 is proved.

Since there are fe elements in the basis (1.10), we also have the following result.

Theorem 5. The product of the ramification index and the degree of inertia is equal to the degree $n = (K : k)$; that is,

$$fe = n.$$

Set $N_{K/k}(\pi) = \pi_0^m u$, where u is a unit of the ring $\mathfrak{o}$. Taking the norm of (1.8), we obtain

$$N_{K/k}(\pi_0) = \pi_0^n = N_{K/k}(\pi^e \varepsilon) = \pi_0^{me} u^e N_{K/k}(\varepsilon) = \pi_0^{me} v,$$

where v is also a unit in $\mathfrak{o}$. It follows that $n = me$ (and $v = 1$), and hence $m = f$. Hence the degree of inertia f of the extension K/k could also be defined by

$$f = v_0(N_{K/k}(\pi)), \tag{1.12}$$

where π is a prime element of the ring of integral elements of K. From this it easily follows that for any α of the field K we have

$$v_0(N_{K/k}(\alpha)) = fv(\alpha). \tag{1.13}$$

Note that Theorem 5 and (1.12) are both immediate corollaries of Theorem 5 and (5.12) of Section 5, Chapter 3.

Definition. If $e = 1$, the extension K/k is called *unramified.* If $e = n$, then K/k is called *totally ramified.*

It follows from Theorem 5 that for unramified extensions the degree of inertia coincides with the degree of the extension. For totally ramified extensions the residue class fields Σ and Σ_0 coincide; that is, every integral element of K is congruent modulo π with an integral element of k.

It can be shown (Problem 12) that if the residue class field Σ of the field K is separable over the residue class field Σ_0 of the field k, then there is a uniquely determined intermediate field T, such that the extension T/k is unramified and K/T is totally ramified. The field T is called the *inertia field* of the extension K/k.

1.5. *Fields of Formal Power Series*

Fields of formal power series are fields, complete with respect to a valuation. These fields are constructed in the following manner.

Let k_0 be any field. The set $\mathfrak{o}$ of all formal power series of the type

$$a_0 + a_1 t + a_2 t^2 + \cdots + a_n t^n + \cdots \qquad (a_n \in k_0) \tag{1.14}$$

becomes a commutative ring with unit when the operations of addition and multiplication are defined as is usual for power series. This ring has no divisors of zero, and the units are precisely those series (1.14) in which $a_0 \neq 0$. The quotient field of $\mathfrak{o}$ is called the *field of formal power series* in t over the field k_0. It is denoted by $k_0\{t\}$. Just as for the field of p-adic numbers (Section 3.3 of Chapter 1), every nonzero element ξ of the field $k_0\{t\}$ has a unique representation in the form

$$\xi = t^m(c_0 + c_1 t + \cdots + c_n t^n + \cdots), \qquad (c_n \in k_0,\ c_0 \neq 0),$$

where m is some integer (positive, negative, or zero). Setting $v(\xi) = m$ for $\xi \neq 0$ and $v(0) = \infty$, we obtain a valuation v, and the field $k_0\{t\}$ is easily shown to be complete under v. The ring of the valuation v coincides with the ring $\mathfrak{o}$ of series of the type (1.14). As prime element in $\mathfrak{o}$ we may take t. Since two series of the form (1.14) are congruent modulo t if and only if their initial terms coincide, then each residue class of modulo t contains a unique element of k_0. Hence the residue class field Σ_0 of the field $k_0\{t\}$ is canonically isomorphic to the field k_0.

It is easily seen that the field of formal power series $k_0\{t\}$ is a completion of the field of rational functions $k_0(t)$, under the valuation corresponding to the irreducible polynomial t of the ring $k_0[t]$ (see Problem 7 of Section 4 of Chapter 1).

Since $k_0 \subset k_0\{t\}$ and $k_0 \approx \Sigma_0$, the characteristic of a field of formal power series coincides with the characteristic of its residue class field. It can be shown that this property characterizes formal power series fields among all fields which are complete under a valuation. Namely, if k is complete under a valuation and its characteristic equals the characteristic of its residue class field, then k contains a subfield k_0, where the elements of k_0 form a complete system of residues modulo the prime element π. But for such a system of residues, the operations with series (1.3) reduce to the usual operations with power series, and hence k will be a field of formal power series in π with coefficients from k_0. The proof of the existence of the subfield k_0 in the general case is rather difficult, and we shall not give it. (Two special cases, for which the proof is relatively easy, are indicated in Problems 7 and 11.)

If k_0' is an extension of k_0, then $k_0'\{t\}$ is an extension of $k_0\{t\}$, and if k_0'/k_0 is finite, then $k_0'\{t\}/k_0\{t\}$ is also finite and has the same degree. Another method for constructing finite extensions of the field $k_0\{t\}$ is to map it isomorphically into the field $k_0\{u\}$, with $t \to u^n$ (n a natural number). If we identify $k_0\{t\}$ with its image under this mapping, that is, if we set $t = u^n$, then $k_0\{u\}$ will be a finite extension of $k_0\{t\}$ of degree n. It is clear that $k_0\{u\}$ is obtained from $k_0\{t\}$ by adjoining an nth root of t.

For fields of characteristic zero, all finite extensions can be reduced to these two types.

Theorem 6. Let k_0 be a field of characteristic zero. If K is a finite extension of the field $k = k_0\{t\}$, with ramification index e, then K is a subfield of an extension of the form $k_0'\{u\}$, where k_0' is a finite extension of k_0 and $u^e = t$.

Proof. Let Σ_0 and Σ denote the residue class fields of k and K, let f denote the degree of inertia of K/k, let π be a prime element of K, and for any $\xi \in K$ let $\bar{\xi}$ be its residue class in Σ. The elements of the field k_0, as we have seen, form a natural system of representatives for the residue classes of Σ_0. We first show that there is a subfield S of K which contains k_0 and is a complete system of representatives for the residue classes of Σ. Since any finite extension of a field of characteristic zero is simple, then $\Sigma = \Sigma_0(\bar{\xi})$, where $\bar{\xi}$ is some residue class of Σ. Let $\bar{F}$ be the minimum polynomial of $\bar{\xi}$ over Σ_0. If we replace all coefficients of F (which are residue classes of Σ_0) by the corresponding elements of k_0, we obtain an irreducible polynomial F over k_0 of degree f, for which

$$F(\xi) \equiv 0 \pmod{\pi} \qquad \text{and} \qquad F'(\xi) \not\equiv 0 \pmod{\pi}.$$

By the second remark at the end of Section 1.2, there is an element θ in the field K with $\bar{\theta} = \bar{\xi}$ and $F(\theta) = 0$. Consider the subfield $S = k_0(\theta)$ of the field K. Since θ is a root of an irreducible polynomial over k_0 of degree f, then $(S : k_0) = f$, and every element of S has a unique representation in the form

$$a_0 + a_1\theta + \cdots + a_{f-1}\theta^{f-1} \qquad (a_i \in k_0).$$

Since $\bar{\theta} = \bar{\xi}$, the corresponding residue classes modulo π are given by $\bar{a}_0 + \bar{a}_1\bar{\xi} + \cdots + \bar{a}_{f-1}\bar{\xi}^{f-1}$. Since $\Sigma = \Sigma_0(\bar{\xi})$ and $(\Sigma : \Sigma_0) = f$, these linear combinations run without repetition through all residue classes of Σ. This shows that the elements of the subfield S (which is a finite extension of the field k_0) form a complete system of representatives for the residue classes of Σ.

By Theorem 1 the field K is the field of formal power series in π with coefficients in S; that is, $K = S\{\pi\}$. Theorem 6 would be proved (in a stronger form) if we could show that π can be chosen so that it is an eth root of t. However, it is not always possible to choose π in this way, and therefore we must pass to some finite extension k_0' of the field of coefficients S.

By (1.8) we have

$$t = \pi^e\varepsilon, \tag{1.15}$$

where ε is a unit in the ring of integral elements of K. Let α denote that element of S for which $\alpha \equiv \varepsilon \pmod{\pi}$ and denote by k_0' the field $S(\sqrt[e]{\alpha})$ (if $\alpha = \gamma^e$ for some $\gamma \in S$, then $k_0' = S$). The field of formal power series $K' = k_0'\{\pi\}$ clearly contains K as a subfield and is a finite extension of k. We show that it can be represented in the form $k_0'\{u\}$, where $u^e = t$. Consider the polynomial $G(X) = X^e - \varepsilon$. Since in K' we have

$$G(\gamma) \equiv 0 \pmod{\pi} \qquad \text{and} \qquad G'(\gamma) \not\equiv 0 \pmod{\pi},$$

where by γ we denote the root $\sqrt[e]{\alpha}$, then in K' there exists a unit η with $\eta \equiv \gamma \pmod{\pi}$ and $\eta^e = \varepsilon$ (here we again apply the remark at the end of Section 2.2). Now replace the prime element π of the field K' by the element $u = \pi\eta$. Then K' can also be considered as the field of formal power series in U over the field k_0', that is, $K' = k_0'\{u\}$, where $u^e = t$ by (1.15). The proof of Theorem 6 is complete.

Remark. Theorem 6 is no longer valid for arbitrary finite extensions of formal power series fields $k = k_0\{t\}$ of characteristic $p \neq 0$. However, it is easily seen that it remains true in the case of extensions K/k, for which the residue class field Σ is separable over Σ_0 and the ramification index e is not divisible by p.

PROBLEMS

1. A nontrivial metric φ of a field k is called *discrete* if the only limit point for the set of values $\varphi(x)$, $x \in k$, is zero. Show that any discrete metric is induced by some valuation ν of k by (1.1).

2. Let k be complete under a valuation and let K/k be a finite extension with a fundamental basis $\theta_1, \ldots, \theta_n$. Show that the elements

$$\theta_i' = \sum_{j=1}^{n} a_{ij}\theta_j, \qquad (a_{ij} \in k)$$

also form a fundamental basis for K over k if and only if all a_{ij} are integral and the determinant $\det(a_{ij})$ is a unit in k.

3. Using the notations of Theorem 4, let $\alpha = \sum_{i=1}^{f} \sum_{j=0}^{e-1} a_{ij}\omega_i\pi_j$ $(a_{ij} \in k)$ be any element of K. Set $m = \min \nu_0(a_{ij})$. Show that if j_0 is the smallest value of the index j for which there exists $i = i_0$ with $\nu_0(a_{i_0j_0}) = m$, then

$$\nu(\alpha) = j_0 + em,$$

where ν is the valuation of the field K.

4. Show that every element of the field of formal power series $k_0\{t\}$, not lying in k_0, is transcendental over k_0.

5. Under the assumptions of Theorem 6, show that the subfield $S \subset K$, which contains k_0 and is a complete set of representatives for the residue class field of K, is uniquely determined.

6. Let k be algebraically closed and of characteristic zero, and let $k = k_0\{t\}$. Show that k has one and only one extension of degree n, for all natural numbers n, namely, $k(\sqrt[n]{t})$ (that is, show that any two extensions of degree n are isomorphic under an isomorphism which is the identity on k).

7. Let K be complete under a valuation with residue class field Σ of characteristic zero. Show that K contains a subfield S which is a full system of representatives for the residue classes of Σ, and hence that $K = S\{\pi\}$, where π is a prime element of K. (Use the fact that any field can be obtained from the prime field by a pure transcendental extension followed by an algebraic extension.)

8. Under the assumptions of Problem 7, assume also that the residue class field Σ is algebraic over the prime field. Show that the subfield S is then unique.

9. Let K be complete under a valuation with residue class field Σ. If Σ is a perfect field of characteristic p (which means that every element is a pth power, and the map which takes every element to its pth power is an automorphism), show that there is a unique "multiplicatively closed" system S of representatives of the residue classes of Σ, so that if $\alpha \in S$ and $\beta \in S$, then $\alpha\beta \in S$. (Determine $\alpha \in S$, representing the class $\xi \in \Sigma$, by $\alpha = \lim_{n\to\infty} \alpha_n^{p^n}$, where α_n belongs to the class $\xi^{p^{-n}}$.)

10. Under the same notations, assume that Σ is a finite field with p^f elements. Show that in the field K the polynomial $t^{p^f} - t$ factors into linear factors and that the set of its roots is a multiplicatively closed system S of representatives of the residue classes of Σ.

11. Assume that the field K of Problem 9 also has characteristic p, and that its residue class field Σ is perfect. Show that then the multiplicatively closed system S will also be additively closed, so that it will be a subfield of K, and $K = S\{\pi\}$, where π is a prime element of K.

12. Let k be complete under a valuation and let K be a finite extension. Assume that the residue class field Σ of K is separable over the residue class field Σ_0 of k. Show that among the intermediate fields L, $k \subset L \subset K$, which are unramified over k, there is a largest such field T (which contains all other intermediate fields which are unramified over k). Verify that the residue class field of T coincides with Σ, and that the degree $(T:k)$ equals $(\Sigma:\Sigma_0)$.

13. Let $f(X) = X^m + a_1X^{m-1} + \cdots + a_m$ be an irreducible polynomial over a field complete with respect to a valuation. If the constant term a_m is integral, show that all other coefficients $a_1, \ldots, a_{m-1}$ are also integral.

14. Let ζ be a primitive root of degree p^s of 1 ($s \geqslant 1$). Show that the degree of the field $R_p(\zeta)$ over the field R_p of p-adic numbers is $(p-1)p^{s-1}$, and that the extension $R_p(\zeta)/R_p$ is totally ramified.

15. Let ζ be a primitive root of degree p of 1. Show that $R_p(\zeta) = R_p(\sqrt[p-1]{-p})$.

16. Let k be complete under a valuation, K/k a finite extension, and Σ and Σ_0 the residue class fields of K and k. Show that if Σ/Σ_0 is separable, then K/k has a fundamental basis consisting of powers of a single element (that is, $\mathfrak{O} = \mathfrak{o}[\theta]$, $\theta \in \mathfrak{O}$, where $\mathfrak{O}$ and $\mathfrak{o}$ are the rings of integral elements of K and k).

Hint: Show that if $\Sigma = \Sigma_0(\bar{\theta})$, then a representative $\theta \in \mathfrak{O}$ can be chosen so that $f(\theta)$ is a prime element in $\mathfrak{O}$. Here $f(t) \in \mathfrak{o}(t)$ is chosen so that $\bar{f}(t) \in \Sigma_0(t)$ is the minimum polynomial of the element $\bar{\theta} \in \Sigma$.

17. In a field complete under a valuation, show that the infinite product $\prod_{n=1}^{\infty}(1+a_n)$, $a_n \neq -1$, converges if and only if $a_n \to 0$ as $n \to \infty$.

2. Finite Extensions of Fields with Valuations

Let k be a field with a valuation $v_{\mathfrak{p}}$ and let K/k be a finite extension. The ring $\mathfrak{o} = \mathfrak{o}_{\mathfrak{p}}$ can be considered as a ring in which we have a theory of divisors with a unique prime divisor $\mathfrak{p}$. By Theorem 1 of Section 5, Chapter 3, if $\mathfrak{O}$ is the integral closure of the ring $\mathfrak{o}$ in the field K, then $\mathfrak{O}$ has a theory of divisors with a finite number of prime divisors $\mathfrak{P}_1, \dots, \mathfrak{P}_m$ (all of which divide $\mathfrak{p}$).

Let $\mathfrak{P}$ be a prime divisor of the ring $\mathfrak{O}$ and let $K_{\mathfrak{P}}$ be the completion of the field K with respect to the valuation $v_{\mathfrak{P}}$. Those elements of $K_{\mathfrak{P}}$ which are limits of elements of k form a subfield, which is topologically isomorphic to the completion $k_{\mathfrak{p}}$ of the field k with respect to the valuation $v_{\mathfrak{p}}$. Using the embedding $k_{\mathfrak{p}} \to K_{\mathfrak{P}}$ we may assume that $k_{\mathfrak{p}}$ is a subfield of the field $K_{\mathfrak{P}}$. Let $K = k(\alpha_1, \dots, \alpha_r)$. The elements $\alpha_i \in K$ also belong to $K_{\mathfrak{P}}$, and since they are algebraic over k, they are also algebraic over $k_{\mathfrak{p}}$. Hence the extension $k_{\mathfrak{p}}(\alpha_1, \dots, \alpha_r)/k_{\mathfrak{p}}$ is finite (with degree not exceeding the degree of K/k), and then by Theorem 2 of Section 1 the field $k_{\mathfrak{p}}(\alpha_1, \dots, \alpha_r)$ is complete. Every element of $K_{\mathfrak{P}}$ is the limit of a sequence of elements of K, and since $K \subset k_{\mathfrak{p}}(\alpha_1, \dots, \alpha_r)$ and $k_{\mathfrak{p}}(\alpha_1, \dots, \alpha_r)$ is complete, then $K_{\mathfrak{P}} \subset k_{\mathfrak{p}}(\alpha_1, \dots, \alpha_r)$. Since the reverse inclusion also holds, we have $K_{\mathfrak{P}} = k_{\mathfrak{p}}(\alpha_1, \dots, \alpha_r)$. We have proved that the extension $K_{\mathfrak{P}}/k_{\mathfrak{p}}$ is finite and that

$$(K_{\mathfrak{P}} : k_{\mathfrak{p}}) \leqslant (K : k).$$

Since the residue class fields of the valuations $v_{\mathfrak{p}}$ and $v_{\mathfrak{P}}$ coincide with the residue class fields of the completions $k_{\mathfrak{p}}$ and $K_{\mathfrak{P}}$ (see Section 1.1), then the degree of inertia $f_{\mathfrak{P}}$ of the divisor $\mathfrak{P}$ relative to $\mathfrak{p}$ coincides with the degree of inertia of the extension $K_{\mathfrak{P}}/k_{\mathfrak{p}}$. It is also clear that the ramification index $e_{\mathfrak{P}}$ of the divisor $\mathfrak{P}$ relative to $\mathfrak{p}$ coincides with the ramification index of $K_{\mathfrak{P}}/k_{\mathfrak{p}}$. By Theorem 5 of Section 1 the numbers $f_{\mathfrak{P}}$ and $e_{\mathfrak{P}}$ are related to the degree $n_{\mathfrak{P}} = (K_{\mathfrak{P}} : k_{\mathfrak{p}})$ by

$$f_{\mathfrak{P}} e_{\mathfrak{P}} = n_{\mathfrak{P}}.$$

For the rest of this section we shall assume that the extension K/k is separable, and we will study the relations between the various completions $K_{\mathfrak{P}}, \dots, K_{\mathfrak{P}_n}$ of the field K under all extensions of the valuation $v_{\mathfrak{p}}$.

Let $\omega_1, \dots, \omega_n$ be a basis for the extension K/k. If for some $\alpha \in K$, all the coefficients a_j in the representation

$$\alpha = a_1\omega_1 + \cdots + a_n\omega_n \qquad (a_j \in k) \tag{2.1}$$

are small relative to $\mathfrak{p}$ (that is, small relative to the valuation $v_\mathfrak{p}$), then the element α will clearly be small relative to each of the prime divisors $\mathfrak{P}_s$. The following converse also holds.

Lemma 1. For any integer N there exist an M such that whenever $v_{\mathfrak{P}_s}(\alpha) \geqslant M$ for all $s = 1, \ldots, m$, then all coefficients a_j in (2.1) satisfy $v_\mathfrak{p}(a_j) \geqslant N$.

Proof. Let $\omega_1^*, \ldots, \omega_n^*$ be the dual basis of the basis $\omega_1, \ldots, \omega_n$ (here we are already using separability; see Section 2.3 of the Supplement). Then

$$a_j = \mathrm{Sp}_{K/k}(\alpha\omega_j^*) = \mathrm{Sp}\ \alpha\omega_j^*.$$

Let e_s be the ramification index of $\mathfrak{P}_s$ relative to $\mathfrak{p}$ and p be a prime element in the ring $\mathfrak{o}_\mathfrak{p}$ of the valuation $v_\mathfrak{p}$, so that $e_s = v_{\mathfrak{P}_s}(p)$. Set

$$M = \max_{s,j}\,(e_s N - v_{\mathfrak{P}_s}(\omega_j^*)).$$

If $v_{\mathfrak{P}_s}(\alpha) \geqslant M$ for all s, then for fixed j we have

$$v_{\mathfrak{P}_s}(\alpha\omega_j^*) \geqslant e_s N = v_{\mathfrak{P}_s}(p^N),$$

and this means that $\alpha\omega_j^* = p^N\gamma$, where $v_{\mathfrak{P}_s}(\gamma) \geqslant 0$ $(1 \leqslant s \leqslant m)$. By Theorem 6 of Section 4 of Chapter 3, the element γ belongs to the integral closure of the ring $\mathfrak{o}_\mathfrak{p}$ in the field K, and therefore $\mathrm{Sp}\ \gamma \in \mathfrak{o}_\mathfrak{p}$, that is, $v_\mathfrak{p}(\mathrm{Sp}\ \gamma) \geqslant 0$, so that

$$v_\mathfrak{p}(a_j) = v_\mathfrak{p}(\mathrm{Sp}(\alpha\omega_j^*)) = v_\mathfrak{p}(p^N\ \mathrm{Sp}\ \gamma) \geqslant N,$$

and Lemma 1 is proved.

Corollary. If a sequence $\{\alpha_k\}$ of elements from the field K is a Cauchy sequence relative to each of the prime divisors $\mathfrak{P}_s$ $(s = 1, \ldots, m)$, then all the sequences $\{a_j^{(k)}\}_{k=1}^{\infty}$, defined by

$$\alpha_k = a_1^{(k)}\omega_1 + \cdots + a_n^{(k)}\omega_n \qquad (a_j^{(k)} \in k),$$

are Cauchy sequences relative to $\mathfrak{p}$.

Now consider the completions $K_{\mathfrak{P}_1}, \ldots, K_{\mathfrak{P}_m}$ of the field K with respect to each of the prime divisors $\mathfrak{P}_1, \ldots, \mathfrak{P}_m$ and form the direct sum $K_{\mathfrak{P}_1} \oplus \cdots \oplus K_{\mathfrak{P}_m}$, which we denote by $K_\mathfrak{p}$. The elements of this direct sum are sequences $\xi = (\xi_1, \ldots, \xi_m)$, where $\xi_1 \in K_{\mathfrak{P}}, \ldots, \xi_m \in K_{\mathfrak{P}_m}$. We define addition and multiplication in $K_\mathfrak{p}$ component-wise, so that $K_\mathfrak{p}$ becomes a ring. For any $\gamma \in k_\mathfrak{p}$, set

$$\gamma(\xi_1, \ldots, \xi_m) = (\gamma\xi_1, \ldots, \gamma\xi_m).$$

The ring $K_\mathfrak{p}$ then becomes a vector space over the field $k_\mathfrak{p}$. If we denote the degree of $K_{\mathfrak{P}_s}$ over $k_\mathfrak{p}$ by n_s, then the dimension of $K_\mathfrak{p}$ over $k_\mathfrak{p}$ is given by

$$n_1 + \cdots + n_m. \tag{2.2}$$

There is a natural way of defining convergence in the ring $K_{\mathfrak{p}}$. We say that the sequence $\{(\xi_1{}^{(k)}, \ldots, \xi_m{}^{(k)})\}_{k=1}^{\infty}$, $\xi_s{}^{(k)} \in K_{\mathfrak{P}_s}$, converges to the element $(\xi_1, \ldots, \xi_m)$ if for each s the sequence $\{\xi_s{}^{(k)}\}$ converges to ξ_s relative to convergence in the field $K_{\mathfrak{P}_s}$. It is easily seen that, under this definition of convergence, the operation of multiplication of elements of $K_{\mathfrak{p}}$ by elements of $k_{\mathfrak{p}}$ is continuous. In other words, if $\gamma = \lim_{k\to\infty} \gamma^{(k)}$, $\gamma^{(k)} \in k_{\mathfrak{p}}$, and $\xi = \lim_{k\to\infty} \xi^{(k)}$, $\xi^{(k)} \in K_{\mathfrak{p}}$, then

$$\lim_{k\to\infty} \gamma^{(k)}\xi^{(k)} = \gamma\xi. \tag{2.3}$$

We now define an imbedding $K \to K_{\mathfrak{p}}$, setting

$$\hat{\alpha} = (\alpha, \ldots, \alpha) \in K_{\mathfrak{p}}, \qquad (\alpha \in K).$$

Since $K \subset K_{\mathfrak{P}_e}$ for all s, the sequence $(\alpha, \ldots, \alpha)$ is an element of $K_{\mathfrak{p}}$. It is clear that the mapping $\alpha \to \hat{\alpha}$ defines an isomorphism of the field K into the ring $K_{\mathfrak{p}}$. We denote the image of K under this isomorphism by $\hat{K}$.

To avoid confusion we note that the components of the product

$$\gamma\hat{\alpha} = (\gamma\alpha, \ldots, \gamma\alpha) \qquad (\gamma \in k_{\mathfrak{p}}),$$

which in this representation look identical, actually may be different, since the product $\gamma\alpha$ depends on the field $K_{\mathfrak{P}_s}$ in which it is taken, even when $\gamma\alpha \in k_{\mathfrak{p}}$.

Theorem 1. If $\omega_1, \ldots, \omega_n$ is a basis for the separable extension K/k, then $\hat{\omega}_1, \ldots, \hat{\omega}_n$ is a basis for the ring $K_{\mathfrak{p}}$ as a vector space over $k_{\mathfrak{p}}$.

Proof. We first show that $\hat{K}$ is everywhere dense in $K_{\mathfrak{p}}$, that is, that every element of $K_{\mathfrak{p}}$ is the limit of a sequence of elements from $\hat{K}$. Let $\xi = (\xi_1, \ldots, \xi_m)$ be any element of $K_{\mathfrak{p}}$, $\xi_s \in K_{\mathfrak{P}_s}$ $(s = 1, \ldots, m)$. Since K is dense in $K_{\mathfrak{P}_s}$, then for any natural number k there exists an element $(\alpha_s{}^{(k)}) \in K$ for which $v_{\mathfrak{P}_s}(\xi_s - \alpha_s{}^{(k)}) \geqslant k$. By Theorem 4 of Section 4, Chapter 3, there is an element $\alpha^{(k)}$ in the field K for which $v_{\mathfrak{P}_s}(\alpha^{(k)} - \alpha_s{}^{(k)}) \geqslant k$ for $s = 1, \ldots, m$. The element $\alpha^{(k)}$ satisfies

$$v_{\mathfrak{P}_s}(\xi_s - \alpha^{(k)}) \geqslant k \qquad (s = 1, \ldots, m),$$

and this means that the sequence $\{\hat{\alpha}^{(k)}\}_{k=1}^{\infty}$ of elements of $K_{\mathfrak{p}}$ converges in the ring $K_{\mathfrak{p}}$ to the element ξ.

Represent each element $\alpha^{(k)}$ in the form

$$\alpha^{(k)} = a_1{}^{(k)}\omega_1 + \cdots + a_n{}^{(k)}\omega_n, \qquad (a_j{}^{(k)} \in k).$$

Since the sequence $\{\alpha^{(k)}\}$ is a Cauchy sequence relative to each prime divisor $\mathfrak{P}_s$, then by the corollary of Lemma 1 the sequences $\{a_j{}^{(k)}\}$ are all Cauchy sequences relative to $\mathfrak{p}$, and hence they converge in $k_{\mathfrak{p}}$. Set $\gamma_j = \lim_{k\to\infty} a_j{}^{(k)}$ $(j = 1, \ldots, n)$. Since for any $a \in k \subset k_{\mathfrak{p}}$ and any $\xi \in K_{\mathfrak{p}}$,

$$a\xi = \hat{a}\xi, \tag{2.4}$$

then

$$\hat{\alpha}_k = \sum_{j=1}^{n} \hat{a}_j^{(k)} \hat{\omega}_j = \sum_{j=1}^{n} a_j^{(k)} \hat{\omega}_j .$$

Passing to the limit as $k \to \infty$ and considering (2.3), we obtain

$$\xi = \lim_{k \to \infty} \hat{\alpha}_k = \sum_{j=1}^{n} \gamma_j \hat{\omega}_j .$$

This proves that the elements $\hat{\omega}_j$ generate the vector space $K_{\mathfrak{p}}$. We still need to show that they are linearly independent over $k_{\mathfrak{p}}$. Let

$$\gamma_1 \hat{\omega}_1 + \cdots + \gamma_n \hat{\omega}_n = 0, \qquad (\gamma_j \in k_{\mathfrak{p}}).$$

Since k is dense in $k_{\mathfrak{p}}$, then $\gamma_j = \lim_{k=\infty} a_j^{(k)}$, where $a_j^{(k)} \in k$. Set

$$\alpha^{(k)} = a_1^{(k)} \omega_1 + \cdots + a_n^{(k)} \omega_n \in K.$$

Then

$$\lim_{k \to \infty} \hat{\alpha}^{(k)} = \lim_{k \to \infty} \sum_j a_j^{(k)} \hat{\omega}_j = \sum_j \gamma_j \hat{\omega}_j = 0.$$

This means that the sequence $\{\alpha^{(k)}\}$ is a null sequence in K relative to all the prime divisors $\mathfrak{P}_s$ $(s = 1, \ldots, m)$. Then by the corollary of Lemma 1 all the sequences $\{a_j^{(k)}\}$ in k are null sequences relative to $\mathfrak{p}$, and this means that $\gamma_1 = 0, \ldots, \gamma_n = 0$.

The proof of Theorem 1 is complete.

Remark. In terms of the tensor products of algebras, Theorem 1 shows that the algebra $K_{\mathfrak{p}}$ over the field $k_{\mathfrak{p}}$ is isomorphic to the tensor product $K \otimes_k k_{\mathfrak{p}}$, that is, that it can be obtained from K (regarded as an algebra over k), by extending the ground field from k to $k_{\mathfrak{p}}$.

We have shown that the dimension of the vector space $K_{\mathfrak{p}}$ over $k_{\mathfrak{p}}$ equals $n = (K : k)$. On the other hand, this dimension is given by the sum (2.2). Since $n_s = n_{\mathfrak{P}_s} = e_{\mathfrak{P}_s} f_{\mathfrak{P}_s}$, we arrive at

$$\sum_{\mathfrak{P}} e_{\mathfrak{P}} f_{\mathfrak{P}} = n$$

($\mathfrak{P}$ running through all prime divisors of the ring $\mathfrak{O}$). Hence we have obtained another proof of Theorem 7 of Section 5, Chapter 3.

Theorem 2. Let $\varphi(X)$ denote the characteristic polynomial of the element $\alpha \in K$, relative to the separable extension K/k, and let $\varphi_{\mathfrak{P}}(X)$ denote the characteristic polynomial of α relative to $K_{\mathfrak{P}}/k_{\mathfrak{p}}$. Then

$$\varphi(X) = \prod_{\mathfrak{P}} \varphi_{\mathfrak{P}}(X).$$

Proof. Consider the linear mapping $\xi \to \hat{\alpha}\xi$ of the vector space $K_{\mathfrak{p}}$ to itself. If $\alpha\omega_k = \sum_{l=1}^{n} a_{kl}\omega_l$, $a_{kl} \in k$, then by (2.4),

$$\hat{\alpha}\hat{\omega}_k = \sum_l a_{kl}\hat{\omega}_l.$$

This means that the characteristic polynomial of our transformation coincides with the characteristic polynomial of the matrix (a_{kl}), that is, coincides with $\varphi(X)$. We now take another basis for $K_{\mathfrak{p}}$ over $k_{\mathfrak{p}}$. Let β_{sj} $(j = 1, \ldots, n_s)$ be any basis for the extension $K_{\mathfrak{P}_s}/k_{\mathfrak{p}}$ $(s = 1, \ldots, m)$. If we denote by $\bar{\beta}_{sj}$ that element of $K_{\mathfrak{p}}$ which is zero in all components except the sth one, and whose sth component equals β_{sj}, then the set of all elements

$$\bar{\beta}_{sj} \qquad (s = 1, \ldots, m; j = 1, \ldots, n_s) \tag{2.5}$$

is a basis for the ring $K_{\mathfrak{p}}$ over $k_{\mathfrak{p}}$. Let

$$\alpha\beta_{sj} = \sum_{l=1}^{n_s} \gamma_{jl}^{(s)}\beta_{sl}, \qquad (\gamma_{jl}^{(s)} \in k_{\mathfrak{p}}),$$

so that $\varphi_{\mathfrak{P}_s}(X)$ is the characteristic polynomial of the matrix (γ_{jl}^{s}). It is now easily seen that the matrix of the linear mapping $\xi \to \hat{\alpha}\xi$ with the basis (2.5) will be a block-diagonal matrix with the blocks (γ_{jl}^{s}) on the main diagonal. Theorem 2 follows immediately.

For elements $\alpha \in K$ we introduce the concepts of local norm $N_{\mathfrak{P}}(\alpha)$ and local trace $\mathrm{Sp}_{\mathfrak{P}}(\alpha)$:

$$N_{\mathfrak{P}}(\alpha) = N_{K_{\mathfrak{P}}/k_{\mathfrak{p}}}(\alpha), \qquad \mathrm{Sp}_{\mathfrak{P}}(\alpha) = \mathrm{Sp}_{K_{\mathfrak{P}}/k_{\mathfrak{p}}}(\alpha).$$

From Theorem 2 we deduce that

$$N_{K/k}(\alpha) = \prod_{\mathfrak{P}/\mathfrak{p}} N_{\mathfrak{P}}(\alpha), \qquad \mathrm{Sp}_{K/k}(\alpha) = \sum_{\mathfrak{P}/\mathfrak{p}} \mathrm{Sp}_{\mathfrak{P}}(\alpha). \tag{2.6}$$

From the first of these formulas and (1.13) we deduce

$$\nu_{\mathfrak{p}}(N_{K/k}(\alpha)) = \sum_{\mathfrak{P}/\mathfrak{p}} f_{\mathfrak{P}}\nu_{\mathfrak{P}}(\alpha), \tag{2.7}$$

which we proved by other methods in Section 5 of Chapter 3.

Theorem 3. Let K/k be a separable extension with primitive element θ, so that $K = k(\theta)$, and let $\varphi(X)$ be the minimum polynomial of θ over k. Let $\mathfrak{p}$ be a prime divisor of k, and let

$$\varphi(X) = \varphi_1(X) \cdots \varphi_m(X)$$

by the factorization into irreducible polynomials in $k_{\mathfrak{p}}[X]$. Let $\mathfrak{P}_1, \ldots, \mathfrak{P}_m$ be the prime divisors of the field K which divide $\mathfrak{p}$. There is a one-to-one correspondence between the divisors $\mathfrak{P}_s$ and the factors $\varphi_s(X)$, such that the polynomial $\varphi_s(X)$ which corresponds to $\mathfrak{P}_s$ coincides with the minimum polynomial of the element $\theta \in K_{\mathfrak{P}_s}$ over the field $k_{\mathfrak{p}}$.

Proof. Since $\varphi(X)$ is the characteristic polynomial of θ relative to K/k, by Theorem 2 we have the factorization $\varphi_1(X)\cdots\varphi_m(X)$, where $\varphi_j(X)$ is the characteristic polynomial for θ relative to $K_{\mathfrak{P}_s}/k_{\mathfrak{p}}$. Hence the factors $\varphi_s(X)$ correspond to the prime divisors $\mathfrak{P}_s$. But we saw in the preceding section that $K_{\mathfrak{P}_s} = k_{\mathfrak{p}}(\theta)$, $\theta \in K \subset K_{\mathfrak{P}_s}$, and therefore each of the polynomials $\varphi_s(X)$ is irreducible over $k_{\mathfrak{p}}$, and Theorem 3 is proved.

Remark. Let $\mathfrak{o}$ be any ring with a theory of divisors (and quotient field k) and let $\mathfrak{p}$ be a prime divisor in $\mathfrak{o}$. If K/k is a finite separable extension, then Theorem 3 gives a description of all prime divisors of the integral closure $\mathfrak{O}$ of $\mathfrak{o}$ in K which divide $\mathfrak{p}$ (at least it gives their number m and the products $e_{\mathfrak{P}}f_{\mathfrak{P}}$).

3. Factorization of Polynomials in a Field Complete with Respect to a Valuation

In view of Theorem 3 of Section 2 it is important to have a method for factoring polynomials into irreducible factors in a field complete under a valuation. In this section we shall show that in such fields the decomposition of a polynomial with integral coefficients is completely determined by its decomposition modulo some power of the prime element.

Lemma. Let $\mathfrak{o}$ be a subring of the field k, and let $g(X)$ and $h(X)$ be polynomials of degree m and n with coefficients in $\mathfrak{o}$. If the resultant $\rho = R(f, g)$ of the polynomials f and g is nonzero, then for any polynomial $l(X) \in \mathfrak{o}[X]$ of degree $\leqslant m+n-1$ there exist polynomials $\varphi(X)$ and $\psi(X)$ in $\mathfrak{o}[X]$ of degrees $\leqslant n-1$ and $\leqslant m-1$ such that

$$\sum_{r+s=i} a_r u_s + \sum_{r+s=i} b_r v_s = \rho c_i \qquad (i = 0, 1, \dots, m+n-1). \tag{3.1}$$

Proof. Set

$$g(X) = \sum_{i=0}^{m} a_i X^{m-i}, \quad h(X) = \sum_{i=0}^{n} b_i X^{n-i}, \quad l(X) = \sum_{i=0}^{m+n-1} c_i X^{m+n-1-i},$$

$$\varphi(X) = \sum_{i=0}^{n-1} u_i X^{n-1-i}, \quad \psi(X) = \sum_{i=0}^{m-1} v_i X^{m-1-i}.$$

To determine the $m+n$ unknowns $u_0, \dots, u_{n-1}; v_0, \dots, v_{m-1}$, we equate the coefficients of like powers of X in (3.1), obtaining a system of $m+n$ equations

$$\sum_{r+s=i} a_r u_s + \sum_{r+s=i} b_r v_s = \rho c_i \qquad (i = 0, 1, \dots, m+n-1).$$

The determinant of this system equals

$$\begin{vmatrix} a_0 & & & & b_0 & & & \\ a_1 & a_0 & & & b_1 & b_0 & & \\ \vdots & & \ddots & & \vdots & & \ddots & \\ \cdot & & & a_0 & \cdot & & & b_0 \\ a_m & a_{m-1} & & a_1 & b_n & b_{n-1} & & b_1 \\ & a_m & & \vdots & & b_n & & \vdots \\ & & \ddots & \vdots & & & \ddots & \vdots \\ & & & a_m & & & & b_n \\ \multicolumn{4}{c}{\underbrace{\hphantom{a_0 \quad a_0 \quad a_0 \quad a_0}}_{n}} & \multicolumn{4}{c}{\underbrace{\hphantom{b_0 \quad b_0 \quad b_0 \quad b_0}}_{m}} \end{vmatrix} \tag{3.2}$$

(with zeros everywhere else); that is, it equals the resultant $\rho = R(g, h)$. Since $\rho \neq 0$, this system has a unique solution, and since all the constant terms ρc_i are divisible by ρ, then the values of u_i and v_i will belong to the ring $\mathfrak{o}$. The lemma is proved.

Now let k be complete under the valuation v, with $\mathfrak{o}$ the ring of integral elements of k and π a prime element of $\mathfrak{o}$. Two polynomials $f(X)$ and $f_1(X)$ of $\mathfrak{o}[X]$ are called *congruent modulo* π^k, and we write $f(X) \equiv f_1(X) \pmod{\pi^k}$, if this congruence holds for the coefficient of each power of X.

Theorem 1. Let $f(X) \in \mathfrak{o}[X]$ have degree $m + n$. Suppose that there exist polynomials $g_0(X)$ and $h_0(X)$ in $\mathfrak{o}[X]$ of degrees m and n such that (1) f and $g_0 h_0$ have the same leading coefficients; (2) the resultant $R(g_0, h_0)$ is nonzero; and (3) if $r = v(R(g_0, h_0))$, then

$$f(X) \equiv g_0(X) h_0(X) \qquad (\mathrm{mod}\ \pi^{2r+1}). \tag{3.3}$$

Then there are polynomials $g(X)$ and $h(X)$ in $\mathfrak{o}[X]$ of degrees m and n, for which

$$f(X) = g(X)h(X),$$

$$g(X) \equiv g_0(X), \quad h(X) \equiv h_0(X) \ (\mathrm{mod}\ \pi^{r+1})$$

and the leading coefficients of $g(X)$ and $h(X)$ coincide with those of $g_0(X)$ and $h_0(X)$, respectively.

Proof. For each $k \geqslant 1$ we construct by induction polynomials $\varphi_k \in \mathfrak{o}[X]$

of degree $\leqslant m-1$ and $\psi_k \in \mathfrak{o}[X]$ of degree $\leqslant n-1$ so that the polynomials

$$g_k = g_0 + \pi^{r+1}\varphi_1 + \cdots + \pi^{r+k}\varphi_k,$$

$$h_k = h_0 + \pi^{r+1}\psi_1 + \cdots + \pi^{r+k}\psi_k$$

will satisfy

$$f \equiv g_k h_k \pmod{\pi^{2r+k+1}}. \tag{3.4}$$

Assume that we have constructed the polynomials $\varphi_1, \ldots, \varphi_{k-1}$ and $\psi_1, \ldots, \psi_{k-1}$ with the required properties, so that

$$f = g_{k-1}h_{k-1} + \pi^{2r+k}l, \tag{3.5}$$

where $l(X) \in \mathfrak{o}[X]$. The polynomials g_0 and g_{k-1}, as well as h_0 and h_{k-1}, have the same leading coefficient, so $l(X)$ has degree $\leqslant m+m+n-1$. Further, $g_{k-1} \equiv g_0$, $h_{k-1} \equiv h_0 \pmod{\pi^{r+1}}$ and therefore

$$R(g_{k-1}, h_{k-1}) \equiv R(g_0, h_0) \pmod{\pi^{r+1}},$$

which means that $v(R(g_{k-1}, h_{k-1})) = r$. By the lemma there exist polynomials φ_k and ψ_k in $\mathfrak{o}[X]$ with degrees $\leqslant m-1$ and $\leqslant n-1$, for which

$$\pi^r l = g_{k-1}\psi_k + h_{k-1}\varphi_k. \tag{3.6}$$

We shall show that φ_k and ψ_k satisfy our requirements. Since

$$g_k = g_{k-1} + \pi^{r+k}\varphi_k, \quad h_{k-1} + \pi^{r+k}\psi_k,$$

then by (3.5) and (3.6)

$$f - g_k h_k = \pi^{2r+k}l - \pi^{r+k}(g_{k-1}\psi_k + h_{k-1}\varphi_k) - \pi^{2r+2k}\varphi_k\psi_k = -\pi^{2r+2k}\varphi_k\psi_k,$$

so (3.4) also holds (since $2k \geqslant k+1$).

Now consider the polynomials

$$g(X) = g_0 + \sum_{k=1}^{\infty} \pi^{r+k}\varphi_k, \quad h(X) = h_0 + \sum_{k=1}^{\infty} \pi^{r+k}\psi_k,$$

whose coefficients (except the leading ones) are defined as the sums of convergent power series. Since $g \equiv g_k$ and $h \equiv h_k \pmod{\pi^{k+r+1}}$, then

$$gh \equiv g_k h_k \pmod{\pi^{r+k+1}},$$

so that by (3.4)

$$f \equiv gh \pmod{\pi^{r+k+1}}.$$

Since this holds for all k, then $f = gh$, and Theorem 1 is proved.

Remark. From the proof of Theorem 1 it easily follows that if g_0 and h_0 satisfy the condition $f \equiv g_0 h_0 \pmod{\pi^s}$, $s \geqslant 2r+1$, instead of (3.3), then g and h can be chosen so that

$$g \equiv g_0, \quad h \equiv h_0 \pmod{\pi^{s-r}}.$$

We consider an important special case of Theorem 1.

The polynomial $f(X) \in \mathfrak{o}[X]$ will be called primitive if at least one of its coefficients is a unit in $\mathfrak{o}$. Let Σ be the residue class field of the ring $\mathfrak{o}$ modulo the prime element π. If we replace the coefficients of f by the corresponding residue classes in Σ, we obtain a polynomial $\bar{f}$ with coefficients in the field Σ. Assume that in the ring $\Sigma[X]$ the polynomial $\bar{f}$ has a decomposition

$$\bar{f} = \bar{g}_0 \bar{h}_0, \tag{3.7}$$

in which the factors $\bar{g}_0$ and $\bar{h}_0$ are relatively prime. We may choose polynomials g_0 and h_0 in the ring $\mathfrak{o}[X]$ so that the degree of g_0 coincides with the degree of $\bar{g}_0$ and the degree and leading coefficient of the polynomials f and $g_0 h_0$ coincide. Consider the resultant $R(g_0, h_0)$ of the polynomials g_0 and h_0, that is, the determinant of (3.2). If we replace each entry in this determinant by the corresponding residue class modulo π, we obtain the resultant $R(\bar{g}_0, \bar{h}_0)$ of the polynomials $\bar{g}_0$ and $\bar{h}_0$ (here the leading coefficient of $\bar{h}_0$ may be zero). The resultant $R(\bar{g}_0, \bar{h}_0)$ is nonzero, since by choice of g_0 the leading coefficient of $\bar{g}_0$ is nonzero, and the polynomials $\bar{g}_0$ and $\bar{h}_0$ are assumed relatively prime. (Recall that the resultant of two polynomials is zero if and only if the two polynomials have a common divisor, or both of them have leading coefficient zero.) Hence $R(g_0, h_0) \not\equiv 0 \pmod{\pi)_{\mathfrak{p}}$, that is, $v(R(g_0, h_0)) = r = 0$. Equation (3.7) means that $f \equiv g_0 h_0 \pmod{\pi}$. Thus we see that g_0 and h_0 fulfill all the conditions of Theorem 1 (with $r = 0$), and we have the following result.

Theorem 2 (Hensel's Lemma). Let $f(X)$ be a primitive polynomial with coefficients in the ring $\mathfrak{o}$ of integral elements of a field complete under a valuation. If in the residue class field Σ the polynomial $\bar{f} \in \Sigma[X]$ has a factorization

$$\bar{f} = \bar{g}_0 \bar{h}_0 \qquad (g_0, h_0 \in \mathfrak{o}[X])$$

with $\bar{g}_0$ and $\bar{h}_0$ relatively prime, then there exist polynomials g_0 and h_0 in $\mathfrak{o}[X]$ such that

$$f(X) = g(X)h(X),$$

with $\bar{g} = \bar{g}_0$, $\bar{h} = \bar{h}_0$, and the degree of g equal to the degree of $\bar{g}_0$.

With the aid of Theorem 1 we can solve the problem of the factorization of polynomials over a field complete under a valuation. We need consider only polynomials with integral coefficients and leading coefficient 1 (if the leading coefficient of a polynomial of degree n in $\mathfrak{o}[X]$ is a, then we can multiply the polynomial by a^{n-1} and take aX as a new variable). Since Gauss's lemma on the factorization of polynomials with integer coefficients also holds for the ring $\mathfrak{o}[X]$, then every irreducible factor of $f(X)$ with leading coefficient 1 will lie in $\mathfrak{o}[X]$.

If the polynomial $f(X)$ does not have multiple roots (in any finite extension of the field k), then its discriminant $D(f) = \pm R(f, f')$ is nonzero. Let $d = v(D(f))$, and consider any factorization

$$f \equiv \varphi_1 \varphi_2 \cdots \varphi_m \pmod{\pi^{d+1}}, \tag{3.8}$$

in which the leading coefficient of each φ_s (as well as of f) equals 1. Set $h_1 = \varphi_2 \cdots \varphi_m$. Since for the discriminant of the product of two polynomials we have the formula

$$D(\varphi\psi) = D(\varphi)D(\psi)R(\varphi, \psi)^2,$$

and $D(f) \equiv D(\varphi_1 h_1) \pmod{\pi^{d+1}}$, so that $v(D(\varphi_1 h_1)) = d$, then $d \geqslant 2r$, where $r = v(R(\varphi_1, h_1))$. By Theorem 1 (see the remark at the end of its proof) there exist polynomials $g_1(X)$ and $f_1(X)$ in the ring $\mathfrak{o}[X]$, with $g = g_1 f_1$ and $f_1 \equiv \varphi_2 \cdots \varphi_m \pmod{\pi^{d-r+1}}$. But $d - r \geqslant d - 2r \geqslant d_1 = v(D(f_1))$, so that for the polynomial f_1 we have the analogous factorization $f_1 = g_2 f_2$, etc. We finally arrive at the decomposition

$$f(X) = g_1(X) \cdots g_m(X), \tag{3.9}$$

in which the polynomial $g_s \in \mathfrak{o}[X]$ has the same degree as φ_s.

If the factorization (3.8) is chosen with m as large as possible, then all the polynomials g_s will be irreducible over the field k, and we have the following result.

Theorem 3. If the factorization (3.8) of the polynomial $f(X)$ is taken with m as large as possible, then f has a factorization of the form (3.9), in which each g_s is irreducible and has the same degree as the corresponding φ_s.

We also note the special case of Theorem 3 when $d = 0$, that is, when $D(f)$ is a unit in $\mathfrak{o}$. In this case the factorization (3.8) coincides (after passage to the residue class field) with the factorization

$$\bar{f} = \bar{\varphi}_1 \cdots \bar{\varphi}_m \tag{3.10}$$

into irreducible polynomials in the ring $\Sigma[X]$. Therefore we have the following

Corollary. Let $f(X) \in \mathfrak{o}[X]$ have discriminant $D(f)$ which is a unit in $\mathfrak{o}$, and let the decomposition of $\bar{f}$ into irreducible polynomials in $\Sigma[X]$ be given by (3.10). Then there exist polynomials $g_1, \ldots, g_m$ in $\mathfrak{o}[X]$ such that $f = g_1 \cdots g_m$ and $\bar{g}_1 = \bar{\varphi}_1, \ldots, \bar{g}_m = \bar{\varphi}_m$.

PROBLEMS

1. Let k be complete under a valuation, K/k a finite separable extension with ramification index e, $\mathfrak{o}$ and $\mathfrak{O}$ the rings of integral elements of k and K, and π_0 and π prime

elements in these rings. Show that if $\alpha \in \mathfrak{O}$ is divisible by π, then $Sp_{K/k}(\alpha)$ is divisible by π_0. Deduce from this that $Sp_{K/k}(\pi^{1-e}\mathfrak{O}) \subset \mathfrak{o}$. Apply Problems 12 and 16 of Section 2, Chapter 2, to this case to show that, if $e > 1$, then for any $\theta \in \mathfrak{O}$ with characteristic polynomial $f(t)$, $f'(\theta)$ is divisible by π.

2. Let k be a finite extension of the field of p-adic numbers, with ramification index e over R_p and let π be a prime element of k. Assume that k contains a primitive pth root of 1, and that e is divisible by $p-1$ (Problem 14 of Section 1). Show that any integral $\alpha \in k$, for which $\alpha \equiv 1 \pmod{\pi^{m+1}}$, where $m = pe/(p-1) = ps = e + s$, is a pth power of some element of k. [Use the fact that if $\beta = 1 + \pi^{e+k}\gamma$ (γ integral), $k > s$, $p = \pi^e \varepsilon^{-1}$, then $\beta \equiv (1 + \pi^k \gamma\varepsilon)^p \pmod{\pi^{e+k+1}}$. Then apply Problem 17 of Section 1.]

3. Under the conditions of Problem 2 assume that α is congruent to 1 modulo π^m but is not a pth power in k. Show that $k(\sqrt[p]{\alpha})/k$ is an unramified extension of degree p. [Find the characteristic polynomial $f(t)$ of the element $\gamma = \pi^{-s}(\sqrt[p]{\alpha} - 1)$ and verify that $f'(\gamma)$ is a unit; now apply the last assertion of Problem 1.]

4. Retaining the conditions of Problem 2, assume that $\alpha \in k$ is integral and satisfies the conditions: $\alpha \equiv 1 \pmod{\pi^h}$, $\alpha \equiv 1 \pmod{\pi^{h+1}}$, $(h, p) = 1$, $h < m = ep/(p-1)$. Show that α is not a pth power in k and that the extension $k(\sqrt[p]{\alpha})/k$ is totally ramified. (Consider the exponent with which the prime element of the field $k(\sqrt[p]{\alpha})$ occurs in $1 - \alpha = \prod_{i=0}^{p-1} (1 - \zeta^{\,i} \sqrt[p]{\alpha})$, where ζ is a primitive pth root of 1.)

4. Metrics on Algebraic Number Fields

4.1. *Description of Metrics*

In Section 4.2 of Chapter 1 we gave a description of all possible completions of the field R of rational numbers, these being the p-adic fields R_p and the real field R_∞. We now do the same for any algebraic number field k. As we saw in Section 1, each prime divisor $\mathfrak{p}$ of the field k corresponds to the $\mathfrak{p}$-adic completion k, that is, to the completion under the metric $\varphi_\mathfrak{p}(x) = \rho^{\nu_\mathfrak{p}(x)}$, $x \in k$ $(0 < \rho < 1)$. We call the metric $\varphi_\mathfrak{p}$ the $\mathfrak{p}$-adic metric of the field k. To classify all possible completions of k, we must find all metrics of the field k other than the $\mathfrak{p}$-adic metrics.

Let φ be any nontrivial metric of the algebraic number field k. Considering the restriction of φ to the rational numbers, we obtain a metric φ_0 of the field R. We first show that the metric φ_0 is also nontrivial. Take any basis $\omega_1, \dots,$ ω_n of k over R. For any $\xi = a_1\omega_1 + \cdots + a_n\omega_n$ $(a_i \in R)$, we have

$$\varphi(\xi) \leqslant \varphi_0(a_1)\varphi(\omega_1) + \cdots + \varphi_0(a_n)\varphi(\omega_n).$$

If the metric φ_0 were trivial, then, since $\varphi_0(a_i) \leqslant 1$, we would have the inequality

$$\varphi(\xi) \leqslant \sum_{i=1}^{n} \varphi(\omega_i)$$

for all $\xi \in k$. But this is impossible, since a nontrivial metric never takes bounded values.

By Theorem 3 of Section 4 of Chapter 1, the metric φ_0 coincides either with a p-adic metric $\varphi_p(x) = \rho^{v_p(x)}$, $0 < \rho < 1$, or with a metric $|x|^\rho$, $0 < \rho \leqslant 1$ ($x \in R$). Consider the first alternative. Let $\mathfrak{o}_p$ denote the ring of the valuation v_p (the ring of p-integral rational numbers), and let $\mathfrak{O}_p$ be its integral closure in k. If $\omega_1, \dots, \omega_n$ is a fundamental basis of the field k, then every $\alpha \in \mathfrak{O}_p$ is represented in the form $\alpha = a_1\omega_1 + \cdots + a_n\omega_n$ with the coefficients $a_i \in \mathfrak{o}_p$. Since $\varphi_p(a_i) \leqslant 1$, then

$$\varphi(\alpha) \leqslant \sum_{i=1}^{n} \varphi(\omega_i),$$

and since all powers α^k ($k \geqslant 0$) of α also lie in $\mathfrak{O}_p$, we must have $\varphi(\alpha) \leqslant 1$. It then follows easily that $\varphi(\varepsilon) = 1$ for all units of the ring $\mathfrak{O}_p$. By Theorem 7 of Section 4 of Chapter 3 each nonzero $\xi \in k$ has a unique representation in the form

$$\xi = \varepsilon\pi_1^{k_1} \cdots \pi_m^{k_m}, \tag{4.1}$$

where ε is a unit in $\mathfrak{O}_p$, and $\pi_1, \dots, \pi_m$ is a fixed system of pairwise-non-associate prime elements. (The number ξ belongs to $\mathfrak{O}_p$ if and only if $k_i \geqslant 0$.) If $\varphi(\pi_i) = 1$ for all i, then $\varphi(\xi)$ would equal 1 for all $\xi \neq 0$ of k. But this would contradict the nontriviality of φ. Suppose that we had $\varphi(\pi_i) < 1$ and $\varphi(\pi_j) < 1$ for two distinct primes π_i and π_j. Choose natural numbers k and l so that $\varphi(\pi_i)^k + \varphi(\pi_j)^l < 1$. The numbers π_i^k and π_j^l are relatively prime in the ring $\mathfrak{O}_p$, so by Lemma 2 of Section 6, Chapter 3, there exist elements α and β in $\mathfrak{O}_p$ such that

$$1 = \alpha\pi_i^k + \beta\pi_j^l.$$

But then

$$1 = \varphi(1) \leqslant \varphi(\alpha)\varphi(\pi_i)^k + \varphi(\beta)\varphi(\pi_j)^l \leqslant \varphi(\pi_i)^k + \varphi(\pi_j)^l < 1,$$

and we have a contradiction. Hence there is only one prime element π_i, for which $\varphi(\pi_i) < 1$. Denote by $\mathfrak{p}$ and $v_\mathfrak{p}$ the corresponding prime divisor and valuation. Since in (4.1) the exponent k_i equals $v_\mathfrak{p}(\xi)$, if we denote the value of $\varphi(\pi_i)$ by ρ_1, we have

$$\varphi(\xi) = \rho_1^{v_\mathfrak{p}(\xi)}. \tag{4.2}$$

Taking $\xi = p$, we find that $\rho = \rho_1^e$, where e is the ramification index of the prime divisor $\mathfrak{p}$. The formula (4.2) shows that the metric φ coincides with the $\mathfrak{p}$-adic metric $\varphi_\mathfrak{p}$, corresponding to the prime divisor $\mathfrak{p}$.

We now consider the case when $\varphi_0(x) = |x|^\rho$, $0 < \rho \leqslant 1$ ($x \in R$).

The completion of the field R under the metric $|x|^\rho$ is the field of real numbers (and does not depend on ρ). As in Section 7.2 of Chapter 1 we denote it by R_∞. The extension of the metric $|x|^\rho$, $x \in R$, to the field R_∞ will clearly be the metric $|\alpha|^\rho$, $\alpha \in R_\infty$. Adjoining to the field R_∞ the root $i = \sqrt{-1}$, we obtain the field C of complex numbers. We shall show that the metric $|\alpha|^\rho$

on the field R can be extended to the field C in only one way, namely, as the metric $|\xi|^\rho$, where $|\xi|$ denotes the absolute value (modulus) of the complex number ξ. Let ψ be some extension. Then we claim that $\psi(\xi) = 1$ for all $\xi \in C$ with $|\xi| = 1$. Otherwise we would have some $\xi \in C$ with $\varphi(\xi) > 1$ and $|\xi| = 1$. Taking a natural number n and setting $\xi^n = \alpha + \beta i$ $(\alpha, \beta \in R_\infty)$, we obtain

$$\psi(\xi^n) \leqslant \psi(\alpha) + \psi(\beta)\psi(i) \leqslant 1 + \psi(i),$$

since $\psi(\alpha) = |\alpha|^\rho \leqslant 1$, and analogously $\psi(\beta) \leqslant 1$. But this is impossible, since $\psi(\xi)^n > 1 + \psi(i)$, if n is sufficiently large. Now let ξ be any nonzero complex number. We have just shown that $\psi(\xi/|\xi|) = 1$. Hence

$$\psi(\xi) = \psi(|\xi|) = |\xi|^\rho,$$

which is what was to be shown.

Every algebraic number field k of degree $n = s + 2t$ (see Section 3.1 of Chapter 2) has n different embeddings in the field C of complex numbers (s real, and t pairs of complex conjugates). Let σ be any such embedding. If for any $\xi \in k$ we set

$$\varphi_\sigma(\xi) = |\sigma(\xi)|^\rho,$$

then the function φ_σ is clearly a metric of the field k, and $\varphi_\sigma(x) = |x|^\rho$ for $x \in R$. If ∞ and $\bar{\sigma}$ are conjugate embeddings, then $|\bar{\sigma}(\xi)| = |\overline{\sigma(\xi)}| = |\sigma(\xi)|$, and this means that the metrics φ_σ and $\varphi_{\bar{\sigma}}$ are the same. Hence we have $s + t$ metrics on k, which coincide on R with the metric $|x|^\rho$.

Now let φ be any metric of the field k which coincides with $|x|^\rho$ on R. On the completion $\bar{k}_\varphi$ of the field k under this metric we have the metric $\bar{\varphi}$ which is the only continuous extension of φ to k. The closure $\bar{R}$ of the field of rational numbers in $\bar{k}_\varphi$ is topologically isomorphic to the real field R_∞. If we denote by σ the (unique) topological isomorphism of $\bar{R}$ to R_∞, then for any $\gamma \in \bar{R}$ we shall have $\bar{\varphi}(\gamma) = |\sigma(\gamma)|^\rho$. Take in k a primitive element θ, so that $k = R(\theta)$, and let $f(X)$ be its minimum polynomial over R. Then $f(X)$ factors over the real field into s linear and t quadratic terms. Hence in the field $\bar{R}$ we have the decomposition

$$f(X) = (X - \theta_1) \cdots (X - \theta_s)(X^2 + p_1 X + q_1) \cdots (X^2 + p_t X + q_t).$$

Since $f(\theta) = 0$, then θ must be a root of one of these polynomials.

Assume first that $\theta = \theta_i$. Since $\theta \in \bar{R}$ and thus $K = R(\theta) \subset \bar{R}$, then the isomorphism $\sigma : \bar{R} \to R_\infty$ induces a real embedding $\sigma : k \to C$, such that if $\xi \in k$, then

$$\varphi(\xi) = \bar{\varphi}(\xi) = |\sigma(\xi)|^\rho.$$

Hence the metric φ coincides with φ_σ. Also we see that in this case $\bar{k}\varphi = \bar{R}$; that is, the completion $\bar{k}_\varphi$ is topologically isomorphic to the real field.

Now let θ be a root of one of the quadratic terms. In this case $(\bar{R}(\theta) : \bar{R}) = 2$, and hence the isomorphism $\sigma : \bar{R} \to R_\infty$ can be extended in two ways to an isomorphism $\sigma : \bar{R}(\theta) \to C$. The induced mapping $\sigma : k \to C$ is clearly a complex embedding of k in the complex field C. We have shown that there is only one metric on C which coincides with the metric $|\alpha|^\rho$ on R_∞, namely, the metric $|\eta|^\rho$, $\eta \in C$. Hence for any $\xi \in k$, we have

$$\varphi(\xi) = \bar{\varphi}(\xi) = |\sigma(\xi)|^\rho;$$

that is, $\varphi = \varphi_\sigma$ for the complex embedding σ. The field $\bar{k}_\varphi$ [which coincides with $\bar{R}(\theta)$] is topologically isomorphic to the field of complex numbers.

Hence we have proved the following theorem.

Theorem 1. Any nontrivial metric φ of the algebraic number field k of degree $n = s + 2t$ coincides either with a $\mathfrak{p}$-adic metric

$$\varphi_\mathfrak{p}(\xi) = \rho^{v_\mathfrak{p}(\xi)} \qquad (0 < \rho < 1,\ \xi \in k),$$

corresponding to a prime divisor $\mathfrak{p}$, or with one of the $s + t$ metrics of the form

$$\varphi_\sigma(\xi) = |\sigma(\xi)|^\rho \qquad (0 < \rho \leqslant 1,\ \xi \in k),$$

where σ is an isomorphism of the field k into the field C of complex numbers.

Definition. The completion $k_\mathfrak{p}$ of the algebraic number field k under the metric $\varphi_\mathfrak{p}$ is called the *field of* $\mathfrak{p}$*-adic numbers*.

From Theorem 1 it follows that every completion of an algebraic number field is either a $\mathfrak{p}$-adic field, the field of real numbers (for $s > 0$), or the field of complex numbers (for $t > 0$).

To emphasize the analogy between the metrics $\varphi_\mathfrak{p}$ and φ_σ of the algebraic number field k of degree $n = s + 2t$, we introduce $s + t = r$ new objects $\mathfrak{p}_{1,\infty}, \dots, \mathfrak{p}_{r,\infty}$, called *infinite prime divisors*, which correspond to the metrics φ_σ. Ordinary prime divisors, distinct from the infinite ones, are then called *finite prime divisors.* The infinite prime divisor $\mathfrak{p} = \mathfrak{p}_{i,\infty}$ is called *real* if it corresponds to a metric φ_σ with a real embedding σ, and is called *complex* if the corresponding metric $\varphi_\sigma = \varphi_{\bar{\sigma}}$ comes from a pair of complex-conjugate embeddings σ and $\bar{\sigma}$.

In the case of the rational field R there is a unique infinite (real) prime divisor p_∞, which we introduced in Section 7.2 of Chapter 1 and denoted by the symbol ∞ Those prime divisors $\mathfrak{p}_1, \dots, \mathfrak{p}_m$ of the field k, which correspond to extensions of the p-adic valuation v_p to k, are the divisors of the number p (considered as a divisor of the field R). In an analogous manner, we call the divisors $\mathfrak{p}_{1,\infty}, \dots, \mathfrak{p}_{m,\infty}$ divisors of p_∞, since the corresponding metrics are extensions of the metric $|x|^\rho$ on the rational field.

The ring $K_\mathfrak{p}$, which we considered in Section 2, when specialized to the case of an extension k/R and a rational prime p, is denoted by $k_\mathfrak{p}$ and consists of all m-tuples $(\xi_1, \dots, \xi_m)$, where $\xi_i \in k_{\mathfrak{p}_i}$. The dimension of the ring $k_\mathfrak{p}$ as a vector space over the p-adic field $R_\mathfrak{p}$ is equal to $n = (k : R)$ (Theorem 1 of Section 2). In an analogous fashion we can construct the ring $k_{\mathfrak{p}\infty}$, consisting of all $(s + t)$-tuples $(\xi_1, \dots, \xi_s, \xi_{s+1}, \dots, \xi_{s+t})$, where ξ_i $(1 \leqslant i \leqslant s)$ belongs to the field of real numbers, and ξ_{s+t} $(1 \leqslant j \leqslant t)$ to the field of complex numbers. The ring $k_{\mathfrak{p}\infty}$, being a vector space of dimension $n = (k : R)$ over the real field R_∞, clearly coincides with the ring $\mathfrak{L}^{s,t}$, which we considered in Chapter 2, which was of such great interest in the study of the group of units and classes of modules of the algebraic number field k. The ring $k_{\mathfrak{p}\infty}$ will again play a large role in Section 1 of Chapter 5.

4.2. *Relations between Metrics*

For any prime divisor $\mathfrak{p}$ of the field k (finite or infinite) we introduce the normed metric $\varphi_\mathfrak{p}$, determined by special choice of ρ. If $\mathfrak{p}$ is a finite prime divisor, then the normed metric $\varphi_\mathfrak{p}$ is determined by

$$\varphi_\mathfrak{p}(\xi) = \left(\frac{1}{N(\mathfrak{p})}\right)^{\nu_\mathfrak{p}(\xi)} \qquad (\xi \in k),$$

where $N(\mathfrak{p})$ is the norm of the divisor $\mathfrak{p}$. For an infinite real prime $\mathfrak{p}$, which corresponds to the real embedding $\sigma : k \to C$, set

$$\varphi_\mathfrak{p}(\xi) = |\sigma(\xi)|, \qquad (\xi \in k).$$

Finally, if $\mathfrak{p}$ is an infinite complex prime divisor, corresponding to the pair of complex embeddings σ and $\bar{\sigma}$, then the normed metric $\varphi_\mathfrak{p}$ is given by

$$\varphi_\mathfrak{p}(\xi) = |\sigma(\xi)|^2 = |\bar{\sigma}(\xi)|^2 = \sigma(\xi)\bar{\sigma}(\xi).$$

Note that the last function $|\sigma(\xi)|^2$, is not, strictly speaking, a metric in the sense of the definition of Section 4.1 of Chapter 1 since the triangle inequality (4.2) does not hold. However, since $|\sigma(\xi)|^2$ is the square of a metric, it can also be used to define convergence in the field k, and therefore we shall consider it a metric.

For any $\xi \neq 0$ of k we clearly have only a finite number of prime divisors $\mathfrak{p}$, for which $\varphi_\mathfrak{p}(\xi) \neq 1$. Therefore, the formally infinite product $\prod_\mathfrak{p} \varphi_\mathfrak{p}(\xi)$ makes sense.

Theorem 2. For any $\xi \neq 0$ of the algebraic number field k, the values of the normed metrics satisfy

$$\prod_\mathfrak{p} \varphi_\mathfrak{p}(\xi) = 1 \tag{4.3}$$

($\mathfrak{p}$ runs through all prime divisors of the field k, finite as well as infinite).

Proof. Let P and P' denote the products of all $\varphi_{\mathfrak{p}}(\xi)$, taken over the infinite and the finite primes, respectively, so that the left side of (4.3) equals PP'. From the definition of the normed metric for infinite $\mathfrak{p}$, we have

$$P = \prod_{\sigma} |\sigma(\xi)| = \left|\prod_{\sigma} \sigma(\xi)\right| = |N(\xi)|$$

(here σ runs through all $n = s + 2t$ embeddings of k in the field C). On the other hand, by formula (7.1) of Chapter 3, the norm of the principal divisor $(\xi) = \prod_{\mathfrak{p}} \mathfrak{p}^{\nu_{\mathfrak{p}}(\xi)}$ (here $\mathfrak{p}$ runs through all finite prime divisors) equals

$$|N(\xi)| = N\left(\prod_{\mathfrak{p}} \mathfrak{p}^{\nu_{\mathfrak{p}}(\xi)}\right) = \prod_{\mathfrak{p}} N(\mathfrak{p})^{\nu_{\mathfrak{p}}(\xi)} = \frac{1}{P'},$$

which proves the theorem.

PROBLEMS

1. Let $\varphi_1, \ldots, \varphi_r$ $(r = s + t)$ be the metrics of the algebraic number field k, of degree $n = s + 2t$, which correspond to the infinite prime divisors. Show that for any $i = 1, \ldots, r$ there exists a number ξ in k for which

$$\varphi_i(\xi_i) > 1, \quad \varphi_j(\xi_i) < 1, \qquad (j \neq i).$$

Show that the metrics $\varphi_1, \ldots, \varphi_r$ define different notions of convergence on k.

2. Show that every relation of the form

$$\prod_{\mathfrak{p}} \varphi_{\mathfrak{p}}(\xi)^{m_{\mathfrak{p}}} = 1, \qquad (\xi \in k^*)$$

between the normed metrics $\varphi_{\mathfrak{p}}$ of an algebraic number field k is a consequence of the relation (4.3), that is, show that this will hold for all $\xi \in k^*$ only if there is some integer m with $m_{\mathfrak{p}} = m$ for all $\mathfrak{p}$.

5. Analytic Functions in Complete Fields

5.1. *Power Series*

Let k be a complete field with valuation ν. We have already studied some properties of series in k (see Section 1.2 of this chapter and Section 3.4 of Chapter 1). We know that the series $\sum_{n=1}^{\infty} a_n$ converges in the field k if and only if $a_n \to 0$, as $n \to \infty$; that for convergent series the operations of addition, subtraction, and multiplication by a constant can be carried out termwise; and that the order of terms in a convergent series may be changed and the sum remains the same. From this it easily follows that if we take all products $a_i b_j$ of terms of the two convergent series $\sum_{i=1}^{\infty} a_i = s$ and $\sum_{j=1}^{\infty} b_j = t$, and write them in any order, then the resulting series will converge and its sum will be st.

We note, for future use, a simple theorem on double series. Recall that the double series

$$\sum_{i,j=1}^{\infty} a_{ij} \tag{5.1}$$

is said to converge to the sum s, if $\sum_{i=1}^{m}\sum_{j=1}^{n} a_{ij} \to s$ as $m, n \to \infty$. The series

$$\sum_{i=1}^{\infty}\left(\sum_{j=1}^{\infty} a_{ij}\right), \quad \sum_{j=1}^{\infty}\left(\sum_{i=1}^{\infty} a_{ij}\right)$$

are called the repeated series of the double series (5.1).

Theorem 1. If, for any N, for all but a finite number of pairs (i, j) we have $v(a_{ij}) > N$, then the double series (5.1) converges and its sum equals the sum of each of the repeated series, which also converge. If we form a simple series of all the terms of the double series (5.1) in any way, then the simple series will also converge, and to the same sum.

The proof of this theorem is completely elementary, and is left to the reader.

Any series in k of the form

$$f(x) = \sum_{n=0}^{\infty} a_n x^n = a_0 + a_1 x + \cdots + a_n x^n + \cdots, \tag{5.2}$$

with $a_n \in k$ is called a *power series*. If (5.2) converges for $x = x_0 \in k$, then we claim that it converges for all $x \in k$ with $v(x) \geqslant v(x_0)$. For any such x we have

$$v(a_n x^n) \geqslant v(a_n x_0{}^n),$$

and therefore the terms $a_n x^n$ also converge to zero as $n \to \infty$. Thus if we set $\mu = \min v(x)$, where x runs through all values of k for which (5.2) converges, then the region of convergence will consist of all x for which $v(x) \geqslant \mu$ [or (5.2) will converge for all x].

If we have two power series $f_1(x) = \sum_{n=0}^{\infty} a_n x^n$ and $f_2(x) = \sum_{n=0}^{\infty} b_n x^n$, then by their product we mean the power series obtained by formal multiplication, that is, the series $h(x) = \sum_{n=0}^{\infty} c_n x^n$, where $c_n = \sum_{i+j=n} a_i b_j$. Let the series $f_1(x)$ and $f_2(x)$ converge for $v(x) \geqslant \mu_1$ and $v(x) \geqslant \mu_2$. It is then clear that the series $h(x)$ will converge for $v(x) \geqslant \max(\mu_1, \mu_2)$, and that its sum will equal $f_1(x) f_2(x)$.

A power series $f(x)$ is a continuous function of x in its region of convergence. Indeed, all terms $a_n x^n$ for $n \geqslant 1$ can be made as small as desired by taking x to be sufficiently small. Hence $f(x) \to a_0 = f(0)$ as $x \to 0$; that is, the function $f(x)$ is continuous at the point $x = 0$. Now let c be any value in the region of convergence of the series $f(x)$. Replacing each term $a_n x^n$ by the expression $a_n(c + y)^n$, expanding each term and taking the sum, we obtain a power series $f_c(y)$. For all values of y from the domain of convergence of $f(x)$, we have

$$f(c + y) = f_c(y). \tag{5.3}$$

It now follows that $f_c(y) \to f_c(0)$ as $y \to 0$, and hence $f(x) \to f(c)$ as $x \to c$, so $f(x)$ is continuous at $x = c$.

A function $f(x)$, defined on some domain in a complete field with valuation, and represented on this domain by a convergent power series, is called an *analytic function.*

Consider a power series

$$g(y) = b_1 y + \cdots + b_n y^n + \cdots$$

without constant term. We claim that it is possible to substitute $g(y)$ for x in a power series $f(x)$, and obtain a series $F(y)$ in y. For if

$$a_n(g(y))^n = c_{nn} y^n + c_{n,n+1} y^{n+1} + \cdots, \tag{5.4}$$

then

$$F(y) = a_0 + c_{11} y + (c_{12} + c_{22}) y^2 + \cdots + (c_{1n} + c_{2n} + \cdots + c_{nn}) y^n + \cdots$$

Theorem 2. (On Substitution of Series in Series). Let the series $f(x)$ converge for $v(x) \geqslant \mu$. If the above series $g(y)$ converges for some $y \in k$ and $v(b_m y^m) \geqslant \mu$ for all $m \geqslant 1$, then the series $F(y)$ also converges (for this value of y) and

$$F(y) = f(g(y)).$$

Proof. Consider the double series

$$\sum_{i,j} c_{ij} y^j. \tag{5.5}$$

From (5.4) we have

$$c_{nm} y^m = \sum_{\substack{\alpha_1, \ldots, \alpha_n \geqslant 1 \\ \alpha_1 + \cdots + \alpha_n = m}} a_n b_{\alpha_1} y^{\alpha_1} \cdots b_{\alpha_n} y^{\alpha_n}.$$

Let $N = \min v(b_m y^m)$. Then

$$v(c_{nm} y^m) \geqslant \min_{\alpha_1, \ldots, \alpha_n} (v(a_n b_{\alpha_1} y^{\alpha_1} \ldots b_{\alpha_n} y^{\alpha_n})) \geqslant v(a_n) + nN.$$

Since $N = v(x_0)$ for some x_0 and for $x = x_0$, the series $f(x)$ converges; then $v(a_n) + nN = v(a_n x_0{}^n) \to \infty$, and this means that $v(c_{nm} y^m) \to \infty$ as $n \to \infty$ uniformly for all m. Further, for fixed n the series (5.4) converges (being the product of convergent series), and therefore $v(c_{nm}y^m) \to \infty$ as $m \to \infty$. This proves that the double series (4.5) satisfies the conditions of Theorem 1. By this theorem both repeated series for (4.5) converge and have the same sum. We now need only note that

$$F(y) = a_0 + \sum_j \Big(\sum_i c_{ij} y^j\Big) \qquad \text{and} \qquad f(g(y)) = a_0 + \sum_i \Big(\sum_j c_{ij} y^j\Big),$$

and Theorem 2 is proved.

In the next two sections we shall also consider analytic functions of n variables, that is, functions which can be represented as power series,

$$f(x_1, \dots, x_n) = \sum_{\alpha_1, \dots, \alpha_n \geqslant 0} a_{\alpha_1 \cdots \alpha_n} x_1^{\alpha_1} \cdots x_n^{\alpha_n}.$$

Suppose that the series $f(x_1, \dots, x_n)$ converges in the region in n-dimensional space over k consisting of all points with $v(x_i) \geqslant N$ $(i = 1, \dots, n)$. If $c = (c_1, \dots, c_n)$ is a point of this region, then, just as in the case of one variable, we easily obtain

$$f(x_1 + c_1, \dots, x_n + c_n) = f_c(x_1, \dots, x_n),$$

for all points of the region $v(x_i) \geqslant N$ [f_c also converges for $v(x_i) \geqslant N$].

5.2. *Exponential and Logarithmic Functions*

In this section we assume that k is a finite extension of the p-adic field R_p. We denote by the valuation of k by v, the ramification index over R_p by e, and a prime element of the ring of integral elements of k by π.

Consider in k the power series

$$\exp x = 1 + \frac{x}{1!} + \frac{x^2}{2!} + \cdots + \frac{x^n}{n!} + \dots, \tag{5.6}$$

$$\log(1 + x) = x - \frac{x^2}{2} + \cdots + (-1)^{n-1}\frac{x^n}{n} + \cdots \tag{5.7}$$

We shall find the region of convergence of the series (5.6). Since the prime number p occurs in $n!$ with exponent $[n/p] + [n/p^2] + \cdots$, then

$$v(n!) = e\left(\left[\frac{n}{p}\right] + \left[\frac{n}{p^2}\right] + \cdots\right) < en \sum_{k=1}^{\infty} \frac{1}{p^k} = \frac{en}{p-1},$$

and hence

$$v\left(\frac{x^n}{n!}\right) = nv(x) - v(n!) > n\left(v(x) - \frac{e}{p-1}\right). \tag{5.8}$$

If $v(x) > e/(p-1)$, then $v(x^n/n!) \to \infty$, and the series (5.6) converges. On the other hand, if $v(x) \leqslant e/(p-1)$, we have for $n = p^s$,

$$v\left(\frac{x^n}{n!}\right) = nv(x) - e(p^{s-1} + \cdots + p + 1)$$

$$= nv(x) - e\frac{n-1}{p-1} = n\left(v(x) - \frac{e}{p-1}\right) + \frac{e}{p-1} \leqslant \frac{e}{p-1},$$

and hence for such x the general term in (5.6) does not converge to zero. This shows that the series (5.6) converges precisely for those x for which $v(x) \geqslant \kappa$, where

$$\kappa = \left[\frac{e}{p-1}\right] + 1.$$

Formal multiplication of the power series $\exp x$ and $\exp y$ is easily seen to give the series $\exp(x+y)$, and hence for $v(x) \geqslant \kappa$ and $v(y) \geqslant \kappa$ we have the formula

$$\exp(x+y) = \exp x \cdot \exp y. \tag{5.9}$$

We now turn to the series (5.7). If $v(x) \leqslant 0$, then $v(x^n/n)$ does not converge to infinity as $n \to \infty$, and hence for such x the series (5.7) does not converge. Now let $v(x) \geqslant 1$. If $n = p^a n_1$, $(n_1, p) = 1$, then $p^a \leqslant n$ and $v(n) = ea \leqslant e(\ln n/\ln p)$, so that

$$v\left(\frac{x^n}{n}\right) = nv(x) - v(n) \geqslant nv(x) - e\frac{\ln n}{\ln p},$$

and this means that $v(x^n/n) \to \infty$ as $n \to \infty$. Hence the series (5.7) converges if and only if $v(x) \geqslant 1$.

If $v(x) \geqslant 1$, then the element $\varepsilon = 1 + x$ is a unit in the ring $\mathfrak{o}$ of integral elements of the field k, with $\varepsilon \equiv 1 \pmod{\pi}$. Conversely, if a unit ε satisfies this congruence, then it has the form $\varepsilon = 1 + x$, where $v(x) \geqslant 1$. We call such a unit of the ring $\mathfrak{o}$ a principal unit of the field k. The series (5.7) hence defines a function $\log \varepsilon$ on the multiplicative group of all principal units of the field k. We shall show that for any two principal units ε_1 and ε_2, we have the formula

$$\log(\varepsilon_1\varepsilon_2) = \log \varepsilon_1 + \log \varepsilon_2. \tag{5.10}$$

Let $\varepsilon_1 = 1 + x$, $\varepsilon_2 = 1 + y$, and let $v(y) \geqslant (x)$, so that $y = tx$ with t integral and

$$(1 + x)(1 + y) = 1 + (t + 1)x + tx^2.$$

We shall consider the expression $(t + 1)x + tx^2$ as a power series in x, for which all terms lie in the region of convergence of the series $\log(1 + z)$. Since the formal substitution of this expression in the series $\log(1 + z)$ gives $\log(1 + x) + \log(1 + tx)$, then by Theorem 2

$$\log(1 + (t + 1)x + tx^2) = \log(1 + x) + \log(1 + tx),$$

which proves (5.10).

The formal substitution of the series (5.7) in (5.6) and of the series $\exp(x - 1)$ in (5.7) give us the following formal identities:

$$\exp \log (1 + x) = 1 + x; \tag{5.11}$$

$$\log \exp x = x. \tag{5.12}$$

Since these are formal identities, to verify them we can assume that x is a complex variable and use the theorem on substitution of series in series for complex power series (see, for example, K. Knopp, "Elements of The Theory of Functions," Sections 41 and 45, Dover (New York, 1952). To see under what conditions the formal identities (5.11) and (5.12) can be considered as equations in k, we turn to Theorem 2. By this theorem (5.11) will hold provided the terms of the series $\log(1 + x)$ satisfy $v(x^n/n) \geqslant \kappa$. For $n = 1$ this gives us $v(x) \geqslant \kappa$. But if $v(x) \geqslant \kappa$, then $v(x^n/n) \geqslant n\kappa \geqslant \kappa$ for $1 \leqslant n \leqslant p - 1$ and

$$v\left(\frac{x^n}{n}\right) - \kappa \geqslant (n-1)\kappa - v(n) > (n-1)\frac{e}{p-1} - e\frac{\ln n}{\ln p}$$

$$= \frac{e(n-1)}{\ln p}\left(\frac{\ln p}{p-1} - \frac{\ln n}{n-1}\right) \geqslant 0$$

for $n \geqslant p \geqslant 2$ [here we are using the fact that the function $\ln t/(t-1)$ for $t \geqslant 2$ is monotone decreasing]. Hence (5.11) is valid under the condition $v(x) \geqslant \kappa$. Further, under this condition, $v(\log(1 + x)) \geqslant \kappa$. We turn to formula (5.12). It follows from (5.8) that if $v(x) \geqslant \kappa$, then every term of the series $\exp(x - 1)$ is contained in the region of convergence of the series $\log(1 + x)$, and this means that (5.12) holds for all x for which $\exp x$ is defined.

We denote by A the additive group of all $x \in k$ for which $v(x) \geqslant \kappa$, and by M the multiplicative group of units of the form $\varepsilon = 1 + x$, $x \in A$. We have shown that the mapping $\varepsilon \to \log \varepsilon$ $(\varepsilon \in M)$ is a homomorphism from the group M to the group A. We now show that the mapping $x \to \exp x$ is a homomorphism from A to M. In view of (5.9) we need only show that $v(x^n/n!) \geqslant n$ for all $x \in A$ and all $n \geqslant 1$. Let $p^s \leqslant n < p^{s+1}$. Then

$$v\left(\frac{x^n}{n!}\right) - \kappa \geqslant (n-1)\kappa - e\left(\left[\frac{n}{p}\right] + \left[\frac{n}{p^2}\right] + \cdots + \left[\frac{n}{p^s}\right]\right)$$

$$\geqslant \frac{(n-1)e}{p-1} - \frac{en}{p^s}\frac{p^s - 1}{p - 1} \geqslant 0,$$

which is what is required. Formulas (5.11) and (5.12) now show that the mappings log: $M \to A$ and exp: $A \to M$ are both one-to-one and are inverses of each other. Hence we have the following result.

Theorem 3. The mapping $x \to \exp x$ is an isomorphism of the additive group of all numbers of the field k which are divisible by π^κ ($\kappa = [e/(p-1)]+1$) onto the multiplicative group of all principal units ε which are congruent to 1 modulo π^κ. The inverse isomorphism is given by $\varepsilon \to \log \varepsilon$ [for $\varepsilon \equiv 1 \pmod{\pi^\kappa}$].

The mapping $\varepsilon \to \log \varepsilon$ is, in general, not an isomorphism on the whole group of principal units (Problem 5). Also, the value of $\log \varepsilon$ is not necessarily integral.

In real analysis one also considers the exponential function $a^x = e^{x \ln a}$. Its analog in the field k is the function

$$\eta^x = \exp(x \log \eta), \tag{5.13}$$

where η is a principal unit of the field k. This function is defined provided that $v(x) \geqslant \kappa - v(\log \eta)$. Therefore if $\eta \equiv 1 \pmod{\pi^\kappa}$, then η^x will make sense for all integral x of k, and the value of η^x will satisfy $\eta^x \equiv 1 \pmod{\pi^\kappa}$. If $\eta \equiv 1 \pmod{\pi^\kappa}$ and x and y are any integral elements, then we have the formulas

$$\eta^{x+y} = \eta^x \eta^y,$$

$$(\eta^x)^y = \eta^{xy}.$$

PROBLEMS

1. Let $f(x)$ be an analytic function in the region $\nu(x) \geqslant \mu$ (in a complete field with valuation ν). If f has an infinite number of zeros in the region $\nu(x) \geqslant \mu$, show that f is identically zero.

2. Let k be a field of characteristic zero, complete under a non-Archimedean metric φ (Problem 4 of Section 4, Chapter 1). Assume that the metric φ satisfies $\varphi(p) < 1$ for some rational prime p. Show that the region of convergence of the series $\log(1+x)$ is the set of all x for which $\varphi(x) < 1$, and the region of convergence of the series $\exp x$ is given by $\varphi(x) < \sqrt[p-1]{\varphi(p)}$.

3. Under the same conditions, determine the regions of convergence of the series

$$\sin x = \sum_{n=1}^{\infty} (-1)^{n-1} \frac{x^{2n-1}}{(2n-1)!}, \quad \cos x = \sum_{n=0}^{\infty} (-1)^n \frac{x^{2n}}{(2n)!}.$$

4. Find the error in the following proof of the irrationality of the number π. The number π is the smallest positive number for which $\sin \pi = 0$. Let π be rational. Since $\pi > 3$, the numerator of π must be divisible either by an odd prime p, or by 2^2 (in the latter case set $p = 2$). From this it follows that the series $\sin x$ and $\cos x$ converge in the p-adic field R_p

for $x = \pi$. But in view of the formula

$$\sin(x + y) = \sin x \cos y + \cos x \sin y$$

it follows from $\sin \pi = 0$ that

$$\sin n\pi = 0$$

for all natural n. The function $\sin x$ thus has an infinite number of zeros in its region of convergence. But then, by Problem 1, it would be identically zero, which is a contradiction.

5. Let k be a finite extension of the p-adic field R_p, and let ε be a principal unit of k. Show that $\log \varepsilon = 0$ if and only if ε is a root of degree p^s $(s \geqslant 0)$ of 1.

6. Keep all notations of Section 5.2. The principal units ε, for which $\varepsilon \equiv 1 \pmod{\pi^k}$, form a multiplicative group, which we call M_k. The integers of k which are divisible by π^k form an additive group A_k. Show that for $k \geqslant \kappa$ the mapping $\varepsilon \to \log \varepsilon$ is an isomorphism of the group M_k onto the group A_k (the inverse mapping being $x \to \exp x$, $x \in A_k$).

7. In any complete field with valuation, show that the region of convergence of a power series $f(x) = \sum_{n=0}^{\infty} a_n x^n$ is contained in the region of convergence of its derivative $f'(x) = \sum_{n=0}^{\infty} n a_n x^{n-1}$. Give an example in which f and f' have different regions of convergence (with the field k of characteristic zero).

8. Show that in the ring of 2-integral rational numbers the sum

$$2 + \frac{2^2}{2} + \frac{2^3}{3} + \cdots + \frac{2^n}{n}$$

can be made divisible by any given power of 2, by taking n large enough.

9. Show that all coefficients a_n of the series

$$E_p(x) = \exp\left(x + \frac{x^p}{p} + \frac{x^{p^2}}{p^2} + \cdots\right) = \sum_{n=0}^{\infty} a_n x^n$$

are p-integral rational numbers.

(*Hint:* Show that the number

$$T_n = a_n n! = \sum_{s \geqslant 1} \sum_{\substack{p^{\alpha_1} + \cdots + p^{\alpha_s} = n \\ \alpha_1 \geqslant 0, \ldots, \alpha_s \geqslant 0}} \frac{n!}{s! p^{\alpha_1} p^{\alpha_2} \cdots p^{\alpha_s}}$$

equals the number of elements in the symmetric group of nth degree which have order a power of p. Use the theorem which says that if d divides the order of the finite group G, then the number of elements $u \in G$ which satisfy the equation $u^d = 1$ is divisible by d.)

10. Show that

$$E_p(x) = \prod_{(m,p)=1} (1 - x^m)^{-\mu(m)/m}$$

(m runs through all natural numbers relatively prime to p; $\mu(m)$ is the Möbius function).

11. Let η be a principal unit in a finite extension field of the field R_p, and let x be a p-adic integer. Choose a sequence of natural numbers $\{a_n\}$ which converges to x. Show that $\lim_{n\to\infty} \eta^{a_n}$ exists and that it is independent of the choice of $\{a_n\}$. Further, show that the function

$$\eta^{\alpha} = \lim_{n\to\infty} \eta^{a_n}$$

of the p-adic integer x coincides with the function (5.13).

6. Skolem's Method

In this section we study the application of Skolem's method to equations of the form

$$F(x_1, \dots, x_m) = c, \tag{6.1}$$

where F is an irreducible, decomposable, nonfull form (Section 1.3 of Chapter 2), and c is a rational number. This method is based on some simple properties of local analytic manifolds over $\mathfrak{p}$-adic fields, which will be proved in the next section. An example which illustrates the idea of Skolem's method was given at the beginning of this chapter.

6.1. *Representation of Numbers by Nonfull Decomposable Forms*

In Section 1.3 of Chapter 2 we saw that (6.1) can be written in the form

$$N(x_1\mu_1 + \cdots + x_m\mu_m) = a \tag{6.2}$$

or

$$N(\alpha) = a, \qquad (\alpha \in M), \tag{6.3}$$

where $\mu_1, \dots, \mu_m$ are numbers of some algebraic number field k, and $M = \{\mu_1, \dots, \mu_m\}$ is the module generated by these numbers (a is a rational number). Replacing, if necessary, the form F by a form integrally equivalent to it, we may assume that the numbers $\mu_1, \dots, \mu_m$ of the module M are linearly independent over the field R of rational numbers. Since M is nonfull, $m < n = (k : R)$.

In Chapter 2 we saw how to find all solutions to (6.3) when M is a full module of k. It is thus natural to embed M in a full module $\overline{M}$, and to use the methods of Chapter 2 to find all solutions of the equation $N(\alpha) = a$, $\alpha \in \overline{M}$, and then to pick out those solutions which lie in M.

It is clear that any module of k can be embedded in a full module. To do this it suffices to extend the linearly independent set $\mu_1, \dots, \mu_m$ to a basis $\mu_1, \dots, \mu_n$ of the field k and to set $\overline{M} = \{\mu_1, \dots, \mu_n\}$.

If all $\alpha \in \overline{M}$ for which $N(\alpha) = a$ have already been found, then we shall obtain all solutions of (6.3) if we can isolate those solutions for which in the representation

$$\alpha = x_1\mu_1 + \cdots + x_n\mu_n$$

the coefficients $x_{m+1}, \dots, x_n$ are equal to zero. To express the conditions $x_{m+1} = \cdots = x_n = 0$ directly in terms of α, it is convenient to use the dual basis $\mu_1^*, \dots, \mu_n^*$ (see Section 2.3 of the Supplement). Since the trace $\mathrm{Sp}\, \mu_j\mu_i^*$ is 0 for $i \neq j$ and 1 for $i = j$, then $x_i = \mathrm{Sp}\, \alpha\mu_i^*$ $(1 \leqslant i \leqslant n)$. It follows that the numbers $\alpha \in \overline{M}$ which lie in the submodule M are characterized by the conditions

$$\operatorname{Sp} \alpha\mu_i^* = 0 \qquad (i = m + 1, \dots, n). \tag{6.4}$$

By Theorem 1 of Section 5, Chapter 2, all solutions of the equation $N(\alpha) = a$, $\alpha \in \overline{M}$, can be written in the form

$$\alpha = \gamma_j \varepsilon_1^{u_1} \cdots \varepsilon_r^{u_r} \qquad (1 \leqslant j \leqslant k), \tag{6.5}$$

where $\gamma_1, \dots, \gamma_k$ is a finite set of numbers of the module $\overline{M}$ with norm a; $\varepsilon_1, \dots, \varepsilon_r$ is a system of independent units of the field k; and $u_1, \dots, u_r$ are arbitrary rational integers. From (6.4) we see that every solution to (6.3) corresponds to a solution in one of the k systems of equations

$$\operatorname{Sp}(\gamma\mu_i^* \varepsilon_1^{u_1} \cdots \varepsilon_r^{u_r}) = 0 \qquad (i = m + 1, \dots, n) \tag{6.6}$$

in rational integers $u_1, \dots, u_r$ (here γ is one of the γ_j).

Let K be an algebraic number field which contains all fields conjugate with k, and let $\sigma_1, \dots, \sigma_n$ be all isomorphisms of k into K. Since $\operatorname{Sp} \xi = \sigma_1(\xi) + \cdots + \sigma_n(\xi)$ for any $\xi \in k$, then the system (6.6) can be written

$$\sum_{j=1}^{n} \sigma_j(\gamma\mu_i^*)\sigma_j(\varepsilon_1)^{u_1} \cdots \sigma_j(\varepsilon_r)^{u_r} = 0 \qquad (i = m + 1, \dots, n). \tag{6.7}$$

It is clear that to prove the finiteness of the number of solutions to the equation (6.3) it suffices to show that each system of the form (6.7) has only a finite number of solutions in rational integers $u_1, \dots, u_r$.

Remark. The set of all numbers of the field k of the form $\varepsilon_1^{u_1} \cdots \varepsilon_r^{u_r}$, where $u_1, \dots, u_r$ run through all rational integers, is a multiplicative subgroup of k which we shall denote by U. The solutions of (6.3) then coincide with numbers of the intersections

$$M \cap \gamma_j U \qquad (j = 1, \dots, k). \tag{6.8}$$

Instead of the set (6.8) we may also consider the similar set $\gamma_j^{-1} M \cap U$. Then the problem of finding all solutions to (6.1) is reduced to the problem of finding the intersection of a module and a multiplicative subgroup of the field k. We note also that we may replace the module M by the vector space L (over R) which is spanned by $\mu_1, \dots, \mu_m$. For $\gamma_j U \subset \overline{M}$ and $L \cap \overline{M} = M$, so $L \cap \gamma_j U = M \cap \gamma_j U$.

6.2. *The Relation to Local Analytic Manifolds*

The idea of Skolem's method is that in some cases we can prove the finiteness of the number of solutions of (6.1) by proving that the system (6.7) has only a finite number of solutions even when the variables $u_1, \ldots, u_r$ take on $\mathfrak{P}$-adic integral values (that is, they take integral values in the completion $K_{\mathfrak{P}}$), where $\mathfrak{P}$ is any prime divisor of the field K. After such an extension of the domain of possible values for the variables, we may consider the set of all solutions of the system (6.7) as a local analytic manifold in r-dimensional space, and then apply properties of such manifolds.

When we allow that variables $u_1, \ldots, u_r$ in (6.7) to take $\mathfrak{P}$-adic values, we encounter the obstacle that the exponential function $\varepsilon^u = \exp(u \log \varepsilon)$ is only defined for all integral $\mathfrak{P}$-adic u when $\varepsilon \equiv 1 \pmod{\mathfrak{P}^\kappa}$ (κ is an integer which depends only on the field $K_{\mathfrak{P}}$; see the end of Section 5). We avoid this difficulty in the following manner. By Problem 6 of Section 7, Chapter 3, there exists a natural number q such that any integer $\alpha \in K$ which is not divisible by $\mathfrak{P}$ satisfies

$$\alpha^q \equiv 1 \pmod{\mathfrak{P}^\kappa}. \tag{6.9}$$

Each exponent u_i in (6.5) can be written

$$u_i = \rho_i + qv_i, \qquad (0 \leqslant \rho_i < q, \qquad v_i \in Z),$$

and hence the unit $\varepsilon = \varepsilon_1^{u_1} \cdots \varepsilon_r^{u_r}$ has the representation

$$\varepsilon = \delta_l \varepsilon_l^{qv_1} \cdots \varepsilon_r^{qv_r} \qquad (l = 1, \ldots, q^r),$$

where δ_l is one of the q^r numbers

$$\varepsilon_1^{\rho_1} \cdots \varepsilon_r^{\rho_r}, \qquad (0 \leqslant \rho_i < q).$$

Hence we obtain a new representation for numbers α of the form (6.5), in which ε_i is replaced by ε_i^q, and the finite set of numbers γ_j by the set of numbers $\gamma_j \delta_l$. Since the ε_i are units, then the congruence (6.9) holds for all of the numbers $\sigma_j(\varepsilon_i)$, and hence the functions $\sigma_j(\varepsilon_i^q)^u$ are defined for all $\mathfrak{P}$-adic integers $u \in K_{\mathfrak{P}}$. We have proved the following result.

Lemma 1. After making new choices, if necessary, for the numbers γ_j and ε_i in (6.5), the functions $\sigma_j(\varepsilon_i)^u$ are defined for all integers of the field $K_{\mathfrak{P}}$.

In the future we shall assume that this condition holds without special mention.

We turn to the system (6.7). In view of (5.9) and (5.13), we can put this system in the form

$$\sum_{j=1}^{n} A_{ij} \exp L_j(u_1, \ldots, u_r) = 0 \qquad (i = m+1, \ldots, n), \tag{6.10}$$

where

$$L_j(u_1, \dots, u_r) = \sum_{k=1}^{r} u_k \log \sigma_j(\varepsilon_k),$$

$$A_{ij} = \sigma_j(\gamma\mu_i{}^*).$$

Since the left side of (6.10) consists of power series which converge for all $\mathfrak{P}$-adic integral $u_1, \dots, u_r$, and hence represent analytic functions, then the set of all solutions to (6.10) can be interpreted as a local analytic manifold (in a neighborhood of any solution) in the sense of the definition of Section 7.

The system (6.10) consists of $n - m$ equations in r variables. It is natural to expect that the manifold defined by this system will consist of only a finite number of points, provided that $n - m \geqslant r$. Recall that the number r comes from the Dirichlet theorem on units, and $r = s + t - 1$, where s is the number of real, and t the number of pairs, of complex embeddings of the field k in the field of complex numbers. Since $n = s + 2t$, then $n - m \geqslant r$ if and only if $t \geqslant m - 1$. In the simplest interesting case $m = 2$, and the condition reduces to $t \geqslant 1$. This means that there should be at least one pair of complex embeddings of k. This case leads to Thue's theorem, and will be considered in the next sections.

Assume that the system (6.10) has an infinite number of solutions $(u_{1s}, \dots, u_{rs})$, $s = 1, 2, \dots$. Since the ring of $\mathfrak{P}$-adic integers is compact (see Theorem 6 of Section 3, Chapter 1, and the second remark at the end of Section 1.2 of this chapter), we can choose from this sequence a convergent subsequence, the limit of which we denote by $(u_1{}^*, \dots, u_r{}^*)$. It is clear that the point $(u_1{}^*, \dots, u_r{}^*)$ is also a solution of (6.10), that is, that it lies on the manifold determined by this system, and that in any neighborhood of this point there are an infinite number of points of the manifold. We now change variables by the formula

$$u_i = u_i{}^* + v_i \qquad (1 \leqslant i \leqslant r).$$

The system (7.10) then becomes

$$\sum_{j=1}^{n} A_{ij}^* \exp L_j(v_1, \dots, v_r) = 0 \qquad (i = m + 1, \dots, n), \tag{6.11}$$

where

$$A_{ij}{}^* = A_{ij} \exp L_j(u_1{}^*, \dots, u_r{}^*).$$

The constant terms of the series on the left of (6.11) are all zero. We denote by V the local analytic manifold (see Section 7) determined by (6.11) [in the neighborhood of the point $(0, \dots, 0)$]. Since this manifold does not consist of a single point (any neighborhood of the origin contains an infinite number of points of the manifold), then by Theorem 2 of Section 7 the manifold V

contains an analytic curve; that is, there is a system of formal power series

$$\omega_1(t), \dots, \omega_r(t)$$

(not identically zero and without constant terms) with coefficients from a finite extension of $K_{\mathfrak{P}}$, such that the series

$$P_j(t) = L_j(\omega_1(t), \dots, \omega_r(t)) \tag{6.12}$$

identically satisfy the relations

$$\sum_{j=1}^{n} A_{ij}^{*} \exp P_j(t) = 0 \qquad (i = m+1, \dots, n),$$

Hence we have the following result.

Theorem 1. If the equation (6.1) has an infinite number of solutions, then at least one local analytic manifold of the type (6.11) [for some $\gamma = \gamma_j$ and some point $(u_1^*, \dots, u_r^*)$] contains an analytic curve.

This theorem is the heart of Skolem's method. It reduces the question of the finiteness of the number of solutions to (6.1) to the proof that the systems (6.11) do not have solutions in formal power series, that is, that the corresponding local analytic manifolds do not contain analytic curves.

Note that there are $n - r$ linear relations on the n series $P_j(t)$ defined by (6.12):

$$\sum_{j=1}^{n} B_{ij}\, P_j(t) = 0. \qquad (1 \leqslant i \leqslant n - r),$$

since they are linear combinations of the r series $\omega_k(t)$. Thus the existence of an analytic curve on V implies the solvability [in power series $P_i(t)$ without constant term] of the system

$$\begin{aligned} &\sum_{j=1}^{n} A_{ij}^{*} \exp P_j(t) = 0 \qquad (m+1 \leqslant i \leqslant n), \\ &\sum_{j=1}^{n} B_{ij} P_j(t) = 0 \qquad (1 \leqslant i \leqslant n - r = t + 1), \end{aligned} \tag{6.13}$$

in which both groups of equations are linearly independent. [The linear independence of the equations of the first group follows from the fact that the determinant det $\sigma_j(\gamma\mu_i^*)$, whose square is the discriminant of the basis $\gamma\mu_i^*$, is nonzero, and hence the rank of the matrix (A_{ij}) $(m+1 \leqslant i \leqslant n$, $1 \leqslant j \leqslant n)$, and, consequently, of the matrix (A_{ij}^*), is $n - m$.] If we assume that $n - m \geqslant r$, then the total number of equations in (6.13) will be $\geqslant n$.

6.3. *Thue's Theorem*

Thue's theorem states that if the form $f(x, y) = a_0x^n + a_1x^{n-1}y + \cdots + a_ny^n$ in two variables with rational integral coefficients is irreducible and has degree $n \geqslant 3$, then the equation

$$f(x, y) = c \tag{6.14}$$

has only a finite number of solutions in integers. Since forms in two variables are always decomposable, and when $n > 2$ they are nonfull, then the equation (6.14) is a special case of (6.1). Here $m = 2$, so to apply Skolem's method we must have $t \geqslant 1$; that is, the equation $f(x, 1) = 0$ must have at least one complex root. In such a case we shall say that the form $f(x, y)$ has complex roots. We shall prove Thue's theorem by Skolem's method under this assumption. In other words, we shall prove the following assertion.

Theorem 2. If the form $f(x, y)$ has integral coefficients, is irreducible, has degree $\geqslant 3$, and has at least one complex root, then the equation

$$f(x, y) = c$$

has only a finite number of solutions in integers.

Proof. We assume that the coefficient a_0 of x^n equals 1 [otherwise we can multiply (6.14) by a_0^{n-1} and replace a_0x by x]. Set $k = R(\theta)$, $K = R(\theta_1, \ldots, \theta_n)$, where $\theta = \theta_1, \theta_2, \ldots, \theta_n$ are determined by the decomposition

$$f(x, 1) = (x + \theta_1) \cdots (x + \theta_n).$$

For each $j = 1, \ldots, n$ we denote by σ_j the isomorphism of k to K which takes θ to θ_j. Since $f(x, y) = N(x + y\theta)$ (N denoting the norm of k/R), then (6.14) can be written in the form (6.3), where by M we mean the module $\{1, \theta\}$. Hence in this case $\mu_1 = 1$, $\mu_2 = \theta$ ($m = 2$).

Assume that the equation (6.3) has an infinite number of solutions $\alpha = x + y\theta$ in the module $M = \{1, \theta\}$. Then for some $\gamma = \gamma_j \in k$, an infinite number of these solutions will be of the form (6.5), where the independent units $\varepsilon_1, \ldots, \varepsilon_r$ of k satisfy Lemma 1. The exponents $u_1, \ldots, u_r$ in (6.5), corresponding to each solution α, will satisfy the system (6.10). We choose among the solutions α a sequence $\alpha_1, \alpha_2, \ldots$ so that the corresponding points

$$(u_{1s}, \ldots, u_{rs}) \qquad (s = 1, 2, \ldots) \tag{6.15}$$

converge to some point $(u_1^*, \ldots, u_r^*)$. In Section 6.2 we saw that the local analytic manifold V, defined by (6.11), contains an analytic curve $\omega_1(t), \ldots, \omega_r(t)$, and for any such curve on V the series (6.12) satisfy some system of the form (6.13).

The rest of the proof of Theorem 2 is based on the following important auxiliary result.

Lemma 2. Let there be given a system of equations

$$\sum_{j=1}^{n} a_{ij} \exp P_j = 0 \qquad (i = 1, \dots, n_1),$$
$$\sum_{j=1}^{n} b_{ij} P_j = 0 \qquad (i = 1, \dots, n_2), \tag{6.16}$$

in which each group of equations is linearly independent. If $n_1 = n - 2$, $n_2 \geqslant 2$ and if the system has a solution in formal power series $P_1(t), \dots, P_n(t)$ without constant term, then $P_k(t) = P_j(t)$ for at least two distinct indices k and j. [The coefficients a_{ij} and b_{ij}, as well as the coefficients of the series $P_j(t)$, lie in some fixed field of characteristic zero.]

We give the proof of this lemma below, but now we show how the lemma implies Theorem 2.

By Lemma 2, for any curve $\omega_1(t), \dots, \omega_r(t)$ on V, $P_k(t) = P_j(t)$ for at least two distinct indices j and k; that is,

$$L_k(\omega_1(t), \dots, \omega_r(t)) = L_j(\omega_1(t), \dots, \omega_r(t)). \tag{6.17}$$

In r-dimensional space consider the points $(v_1, \dots, v_r)$ of the manifold W which are determined by

$$\prod_{1 \leqslant k < j \leqslant n} (L_k(v_1, \dots, v_r) - L_j(v_1, \dots, v_r)) = 0.$$

It follows from (6.17) that any curve which lies on the local analytic manifold V also lies on W. But then by Theorem 3 of Section 7, $V \subset W$; that is, all points of the manifold V, contained is some sufficiently small neighborhood of the origin, also belong to W.

But on the other hand, we shall now show that among the points $(v_{1s}, \dots, v_{rs}) \in V$, which are obtained from the points (6.15) by $u_{is} = u_i^* + v_{is}$ and which converge to the origin, only a finite number lie on the manifold W. This contradiction will prove Theorem 2.

Let $\alpha = x + y\theta$ and $\alpha' = x' + y'\theta$ be two points of the sequence $\{\alpha_s\}$, for which the corresponding points of V lie in the manifold determined by $L_k = L_j$. If $\alpha = \gamma\varepsilon_1^{u_1} \cdots \varepsilon_r^{u_r}$ and $u_i = u_i^* + v_i$, then

$$u_i = u_i^* + v_i,$$
$$\sigma_j(\alpha) = \sigma_j(\gamma)\sigma_j(\varepsilon_1)^{u_1^*} \cdots \sigma_j(\varepsilon_r)^{u_r^*}\sigma_j(\varepsilon_1)^{v_1} \cdots \sigma_j(\varepsilon_r)^{v_r}$$
$$= c_j \exp L_j(v_1, \dots, v_r)$$

(with c_j independent of α) and analogously

$$\sigma_k(\alpha) = c_k \exp L_k (v_1, \cdots, v_r),$$

so that

$$\frac{\sigma_j(\alpha)}{c_j} = \frac{\sigma_k(\alpha)}{c_k}.$$

In precisely the same fashion we find that

$$\frac{\sigma_j(\alpha')}{c_j} = \frac{\sigma_k(\alpha')}{c_k}.$$

From the last two equations we obtain

$$\frac{x + y\theta_j}{x' + y'\theta_j} = \frac{x + y\theta_k}{x' + y'\theta_k},$$

so that

$$(xy' - x'y)(\theta_k - \theta_j) = 0,$$

and since $\theta_j \neq \theta_k$, then

$$xy' - x'y = 0.$$

This means that $x + y\theta = d(x' + y'\theta)$ for some rational d. Taking norms and using the fact that $N(\alpha) = N(\alpha')$, we obtain $d^n = 1$, so that $d = \pm 1$, and $\alpha' = \pm\alpha$.

Thus each of the $n(n-1)/2$ manifolds given by $L_k = L_j$, whose union is W, contains not more than two points of V which correspond to numbers of the sequence $\{\alpha_s\}$. Then W contains at most $n(n-1)$ such points. Thus any neighborhood of the origin contains points of V not lying on W, so V (as a local analytic manifold) is not contained in W, which contradicts our earlier conclusion that $V \subset W$. As we have noted, this contradiction proves Theorem 2.

Proof of Lemma 2. Since the first group of equations linearly independent (and $n_1 = n - 2$), we can, after changing the numbering if necessary, express $\exp P_i$ $(i = 1, \dots, n-2)$ in terms of $\exp P_{n-1}$ and $\exp P_n$:

$$\exp P_i = a_i \exp P_{n-1} + b_i \exp P_n. \tag{6.18}$$

If $a_i = 0$, then from the absence of constant terms and the equation $\exp P_i = b_i \exp P_n$, we deduce that $b_i = 1$ and $P_i = P_n$. Hence we may assume that all a_i are nonzero. Set

$$P_i - P_n = Q_i \qquad (i = 1, \dots, n-1)$$

and assume that all Q_i are nonzero. By (6.18) we have

$$\exp Q_i = a_i \exp Q_{n-1} + b_i, \tag{6.19}$$

so that by differentiation with respect to t (Problem 10) we obtain

$$Q_i' \exp Q_i = a_i Q_{n-1}' \exp Q_{n-1}. \tag{6.20}$$

From (6.19) and (6.20) we deduce

$$Q_i' = Q_{n-1}' \exp Q_{n-1} \frac{1}{c_i + \exp Q_{n-1}} \qquad (i = 1, \dots, n-2), \tag{6.21}$$

where $c_i = b_i a_i^{-1}$.

We now use the second group of equations of (6.16). By assumption there are at least two linearly independent equations in this group. Hence we can find a nontrivial relation among the Q_i:

$$\sum_{i=1}^{n-1} d_i Q_i = 0.$$

Differentiating this identity and replacing Q_i' by (6.21), we obtain

$$Q_{n-1}' \exp Q_{n-1} \left(\sum_{i=1}^{n-2} \frac{d_i}{c_i + \exp Q_{n-1}} + \frac{d_{n-1}}{\exp Q_{n-1}} \right) = 0,$$

and since $Q_{n-1}' \neq 0$ and $\exp Q_{n-1} \neq 0$, then

$$\sum_{i=1}^{n-1} \frac{d_i}{c_i + \exp Q_{n-1}} = 0 \tag{6.22}$$

(here we take $c_{n-1} = 0$).

We claim that (6.22) can hold only if the rational function

$$\sum_{i=1}^{n-1} \frac{d_i}{c_i + z} \tag{6.23}$$

is identically zero. Assume that it is not, that is, assume that (6.23) equals $\varphi(z)/\psi(z)$, where $\varphi(z) \neq 0$. Then since $\varphi(\exp Q_{n-1}) = 0$, the nonconstant formal power series $\exp Q_{n-1}$ is the root of a polynomial, which contradicts the assertion of Problem 4 of Section 1. It is clear that the function (6.4) can vanish identically only when $c_k = c_j$ for at least two distinct indices j and k. Since $c = ba$, we then find from (6.19) that

$$\exp P_k = \frac{a_k}{a_j} \exp P_j,$$

from which it easily follows that $P_k = P_j$. Lemma 2 is proved.

Remark. Skolem's method allows us to prove that (6.14) has only a finite number of integral solutions. But it does not give an algorithm for finding these solutions. The reason is as follows. After proving that the system (6.7) has only a finite number of $\mathfrak{P}$-adic solutions, it is easy to find an algorithm for the computation of the coefficients in the expansions of these solutions in power series of the prime element. However, there is no algorithm which allows us to judge from a finite number of coefficients whether or not we are dealing with a rational solution.

This defect is shared by all known proofs of Thue's theorem. Even when the equation (6.14) is of third degree, there is no known algorithm for finding all integral solutions, or even for determining if there exist any solutions.

6.4. *Remarks on Forms in More Variables*

The following question now arises: Under what conditions does an equation of the type (6.1) with a nonfull decomposable form have only a finite number of solutions in integers? Such equations sometimes have an infinite number of solutions. As an example consider the equation

$$x^4 + 4y^4 + 9z^4 - 4x^2y^2 - 6x^2z^2 - 12y^2z^2 = N(x + y\sqrt{2} + z\sqrt{3}) = 1$$

[the norm is taken in the extension $R(\sqrt{2}, \sqrt{3})/R$]. This equation has two infinite sets of solutions, given by the formulas

$$x + y\sqrt{2} = \pm(1 + \sqrt{2})^n \qquad (z = 0),$$

$$x + z\sqrt{3} = \pm(2 + \sqrt{3})^n \qquad (y = 0).$$

The reason for this is that, setting $z = 0$ or $y = 0$, we obtain from our form the square of a full form: $(x^2 - 2y^2)^2$ or $(x^2 - 3z^2)^2$. This occurs because the module $\{1, \sqrt{2}, \sqrt{3}\}$, which corresponds to our form, contains full modules of smaller fields, namely, $\{1, \sqrt{2}\} \subset R(\sqrt{2})$ and $\{1, \sqrt{3}\} \subset R(\sqrt{3})$.

We describe a general class of forms with analogous properties. We write (6.1) in the form (6.3) and consider the vector space L (over R) which is generated by the numbers of the module M. The module M is called *degenerate* if the corresponding space L contains a subspace L' which is similar to some subfield $k' \subset k$, where k' is neither the field of rational numbers nor an imaginary quadratic field.

We show that for a degenerate module the equation (6.3) has an infinite number of solutions (at least for some a). For if $L' = \gamma k'$ $(\gamma \in k)$ and $M' = L' \cap M$, then $\gamma^{-1}M'$ is a full module of the field k'. By the assumptions on the field k' the number of fundamental units in any order is nonzero, and therefore the equation

$$N_{k'/R}(\xi) = a, \qquad (\xi \in \gamma^{-1}M') \tag{6.24}$$

has an infinite number of solutions (provided it has at least one solution). Set

$a_1 = N_{k/R}(\gamma)a^r$, where $r = (k : k')$. Since

$$N_{k/R}(\xi\gamma) = (N_{k'/R}(\xi))^r N_{k/R}(\gamma) = a$$

and $\xi\gamma \in M' \subset M$ [for any ξ which satisfies (6.24)], then the equation $N_{k/R}(\eta) = a_1$, $\eta \in M$, has an infinite number of solutions.

The basic conjecture on equations of the form (6.1) is that any such equation has only a finite number of integral solutions, provided that the associated module is not degenerate.

Apparently the only known approach to this hypothesis lies in Skolem's method (application of which, as we have seen, requires the additional restriction that $t \geqslant m - 1$).

The basic stage at which the condition $m = 2$ was used in the proof of Theorem 2 was Lemma 2. The generalization of Lemma 2 to the case $n_1 + n_2 \geqslant n$ (instead of $n_1 = n - 2$ and $n_2 \geqslant 2$) is apparently the principal obstacle to a proof of the above hypothesis (in the case $t \geqslant m - 1$). Skolem proved this generalization in the case $n = 5$, $n_1 = 2$, $n_2 = 3$ and thus deduced that the number of solutions to (6.1) is finite when $n = 5$, $m = 3$, $t = 2$ [T. Skolem, Einige Satze uber $\mathfrak{p}$-adische Potenzreihen mit Anwendung auf gewisse exponentielle Gleichungen, *Math. Ann.* **111**, No. 3, 399/424 (1935)]. This indicates the validity of our hypothesis in the case $n = 5$ (under the condition $t \geqslant m - 1$; the nondegeneracy of the module does not arise here since the field k has prime degree and hence has no subfields).

The validity of the hypothesis was proved for $m = 3$ (and hence under the restriction that $t \geqslant 2$) by Chabauty [C. Chabauty, Sur les equations diophantiennes liees aux unites d'un corps de nombres algebriques fini, *Ann. Mat. Pura Appl.* **17**, 127/168 (1938)]. His method, however, avoids consideration of the system (6.16) by introducing other more refined techniques. Hence the generalization of Lemma 2 remains unproved even in the case $n_1 = n - 3$ (except for the case $n = 5$ considered by Skolem).

PROBLEMS

1. Let the series $f(t) = a_0 + a_1 t + a_2 t^2 + \cdots$ with p-adic integral coefficients converge for all p-adic integral values of t. If

$$\nu_p(a_1) < \nu_p(a_k), \qquad (k = 2, 3, \ldots),$$

show that the equation $f(t) = 0$ has precisely one solution in p-adic integers if $\nu_p(a_0) \geqslant \nu_p(a_1)$, and has no solution in p-adic integers if $\nu_p(a_0) < \nu_p(a_1)$.

2. Let $d > 1$ be a square-free natural number, and let (a, b) and (a, b) be two nontrivial (distinct from $(1, 0)$) integral solutions of the equation

$$x^3 + dy^3 = 1.$$

Set $\varepsilon = a + b\sqrt[3]{d}$ and $\varepsilon_1 = a_1 + b_1\sqrt[3]{d}$ in the cubic field $K = R(\sqrt[3]{d})$. Show that then

$$\varepsilon^u = \varepsilon_1{}^v$$

for rational integers u and v, at least one of which is not divisible by 3.

3. Keeping the notations of the preceding problem, assume that $d \not\equiv \pm 1 \pmod 9$. Then in K we have the decomposition $3 = \mathfrak{p}^3$ (Problem 24 of Section 7, Chapter 3), and hence the completion $K_{\mathfrak{p}}$ of the field K is of degree 3 over the field R_3 of 3-adic numbers. Assuming that $v \not\equiv 0 \pmod 3$, set $t = u/v$. Show that the number t, considered as a 3-adic integer, is a root of the equation

$$\sum_{n=2}^{\infty} a_n t^n = 0, \tag{*}$$

where $a_n = (1/n!)\,\mathrm{Sp}((\log \eta)^n)$, $\eta = \varepsilon^3$. (Here Sp denotes the trace for the extension $K_{\mathfrak{p}}/R_3$.) Show that the series in (*) converges when t is any 3-adic integer.

(*Hint:* Show that $\mathrm{Sp}(\log \eta) = 0$ and $\mathrm{Sp}\ \eta_1 = 3$, $\eta_1 = \varepsilon_1{}^3$.)

4. Show that the coefficients a_n in (*) satisfy

$$\nu_3(a_2) = \nu_3(a_3) = \mu + 3, \quad \nu_3(a_n) > \mu + 3 \qquad \text{for } n > 3,$$

where $\mu = \nu_3(a^3b^3d)$ (ν_3 being the 3-adic valuation).

[*Hint:* Use the fact that if $\eta = 1 + 3x$, $x = ab\sqrt[3]{d}\,\varepsilon$, then

$$\log \eta \equiv 3x - \frac{9}{2}x^2 + 9x^3 \pmod{3^{4+\mu}},$$

and also the fact that the trace of any element of the ring $O_3[\sqrt[3]{d}]$ is divisible by 3 (O_3 is the ring of 3-adic integers).]

5. Using Problems 1 to 4, show that the equation $x^3 + dy^3 = 1$, with $d \not\equiv \pm 1 \pmod 9$, has at most one nontrivial solution in rational integers.

6. Prove the assertion of the preceding problem in the case $d \equiv \pm 1 \pmod 9$.

[*Hint:* Recall that the number 3 factors in $K = R(\sqrt[3]{d})$ in the form $3 = \mathfrak{p}^2\mathfrak{q}$ (Problem 24 of Section 7, Chapter 3), and carry over the considerations of Problems 3 and 4 to the direct sum $K_3 = K_{\mathfrak{p}} \oplus K_{\mathfrak{q}}$ (see Section 2). The logarithmic function on K_3 is defined just as on a field; the series converges for all $\xi = (\alpha, \beta) \in K_3$, where α and β are principal units in $K_{\mathfrak{p}}$ and $K_{\mathfrak{q}}$. The trace $\mathrm{Sp}(\xi)$ is defined as the trace of the matrix of the linear mapping $\xi' \to \xi\xi'$ ($\xi' \in K_3$), and therefore for any elements of $\hat{K}$ (Section 2) it coincides with the trace of the corresponding number of K.]

7. Let the series $f(t) = a_0 + a_1 t + a_2 t^2 + \cdots$ have p-adic integral coefficients, and converge for all p-adic integral values of t. If a_n is a p-adic unit and $a_s \equiv 0 \pmod p$ for all $s > n$, show that the equation $f(t) = 0$ has at most n solutions in p-adic integers.

8. Let the sequence of integers

$$u_0, u_1, \ldots, u_n, \ldots \tag{**}$$

satisfy the recurrence relation $u_n = a_1u_{n-1} + \cdots + a_m u_{n-m}$ ($a_m \neq 0$) with rational integer coefficients $a_1, \ldots, a_m$. Assume that the polynomial $\varphi(x) = x^m - a_1x^{m-1} - \cdots - a_m$ has no multiple roots. Show that there exists a natural number M with the following property: For each residue class modulo M, either all u_n (with n in that class) are equal, or no number occurs infinitely often among the u_n (that is, show that the sequence (**) is either periodic or else assumes any given value only a finite number of times).

[*Hint:* Use the formula $u_n = A_1\alpha_1{}^n + \cdots + A_m\alpha_m{}^n$ [α_i are the roots of $(\varphi(x)$] and the fact that for any prime p and natural number M the function $\alpha_i{}^{Mx} = \exp(x \log \alpha_i{}^M)$ will be an analytic function for all p-adic integral values of x.]

9. Using the notations of the preceding problem, assume that all roots α_i $(1 \leqslant i \leqslant m)$ of the polynomial $\varphi(x)$ and all expressions α_i/α_j $(i \neq j)$ are not roots of 1. Show that no integer occurs in the sequence (**) infinitely often (unless it consists solely of zeros).

10. Let $f(y)$ be any power series, and let $g(x)$ be a power series without constant term (the coefficients are taken from some field). Set $F(x) = f(g(x))$. Show that

$$F'(x) = f'(g(x))g'(x).$$

11. Let $P(t)$ be a nonzero formal power series without constant term (over an arbitrary field). If

$$\sum_{i=1}^{n} a_i \exp \gamma_i P(t) = 0,$$

where not all a_i equal zero, show that $\gamma_k = \gamma_j$ for at least two indices $j \neq k$.

12. Prove the assertion of Lemma 2 under the conditions $n_1 = n - 1$, $n_2 = 1$, and also $n_1 = 1$, $n_2 = n - 1$.

7. Local Analytic Manifolds

Let k be a field of characteristic zero, complete under the valuation v, and let φ be a metric corresponding to v. By the n-dimensional space $\tilde{k}^n$ we mean the set of all n-tuples $(\alpha_1, \ldots, \alpha_n)$, called *points*, where the components α_i lie either in k or in a finite extension of k. By the ε-neighborhood or the origin in k we mean the set of all points $(\alpha_1, \ldots, \alpha_n)$ which satisfy $\varphi(\alpha_i) < \varepsilon$ $(i = 1, \ldots, n)$ (ε a positive real number).

Consider the set of all power series $f(x_1, \ldots, x_n)$ in n variables which converge in some ε-neighborhood of the origin. It is easily seen that the set of all such series is a ring, which we denote by $\mathfrak{O}$. We will sometimes write $f(X)$ instead of $f(x_1, \ldots, x_n)$.

Definition. If $f_1(X), \ldots, f_m(X)$ are power series in the ring $\mathfrak{O}$ without constant term, then the set V of all points $(\alpha_1, \ldots, \alpha_n) \in \tilde{k}^n$ in some ε-neighborhood of the origin which satisfy the equations

$$f_1(X) = 0, \ldots, f_m(X) = 0, \tag{7.1}$$

is called a *local analytic manifold*, or a *local manifold*.

Two local manifolds are considered equal if they coincide in some ε-neighborhood of the origin.

Local manifolds could, of course, be considered in the neighborhood of any point in k. We choose the origin for simplicity of notation.

Let V be a local manifold. The set of all power series $f(X) \in \mathfrak{O}$ which vanish on V in some ε-neighborhood of the origin is an ideal in the ring $\mathfrak{O}$. We denote this ideal by $\mathfrak{A}_V$. It is clear that an element of $\mathfrak{O}/\mathfrak{A}_V = \bar{\mathfrak{O}}$ can be considered as a function on the points of V lying in some ε-neighborhood of the origin (the neighborhood depending on the function). Hence we call the factor ring $\bar{\mathfrak{O}}$ the *ring of analytic functions* on V.

Definition. The local manifold V is called *irreducible* if the ring of functions $\mathfrak{O}/\mathfrak{A}_V$ on V has no divisors of zero. Otherwise it is called *reducible*.

The investigation of local analytic manifolds is based on three simple facts, one from algebra and two dealing with the properties of power series. We state them without proof, giving references.

Lemma 1. Let $g_1(t), \ldots, g_m(t)$ be polynomials of $k[t]$ with leading coefficient 1. There is a system $h_1, \ldots, h_r$ of polynomials in several variables, one for each coefficient of the g_j, and with integer coefficients, such that if the coefficients of $g_1(t), \ldots, g_m(t)$ are substituted for the corresponding variables in $h_1, \ldots, h_r$, then $h_1 = \cdots = h_r = 0$ if and only if $g_1, \ldots, g_m$ have a common root in some extension of k.

If $m = 2$, then $r = 1$ and h_1 is the resultant of the polynomials g_1 and g_2. The general case is easily reduced to this case. The proof is given in "Modern Algebra" by B. L. van der Waerden, Vol. II, Section 77, Ungur, New York, 1950.

Lemma 2. Suppose that the power series $f(x_1, \ldots, x_n)$ is such that all terms of degree $<k$ have zero coefficient, and the coefficient of $x_n{}^k$ is nonzero. Then there is a power series $e(x_1, \ldots, x_n)$ in $\mathfrak{O}$ with nonzero constant term such that

$$f(X)e(X) = x_n{}^k + \varphi_1(x_1, \ldots, x_{n-1})x_n{}^{k-1} + \cdots + \varphi_k(x_1, \ldots, x_{n-1}),$$

where $\varphi_1, \ldots, \varphi_k$ are power series in $x_1, \ldots, x_{n-1}$ with zero constant term.

This corollary to the Weierstrass preparation theorem is proved by O. Zariski and P. Samuel, in "Commutative Algebra," Vol. II, p. 145, Princeton University Press, Princeton, N.J., 1960.

Note that the condition that the coefficient of $x_n{}^k$ be nonzero always can be obtained after a linear change of variables. Further, it is easily checked that if we have a finite set $f_1, \ldots, f_m$ of power series, then a linear change of variables can be found so that they all satisfy this condition simultaneously.

Lemma 3. Any ideal $\mathfrak{A}$ of the ring $\mathfrak{O}$ has a finite set of generators; that is there exist power series $h_1, \ldots, h_s$ in $\mathfrak{A}$ such that any $h \in \mathfrak{A}$ can be represented in the form

$$h = g_1 h_1 + \cdots + g_s h_s,$$

for some $g_1, \ldots, g_s$ of $\mathfrak{O}$.

The proof of this lemma can also be found in "Commutative Algebra," Vol. II, p. 148.

We need Lemma 3 for the proof of the following theorem.

Theorem 1. Every local manifold is a finite union of irreducible manifolds.

Proof. Let the manifold V be determined by (7.1). If V is reducible, then there exist power series f and g in $\mathfrak{O}$, which do not vanish on the points of V in any neighborhood of the origin, but such that fg is identically zero on V in some neighborhood of the origin. Let V_1 and V_1' be the manifolds obtained by adjoining to the system (7.1) the equations $f(X) = 0$ and $g(X) = 0$, respectively. It is clear that V_1 and V_1' are submanifolds of V and that

$$V = V_1 \cup V_1'.$$

If the manifolds V_1 and V_1' are irreducible, then the theorem is proved. If one of them is reducible, then we can, in the same way, represent it as the union of two proper submanifolds. Continuing this process, we either represent V as a finite union of irreducible submanifolds, or we obtain an infinite sequence of manifolds

$$V = V_0 \underset{\neq}{\supset} V_1 \underset{\neq}{\supset} V_2 \underset{\neq}{\supset} \cdots. \tag{7.2}$$

We show that the second case is impossible. Let $\mathfrak{A}_{V_i}$ be the ideal of the variety V_i. From (7.2) it follows that

$$\mathfrak{A}_V \underset{\neq}{\subset} \mathfrak{A}_{V_1} \underset{\neq}{\subset} \mathfrak{A}_{V_2} \underset{\neq}{\subset} \cdots. \tag{7.3}$$

Denote by $\mathfrak{A}$ the union of the ideals $\mathfrak{A}_{V_i}$. By Lemma 3 the ideal $\mathfrak{A}$ is generated by a finite system of series $h_1, \ldots, h_s$. Since each series of $\mathfrak{A}$ is contained in some ideal $\mathfrak{A}_{V_i}$, then there is an integer k such that all the series $h_1, \ldots, h_s$ are contained in $\mathfrak{A}_{V_k}$. But then $\mathfrak{A} \subset \mathfrak{A}_{V_k}$ and hence $\mathfrak{A}_{V_k} = \mathfrak{A}_{V_{k+1}} = \cdots$, which contradicts (7.3). Theorem 1 is proved.

We now describe a general method for studying local manifolds, based on a reduction to manifolds in spaces of lowest possible dimension.

Let the manifold V in the space $\tilde{k}^n$ be defined by the equations (7.1). Assume that V is different from $\tilde{k}^n$, so that the series $f_1, \ldots, f_m$ $(m \geqslant 1)$ are not identically zero. Also assume that we have made a linear change of variables so that the polynomials f_i all satisfy the conditions of Lemma 2. Then, by this lemma, we can find power series $e_1(X), \ldots, e_m(X)$ in $\mathfrak{O}$ with nonzero constant term, such that

$$f_i e_i = g_i = x_n^{k_i} + \varphi_{i1} x_n^{k_i - 1} + \cdots + \varphi_{ik_i}, \tag{7.4}$$

where $\varphi_{ij} = \varphi_{ij}(x_1, \ldots, x_{n-1})$ are power series in $n - 1$ variables with zero constant term. Since $e_i(X) \neq 0$ in some ε-neighborhood of the origin, then the local manifold V is also given by the system of equations

$$g_1(X) = 0, \dots, g_m(X) = 0, \tag{7.5}$$

where each g_j is a polynomial in x_n with leading coefficient 1. We now apply Lemma 1 to these polynomials. The corresponding polynomials $h_1, \dots, h_r$ in the coefficients of the polynomials $g_1, \dots, g_m$ will be power series in $x_1, \dots, x_{n-1}$ without constant term, and since all the φ_{ij} converge in some ε-neighborhood of the origin, then the series $h_1, \dots, h_r$ will converge in the same neighborhood.

Consider now the local manifold W in the space $\tilde{k}^{n-1}$ defined by the equations

$$h_1(x_1, \dots, x_{n-1}) = 0, \dots, h_r(x_1, \dots, x_{n-1}) = 0.$$

It is clear that a point $(\alpha_1, \dots, \alpha_{n-1}) \in \tilde{k}^{n-1}$ belongs to W if and only if there exists an α_n such that $(\alpha_1, \dots, \alpha_{n-1}, \alpha_n) \in V$. Thus W is a projection of the manifold V into the hyperplane $x_n = 0$. Here each point $(\alpha_1, \dots, \alpha_{n-1}) \in W$ is the projection of a finite set of points $(\alpha_1, \dots, \alpha_{n-1}, \alpha_n) \in V$, since α_n is a common root of the polynomials $g_i(\alpha_1, \dots, \alpha_{n-1}, x_n)$. The passage from the manifold V to its projection W gives us a method for investigating local manifolds.

Definition. By a curve in the space $\tilde{k}^n$ we mean a system of n integral formal power series $\omega_1(t), \dots, \omega_n(t)$ which have zero constant term and have coefficients in k or in some finite extension of k (and are not all identically zero).

For our purposes it is not necessary to assume that the series $\omega_i(t) = \alpha_{i_1}t + \alpha_{i_2}t^2 + \cdots$ converge, and it is simpler not to do so. Thus a curve does not consist of a set of points, but only of a collection of series $\omega_i(t)$.

Definition. We shall say that the curve $\omega_1(t), \dots, \omega_n(t)$ lies on the manifold V if for any series $f(x_1, \dots, x_n)$ of the ideal $\mathfrak{A}_V$ the power series $f(\omega_1(t), \dots, \omega_n(t))$ is identically zero.

Our basic result on local analytic manifolds is the following.

Theorem 2. Any local manifold either coincides with the origin or contains some curve.

The proof will be given by induction on the dimension n.

By Lemma 3 the ideal $\mathfrak{A}_V$ has a finite set of generators. Hence we may assume that the system (7.1), which defines the variety V, consists of a set of generators for the ideal $\mathfrak{A}_V$. For $n = 1$ the variety V consists only of the origin if at least one of the series f_i is not identically zero, and coincides with $\tilde{k}^1$

if all f_i are identically zero. In the second case any series $\omega(t)$ satisfies the system (7.1).

Now let $n > 1$. The assertion of the theorem is clear if all f_i are identically zero (or if $m = 0$). Therefore we assume that none of the series $f_1, \dots, f_m$ ($m > 0$) are equal to zero. Further, assume that these series are in the form given by Lemma 2, so that instead of the equations (7.1) we have V given by (7.5), where g_i is determined by (7.4). Let W be the projection of V in $\tilde{k}^{n-1}$. By induction, we assume the theorem valid for W. If W coincides with the origin, then the local manifold V will be defined by the equations

$$g_i(0, \dots, 0, x_n) = 0 \qquad (1 \leqslant i \leqslant m),$$

and it will also coincide with the origin. If W is different from the origin, then W contains a curve $\omega_1(t), \dots, \omega_{n-1}(t)$. Let k_1 denote a finite extension of the field k which contains all coefficients of the power series $\omega_1, \dots, \omega_{n-1}$. From the definition of W it follows that if we substitute the series $\omega_1(t), \dots, \omega_{n-1}(t)$ in the series $g_1, \dots, g_m$ for $x_1, \dots, x_{n-1}$ then we obtain m polynomials in x_n,

$$g_i(\omega_1(t), \dots, \omega_{n-1}(t), x_n) \qquad (1 \leqslant i \leqslant m), \tag{7.6}$$

whose coefficients lie in the field $k_1\{t\}$ of formal power series in t over k_1. Further, these polynomials have a common root $x_n = \xi$ in some finite extension Ω of the field $k_1\{t\}$. By Theorem 6 of Section 1 the field Ω is contained in the field of formal power series $k'\{u\}$, where $u^e = t$ for some natural number e, and k' is a finite extension of k_1. Hence the element ξ can be represented as a power series $\xi = \omega(u)$ with coefficients in k'. Since ξ is a root of the polynomials (7.6), which have leading coefficient 1 and integral coefficients in the field $k_1\{t\}$, then the series $\omega(u)$ is an integral element of the field $k'\{u\}$, that is, it does not contain any term with negative exponent. In the representation (7.4) all the series φ_{ij} have zero constant term. Substituting the series $\omega_1(u^e), \dots, \omega_{n-1}(u^e)$ for $x_1, \dots, x_{n-1}$ in (7.4), and substituting $\omega(u)$ for x_n, we see that the series $\omega(u)$ has zero constant term and that

$$g_i(\omega_1(u^e), \dots, \omega_{n-1}(u^e), \omega(u)) = 0 \qquad (1 \leqslant i \leqslant m).$$

Since the series $\omega_1, \dots, \omega_{n-1}$ are not all zero, then the set of series $\omega_1(u^e), \dots, \omega_{n-1}(u^e), \omega(u)$ is a curve in $\tilde{k}^n$. By assumption the series $f_1, \dots, f_m$, and thus also the series $g_1, \dots, g_m$ generate the ideal $\mathfrak{A}_V$. Hence for any series $f(x_1, \dots, x_n)$ of $\mathfrak{A}_V$ we have

$$f(\omega_1(u^e), \dots, \omega_{n-1}(u^e), \omega(u)) = 0,$$

and this means that the curve $\omega_1(u^e), \dots, \omega_{n-1}(u^e), \omega(u)$ lies on the manifold V. Theorem 2 is proved.

Theorem 3. If V and V' are two local manifolds in $\tilde{k}^n$, where V is not contained in V', then there is a curve in $\tilde{k}^n$ which lies on V and does not lie on V'.

Proof. We can assume that the manifold V is irreducible, since otherwise we may replace V by one of its irreducible components.

Let the manifold V' be defined by the equations

$$F_1(X) = 0, \dots , F_l(X) = 0,$$

where F_j is a series of the ring $\mathfrak{O}$. Since $V \not\subset V'$, then at least one of the series F_j does not vanish on the points of V (in any neighborhood of the origin). We denote this series by $F(X)$ and will show that there is a curve $\omega_1(t), \dots ,$ $\omega_n(t)$ on V for which

$$F(\omega_1(t), \dots , \omega_n(t)) \neq 0.$$

The proof will proceed by induction on n.

We can clearly assume that the series $F(X)$ satisfies the conditions of Lemma 2, so that there exists a series $e(X) = e(x_1, \dots , x_n) \in \mathfrak{O}$ with nonzero constant term so that

$$e(X)F(X) = G(x_1, \dots , x_n) = x_n{}^k + \psi_1 x_n{}^{k-1} + \cdots + \psi_k, \tag{7.7}$$

where $\psi_1, \dots , \psi_k$ are series in $x_1, \dots , x_{n-1}$.

In the case $V = \tilde{k}^n$ (in particular, if $n = 1$) Theorem 3 is proved, for example, by taking $\omega_1(t) = \cdots = \omega_{n-1}(t) = 0$, $\omega_n(t) = t$. If $V \neq \tilde{k}^n$, then we consider the projection $W \subset \tilde{k}^{n-1}$ of the manifold V (here we assume that the series $f_1, \dots , f_m$, as well as $F(X)$, satisfy the conditions of Lemma 2; as we have seen, this can be achieved by a linear change of variables). The manifold W is also irreducible, since the ring of functions on it, that is, the factor ring $\mathfrak{O}_{n-1}/\mathfrak{A}_W = \bar{\mathfrak{O}}_{n-1}$, is a subring of the ring of functions $\mathfrak{O}/\mathfrak{A}_V = \bar{\mathfrak{O}}$ on V (as $\mathfrak{O}_{n-1} \subset \mathfrak{O}$ and $\mathfrak{A}_W \subset \mathfrak{A}_V$). For each series $f \in \mathfrak{O}$ we denote by $\bar{f}$ the corresponding function of $\bar{\mathfrak{O}}$. It follows from (7.4) that

$$\bar{x}_n^{k_i} + \bar{\varphi}_{i1}\bar{x}_n{}^{k_i-1} + \cdots + \bar{\varphi}_{ik_i} = 0,$$

and this means that the function $\bar{x}_n$ of the ring $\bar{\mathfrak{O}}$ is integral over the subring $\bar{\mathfrak{O}}_{n-1}$. It follows that the functions

$$\bar{G} = \bar{x}_n{}^k + \bar{\psi}_1\bar{x}_n{}^{k-1} + \cdots + \bar{\psi}_k \qquad (\bar{\psi}_i \in \bar{\mathfrak{O}}_{n-1})$$

also are integral over $\bar{\mathfrak{O}}_{n-1}$.

We take an equation

$$\bar{G}^s + \bar{L}_1\bar{G}^{s-1} + \cdots + \bar{L}_s = 0 \qquad (L_j \in \mathfrak{O}_{n-1}) \tag{7.8}$$

with smallest possible s. It is clear that $\bar{L}_s \neq 0$, since then we could obtain an equation for $\bar{G}$ with smaller s. Hence the series $L_s \in \mathfrak{D}_{n-1}$ does not vanish on the points of W (in any neighborhood of the origin). By the induction hypothesis there exists a curve $\omega_1(t), \dots, \omega_{n-1}(t)$ in the space $\tilde{k}^{n-1}$ which lies on W and is such that $L_s(\omega_1(t), \dots, \omega_{n-1}(t)) \neq 0$. In the proof of Theorem 2 we saw that there exist curves of the form $\omega_1(u^e), \dots, \omega_{n-1}(u^e), \omega(u)$ which lie on the manifold V. We shall show that for such a curve

$$G(\omega_1(u^e), \dots, \omega_{n-1}(u^e) \qquad (\omega(u)) \neq 0)$$

and hence that this curve does not lie on the manifold V'. For if the series on the left were identically zero, then by (7.8) we would have

$$L_s(\omega_1(u^e), \dots, \omega_{n-1}(u^e)) = 0,$$

and after replacing u^e by t,

$$L_s(\omega_1(t), \dots, \omega_{n-1}(t)) = 0,$$

which is impossible by choice of the curve $\omega_1(t), \dots, \omega_{n-1}(t)$. Theorem 3 is proved.

CHAPTER 5

Analytic Methods

In Chapter 3 we saw how important a role the number h of divisor classes of an algebraic number field played in the arithmetic of the field. Thus one would like to have an explicit formula for the number h, in terms of simpler values which depend on the field K. Although this has not been accomplished for arbitrary algebraic number fields, for certain fields of great interest (such as quadratic fields and cyclotomic fields) such formulas have been found.

The number of divisor classes is a characteristic of the set of all divisors of the field K. Since all divisors are products of prime divisors and the number of prime divisors is infinite, then to compute the number h in a finite number of steps we must use some infinite processes. This is why, in the determination of h, we shall have to consider infinite products, series, and other analytic concepts. The apparatus of mathematical analysis can be applied to solve many problems of the theory of numbers. In this chapter we give an example of the application of this apparatus by using it to compute the number of divisor classes.

1. Analytic Formulas for the Number of Divisor Classes

1.1. *The Dedekind Zeta Function*

The determination of the number h of divisor classes of the algebraic number field K is based on consideration of the Dedekind zeta function $\zeta_K(s)$, defined by

$$\zeta_K(s) = \sum_{\mathfrak{a}} \frac{1}{N(\mathfrak{a})^s}, \tag{1.1}$$

where $\mathfrak{a}$ runs through all integral divisors of the field K, and $N(\mathfrak{a})$ denotes the norm of the divisor $\mathfrak{a}$. We shall show that the series on the left side of (1.1) converges for $1 < s < \infty$, and is a continuous function of the real variable s on this interval. Further, we shall obtain the formula

$$\lim_{s \to 1+0} (s-1)\zeta_K(s) = h\kappa, \tag{1.2}$$

where κ is a constant which depends on the field K in a simple manner, and which will be computed in the course of the proof.

Formula (1.2) becomes valuable because the function $\zeta_K(s)$ also has a representation as an infinite product

$$\zeta_K(s) = \prod_{\mathfrak{p}} \frac{1}{1 - [1/N(\mathfrak{p})^s]}, \tag{1.3}$$

carried out over all prime divisors $\mathfrak{p}$ of the field K, this representation being called *Euler's identity*. If for the field K we have a good knowledge of the prime divisors (that is, if we know how rational primes factor into prime divisors in K), then we can obtain an explicit expression for h from formulas (1.2) and (1.3). By this route we shall obtain formulas for h in later sections when K is a quadratic or a cyclotomic field.

We break the series (1.2) into the sum of h series

$$\zeta_K(s) = \sum_C \left(\sum_{\mathfrak{a} \in C} \frac{1}{N(\mathfrak{a})^s} \right),$$

where $\mathfrak{a}$ runs through all integral divisors of the given divisor class C, and the exterior summation is taken over all h classes C. To prove that the series (1.1) converges, it suffices to show that each of the series

$$f_C(s) = \sum_{\mathfrak{a} \in C} \frac{1}{N(\mathfrak{a})^s} \tag{1.4}$$

converges for $s > 1$. Further, if we show that for each class C the limit

$$\lim_{s \to 1+0} (s-1) f_C(s)$$

exists and has the same value κ for each divisor class C, then we will have obtained formula (1.2).

We now transform the series (1.4) into a series over certain integers of the field K. In the inverse divisor class C^{-1} we choose an integral divisor $\mathfrak{a}'$. Then for any $\mathfrak{a} \in C$ the product $\mathfrak{a}\mathfrak{a}'$ will be a principal divisor:

$$\mathfrak{a}\mathfrak{a}' = (\alpha), \qquad (\alpha \in K).$$

It is clear that the mapping

$$\mathfrak{a} \to (\alpha) \qquad (\mathfrak{a} \in C)$$

establishes (for fixed $\mathfrak{a}'$) a one-to-one correspondence between integral divisors $\mathfrak{a}$ of the class C and principal divisors (α) divisible by $\mathfrak{a}'$. Using the equality

$$N(\mathfrak{a})N(\mathfrak{a}') = |N(\alpha)|,$$

we obtain

$$f_C(s) = N(\mathfrak{a}')^s \sum_{\substack{(\alpha) \\ \alpha \equiv 0 (\mathrm{mod}\ \mathfrak{a}')}} \frac{1}{|N(\alpha)|^s}, \tag{1.5}$$

where the summation is taken over all principal divisors of the field K which are divisible by $\mathfrak{a}'$. Since two principal divisors (α_1) and (α_2) are equal if and only if the numbers α_1 and α_2 are associate, then we may consider that the summation in (1.5) is taken over a complete set of nonzero pairwise-non-associate numbers of the field K which are divisible by $\mathfrak{a}'$.

To put the series (1.5) in a still more convenient form, we use the geometric representation of points of the field K by points in the n-dimensional space $\mathfrak{R}^n = \mathfrak{L}^{s,t}$ and in the logarithmic space $\mathfrak{R}^{s+t}$ [here $n = s + 2t$ is the degree of the field K; see Section (3.3), Chapter 2]. We shall determine a cone X in $\mathfrak{R}^n$ such that in each class of associate numbers of the field K there is one and only one whose geometric representation lies in X (by a cone we mean a subset of $\mathfrak{R}^n$ which, whenever it contains any nonzero point x, also contains the whole ray ξx, $0 < \xi < \infty$).

In Section 3 of Chapter 2 (all notations of which we preserve), we defined a homomorphism $x \to l(x)$ of the multiplicative group of points $x \in \mathfrak{R}^n$ with nonzero norm $N(x)$ to the additive group of vectors of the logarithmic space $\mathfrak{R}^{s+t}$ by formula (3.13). If $\varepsilon_1, \dots, \varepsilon_r$ is some system of fundamental units of the field K, then we showed that the vectors $l(\varepsilon_1), \dots, l(\varepsilon_r)$ formed a basis for the subspace of dimension $r = s + t - 1$ consisting of all points $(\lambda_1, \dots, \lambda_{s+t}) \in \mathfrak{R}^{s+t}$, for which $\lambda_1 + \cdots + \lambda_{s+t} = 0$. Since the vector

$$l^* = (\underbrace{1, \dots, 1}_{s}; \underbrace{2, \dots, 2}_{t})$$

does not lie in this subspace, then the set of vectors

$$l^*, l(\varepsilon_1), \dots, l(\varepsilon_r) \tag{1.6}$$

is a basis for $\mathfrak{R}^{s+t}$. Any vector $l(x) \in \mathfrak{R}^{s+t}$ [$x \in \mathfrak{R}^n$, $N(x) \neq 0$] can be represented in the form

$$l(x) = \xi l^* + \xi_1 l(\varepsilon_1) + \cdots + \xi_r l(\varepsilon_r), \tag{1.7}$$

where $\xi, \xi_1, \dots, \xi_r$ are real numbers.

Let m denote the order of the group of roots of 1 contained in the field K.

Definition. A subset X of the space $\mathfrak{R}^n$ is called a *fundamental domain* for the field K if it consists of all points x which satisfy the following conditions:

(1) $N(x) \neq 0$.

(2) In the representation (1.7) the coefficients ξ_i $(i = 1, \ldots, r)$ satisfy the inequality $0 \leqslant \xi_i < 1$.

(3) $0 \leqslant \arg x_1 < 2\pi/m$, where x_1 is the first component of the point x.

Note that for $s \geqslant 1$ the number m equals 2, so that condition (3) in this case simply means that $x_1 > 0$.

In the next section we shall see that the fundamental domain X is a cone in $\mathfrak{R}^n$, and we shall use this fact to prove the following theorem.

Theorem 1. In every class of associate numbers ($\neq 0$) of the field K there is one and only one number whose geometric representation in the space $\mathfrak{R}^n$ lies in the fundamental domain X.

We turn to the series (1.5). If we denote by $\mathfrak{M}$ the n-dimensional lattice in $\mathfrak{R}^n$ which consists of all images $x(\alpha)$, where α is an integer of K divisible by $\mathfrak{a}'$, then since $|N(\alpha)| = |N(x(\alpha))|$ we can write (1.5) in the form

$$f_C(s) = N(\mathfrak{a}')^s \sum_{x \in \mathfrak{M} \cap X} \frac{1}{|N(x)|^s}, \tag{1.8}$$

where the summation is taken over all points $x = x(\alpha)$ in the lattice $\mathfrak{M}$ which are contained in X.

In Section 1.4 we shall prove a general result on series, in which the summation is carried out over all points of a lattice which lie in some cone (Theorem 3). Applying this result to our case, we find that the series (1.8) converges for $s > 1$ and

$$\lim_{s \to 1+0} (s-1) \sum_{x \in \mathfrak{M} \cap X} \frac{1}{|N(x)|^s} = \frac{v}{\Delta}, \tag{1.9}$$

where Δ is the volume of a fundamental parallelepiped of the lattice $\mathfrak{M}$ and v is the volume of the set T which consists of all points x of the fundamental domain X for which $|N(x)| \leqslant 1$.

By Theorem 2 of Section 4, Chapter 2, and (6.3), Chapter 2, Δ is given by

$$\Delta = \frac{1}{2^t} N(\mathfrak{a}') \sqrt{|D|}, \tag{1.10}$$

where D is the discriminant of the field K. We shall compute the volume v of T in Section 1.3, where we will show that

$$v = \frac{2^s \pi^t R}{m}, \tag{1.11}$$

where R is the regulator of the field K. From (1.9), (1.10), and (1.11) it easily follows that

$$\lim_{s\to 1+0} (s-1)f_C(s) = \frac{2^{s+t}\pi^t R}{m\sqrt{|D|}},$$

and since $\zeta_K(s) = \sum_C f_C(s)$, we have established the following basic result.

Theorem 2. If K is an algebraic number field of degree $n = s + 2t$, the series

$$\zeta_K(s) = \sum_{\mathfrak{a}} \frac{1}{N(\mathfrak{a})^s}$$

converges for all $s > 1$. Further, we have the formula

$$\lim_{s\to 1+0} (s-1)\zeta_K(s) = \frac{2^{s+t}\pi^t R}{m\sqrt{|D|}}\, h,$$

where h, D, and R denote the number of divisor classes, the discriminant, and the regulator of the field K, and m is the number of roots of 1 contained in K.

We now turn to the verification of those assertions used in the derivation of Theorem 2.

1.2. *Fundamental Domains*

If ξ is a positive real number, we shall compute $l(\xi x) \in \mathfrak{L}^{s,t}$, where $x \in \mathfrak{R}^n$, $N(x) \neq 0$. From (3.12) of Chapter 2 we have

$$l_k(\xi x) = \ln \xi + l_k(x) \qquad (1 \leqslant k \leqslant s),$$
$$l_{s+j}(\xi x) = 2\ln \xi + l_{s+j}(x) \quad (1 \leqslant j \leqslant t).$$

It follows that

$$l(\xi x) = \ln \xi \cdot l^* + l(x),$$

and this means that the vectors $l(x)$ and $l(\xi x)$ will have the same coefficients for $l(\varepsilon_1), \ldots, l(\varepsilon_r)$ in terms of the basis (1.6). Since $N(\xi x) = \xi^n N(x) \neq 0$ and $\arg(\xi x) = \arg x$, then if x lies in the fundamental domain X, so does ξx; that is, the domain X is a cone in $\mathfrak{R}^n$ [X is nonempty, since it contains the point $x(1)$, the image of the number $1 \in K$].

Lemma 1. If $y \in \mathfrak{R}^n$ and $N(y) \neq 0$, then y has a unique representation in the form

$$y = x x(\varepsilon), \tag{1.12}$$

where x is a point of the fundamental domain X and ε is a unit of the field K.

Proof. We represent the vector $l(y)$ in terms of the basis (1.6):

$$l(y) = \gamma l^* l + \gamma_1 l(\varepsilon_1) + \cdots + \gamma_r l(\varepsilon_r),$$

and for $j = 1, \ldots, r$ we set

$$\gamma_j = k_j + \xi_j,$$

where k_j is a rational integer and $0 \leqslant \xi_j < 1$. We set $\eta = \varepsilon_1^{k_1} \cdots \varepsilon_r^{k_r}$ and consider the point $z = yx(\eta^{-1})$. We have

$$\begin{aligned} l(z) = l(y) + l(\eta^{-1}) &= l(y) - k_1 l(\varepsilon_1) - \cdots - k_r l(\varepsilon_r) \\ &= \gamma l^* + \xi_1 l(\varepsilon_1) + \cdots + \xi_r l(\varepsilon_r). \end{aligned}$$

Now let $\arg z_1 = \varphi$. For some integer k,

$$0 \leqslant \varphi - \frac{2\pi k}{m} < \frac{2\pi}{m}.$$

Under the isomorphism $\alpha \to \sigma_1(\alpha)$ $(\alpha \in K)$, the mth roots of 1 in K are mapped to the mth roots of 1 in the field C of complex numbers. Denote by ζ that mth root of 1 (which must be primitive) for which $\sigma_1(\zeta) = \cos(2\pi/m) + i \sin(2\pi/m)$.

We shall show that the point $x = zx(\zeta^{-k})$ belongs to the fundamental domain X. We have

$$l(x) = l(z) + l(\zeta^{-k}) = l(z) = \gamma l^* + \xi_1 l(\varepsilon_1) + \cdots + \xi_r l(\varepsilon_r),$$

where $0 \leqslant \xi_j < 1$, so that conditions (1) and (2) are fulfilled. Further, $x_1 = z_1 x(\zeta^{-k})_1 = z_1 \sigma_1(\zeta)^{-k}$, so that

$$\arg x_1 = \arg z_1 - k\frac{2\pi}{m} = \varphi - \frac{2\pi k}{m}$$

and hence

$$0 \leqslant \arg x_1 < \frac{2\pi}{m}.$$

Thus $x \in X$. Now note that $x(\alpha)^{-1} = x(\alpha^{-1})$, so that

$$y = zx(\eta) = xx(\zeta^k)x(\eta) = xx(\varepsilon),$$

where $\varepsilon = \zeta^k \eta$. Hence we have represented y in the form (1.12). We now must show the uniqueness of this representation. Assume that also $y = x'x(\varepsilon')$, where $x' \in X$ and ε' is a unit in K. Since $xx(\varepsilon) = x'x(\varepsilon')$, then

$$l(x) + l(\varepsilon) = l(x') + l(\varepsilon').$$

The vectors $l(\varepsilon)$ and $l(\varepsilon')$ are integral linear combinations of the vectors

$l(\varepsilon_1), \ldots, l(\varepsilon_r)$. The coefficients of the vectors $l(\varepsilon_i)$ in the expansions in the basis (1.6) of $l(x)$ and $l(x')$ are nonnegative and less than 1 [by condition (2)]. Hence $l(\varepsilon') = l(\varepsilon)$ and this means that $\varepsilon' = \varepsilon\zeta_0$, where ζ_0 is an mth root of 1 (see Section 3.4 of Chapter 2). From the equation $x(\varepsilon') = x(\varepsilon)x(\zeta_0)$ it follows that $x = x'x(\zeta_0)$, and hence

$$x_1 = x_1'\sigma_1(\zeta_0).$$

By condition (3) we have

$$0 \leqslant \arg x_1 < \frac{2\pi}{m}, \qquad 0 \leqslant \arg x_1' < \frac{2\pi}{m},$$

and hence $0 \leqslant |\arg \sigma_1(\zeta_0)| < 2\pi/m$, and since $\sigma_1(\zeta_0)$ is an mth root of 1, this is possible only when $\arg \sigma_1(\zeta_0) = 0$, so that $\sigma_1(\zeta_0) = 1$ and $\zeta_0 = 1$. Hence $x' = x$ and $\varepsilon' = \varepsilon$. Lemma 1 is proved.

Proof of Theorem 1. Let β be any nonzero number of K. By Lemma 1 we can write $x(\beta) = xx(\varepsilon)$, where $x \in X$ and ε is a unit. The number $\alpha = \beta\varepsilon^{-1}$ is associate with β, and its geometric image $x(\alpha)$ (coinciding with the point x) lies in the domain X. Since the decomposition (1.12) is unique, the number α which satisfies $\beta = \alpha\varepsilon$ and $x(\alpha) \in X$ is uniquely determined, and this proves Theorem 1.

As an example we shall find the fundamental domains for quadratic fields.

First, we take the case where K is a real quadratic field, so that $n = s = 2$, $t = 0$, $r = s + t - 1 = 1$. We shall assume that K is a subfield of the field C of complex numbers, and that the first isomorphism $\sigma_1 : K \to C$ is the inclusion mapping (see Section 3.1 of Chapter 2). If ε is a fundamental unit of the field K, then $-\varepsilon$, $1/\varepsilon$, $-1/\varepsilon$ are also fundamental units, so we may assume that $\varepsilon > 1$. If $x = (x_1, x_2) \in \mathfrak{R}^2$, with $N(x) = x_1x_2 \neq 0$, then $l(x) = (\ln|x_1|, \ln|x_2|)$. The decomposition (1.7) then has the form

$$l(x) = \xi(1, 1) + \xi_1(\ln \varepsilon, -\ln \varepsilon).$$

The fundamental domain X is determined by the conditions

$$x_1 > 0, \quad x_2 \neq 0, \qquad (0 \leqslant \xi_1 < 1).$$

It is easily seen that

$$\ln|x_1| = \ln|x_2| + 2\xi_1 \ln \varepsilon,$$

and hence

$$|x_1| = |x_2|\varepsilon^{2\xi_1}.$$

The condition $0 \leqslant \xi_1 < 1$ then leads to

$$1 \geqslant \frac{|x_2|}{|x_1|} > \varepsilon^{-2}.$$

The fundamental domain X hence consists of the points indicated in Figure 7 (the boundary rays which lie closest to the positive x-axis are not included in X).

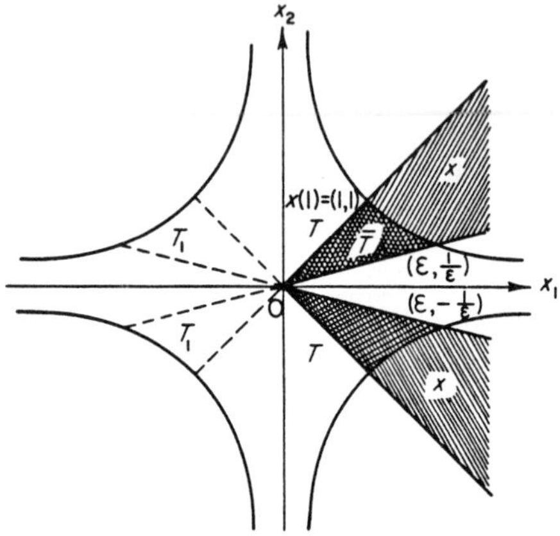

FIG. 7

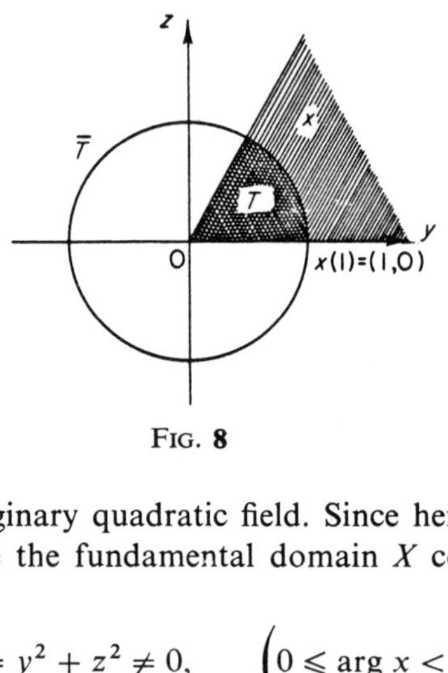

FIG. 8

Now let K be an imaginary quadratic field. Since here $s = 0$, $t = 1$, then $r = s + t - 1 = 0$. Hence the fundamental domain X consists of all points $x = y + iz$ for which

$$N(x) = y^2 + z^2 \neq 0, \qquad \left(0 \leqslant \arg x < \frac{2\pi}{m}\right)$$

[see Figure 8 for the case $K = R(\sqrt{-3})$, with $m = 6$].

1.3. *Computation of the Volume*

We now turn to the computation of the n-dimensional volume of the set T, which consists of all points of the fundamental domain X for which $|N(x)| \leqslant 1$.

It will be shown in the course of the computations that the volume exists and is nonzero. (For quadratic fields, the set T is indicated in Figures 7 and 8.)

We first show that the set T is bounded. In every ray which is contained in the cone X, there is one and only one point x for which $|N(x)| = 1$. Denote the set of all such points by S. It is clear that T consists of all points ξx $(0 < \xi \leqslant 1)$, where x runs through all points of S.

Consider the formula (1.7) for any point $x \in \mathfrak{R}^n$ with nonzero norm. We compute the sum of the components of this vector. By (3.15) of Chapter 2 the sum on the left equals $\ln|N(x)|$. By (3.18) of Chapter 2 the sum on the right is $\xi(s + 2t) = n\xi$. This means that $\xi = (1/n) \ln|N(x)|$, and (1.7) can be written in the form

$$l(x) = \frac{1}{n} \ln|N(x)| \cdot l^* + \xi_1 l(\varepsilon_1) + \cdots + \xi_r l(\varepsilon_r). \tag{1.13}$$

Now if $x \in S$, then $\ln|N(x)| = 0$, and hence the point $l(x) = (l_1(x), \ldots, l_{s+t}(x)) \in \mathfrak{R}^{s+t}$ is represented in the form $l(x) = \xi_1 l(\varepsilon_1) + \cdots + \xi_r l(\varepsilon_r)$, where $0 \leqslant \xi_i < 1$. It follows that there is a constant ρ such that $l_j(x) < \rho$, and then $|x_k| < e^\rho$ for $1 \leqslant k \leqslant s$ and $|x_{s+j}| < e^{\rho/2}$ for $1 \leqslant j \leqslant t$ for all $x \in S$ [see (3.13) and (3.12) of Chapter 2]. This shows that the set S, and hence also the set T, is bounded.

We shall replace the set T by another set which is easily obtained from T, and which has the advantage that it is defined by a simpler set of conditions. We first note the following almost obvious lemma.

Lemma 2. If ε is a unit of the field K, then the linear transformation of the space $\mathfrak{R}^n$ given by $x \to xx(\varepsilon)$ is volume-preserving.

Under any nonsingular linear transformation the volume of a set is multiplied by the absolute value of the determinant of the matrix of the transformation [see (4.2) of Chapter 2]. We showed in Section 3.1 of Chapter 2 that the determinant of the transformation $x \to xx(\varepsilon)$ equals $N(x(\varepsilon))$, that is, equals $N(\varepsilon) = \pm 1$.

As before, let ζ denote that mth root of 1 for which $\sigma_1(\zeta) = \cos(2\pi/m) + i \sin(2\pi/m)$. Consider the sets T_k $(k = 0, 1, \ldots, m-1)$, obtained from T by the linear transformation $x \to xx(\zeta^k)$ $(T_0 = T)$. By Lemma 2 we have $v(T_k) = v(T)$ (provided the volume of one of the sets exists). Since

$$|N(xx(\zeta^k))| = |N(x)N(\zeta^k)| = |N(x)|,$$

$$l(xx(\zeta^k)) = l(x) + l(\zeta^k) = l(x),$$

$$\arg\,(xx(\zeta^k))_1 = \arg x_1 + \frac{2\pi}{m} k,$$

then (by the definition of the fundamental domain X) the set T_k consists of all points $x \in \mathfrak{R}^n$ for which:

(1) $0 < |N(x)| \leqslant 1$.
(2) The coefficients in (1.13) satisfy $0 \leqslant \xi_i < 1$.
(3) $2\pi k/m \leqslant \arg x_1 < (2\pi/k)(k+1)$.

Thus $T_0, T_1, \ldots, T_{m-1}$ are pairwise-nonintersecting and their union $\bigcup_{k=0}^{m-1} T_k$ is defined by conditions (1) and (2) [without condition (3)].

Let $\overline{T}$ denote the set of all points $x \in \bigcup_{k=0}^{m-1} T_k$, for which $x_1 > 0, \ldots, x_s > 0$ [see (3.2) of Chapter 2]. We fix a set of s signs $\delta_1, \ldots, \delta_s$ ($\delta_i = \pm 1$). If we multiply all points in $\mathfrak{R}^n$ by the point $(\delta_1, \ldots, \delta_s; 1, \ldots, 1) \in \mathfrak{L}^{s,t} = \mathfrak{R}^n$, we obtain a volume-preserving linear transformation of $\mathfrak{R}^n$. If we apply all 2^s such linear transformations to $\overline{T}$, we obtain 2^s pairwise-nonintersecting sets whose union coincides with $\bigcup_{k=0}^{m-1} T_k$. If we can show that $\overline{T}$ has nonzero volume $\bar{v}$, then it will follow that T has a well-defined volume, which is given by

$$v(T) = \frac{2^s}{m}\bar{v}. \tag{1.14}$$

(For real quadratic fields $\overline{T}$ is that part of T which is contained in the first quadrant, and for imaginary quadratic fields $\overline{T}$ coincides with the unit disc minus the origin, see Figures 7 and 8.)

The vector equation (1.13) yields the following system:

$$l_j(x) = \frac{e_j}{n} \ln|N(x)| + \sum_{k=1}^{r} \xi_k l_j(\varepsilon_k) \qquad (j = 1, \ldots, s+t),$$

where $e_j = 1$ if $1 \leqslant j \leqslant s$, and $e_j = 2$ if $s+1 \leqslant j \leqslant s+t$. We change variables by the formulas

$$x_k = \rho_k \qquad (k = 1, \ldots, s),$$

$$\left.\begin{aligned} y_j &= \rho_{s+j} \cos \varphi_j \\ z_j &= \rho_{s+j} \sin \varphi_j \end{aligned}\right\} \qquad (j = 1, \ldots, t).$$

(Here the real numbers y_j and z_j are given by $x_{s+j} = y_j + iz_j$, $1 \leqslant j \leqslant t$, see Section 3.1 of Chapter 2.) The Jacobian of this transformation is easily computed to be $\rho_{s+1} \cdots \rho_{s+t}$. Since $l_j(x) = \ln \rho_j^{e_j}$ and $N(x) = \prod_{j=1}^{s+t} \rho_j^{e_j}$ (we assume that $x_1 > 0, \ldots, x_s > 0$), then in terms of the variables $\rho_1, \ldots, \rho_{s+t}$, $\varphi_1, \ldots, \varphi_t$, the set $\overline{T}$ is given by the conditions:

(1) $\rho_1 > 0, \ldots, \rho_{s+t} > 0$, $\prod_{j=1}^{s+t} \rho_j^{e_j} \leqslant 1$.
(2) In the equations

$$\ln \rho_j^{e_j} = \frac{e_j}{n} \ln\left(\prod_{i=1}^{s+t} \rho_i^{e_i}\right) + \sum_{k=1}^{r} \xi_k l_j(\varepsilon_k)$$

($j = 1, \ldots, x+t$) the coefficients ξ_k satisfy $0 \leqslant \xi_k < 1$ ($k = 1, \ldots, r$).

Since these conditions do not impose any restrictions on the variables $\varphi_1, \dots, \varphi_t$, then they independently take on all values in $[0, 2\pi)$. We now replace $\rho_1, \dots, \rho_{s+t}$ by the new variables $\xi, \xi_1, \dots, \xi_r$ by the formulas

$$\ln \rho_j^{e_j} = \frac{e_j}{n} \ln \xi + \sum_{k=1}^{r} \xi_k l_j(\varepsilon_k) \qquad (j = 1, \dots, s+t). \tag{1.15}$$

Adding these equations and noting that

$$\sum_{j=1}^{s+t} e_j = n, \qquad \sum_{j=1}^{s+t} l_j(\varepsilon_k) = 0, \tag{1.16}$$

we obtain

$$\xi = \prod_{j=1}^{s+t} \rho_j^{e_j}. \tag{1.17}$$

The set $\overline{T}$ is determined by the conditions

$$0 < \xi \leqslant 1, \quad 0 \leqslant \xi_k < 1 \qquad (k = 1, \dots, r).$$

It is now clear that the volume $\bar{v} = v(\overline{T})$ exists. Since

$$\frac{\partial \rho_j}{\partial \xi} = \frac{\rho_j}{n\xi}, \quad \frac{\partial \rho_j}{\partial \xi_k} = \frac{\rho_j}{e_j} l_j(\varepsilon_k),$$

the Jacobian of the transformation (1.17) equals

$$J = \begin{vmatrix} \dfrac{\rho_1}{n\xi} & \dfrac{\rho_1}{e_1} l_1(\varepsilon_1) & \cdots & \dfrac{\rho_1}{e_1} l_1(\varepsilon_r) \\ \cdot & \cdot & \cdot & \cdot \\ \dfrac{\rho_{s+t}}{n\xi} & \dfrac{\rho_{s+t}}{e_{s+t}} l_{s+t}(\varepsilon_1) & \cdots & \dfrac{\rho_{s+t}}{e_{s+t}} l_{s+t}(\varepsilon_r) \end{vmatrix}$$

$$= \frac{\rho_1 \cdots \rho_{s+t}}{n\xi 2^t} \begin{vmatrix} e_1 & l_1(\varepsilon_1) & \cdots & l_1(\varepsilon_r) \\ \cdot & \cdot & \cdot & \cdot \\ e_{s+t} & l_{s+t}(\varepsilon_1) & \cdots & l_{s+t}(\varepsilon_r) \end{vmatrix}.$$

In the last determinant we add all rows to the first row. Considering (1.16) and (1.17) and recalling the definition of the regulator R of the field K (see Section 4.4 of Chapter 2), we obtain

$$|J| = \frac{R}{2^t \rho_{s+1} \cdots \rho_{s+t}}.$$

It is now easy to compute the volume $\bar{v}$:

$$\begin{aligned}\bar{v} &= \int_{(T)}\cdots\int dx_1 \cdots dx_s dy_1 dz_1 \cdots dy_t dz_t \\ &= \int_{(T)}\cdots\int \rho_{s+1} \cdots \rho_{s+t}\, d\rho_1 \cdots d\rho_{s+t}\, d\varphi_1 \cdots d\varphi_t \\ &= \int_0^{2\pi} d\varphi_1 \cdots \int_0^{2\pi} d\varphi_t \int\cdots\int \rho_{s+1} \cdots \rho_{s+t}\, d\rho_1 \cdots d\rho_{s+t} \\ &= 2^t\pi^t \int\cdots\int |J| \rho_{s+1} \cdots \rho_{s+t}\, d\xi\, d\xi_1 \cdots d\xi_r \\ &= \pi^t R \int_0^1 d\xi \int_0^1 d\xi_1 \cdots \int_0^1 d\xi_r = \pi^t R.\end{aligned}$$

Substituting this value of $\bar{v}$ in (1.14) we finally obtain,

$$v(T) = \frac{2^s \pi^t R}{m}.$$

1.4. *Dirichlet's Principle*

We first consider the function $\zeta_K(s)$ when K is the rational field R. Since integral divisors in R can be identified with natural numbers and $N(n) = n$, then

$$\zeta_R(s) = \sum_{n=1}^{\infty} \frac{1}{n^s}. \tag{1.18}$$

Hence the Dedekind ζ-function for R coincides with the Riemann ζ-function $\zeta(s)$. We shall show that the series (1.18) converges for $s > 1$. Since the function $1/x^s$ is decreasing for $X > 0$, then

$$\int_n^{n+1} \frac{dx}{x^s} < \frac{1}{n^s} < \int_{n-1}^n \frac{dx}{x^s},$$

where the inequality on the left holds for $n \geqslant 1$, and that on the right for $n \geqslant 2$. Hence for any natural number $N > 1$ we have

$$\int_1^{N+1} \frac{dx}{x^s} < \sum_{n=1}^{N} \frac{1}{n^s} < 1 + \int_1^N \frac{dx}{x^s}.$$

Since the integral $\int_1^\infty (dx/x^s)$ converges for $s > 1$, the inequality on the right shows that the series (1.18) converges. Further, for $s > 1$, we have

$$\int_1^\infty \frac{dx}{x^s} < \zeta(s) < 1 + \int_1^\infty \frac{dx}{x^s},$$

or

$$\frac{1}{s-1} < \zeta(S) < 1 + \frac{1}{s-1}.$$

Multiplying this inequality by $s - 1$ and letting s tend to 1, we obtain

$$\lim_{s \to 1+0} (s-1)\zeta(s) = 1, \tag{1.19}$$

which indicates the order of growth of the function $\zeta(s)$ as $s \to 1$.

We now turn to the proof of a general theorem of Dirichlet on series.

Let X be a cone in the space $\mathfrak{R}^n$, and let $F(x)$ be a positive real-valued function on X. (We assume that the origin is not contained in X.) The function F on X is assumed to satisfy the following conditions:

(1) For any $x \in X$ and any real number $\xi > 0$, the equation $F(\xi x) = \xi^n F(x)$ holds.

(2) The set T, which consists of all points of X for which $F(x) \leqslant 1$, is bounded and has nonzero n-dimensional volume $v = v(T)$.

The points of the cone at which $F(x) = 1$ form a surface which intersects each ray of the cone in precisely one point, and which separates from the rest of the cone a bounded subset with nonzero volume. It is clear that the giving of such a surface in X is equivalent to the definition of such a function $F(x)$.

Let $\mathfrak{M}$ be an n-dimensional lattice in $\mathfrak{R}^n$ with the volume of a fundamental parallelepiped denoted by Δ. Consider the series

$$\bar{\zeta}(S) = \sum_{x \in \mathfrak{M} \cap X} \frac{1}{F(x)^s} \qquad (s > 1), \tag{1.20}$$

taken over all points x of the lattice $\mathfrak{M}$ which are contained in the cone X. This series depends on the cone X, the function F, and the lattice $\mathfrak{M}$.

Theorem 3. Under the assumptions given above, the series (1.20) converges for all $s > 1$ and

$$\lim_{s \to 1+0} (s-1)\bar{\zeta}(s) = \frac{v}{\Delta}. \tag{1.21}$$

Proof. For any real $r > 0$ we denote by $\mathfrak{M}_r$ the lattice which is obtained by contracting $\mathfrak{M}$ by a factor of r. The volume of a fundamental parallelepiped of $\mathfrak{M}$ is then given by Δ/r^n. If $N(r)$ is the number of points of the lattice $\mathfrak{M}_r$ which are contained in the set T, then by the definition of volume we have

$$v = v(T) = \lim_{r \to \infty} N(r)\frac{\Delta}{r^n} = \Delta \lim_{r \to \infty} \frac{N(r)}{r^n}. \tag{1.22}$$

Consider the set rT, obtained by expanding T by a factor of r. It is clear that

$N(r)$ also equals the number of points of the lattice $\mathfrak{M}$ contained in rT, and that this is equal to the number of points $x \in \mathfrak{M} \cap X$, for which $F(x) \leqslant r^n$. The points of $\mathfrak{M} \cap X$ can be arranged in a sequence $\{x_k\}$ so that

$$0 < F(x_1) \leqslant F(x_2) \leqslant \cdots \leqslant F(x_k) \leqslant \cdots.$$

Set

$$\sqrt[n]{F(x_k)} = r_k.$$

The points $x_1, \ldots, x_k$ belong to the set r_kT, so $N(r_k) \geqslant k$. But for any $\varepsilon > 0$, the point x_k does not belong to the set $(r_k - \varepsilon)T$, so $N(r_k - \varepsilon) < k$. Thus

$$N(r_k - \varepsilon) < k \leqslant N(r_k).$$

so that

$$\frac{N(r_k - \varepsilon)}{(r_k - \varepsilon)^n}\left(\frac{r_k - \varepsilon}{r_k}\right)^n < \frac{k}{r_k^{\ n}} \leqslant \frac{N(r_k)}{r_k^{\ n}}.$$

Taking the limit as $k \to \infty$, that is, as $r_k \to \infty$, and considering (1.22), we obtain

$$\lim_{k \to \infty} \frac{k}{F(x_k)} = \frac{v}{\Delta}. \tag{1.23}$$

We compare the series $\tilde{\zeta}(s) = \sum_{k=1}^{\infty} 1/F(x_k)^s$ with the series (1.18). Since $\lim_{k \to \infty} [k^s/F(x_k)^s] = (v/\Delta)^s \neq 0$, then along with the series (1.18) the series (1.20) also converges (if, of course, $s > 1$). Let ε be any positive number. By (1.23) we have

$$\left(\frac{v}{\Delta} - \varepsilon\right)\frac{1}{k} < \frac{1}{F(x_k)} < \left(\frac{v}{\Delta} + \varepsilon\right)\frac{1}{k}$$

for all k greater than some sufficiently large k_0. Hence

$$\left(\frac{v}{\Delta} - \varepsilon\right)^s \sum_{k=k_0}^{\infty} \frac{1}{k^s} < \sum_{k=k_0}^{\infty} \frac{1}{F(x_k)^s} < \left(\frac{v}{\Delta} + \varepsilon\right)^s \sum_{k=k_0}^{\infty} \frac{1}{k^s}$$

for all $s > 1$. We multiply this inequality by $s - 1$ and let s tend to 1 from the right. Since $\lim_{s \to 1}(s-1)\sum_{k=1}^{k_0-1}(1/k^s) = 0$, then by (1.19), $\lim_{s \to 1+0}(s-1) \sum_{k=k_0}^{\infty} (1/k^s) = 1$. Since also $\lim_{s \to 1}(s-1)\sum_{k=1}^{k_0-1}[1/F(x_k)^s] = 0$, we obtain the inequality

$$\frac{v}{\Delta} - \varepsilon \leqslant \varliminf_{s \to 1+0} (s-1)\tilde{\zeta}(s) \leqslant \varlimsup_{s \to 1+0} (s-1)\tilde{\zeta}(s) \leqslant \frac{v}{\Delta} + \varepsilon.$$

and since ε was arbitrary, this proves Theorem 3.

Remark. There is a certain similarity between (1.21) and (1.22). To make this similarity more precise, we assume that the volume Δ of a fundamental parallelepiped of the lattice $\mathfrak{M}$ is equal to 1, and we write them in the form

$$\lim_{s\to 1+0}(s-1)\tilde{\zeta}(s)=v, \tag{1.21'}$$

$$\lim_{r\to\infty}\frac{1}{r^n}N(r)=v. \tag{1.22'}$$

Both limits have the same value, the volume of the set T. The volume is determined in (1.22') in the following way. The lattice $\mathfrak{M}$ is shrunk by a factor of r, and the number $N(r)$ of points of $\mathfrak{M}_r$ which are contained in T is determined. Then the number $N(r)$ is multiplied by the volume $1/r^n$ of a fundamental parallelepiped of the lattice $\mathfrak{M}_n$, and finally we pass to the limit as $r\to\infty$. The same idea is involved in (1.21'). Here the sum $\tilde{\zeta}(s)$ plays the role of the number $N(r)$, and the factor $(s-1)$ corresponds to the factor $1/r^n$. We take the limit as $s\to 1$ from the right instead of as $r\to\infty$.

Turning to the fundamental domain X of an algebraic number field K, we see that the function $F(x)=|N(x)|$ satisfies conditions (1) and (2). Hence we may apply Theorem 3 to the series (1.8), and this means that it converges for $s>1$ and that relation (1.9) holds.

We have now proved all assertions which were used in the first paragraph, and hence have completed the proof of Theorem 2.

1.5. *Euler's Identity*

To use the formula (1.2) to compute the number h, we must have some other method for determining the limit $\lim_{s\to 1}(s-1)\zeta_K(s)$. In some cases this can be done by using the representation of $\zeta_K(s)$ as an infinite product, known as Euler's identity.

Theorem 4. For $s>1$, the function $\zeta_K(s)$ can be represented as a convergent infinite product

$$\zeta_K(s)=\prod_{\mathfrak{p}}\frac{1}{1-[1/N(\mathfrak{p})^s]},$$

where $\mathfrak{p}$ runs through all prime divisors of the field K.

Proof. For every prime divisor $\mathfrak{p}$ we have

$$\frac{1}{1-[1/N(\mathfrak{p})^s]}=1+\frac{1}{N(\mathfrak{p})^s}+\frac{1}{N(\mathfrak{p})^{2s}}+\cdots. \tag{1.24}$$

Let N be any natural number, and $\mathfrak{p}_1,\dots,\mathfrak{p}_r$ all prime divisors with norms not exceeding N. Multiplying the absolutely convergent series (1.24) for $\mathfrak{p}=\mathfrak{p}_1,\dots,\mathfrak{p}_r$, we obtain

$$\prod_{N(\mathfrak{p})\leqslant N}\left(1-\frac{1}{N(\mathfrak{p})^s}\right)^{-1}=\sum_{k_1,\dots,k_r=0}^{\infty}\frac{1}{N(\mathfrak{p}_1^{k_1}\cdots\mathfrak{p}_r^{k_r})^s}=\sum_{\mathfrak{a}}{}'\frac{1}{N(\mathfrak{a})^s},$$

where in the sum $\sum'$, $\mathfrak{a}$ runs through all integral divisors of the field K which are not divisible by a prime divisor with norm exceeding N. Comparing the series $\sum'$ with the series $\zeta_K(s) = \sum 1/N(\mathfrak{a})^s$, we obtain

$$\left| \prod_{N(\mathfrak{p}) \leqslant N} \left(1 - \frac{1}{N(\mathfrak{p})^s}\right)^{-1} - \zeta_K(s) \right| < \sum_{N(\mathfrak{a}) > N} \frac{1}{N(\mathfrak{a})^s}$$

since the series $\sum'$ contains all terms corresponding to integral divisors with norm $\leqslant N$. Since for $s > 1$ the series (1.1) converges, then

$$\sum_{N(\mathfrak{a}) > N} \frac{1}{N(\mathfrak{a})^s} \to 0$$

as $N \to \infty$, and this proves the theorem.

Theorem 4 will be valuable because, along with Theorem 2, it establishes a connection between the number h and prime divisors of the field K. As we remarked in Section 1.1, if we know all prime divisors of the field K, then using Theorem 4, the left side of (1.2) can be computed, and this allows us to obtain a formula for h. On the other hand, since $\kappa h \neq 0$, Theorem 4 gives an important property of prime divisors in a field K. For example, we shall use it in Section 3 to obtain the celebrated theorem of Dirichlet on the distribution of rational prime numbers in arithmetic progressions.

PROBLEMS

1. Use the convergence of the series $\sum_{n=1}^{\infty} (1/n^s)$ $(s > 1)$ to show that when $s > 1$, the series

$$\sum_{\mathfrak{p}} \frac{1}{N(\mathfrak{p})^s},$$

where $\mathfrak{p}$ runs through all prime divisors of the field K, also converges.

2. Use Problem 1 to prove the convergence of the product

$$\prod_{\mathfrak{p}} \frac{1}{1 - [1/N(\mathfrak{p})^s]} \qquad (s > 1).$$

Deduce that the series

$$\sum_{\mathfrak{a}} \frac{1}{N(\mathfrak{a})^s}$$

converges.

3. Let a_k and b_k $(k \geqslant 1)$ be positive real numbers with $\lim_{k \to \infty} b_k/a_k = c$. Show that if the series $\sum_{k=1}^{\infty} a_k^s$ converges for $s > 1$ and $\lim_{s \to 1+0} (s-1)\sum_{k=1}^{\infty} a_k^s = A$, then the series $\sum_{k=1}^{\infty} b_k^s$ also converges for $s > 1$ and

$$\lim_{s \to 1+0} (s-1) \sum_{k=1}^{\infty} b_k^s = cA.$$

4. Let C be any divisor class of the algebraic number field K. Denote by $Z(\xi, C)$ the number of integral divisors $\mathfrak{a}$ of the class C for which $N(\mathfrak{a}) \leqslant \xi$. Show that

$$\lim_{\xi \to \infty} \frac{Z(\xi, C)}{\xi} = x = \frac{2^{s+t}\pi^t R}{m\sqrt{|D|}} .$$

5. Let $\psi(a)$ denote the number of integral divisors of the algebraic number field K with norm a. Show that

$$\frac{\zeta_K(s)}{\zeta(s)} = \sum_{n=1}^{\infty} \frac{c_n}{n^s},$$

where

$$c_n = \sum_{d \mid n} \mu(d)\psi\left(\frac{n}{d}\right)$$

[$\mu(a)$ is the Möbius function].

2. The Number of Divisor Classes of Cyclotomic Fields

Let m be a natural number, and let ζ be a primitive mth root of 1. Since the mth roots of 1 in the complex plane divide the unit circle into m equal parts, the field $R(\zeta)$ is called the mth cyclotomic (circle-dividing) field. In this section we shall use Theorems 2 and 4 of Section 1 to find a formula for h, the number of divisor classes in a cyclotomic field. To do this we must determine the factorization of rational primes into prime divisors in such fields. We first determine the degree of the field $R(\zeta)$.

2.1. *Irreducibility of the Cyclotomic Polynomial*

The degree of the field $R(\zeta)$ equals the degree of the minimum polynomial of the number ζ over the rational field R. In this section we shall show that the minimum polynomial of ζ is the polynomial

$$\Phi_m = \Phi_m(t) = \prod_{(k,\, m)=1} (t - \zeta^k)$$

(the product is taken over the indicated residue classes modulo m), which has as roots all primitive mth roots of 1. Since the degree of Φ_m equals the value of the Euler function $\varphi(m)$, it will follow that $(R(\zeta) : R) = \varphi(m)$.

The polynomial $\Phi_m(t)$ is called the mth cyclotomic polynomial.

We first show that the coefficients of $\Phi_m(t)$ are rational integers. For $m = 1$, this is clear ($\Phi_1 = t - 1$). We proceed by induction on m. Since every mth root of 1 is a primitive root of some degree $d \mid m$, then

$$t^m - 1 = \prod_d \Phi_d ,$$

where d runs through all divisors of the number m. By the induction assumption the polynomial $F = \prod_{d \neq m} \Phi_d$ has rational integral coefficients and its leading coefficient is 1. Hence the coefficients of $\Phi_m = (t^m - 1)/F$ are also rational integers.

As usual, Z denotes the ring of rational integers, Z_p the field of residue classes modulo the prime number p, and for $a \in Z$ we denote the corresponding residue class in Z_p by $\bar{a}$. If $f(t)$ is a polynomial with rational integer coefficients, we denote by $\bar{f}(t)$ the polynomial obtained from f by replacing all coefficients by their residue classes modulo p. It is clear that the mapping $f \to \bar{f}$ is a homomorphism of the ring $Z[t]$ onto the ring $Z_p[t]$. Since $(\bar{f} + \bar{g})^p = \bar{f}^p + \bar{g}^p$, and $\bar{a}^p = \bar{a}$, then in the ring $Z_p[t]$ we have the formula

$$(\bar{f}(t))^p = \bar{f}(t^p). \tag{2.1}$$

Set $h = t^m - 1$. If p does not divide m, then the polynomial $\bar{h}$ of $Z_p[t]$ is relatively prime to its derivative and hence has distinct roots. Noting that $\bar{\Phi}_m$ divides $\bar{h}$, we have the following assertion.

Lemma 1. If the prime number p does not divide m, then the polynomial $\bar{\Phi} \in Z_p[t]$ has no multiple roots.

If $f(t)$ is the minimum polynomial of ζ, then $\Phi_m = fG$, where G and f both belong to the ring $Z[t]$. If p is any prime not dividing m, then ζ^p is a primitive mth root of 1; that is, $\Phi_m(\zeta^p) = 0$. We shall show that ζ^p is a root of f. Otherwise we would have $G(\zeta^p) = 0$. Then consider the polynomial $H(t) = G(t^p)$. Since $H(\zeta) = G(\zeta^p) = 0$, then H is divisible by f, that is, $H = fQ$, where $Q \in Z[t]$. Passing to Z_p we obtain $\bar{H} = \bar{f}\bar{Q}$. But by (2.1), $\bar{H}(t) = \bar{G}(t^p) = (\bar{G}(t))^p$, so that

$$\bar{G}^p = \bar{f}\bar{Q}.$$

Let $\bar{\psi}$ be any irreducible factor of $\bar{f}$ (in the ring $Z_p[t]$). It follows from the last equation that $\bar{G}$ is divisible by $\bar{\psi}$. But then it follows from $\bar{\Phi}_m = \bar{f}\bar{G}$ that $\bar{\Phi}_m$ is divisible by $\bar{\psi}^2$, contradicting Lemma 1. Thus ζ^p cannot be a root of $G(t)$, and hence is a root of $f(t)$.

If ζ' is any root of Φ_m, then $\zeta' = \zeta^k$, where k is relatively prime to m. Let $k = p_1 p_2 \cdots p_s$. We have just shown that ζ^{p_1} is a root of $f(t)$. Analogously, replacing ζ by ζ^{p_1}, we find that $\zeta^{p_1 p_2}$ is a root of $f(t)$. Continuing this process, we find that ζ^k is a root of $f(t)$.

We have shown that any root of Φ_m is also a root of f, and hence $\Phi_m = f$. We formulate this result in the following theorem.

Theorem 1. For any natural number m, the cyclotomic polynomial Φ_m is irreducible over the field of rational numbers.

Corollary. The degree of the mth cyclotomic field is $\varphi(m)$ [where $\varphi(m)$ is Euler's function].

2.2. *Decomposition of Primes in Cyclotomic Fields*

Since the mth cyclotomic field $R(\zeta)$ has degree $\varphi(m)$, then the numbers

$$1, \zeta, \dots, \zeta^{\varphi(m)-1} \tag{2.2}$$

form a basis for $R(\zeta)$ over R.

Lemma 2. If the prime number p does not divide m, then it does not divide the discriminant $D = D(1, \zeta, \dots, \zeta^{\varphi(m)-1})$ of the basis (2.2).

Proof. The discriminant D equals the discriminant $D(\Phi_m)$ of the cyclotomic polynomial Φ_m. The residue class $\overline{D(\Phi_m)} \in Z_p$ of the number $D(\Phi_m)$ clearly coincides with the discriminant $D(\overline{\Phi}_m)$ of the polynomial $\overline{\Phi}_m \in Z_p[t]$. But $\overline{\Phi}_m(t)$ has no multiple roots (Lemma 1), and hence $D(\overline{\Phi_m}) \neq 0$, which means that $D = D(\Phi_m)$ is not divisible by p.

Lemma 3. If the algebraic number field K contains a primitive mth root of 1 and $\mathfrak{p}$ is any prime divisor of K which is relatively prime to m, then

$$N(\mathfrak{p}) \equiv 1 \pmod{m}.$$

Proof. Let $\mathfrak{O}$ be the ring of integers of K, p the rational prime which is divisible by $\mathfrak{p}$, and ζ a primitive mth root of 1 ($\zeta \in \mathfrak{O}$). In Section 2.1 we saw that the polynomial $t^m - 1$ has no multiple roots in the extension field $\mathfrak{O}/\mathfrak{p}$ of Z_p (since $p \nmid m$). Hence the residue classes $1, \tilde{\zeta}, \dots, \tilde{\zeta}^{m-1}$ of $\mathfrak{O}/\mathfrak{p}$ are pairwise-distinct. These classes form a group under multiplication, a subgroup of the multiplicative group of the field $\mathfrak{O}/\mathfrak{p}$. But the order of this group is $N(\mathfrak{p}) - 1$, and since the order of a subgroup divides the order of the group, m divides $N(\mathfrak{p}) - 1$. The lemma is proved.

Theorem 2. If p is a prime number not dividing m, let f be the smallest natural number such that $p^f \equiv 1 \pmod{m}$, and set $g = \varphi(m)/f$. Then the prime p has the factorization

$$p = \mathfrak{p}_1 \cdots \mathfrak{p}_g, \tag{2.3}$$

in the mth cyclotomic field, where the prime divisors $\mathfrak{p}_1, \dots, \mathfrak{p}_g$ are distinct and $N(\mathfrak{p}_i) = p^f$.

Proof. Since $(p, m) = 1$, then by Lemma 2, p does not divide the discriminant of the basis (2.2). It now follows from Theorem 8 of Section 5, Chapter 3, that p has a decomposition of the type (2.3). We need only determine the degree of each prime divisor $\mathfrak{p}_i$, and show that there are $\varphi(m)/f$ of them.

Let $\mathfrak{p}$ be any of the prime divisors $\mathfrak{p}_i$ and let s be its degree, so that $N(\mathfrak{p}) = p^s$. By Lemma p^s 3, $\equiv 1 \pmod{m}$, and hence $s \geqslant f$. To prove the opposite inequality, we consider the residue class field $\mathfrak{O}/\mathfrak{p}$, where $\mathfrak{O}$ is the ring of integers of the field $R(\zeta)$. By the corollary of the lemma in Section 7.4 of Chapter 3 every residue class of $\mathfrak{O}/\mathfrak{p}$ contains a representative of the form

$$\xi = \sum_{j=0}^{\varphi(m)-1} a_j \zeta^j, \tag{2.4}$$

where a_j are rational integers. We raise (2.4) to the p^fth power. Since $p^f \equiv 1 \pmod{m}$, then $\zeta^{p^f} = \zeta$. But also $(\alpha + \beta)^{p^f} \equiv \alpha^{p^f} + \beta^{p^f} \pmod{\mathfrak{p}}$ for any α and β in $\mathfrak{O}$, and thus $a^{p^f} \equiv a \pmod{\mathfrak{p}}$ for any rational integer a. Hence from (2.4) we obtain the congruence

$$\xi^{p^f} \equiv \xi \pmod{\mathfrak{p}}.$$

Thus any residue class $\bar{\xi} \in \mathfrak{O}/\mathfrak{p}$ is a root of the polynomial $t^{p^f} - t$. But in any field the number of roots of a polynomial does not exceed its degree, so $p^s \leqslant p^f$, and $s \leqslant f$. Hence we have $s = f$.

We have shown that all prime divisors $\mathfrak{p}$ in (2.3) have the same degree f, which is equal to the order of the number p modulo m. Now applying Theorem 8 of Section 5 of Chapter 3, we see that the number of prime divisors $\mathfrak{p}_i$ equals $\varphi(m)/f$. Theorem 2 is proved.

2.3. *The Expression of h in Terms of L-Series*

Let $\zeta_K(s)$ be the ζ-function of the mth cyclotomic field $K = R(\zeta)$, $\zeta^m = 1$. If we group together those terms in Euler's identity which involve the prime divisors $\mathfrak{p}$ of each rational prime p, we obtain

$$\zeta_K(s) = \prod_p \prod_{\mathfrak{p}|p} \frac{1}{1 - [1/N(\mathfrak{p})^s]} \tag{2.5}$$

(the product being taken over all rational primes p). Only a finite number of terms correspond to prime divisors $\mathfrak{p}$ which divide m. We denote the product of these terms by

$$G(s) = \prod_{\mathfrak{p}|m} \left(1 - \frac{1}{N(\mathfrak{p})^s}\right)^{-1}. \tag{2.6}$$

If $(p, m) = 1$ and $\mathfrak{p}$ is any prime divisor of p, then $N(\mathfrak{p}) = p^{f_p}$, where f_p is the order of the number p modulo m. Since the number of distinct $\mathfrak{p}$ dividing p is $\varphi(m)/f_p$ (Theorem 2), then

$$\zeta_K(s) = G(s) \prod_{(p,m)=1} \left(1 - \frac{1}{p^{f_p s}}\right)^{-\varphi(m)/f_p} \tag{2.7}$$

Each factor in this product can be put in more convenient form. We use the expansion

$$1 - \left(\frac{1}{p^s}\right)^{f_p} = \prod_{k=0}^{f_p-1} \left(1 - \frac{\varepsilon^k}{p^s}\right), \tag{2.8}$$

where $\varepsilon = \varepsilon_p = \cos(2\pi/f_p) + i \sin(2\pi/f_p)$. Then the product

$$\prod_{k=0}^{f_p-1}\left(1 - \frac{\varepsilon_p{}^k}{p^s}\right)^{-\varphi(m)/f_p}$$

has $\varphi(m)$ terms, and the number of terms is independent of p. We shall show that the products, corresponding to different p, can be associated in such a fashion that the infinite product (2.7) will factor into $\varphi(m)$ products, each having a simple form. We use here the concept of a character modulo m, and the results on characters obtained in Section 5 of the Supplement.

Let G_m denote the group of residue classes of rational integers modulo m, which consists of all classes of numbers relatively prime to m. The class $\bar{p} \in G_m$ which contains p has order f_p. Hence, if χ is any character of the group G_m, the value of $\chi(\bar{p})$, which is an f_pth root of 1, must coincide with some ε^k. Conversely, if we take any root ε^k, then there is one and only one character χ_1 of the cyclic subgroup $\{\bar{p}\}$ of G_m such that $\chi_1(\bar{p}) = \varepsilon^k$. By Theorem 3 of Section 5 of the Supplement this character can be extended in $\varphi(m)/f_p$ ways to a character of the group G_m. Thus as χ runs through all characters of the group G_m, $\chi(\bar{p})$ takes on each value ε^k ($k = 0, 1, \ldots, f_p - 1$) precisely $\varphi(m)/f_p$ times. Now we may substitute (2.8) in (2.7) and obtain

$$\zeta_K(s) = G(s) \prod_{(p,m)=1} \prod_{\chi}\left(1 - \frac{\chi(\bar{p})}{p^s}\right)^{-1} \tag{2.9}$$

(the second product being over all characters χ of the group G).

In the place of characters of the group G_m, we may consider numerical characters modulo m (see Section 5.3 of the Supplement). If χ is a numerical character modulo m, and p is a prime which divides m, then $\chi(p) = 0$, and hence (2.9) takes the form

$$\zeta_K(s) = G(s) \prod_p \prod_{\chi}\left(1 - \frac{\chi(p)}{p^s}\right)^{-1}$$

(here p runs through all prime numbers and χ runs through all numerical characters modulo m). Reversing the order of multiplication, we arrive at the formula

$$\zeta_K(s) = G(s) \prod_{\chi} L(s, \chi), \tag{2.10}$$

where

$$L(s, \chi) = \prod_p \frac{1}{1 - [\chi(p)/p^s]}. \tag{2.11}$$

Note that all the products converge for $s > 1$, and hence all the operations made on infinite products are easily justified.

Remark. In formula (2.10) the term $G(s)$ can be dropped if we let χ run through all primitive characters modulo d, for all d which divide m; see Problems 13 to 16.

The term $L(s, \chi_0)$ in the product (2.10), which corresponds to the unit character χ_0, differs only slightly from the Riemann ζ-function $\zeta(s)$. Since $\chi_0(p) = 1$ for $(p, m) = 1$ and $\chi_0(p) = 0$ for $(p, m) > 1$, then

$$L(s, \chi_0) = \prod_{(p, m)=1} \frac{1}{1 - (1/p^s)} \qquad (s > 1).$$

On the other hand, applying Theorem 4 of Section 1 to the rational field R, we obtain

$$\zeta(s) = \prod_p \frac{1}{1 - (1/p^s)}.$$

Thus

$$L(s, \chi_0) = \left(\prod_{p|m} \frac{1}{1 - (1/p^s)}\right)^{-1} \zeta(s).$$

Substituting this expression in (2.10), we obtain the following formula for $\zeta_K(s)$:

$$\zeta_K(s) = F(s)\zeta(s) \prod_{\chi \neq \chi_0} L(s, \chi) \qquad (s > 1), \tag{2.12}$$

where [see (2.6)]

$$F(s) = \prod_{\mathfrak{p}|m} \left(1 - \frac{1}{N(\mathfrak{p})^s}\right)^{-1} \cdot \prod_{p|m} \left(1 - \frac{1}{p^s}\right).$$

We now simplify the functions $L(s, \chi)$. Since the series $\sum \chi(n)/n^s$ converges absolutely for $s > 1$, we obtain, as in (1.24), the equation

$$\frac{1}{1 - [\chi(p)/p^s]} = \sum_{k=0}^{\infty} \left(\frac{\chi(p)}{p^s}\right)^k.$$

By an almost verbatim repetition of the proof of Theorem 4 of Section 1 (using only the multiplicative property of the character χ), we easily find that

$$L(s, \chi) = \sum_{n=1}^{\infty} \frac{\chi(n)}{n^s} \qquad (s > 1). \tag{2.13}$$

The series on the right in (2.13) is called the L-series or the Dirichlet series for the numerical character χ. Our first goal is now to show that the L-series of a nonunit character converges not only for $s > 1$, but even for $s > 0$ (however, convergence in the interval $0 < s \leqslant 1$ will be nonabsolute). For this we prove the following lemma.

Lemma 4. Let the sequence of complex numbers $\{a_n\}$ $(n = 1, 2, \ldots)$ be such that the sums $A_n = \sum_{k=1}^{n} a_k$ are bounded; that is, $|A_n| \leqslant C$ for all $n \geqslant 1$. Then the series

$$f(s) = \sum_{n=1}^{\infty} \frac{a_n}{n^s}$$

converges for all real $s > 0$. For any $\sigma > 0$, convergence is uniform in the interval $[\sigma, \infty)$, so that the sum $f(s)$ is continuous in s.

Proof. Fix $\sigma > 0$. For any $\varepsilon > 0$ we can pick n_0 so that $1/n^\sigma < \varepsilon$ for all $n > n_0$. For all such $n > n_0$, we also have $1/n^s < \varepsilon$, provided that $s \geqslant \sigma$. Let $M > N > n_0$. Then

$$\sum_{k=N}^{M} \frac{a_k}{k^s} = \sum_{k=N}^{M} \frac{A_k - A_{k-1}}{k^s} = \sum_{k=N}^{M} \frac{A_k}{k^s} - \sum_{k=N-1}^{M-1} \frac{A_k}{(k+1)^s}$$

$$= -\frac{A_{N-1}}{N^s} + \sum_{k=N}^{M-1} A_k\left(\frac{1}{k^s} - \frac{1}{(k+1)^s}\right) + \frac{A_M}{M^s},$$

so that

$$\left|\sum_{k=N}^{M} \frac{a_k}{k^s}\right| \leqslant \frac{C}{N^s} + C\sum_{k=N}^{M-1}\left(\frac{1}{k^s} - \frac{1}{(k+1)^s}\right) + \frac{C}{M^s} = \frac{2C}{N^s} < 2C\varepsilon$$

for all s in the interval $[\sigma, \infty)$. Lemma 4 is proved.

Corollary. If χ is a nonunit character, the series $L(s, \chi)$ converges for $s > 0$ and represents a continuous function on the interval $(0, \infty)$.

For if $\chi \neq \chi_0$, then $\sum \chi(k) = 0$, where k runs through a complete set of residues modulo m. Representing the natural number n in the form $n = mq + r$, $0 \leqslant r < m$, we find $A_n = \sum_{k=1}^{n} \chi(k) = \sum_{k=1}^{r} \chi(k)$, so that $|A_n| \leqslant r < m$.

Turning to the function $\zeta_K(s)$, we multiply (2.12) by $s - 1$ and take the limit as $s \to 1$ from above. By (1.19) we find that

$$\lim_{s \to 1+0} (s-1)\zeta_K(s) = F(1) \prod_{\chi \neq \chi_0} L(1, \chi), \tag{2.14}$$

where

$$L(1, \chi) = \sum_{n=1}^{\infty} \frac{\chi(n)}{n}. \tag{2.15}$$

Note that since the series (2.15) does not converge absolutely, we must keep in mind that its terms appear in order of increasing n. Comparing (2.14) with Theorem 2 of Section 1, we obtain the following formula:

$$h = \frac{w\sqrt{|D|}}{2^{s+t}\pi^t R} F(1) \prod_{\chi \neq \chi_0} L(1, \chi) \tag{2.16}$$

(here w denotes the number of roots of 1 contained in K). The expression (2.16) for the number of divisor classes of a cyclotomic field is not definitive, since it still contains the infinite series $L(1, \chi)$. The summation of these series will be carried out in the next section.

2.4. *Summation of the Series* $L(1, \chi)$.

Assuming that χ is a nonunit character modulo m, we turn to the series (2.13). Omitting those summands which are zero and noting that if $n_1 \equiv n_2 \pmod{m}$, then $\chi(n_1) = \chi(n_2)$, we arrive at the following form (valid for $s > 1$):

$$L(s, \chi) = \sum_{(x, m)=1} \chi(x) \sum_{n \equiv x \pmod{m}} \frac{1}{n^s}.$$

The inner series can be written in the form

$$\sum_{n=1}^{\infty} \frac{c_n}{n^s},$$

where

$$c_n = \begin{cases} 1 & \text{for } n \equiv x \pmod{m}, \\ 0 & \text{for } n \not\equiv x \pmod{m}. \end{cases}$$

To find a convenient way of writing the coefficients c_n, we consider the following formula:

$$\sum_{k=0}^{m-1} \zeta^{rk} = \begin{cases} m & \text{for } r \equiv 0 \pmod{m}, \\ 0 & \text{for } r \not\equiv 0 \pmod{m}, \end{cases}$$

where

$$\zeta = \cos \frac{2\pi}{m} + i \sin \frac{2\pi}{m}$$

is a primitive mth root of 1. We stress that when considering the algebraic properties of a cyclotomic field, it does not matter which primitive mth root of 1 is denoted by ζ, but for analytic computations we must fix a definite complex root. Hence we have

$$c_n = \frac{1}{m} \sum_{k=0}^{m-1} \zeta^{(x-n)k}.$$

Thus

$$\begin{aligned} L(s, \chi) &= \sum_{(x,m)=1} \chi(x) \sum_{n=1}^{\infty} \frac{1}{m} \sum_{k=0}^{m-1} \zeta^{(x-n)k} \frac{1}{n^s} \\ &= \frac{1}{m} \sum_{k=0}^{m-1} \left(\sum_{(x,m)=1} \chi(x) \zeta^{xk} \right) \sum_{n=1}^{\infty} \frac{\zeta^{-nk}}{n^s}. \end{aligned}$$

We have already encountered the expression in parentheses in the case $m = p$ in Section 2 of Chapter 1, where it was called a Gaussian sum. We now define Gaussian sums for arbitrary m.

Definition. Let ζ be a fixed primitive mth root of 1, and let χ be a numerical character modulo m. The expression

$$\tau_a(\chi) = \sum_{x \bmod m} \chi(x)\zeta^{ax},$$

where x runs through a full (or a reduced) system of residues modulo m, is called the Gaussian sum corresponding to the character χ and the rational integer a.

The Gaussian sum $\tau_a(\chi)$ depends not only on χ and the residue of a modulo m, but also on the choice of the root ζ. In the future we shall always assume that ζ is $\cos(2\pi/m) + i\sin(2\pi/m)$. The Gaussian sum with this choice of ζ is called *normed*.

The sum $\tau_1(\chi)$ will also be denoted by $\tau(\chi)$.

If χ is not the unit character, then

$$\tau_0(\chi) = \sum_{(x,m)=1} \chi(x) = 0.$$

Hence our expression for $L(s, \chi)$ can be written in the form

$$L(s, \chi) = \frac{1}{m} \sum_{k=1}^{m-1} \tau_k(\chi) \sum_{n=1}^{\infty} \frac{\zeta^{-nk}}{n^s}.$$

We can apply Lemma 4 to the series $\sum_{n=1}^{\infty}(\zeta^{-nk}/n^s)(\zeta^{-k} \neq 1$ for $k \neq 0$, so $\sum_{n=1}^{mr} \zeta^{-nk} = 0)$. By this lemma our series converges for $0 < s < \infty$ and represents a continuous function of s. Hence we may set $s = 1$ in this last equation and obtain

$$L(1, \chi) = \frac{1}{m} \sum_{k=1}^{m-1} \tau_k(\chi) \sum_{n=1}^{\infty} \frac{\zeta^{-nk}}{n}.$$

To find the sum of the inner series, we turn to the power series $\sum_{n=1}^{\infty}(z^n/n)$. It is well known that it converges for $|z| < 1$ and represents there that branch of the function $-\ln(1 - z)$, the imaginary part of which (that is, the coefficient of i) is contained in the interval $(-\pi/2, \pi/2)$. Since this series also converges at the point $z = \zeta^{-k}$ (on the unit circle), then by Abel's theorem

$$\sum_{n=1}^{\infty} \frac{\zeta^{-nk}}{n} = -\ln(1 - \zeta^{-k}),$$

and hence

$$L(1, \chi) = -\frac{1}{m} \sum_{k=1}^{m-1} \tau_k(\chi) \ln(1 - \zeta^{-k}). \tag{2.17}$$

Hence we have obtained a finite expression for the series $L(1, \chi)$. Substituting it in (2.16), we obtain a formula for the number of divisor classes of a cyclotomic field which does not contain any infinite series.

The formula (2.17) can be further investigated and considerably simplified. In the next section we shall do this, but only for the case when χ is a primitive character. In Section 5 we shall apply these results to the further study of the formula for h in the case of the lth cyclotomic field, with l a prime. In this case the formula has particularly important applications.

2.5. *The Series $L(1, \chi)$ for Primitive Characters*

We shall show that if χ is a primitive character modulo m and $(a, m) = r > 1$, then

$$\tau_a(\chi) = 0.$$

Set $m = rd$. It is clear that ζ^a is a primitive dth root of 1, and therefore $\zeta^{az} = \zeta^a$, provided that $z \equiv 1 \pmod{d}$. We take as z a number for which $(z, m) = 1$, $z \equiv 1 \pmod{d}$ and $\chi(z) \neq 1$ (the existence of such a d is guaranteed by Theorem 4 of Section 5 of the Supplement). As x runs through a complete system of residues modulo m so does zx, so that

$$\tau_a(\chi) = \sum_{x \bmod m} \chi(zx)\zeta^{azx} = \chi(z) \sum_{x \bmod m} \chi(x)\zeta^{ax} = \chi(z)\tau_a(\chi).$$

Since $\chi(z) \neq 1$, it follows that $\tau_a(\chi) = 0$.

Further, if $(a, m) = 1$, then

$$\tau_a(\chi) = \chi(a)^{-1}\tau(\chi).$$

Indeed, as x runs through a complete system of residues modulo m, so does ax and thus

$$\chi(a)\tau_a(\chi) = \sum_{x \bmod m} \chi(ax)\zeta^{ax} = \tau_1(\chi) = \tau(\chi).$$

Hence if χ is a primitive character we can write (2.17) in the form

$$L(1, \chi) = -\frac{\tau(\chi)}{m} \sum_{(k,m)=1} \bar{\chi}(k)\ln(1 - \zeta^{-k}). \tag{2.18}$$

We turn to the study of the sum

$$S_\chi = \sum_{(k,m)=1} \bar{\chi}(k)\ln(1 - \zeta^{-k}) \tag{2.19}$$

(k running through a reduced system of residues modulo m). The study of the sum S_χ leads to two essentially different types of behavior. To distinguish these cases we need the following definition.

Definition. The numerical character χ is called *even* if $\chi(-1) = 1$ [and hence $\chi(-x) = \chi(x)$ for all integers x], and is called *odd* if $\chi(-1) = -1$ [and $\chi(-x) = -\chi(x)$].

Since

$$(\chi(-1))^2 = \chi((-1)^2) = \chi(1) = 1,$$

then $\chi(-1) = \pm 1$, and thus every character χ is either even or odd.

The number $1 - \zeta^{-k}$ (for $0 < k < m$) can be represented as

$$1 - \zeta^{-k} = 2 \sin \frac{\pi k}{m}\left(\cos\left(\frac{\pi}{2} - \frac{\pi k}{m}\right) + i \sin\left(\frac{\pi}{2} - \frac{\pi k}{m}\right)\right),$$

where $-\pi/2 < \pi/2 - \pi k/m < \pi/2$; therefore

$$\ln(1 - \zeta^{-k}) = \ln|1 - \zeta^{-k}| + i\pi\left(\frac{1}{2} - \frac{k}{m}\right).$$

Further, since $1 - \zeta^{-k}$ and $1 - \zeta^k$ are conjugate, then

$$\ln(1 - \zeta^k) = \ln|1 - \zeta^k| - i\pi\left(\frac{1}{2} - \frac{k}{m}\right).$$

(Note that the last two formulas are valid only when k lies between 0 and m.)

Now assume that the character χ (and hence also $\bar{\chi}$) is even. Interchanging k and $-k$ in (2.19), we obtain

$$S_\chi = \sum_{(k,m)=1} \bar{\chi}(k)\ln(1 - \zeta^k),$$

and with (2.19) this yields

$$\begin{aligned} 2S_\chi &= \sum_{(k,m)=1} \bar{\chi}(k)[\ln(1 - \zeta^{-k}) + \ln(1 - \zeta^k)] \\ &= 2 \sum_{(k,m)=1} \bar{\chi}(k)\ln|1 - \zeta^k| = 2 \sum_{\substack{(k,m)=1 \\ 0<k<m}} \bar{\chi}(k)\ln 2 \sin \frac{\pi k}{m}. \end{aligned}$$

If the character χ is odd, then when we interchange k and $-k$ in (2.19) we obtain

$$S_\chi = -\sum_{(k,m)=1} \bar{\chi}(k)\ln(1 - \zeta^k),$$

so that

$$\begin{aligned} 2S_\chi &= \sum_{(k,m)=1} \bar{\chi}(k)[\ln(1 - \zeta^{-k}) - \ln(1 - \zeta^k)] \\ &= 2 \sum_{\substack{(k,m)=1 \\ 0<k<m}} \bar{\chi}(k)\pi i\left(\frac{1}{2} - \frac{k}{m}\right). \end{aligned}$$

Since $\sum_{(k,m=1)} \bar{\chi}(k) = 0$ ($\bar{\chi}$ is not the unit character), then we obtain the following result from (2.18).

Theorem 3. Let χ be a primitive character with modulus $m > 1$. If χ is even, then

$$L(1, \chi) = -\frac{\tau(\chi)}{m} \sum_{(k,m)=1} \bar{\chi}(k)\ln|1 - \zeta^k|$$

$$= -\frac{\tau(\chi)}{m} \sum_{\substack{(k,m)=1 \\ 0<k<m}} \bar{\chi}(k)\ln \sin\frac{\pi k}{m}. \tag{2.20}$$

If χ is odd, then

$$L(1, \chi) = \frac{\pi i\tau(\chi)}{m^2} \sum_{\substack{(k,m)=1 \\ 0<k<m}} \bar{\chi}(k)k. \tag{2.21}$$

PROBLEMS

1. If χ is a primitive character modulo m, show that

$$|\tau(\chi)| = \sqrt{m}.$$

2. Let p be an odd prime and set $p^* = (-1)^{(p-1)/2}p$. Show that the quadratic field $R(\sqrt{p^*})$ is contained in the pth cyclotomic field (use Problem 5 of Section 2, Chapter 1, with $a = b = 1$).

3. Show that every quadratic field is contained in some cyclotomic field.

4. Using the notation of Problem 6 of Section 5 of the Supplement, show that

$$\tau_a(\chi) = \tau_a(\chi_1) \cdots \tau_a(\chi_k)\chi_1\left(\frac{m}{m_1}\right) \cdots \chi_k\left(\frac{m}{m_k}\right)$$

[assume that the m_ith root of 1 used to define the Gaussian sum $\tau_a(\chi_i)$ is ζ^{m/m_i}, where ζ is the primitive mth root of 1 which is used to define the sum $\tau_a(\chi)$].

5. Let p be a prime number which does not divide m, and let f be the smallest natural number such that $p^f \equiv 1 \pmod{m}$. Show that the polynomial $\bar{\Phi}_m(t)$ with coefficients in Z_p (see Section 2.1) factors in $Z_p[t]$ as a product of $\varphi(m)/f$ irreducible polynomials, each of degree f. (In view of Theorem 8 of Section 5, Chapter 3, this gives another proof of Theorem 2.)

6. Let p be an odd prime. By applying Theorem 1 of Section 8, Chapter 3, and Theorem 2 to the field $R(\sqrt{-1})$, show that

$$\left(\frac{-1}{p}\right) = (-1)^{(p-1)/2}$$

(this is the first supplement to the law of quadratic reciprocity).

7. Let p and $q \neq 2$ be distinct primes, let K be the qth cyclotomic field, and let g be the number of distinct prime divisors of K which divide p. Using the Euler criterion $(a/q) \equiv a^{(q-1)/2} \pmod{q}$, show that

$$\left(\frac{p}{q}\right) = (-1)^g.$$

8. Using the same notations, consider the quadratic subfield $k = R(\sqrt{q^*})$ of the field K, where $q^* = (-1)^{(q-1)/2}q$, and set $f = (q-1)/g$. If p factors as the product of two prime divisors in k, show that g is even, and if p remains prime in k, show that f is even. If $p \neq 2$, use Theorem 1 of Section 8, Chapter 3, to show that

$$\left(\frac{q^*}{p}\right) = (-1)^g.$$

Thus p factors in k if and only if g is even.

[*Hint:* If $q \equiv 1 \pmod 4$, use Problem 7 and show that from $(p/q) = (p^*/q) = 1$ it follows that $(q/p) = (q^*/p) = 1$.]

9. Use the preceding two problems to prove the law of quadratic reciprocity:

$$\left(\frac{p}{q}\right)\left(\frac{q}{p}\right) = (-1)^{[(p-1)/2][(q-1)/2]}.$$

10. Let q be a prime which factors as the product of two distinct prime divisors in the field $R(\sqrt{2})$, and let $q \equiv 1 \pmod 4$. Show that $q \equiv 1 \pmod 8$. [Consider the factorization of q in the field $R(\sqrt{2}, \sqrt{-1})$, the 8th cyclotomic field.]

11. Using the notations of Problems 7 and 8, show that the prime $p = 2$ factors into two distinct prime divisors in the field k if and only if g is even.

12. Comparing the result of the preceding problem with Theorem 1 of Section 8, Chapter 3, show that $(2/q) = +1$ if and only if $q^* \equiv 1 \pmod 8$; that is, show that

$$\left(\frac{2}{q}\right) = (-1)^{(q^2-1)/8}$$

(this is the second supplement to the law of quadratic reciprocity).

13. Show that the prime number p has the factorization

$$p = \mathfrak{p}^g, \quad g = \varphi(p^k) = p^{k-1}(p-1), \quad N(\mathfrak{p}) = p,$$

in the pth cyclotomic field.

14. Let $m = p^k m'$, $(p, m') = 1$, and let f be the smallest natural number for which $p^f \equiv 1 \pmod{m'}$. Show that the prime number p has the factorization

$$p = (\mathfrak{p}_1 \cdots \mathfrak{p}_g)^e, \quad N(p_i) = p^f,$$

in the mth cyclotomic field, where $e = \varphi(p^k)$, $fg = \varphi(m')$ (φ is Euler's function).

15. If $G(s)$ is the function determined by (2.6), show that

$$G(s) = \prod_{p \mid m} \prod_{\chi \bmod m'} \left(1 - \frac{\chi(p)}{p^s}\right)^{-1},$$

where p runs through all prime divisors of m, and χ (for given p) runs through all numerical characters modulo m', where $m = p^k m'$, $p \nmid m'$.

16. Using Problem 9 of Section 5 of the Supplement, formula (2.10), and the preceding problem, show that the ζ-function $\zeta_K(s)$ of the mth cyclotomic field has the representation

$$\zeta_K(s) = \prod_{d \mid m} \prod_{\substack{\chi \bmod d \\ \chi \text{ prim}}} L(s, \chi),$$

where d runs through all divisors of m (including 1 and m), and χ (for given d) runs through all primitive characters modulo d. Deduce that

$$\lim (s-1)\zeta_K(s) = \prod_{\substack{d \mid m \\ d \neq 1}} \prod_{\substack{\chi \bmod d \\ \chi \text{ prim}}} L(1, \chi).$$

3. Dirichlet's Theorem on Prime Numbers in Arithmetic Progressions

In Section 2 we used Theorems 2 and 4 of Section 1 to compute the number of divisor classes of a cyclotomic field. In this section we shall show that from the existence of formula (1.2), with a nonzero constant on the right, we can deduce important results on prime divisors of first degree and prime numbers in arithmetic progressions.

3.1. *Prime Divisors of First Degree*

Theorem 1. Any algebraic number field K has an infinite number of prime divisors of first degree.

Proof. By Theorem 4 of Section 1 the function $\zeta_K(s)$ has the expansion

$$\zeta_K(s) = \prod_{\mathfrak{p}} \left(1 - \frac{1}{N(\mathfrak{p})^s}\right)^{-1}. \tag{3.1}$$

Since convergent infinite products are nonzero, $\zeta_K(s) \neq 0$ for $s > 1$. Taking logarithms in (3.1), we obtain

$$\ln \zeta_K(s) = \sum_{\mathfrak{p}} \sum_{m=1}^{\infty} \frac{1}{mN(\mathfrak{p})^{ms}}. \tag{3.2}$$

We isolate the following summands:

$$P(s) = \sum_{\mathfrak{p}_1} \frac{1}{N(\mathfrak{p}_1)^s}, \tag{3.3}$$

the summation being taken over all prime divisors $\mathfrak{p}_1$ of K of first degree. If we denote the sum of all remaining terms by $G(s)$, then (3.2) can be put in the form

$$\ln \zeta_K(s) = P(s) + G(s). \tag{3.4}$$

Let f denote the degree of the prime divisor $\mathfrak{p}$, so that $N(\mathfrak{p}) = p^f$. If $f \geqslant 2$, then

$$\sum_{m=1}^{\infty} \frac{1}{mN(\mathfrak{p})^{ms}} < \sum_{m=1}^{\infty} \frac{1}{p^{2sm}} = \frac{1}{p^{2s}-1} < \frac{2}{p^{2s}}.$$

If $f = 1$, then

$$\sum_{m=2}^{\infty} \frac{1}{mN(\mathfrak{p})^{ms}} < \sum_{m=2}^{\infty} \frac{1}{p^{sm}} = \frac{1}{p^s(p^s-1)} < \frac{2}{p^{2s}}.$$

For each rational prime p there are at most $n = (K : R)$ prime divisors of the field K which divide p, so we have the following estimate for $G(s)$:

$$G(s) < \sum_p \frac{2n}{p^{2s}} < 2n \sum_{m=1}^{\infty} \frac{1}{m^{2s}}.$$

It follows that the function $G(s)$ is bounded as $s \to 1 + 0$. But since $\kappa h \neq 0$ in (1.2), both $\zeta_K(s)$ and $\ln \zeta_K(s)$ must go to infinity as $s \to 1 + 0$. We have seen that $G(s)$ is bounded, and hence by (3.4) the sum (3.3) must contain an infinite number of terms. Theorem 1 is proved.

We note that this proof uses the idea of one of the proofs of the existence of infinitely many prime numbers (see Problem 1).

3.2. *Dirichlet's Theorem*

Theorem 2 (Dirichlet's Theorem). Every residue class modulo m which consists of numbers relatively prime to m contains an infinite number of prime numbers.

Proof. The proof of Section 3.1 was based on the nonvanishing of the limit (1.2). Analogously, the proof of Dirichlet's theorem uses the fact that $L(1, \chi) \neq 0$ for any nonunit character χ modulo m, which is an immediate consequence of (2.16).

Consider the representation of $L(s, \chi)$ as an infinite product,

$$L(s, \chi) = \prod_p \left(1 - \frac{\chi(p)}{p^s}\right)^{-1}. \tag{3.5}$$

From the convergence of this infinite product it follows that $L(s, \chi)$ is nonzero for all $s > 1$. (Here χ may be any numerical character modulo m, including the unit character χ_0.) Therefore we can consider the complex function $\ln L(s, \chi)$ on the interval $(1, \infty)$. We choose a fixed branch of the logarithm function as follows. In each factor of the infinite product (3.5) we choose the value of the logarithm so that

$$-\ln\left(1 - \frac{\chi(p)}{p^s}\right) = \sum_{n=1}^{\infty} \frac{\chi(p)^n}{np^{sn}}. \tag{3.6}$$

Summing the series (3.6) over all p, we obtain

$$\sum_p -\ln\left(1-\frac{\chi(p)}{p^s}\right)=\sum_p \frac{\chi(p)}{p^s}+R(s,\chi)$$

where

$$R(s,\chi)=\sum_p\left(\frac{1}{2}\frac{\chi(p)^2}{p^{2s}}+\frac{1}{3}\frac{\chi(p)^3}{p^{3s}}+\cdots\right)$$

(it is clear that all series involved are absolutely convergent for $s>1$). The value for $\ln L(s,\chi)$ is now chosen so that

$$\ln L(s,\chi)=\sum_p \frac{\chi(p)}{p^s}+R(s,\chi) \tag{3.7}$$

for all $s>1$. Note that the values of $\ln L(s,\chi_0)$ will be real for the unit character χ_0.

We estimate the function $R(s,\chi)$:

$$|R(s,\chi)|<\sum_p\sum_{n=2}^{\infty}\frac{1}{p^{sn}}<\sum_p\frac{1}{p(p-1)}<\sum_{n=1}^{\infty}\frac{1}{n(n+1)}=1.$$

Thus $|R(s,\chi)|<1$ for all $s>1$.

Along with the numerical character χ we consider the corresponding character of the group G_m (which consists of all residue classes modulo m which consist of numbers relatively prime to m), which we also denote by χ. Let C run through all classes of the group G_m. Since $\chi(p)=\chi(C)$ for $p\in C$, then

$$\sum_p\frac{\chi(p)}{p^s}=\sum_C\chi(C)\sum_{p\in C}\frac{1}{p^s}$$

[recall that $\chi(p)=0$ if p divides m]. Setting

$$f(s,C)=\sum_{p\in C}\frac{1}{p^s},$$

we can put (3.7) in the form

$$\ln L(s,\chi)=\sum_C\chi(C)f(s,C)+R(s,\chi). \tag{3.8}$$

Since there are $\varphi(m)$ characters modulo m, we may regard the equations of the form (3.8) as a system of $\varphi(m)$ linear equations in the $\varphi(m)$ variables $f(x,C)$ [the constant terms are $\ln L(s,\chi)-R(s,\chi)$]. To use this system to find $f(s,A)(A\in G_m)$, multiply (3.8) by $\chi(A^{-1})$, and then sum over all characters χ. We obtain

$$\sum_\chi\chi(A^{-1})\ln L(s,\chi)=\sum_C\sum_\chi\chi(CA^{-1})f(s,C)+R_A(s), \tag{3.9}$$

where we have the estimate $|R_A(s)| = \left|\sum_\chi \chi(A^{-1})R(s, \chi)\right| < \varphi(m)$ for all $s > 1$. By formula (5.6) of the Supplement the sum $\sum_\chi \chi(CA^{-1})$ equals $\varphi(m)$ for $C = A$ and equals zero for $C \neq A$. Therefore (3.9) takes the form

$$\ln L(s, \chi_0) + \sum_{\chi \neq \chi_0} \chi(A^{-1})\ln L(s, \chi) = \varphi(m)f(s, A) + R_A(s). \tag{3.10}$$

This gives the value of $f(s, A)$ in terms of the system (3.8).

Now we let s approach 1 from the right. If $\chi \neq \chi_0$, then $L(s, \chi) \to L(1, \chi)$, with $L(1, \chi) \neq 0$, as was noted at the beginning of the proof. Hence the sum on the left in (3.10) (over all nonunit characters) has a finite limit. Taking this sum to the right side and combining it with $R_A(s)$, we obtain

$$\ln L(s, \chi_0) = \varphi(m)f(s, A) + T_A(s), \tag{3.11}$$

where T_A remains bounded as $s \to 1 + 0$.

Now we assume that the number of primes in the class A is finite. Then the function $f(s, A) = \sum_{p\in A} 1/p^s$ will have a finite limit as $s \to 1$, and therefore the right side of (3.11) will remain bounded as $s \to 1 + 0$. But this is impossible, since

$$\lim_{s\to 1+0} L(s, \chi_0) = \infty,$$

since

$$L(s, \chi_0) = \zeta(s) \prod_{p|m} \left(1 - \frac{1}{p^s}\right).$$

This contradiction proves Theorem 2.

Dirichlet's theorem can be strengthened as follows. Set

$$f(s) = \sum_A f(s, A) = \sum_{(p,m)=1} \frac{1}{p^s}.$$

Dividing (3.11) by $\varphi(m)$ and summing over all $A \in G_m$, we obtain

$$\ln L(s, \chi_0) = f(s) + T(s), \tag{3.12}$$

where $T(s)$ is bounded as $s \to 1 + 0$. Comparing the right sides of (3.11) and (3.12) and taking the limit as $s \to 1 + 0$, we arrive at the formula

$$\lim_{s\to 1+0} \left(\sum_{p\in A} \frac{1}{p^s} \Big/ \sum_{(p,m)=1} \frac{1}{p^s}\right) = \frac{1}{\varphi(m)}.$$

This formula says that, in a certain sense, the prime numbers which are relatively prime to m are uniformly distributed in the residue classes of G_m.

PROBLEMS

1. Show that the difference between the functions $\ln \zeta(s)$ and $g(s) = \sum_p 1/p^s$ (p running through all rational primes) remains bounded as $s \to 1 + 0$.

2. Let $P(s)$ be the function determined by (3.3). Show that the difference

$$P(s) - \ln \frac{1}{s-1}$$

remains bounded as $s \to 1 + 0$.

3. The rational integer a is called an nth *power residue* modulo the prime p, if the congruence $x^n \equiv a \pmod p$ is solvable. For any a and any n, show that there are infinitely many p such that a is an nth power residue.

4. Let the integers $a_1, \ldots, a_n$ be such that $a_1^{x_1} \cdots a_n^{x_n}$ is a square if and only if all x_i are even. For any choice of $\varepsilon_1, \ldots, \varepsilon_n$ ($\varepsilon_i = \pm 1$), show that there exist infinitely many primes p (not dividing $a_1, \ldots, a_n$) for which

$$\left(\frac{a_1}{p}\right) = \varepsilon_1, \ldots, \left(\frac{a_n}{p}\right) = \varepsilon_n.$$

Hint: Consider the sum

$$\sum_p \left(\prod_i \left(1 + \varepsilon_i \left(\frac{a_i}{p}\right)\right)\right)\frac{1}{p^s}.$$

4. The Number of Divisor Classes of Quadratic Fields

4.1. *A Formula for the Number of Divisor Classes*

Let $K = R(\sqrt{d})$ be a quadratic field (d a square-free rational integer). By Theorem 2 of Section 8, Chapter 3, a rational prime p has the following factorization into prime divisors in K:

$$\begin{aligned}
&(1)\ p = \mathfrak{p}\mathfrak{p}', \quad \mathfrak{p} \neq \mathfrak{p}', && N(\mathfrak{p}) = N(\mathfrak{p}') = p, && \text{if } \chi(p) = 1;\\
&(2)\ p = \mathfrak{p}, && N(\mathfrak{p}) = p^2, && \text{if } \chi(p) = -1;\\
&(3)\ p = \mathfrak{p}^2, && N(\mathfrak{p}) = p, && \text{if } \chi(p) = 0;
\end{aligned}$$

where χ is the character of the quadratic field K (see the definition of Section 8.2 of Chapter 3). Hence in the product

$$\zeta_K(s) = \prod_{\mathfrak{p}} \left(1 - \frac{1}{N(\mathfrak{p})^s}\right)^{-1}$$

the factor corresponding to p will be one of the following:

$$(1)\quad \left(1-\frac{1}{p^s}\right)^{-1}\left(1-\frac{1}{p^s}\right)^{-1};$$

$$(2)\quad \left(1-\frac{1}{p^{2s}}\right)^{-1}=\left(1-\frac{1}{p^s}\right)^{-1}\left(1+\frac{1}{p^s}\right)^{-1};$$

$$(3)\quad 1-\frac{1}{p^s}.$$

In all three cases this can be written

$$\left(1-\frac{1}{p^s}\right)^{-1}\left(1-\frac{\chi(p)}{p^s}\right)^{-1}.$$

Since $\prod_p(1-1/p^s)^{-1}=\zeta(s)$ (Theorem 4 of Section 1), then $\zeta_K(s)$ has the representation

$$\zeta_K(s)=\zeta(s)\prod_p\left(1-\frac{\chi(p)}{p^s}\right)^{-1}. \tag{4.1}$$

The infinite product on the right is the L-series $L(s, \chi)$ for the character χ (with modulus $|D|$, where D is the discriminant of the field K), and since this character is not the unit character, then $L(s, \chi)$ is a continuous function on the interval $0 < s < \infty$ (corollary of Lemma 4 of Section 2). Multiplying (4.1) by $s-1$ and taking the limit as $s \to 1+0$, we obtain [by (1.19)]

$$\lim_{s\to 1+0}(s-1)\zeta_K(s)=L(1,\chi).$$

Now we use Theorem 2 of Section 1. For real quadratic fields $s=2$, $t=0$, $m=2$, and $R=\ln\varepsilon$ ($\varepsilon>1$ a fundamental unit of the field); for imaginary quadratic fields $s=0$, $t=1$, and $R=1$, and hence

$$h=\begin{cases}\dfrac{\sqrt{D}}{2\ln\varepsilon}L(1,\chi) & \text{for } d>0,\\[2ex] \dfrac{m\sqrt{|D|}}{2\pi}L(1,\chi) & \text{for } d<0.\end{cases}$$

[By Section 7.3 of Chapter 2 the number m of roots of 1 contained in K equals 4 for $K=R(\sqrt{-1})$, equals 6 for $K=R(\sqrt{-3})$, and equals 2 for all other imaginary quadratic fields.]

In the next section we shall show that the character of a quadratic field with discriminant D is a primitive character modulo $|D|$ (see the definition of Section 5.3 of the Supplement), and also that it is even for real fields and odd for imaginary fields. Therefore we can use formulas (2.20) and (2.21) to find

$L(1, \chi)$. To find a closed formula for h we must still find the value of the normed Gaussian sum $\tau(\chi) = \tau_1(\chi)$. In Section 4.3 we shall see that the sum $\tau(\chi)$ equals $\sqrt{D}$ for real fields and equals $i\sqrt{|D|}$ for imaginary fields. Noting also that for real fields $\chi(D - x) = \chi(x)$, we can formulate the following theorem [to simplify the formulas we eliminate the fields $R(\sqrt{-1})$ and $R(\sqrt{-3})$, which have discriminants -4 and -3, and for which m equals 4 and 6; for these fields $h = 1$].

Theorem 1. The number of divisor classes of a real quadratic field with discriminant D is given by

$$h = -\frac{1}{\ln \varepsilon} \sum_{\substack{(x,D)=1 \\ 0<x<D/2}} \chi(x) \ln \sin \frac{\pi x}{D}, \tag{4.2}$$

where $\varepsilon > 1$ is a fundamental unit of the field; for an imaginary quadratic field with discriminant $D < -4$ we have the formula

$$h = -\frac{1}{|D|} \sum_{\substack{(x,D)=1 \\ 0<x<|D|}} \chi(x)x. \tag{4.3}$$

In both cases χ denotes the character of the given field, defined in Section 8.2 of Chapter 3 [formula (8.5)].

We note some number-theoretic consequences of Theorem 1. We begin with formula (4.2). Consider the number

$$\eta = \prod_b \sin \frac{\pi b}{D} \Big/ \prod_a \sin \frac{\pi a}{D}, \tag{4.4}$$

where a and b run through all natural numbers in $(0, D/2)$ which are relatively prime to D and satisfy $\chi(a) = +1$, $\chi(b) = -1$. Then formula (4.2) can be written in the form $\varepsilon^h = \eta$. Hence η is a unit of the quadratic field in question, with $\eta > 1$ (since $\varepsilon > 1$). Hence we have the following theorem.

Theorem 2. Let K be a real quadratic field with discriminant D and character χ. The number η given by (4.4) is a unit in K, and is related to the fundamental unit $\varepsilon > 1$ by

$$\varepsilon^h = \eta,$$

where h is the number of divisor classes of K.

In spite of its simple formulation, there has never been an elementary proof of Theorem 2. Further, it has not been proved by purely arithmetic methods even that $\eta > 1$. From the inequality $\eta > 1$ we can deduce some consequences on the distribution of quadratic residues modulo a prime $p \equiv 1 \pmod 4$.

The quadratic field $R(\sqrt{p})$ has discriminant p and its character $\chi(x)$ coincides with the Legendre symbol (x/p). Therefore we have the inequality

$$\prod_b \sin \frac{\pi b}{p} > \prod_a \sin \frac{\pi a}{p},$$

where a and b run through the quadratic residues and nonresidues, respectively, in the interval $(0, p/2)$. Since the function $\sin x$ is monotone on the interval $(0, \pi/2)$, it follows from this inequality that the values of the $\pi b/p$ are "on the average" greater than the values $\pi a/p$, that is, that the quadratic residues modulo p "cluster" at the beginning of the interval $(0, p/2)$, and the nonresidues at the end [when $p \equiv 1 \pmod 4$ precisely half of the numbers in the interval $(0, p/2)$ are quadratic residues].

For prime numbers $p \equiv 3 \pmod 4$ we can obtain information on the distribution of residues and nonresidues by considering formula (4.3) for the field $R(\sqrt{-p})$.

First, we put formula (4.3) in simpler form in the general case. We denote $|D|$ by m.

First, assume that m is even. It is easily verified (Problem 9) that in this case $\chi(x+m/2) = -\chi(x)$ and formula (4.3) gives us

$$\begin{aligned} hm &= -\sum_{0<x<m/2} \chi(x)x - \sum_{0<x<m/2} \chi\left(x+\frac{m}{2}\right)\left(x+\frac{m}{2}\right) \\ &= -\sum_{0<x<m/2} \chi(x)x + \sum_{0<x<m/2} \chi(x)\left(x+\frac{m}{2}\right) \\ &= \frac{m}{2} \sum_{0<x<m/2} \chi(x), \end{aligned}$$

so that

$$h = \frac{1}{2} \sum_{0<x<m/2} \chi(x).$$

Note that since m is even, $\chi(2) = 0$.

Now let m be odd. Since the character χ of an imaginary quadratic field is odd, that is, $\chi(-1) = -1$ (we have already noted that this will be proved in the following section in Theorem 6), then it follows from (4.3) that

$$\begin{aligned} hm &= -\sum_{0<x<m/2} \chi(x)x - \sum_{0<x<m/2} \chi(m-x)(m-x) \\ &= -2\sum_{0<x<m/2} \chi(x)x + m \sum_{0<x<m/2} \chi(x). \end{aligned} \tag{4.5}$$

On the other hand,

$$hm = -\sum_{\substack{0<x<m \\ x \text{ even}}} \chi(x)x - \sum_{\substack{0<x<m \\ x \text{ even}}} \chi(m-x)(m-x)$$

$$= -4\sum_{0<x<m/2} \chi(2x)x + m\sum_{0<x<m/2} \chi(2x),$$

so that

$$hm\chi(2) = -4\sum_{0<x<m/2} \chi(x)x + m\sum_{0<x<m/2} \chi(x). \tag{4.6}$$

Equating the sums $\sum \chi(x)x$ in (4.5) and (4.6), we obtain the equation

$$h(2-\chi(2)) = \sum_{0<x<m/2} \chi(x).$$

Since this equation also holds for even m [when $2 \mid m$, then $\chi(2) = 0$], we have the following theorem.

Theorem 3. For an imaginary quadratic field with discriminant $D < -4$ and character χ we have the following formula:

$$h = \frac{1}{2-\chi(2)} \sum_{\substack{0<x<|D|/2 \\ (x,D)=1}} \chi(x). \tag{4.7}$$

We now apply Theorem 3 to the case of the field $R(\sqrt{-p})$, where p is a prime of the form $4n+3$. Since $-p \equiv 1 \pmod 4$, in this case $D = -p$ and the value of the character $\chi(x)$ coincides with that of the Legendre symbol (x/p). The number of summands in $\sum_{0<x<p/2}(x/p)$ is odd $[(p-1)/2 = 2n+1]$, and hence the sum itself is odd. Further, $\chi(2) = +1$ if $p \equiv 7 \pmod 8$, and $\chi(2) = -1$ if $p \equiv 3 \pmod 8$, so that we deduce from Theorem 3 the following result.

Theorem 4. Let p be a prime number of the form $4n+3$ and let V and N denote the number of quadratic residues and nonresidues in the interval $(0, p/2)$. The number of divisor classes of the field $R(\sqrt{-p})$ is odd and is given by

$$h = V - N \qquad \text{for } p \equiv 7 \pmod 8,$$

$$h = \frac{1}{3}(V - N) \qquad \text{for } p \equiv 3 \pmod 8.$$

It clearly follows from Theorem 4 that $V > N$. Thus if p is a prime of the form $4n+3$, then the quadratic residues outnumber the nonresidues on the interval $(0, p/2)$ [by a number divisible by 3 if $p \equiv 3 \pmod 8$ and $p \neq 3$].

This assertion, despite its simplicity, lies among some very deep results of number theory. It was obtained by us as a simple corollary of the fact that the number h, and hence the expression on the right in (4.7) is positive. However, the sign of this expression depends on knowledge of the value of the Gaussian sum $\tau_1(\chi)$, and we shall see in Section 4.3 that the determination of the sign $\tau_1(\chi)$ is a very difficult problem.

If $D \equiv 1 \pmod 8$, the formula for the number h for an imaginary quadratic field can be proved by purely arithmetic methods. This was done by B. A. Venkov. His proof was based on the theory of the representation of binary forms by sums of squares of linear forms and on some delicate properties of continued fractions [B. A. Venkov, On the number of classes of binary quadratic forms with negative determinant. I and II, *Izv. Akad. Nauk SSSR Ser. VII*, No. 4-5, 375–392 (1928); No. 6-7, 455–480 (1928)]. In the case $D \equiv 1 \pmod 8$, as in the case of real quadratic fields, a purely arithmetic derivation of the formula for h has never been obtained. Also there is no known elementary proof of the fact that for a prime p of the form $8n + 7$ the interval $(0, p/2)$ contains more quadratic residues than nonresidues.

Remark. Let p be a prime of the form $8n + 7$. It can be shown by elementary means (Problem 7) that the interval $(0, p/2)$ contains just as many odd quadratic residues as odd nonresidues. Hence the number h for the field $R(\sqrt{-p})$, $p \equiv 7 \pmod 8$, is also given by

$$h = V^* - N^*,$$

where V^* and N^* denote the number of quadratic residues and nonresidues among the even integers in the interval $(0, p/2)$.

4.2. *The Character of a Quadratic Field*

We shall prove those assertions about the character of a quadratic field which were used in Section 4.1.

Theorem 5. The character χ of a quadratic field with discriminant D is primitive (with modulus $|D|$).

Proof. By Theorem 4 of Section 5 of the Supplement it suffices to show that for any prime number p which divides D there is an x such that $(x, D) = 1$, $x \equiv 1 \pmod{|D|/p}$ and $\chi(x) = -1$. First, consider the case $p \neq 2$. Choose any quadratic nonresidue s modulo p and pick x as a solution to the system of congruences

$$x \equiv s \pmod p,$$

$$x \equiv 1 \left(\bmod \frac{2|D|}{p}\right).$$

Using formula (8.5) of Chapter 3 it is easily checked that $\chi(x) = (x/p) = (s/p) = -1$.

Now let $p = 2$. If $d \equiv 3 \pmod 4$, $D = 4d$, then, solving the congruences

$$x \equiv 3 \pmod 4,$$

$$x \equiv 1 \pmod{2|d|},$$

we shall also have $\chi(x) = (-1)^{(x-1)/2} = -1$. If $d = 2d'$, $D = 4d = 8d'$, then for the number x, given by

$$x \equiv 5 \pmod 8,$$

$$x \equiv 1 \pmod{4|d'|},$$

we shall have $\chi(x) = (-1)^{(x^2-1)/8} = -1$.

We have shown that χ is primitive.

Theorem 6. The characters of real quadratic fields are even and the characters of imaginary quadratic fields are odd.

Proof. Let χ be the character of the quadratic field $R(\sqrt{d})$. We compute $\chi(-1)$, using (8.5) of Chapter 3. If $d \equiv 1 \pmod 4$, then

$$\chi(-1) = \left(\frac{-1}{|d|}\right) = (-1)^{(|d|-1)/2} = (-1)^{[(d-1)/2]+[(|d|-1)/2]}.$$

If $d \equiv 3 \pmod 4$, then

$$\chi(-1) = -\left(\frac{-1}{|d|}\right) = -(-1)^{(|d|-1)/2} = (-1)^{[(d-1)/2]+[(|d|-1)/2]}.$$

Finally, if $d = 2d'$, then

$$\chi(-1) = (-1)^{(d'-1)/2}\left(\frac{-1}{|d'|}\right) = (-1)^{[(d'-1)/2]+[(|d'|-1)/2]}.$$

But if a is odd, then

$$\frac{a-1}{2} + \frac{|a|-1}{2} = \begin{cases} a-1 \equiv 0 \pmod 2 & \text{for } a > 0, \\ -1 & \text{for } a < 0. \end{cases}$$

Hence in all cases

$$\chi(-1) = \begin{cases} 1 & \text{for } d > 0, \\ -1 & \text{for } d < 0. \end{cases}$$

Theorem 6 is proved.

4.3. *Gaussian Sums for Quadratic Characters*

In deriving formulas for the number of divisor classes of a quadratic field, we used a formula for the value of the normed Gaussian sum $\tau(\chi)$. Recall that the Gaussian sum $\tau_a(\chi)$ of the character χ modulo m is called *normed* if $\zeta = \cos 2\pi/m + i \sin 2\pi/m$ is taken as the primitive mth root of 1 in its definition (see Section 2.4). We now consider the computation of the value of $\tau(\chi)$.

By Theorem 5 the character χ of the quadratic field $R(\sqrt{d})$ with discriminant D is a primitive numerical character modulo $|D|$. Also it satisfies the condition $\chi^2 = \chi_0$, where χ_0 is the unit character. This simply means that the character χ takes the values ± 1 (and, of course, zero).

Definition. A nonunit numerical character χ is called *quadratic* if $\chi^2 = \chi_0$.

We shall show that every primitive quadratic character is the character of a quadratic field. By Problem 8, primitive quadratic characters occur only for moduli of the form r and $4r$ (one character for each) and $8r$ (two characters), where r is an odd square-free natural number. The set of these moduli hence coincides with the set of numbers of the form $|D|$, where D is the discriminant of a quadratic field. We note that when $|D| = 8r$ there are two quadratic fields; $R(\sqrt{2r})$ and $R(\sqrt{-2r})$, which have distinct characters, since one is even and the other odd. Hence each primitive quadratic character is the character of a quadratic field.

The value of the Gaussian sum for primitive quadratic characters is determined by the following theorem.

Theorem 7. Let χ be a primitive quadratic character modulo m. Then the normed Gaussian sum $\tau_1(\chi) = \tau(\chi)$ satisfies

$$\tau(\chi) = \begin{cases} \sqrt{m} & \text{if } \chi(-1) = 1, \\ i\sqrt{m} & \text{if } \chi(-1) = -1. \end{cases}$$

Proof. We shall give the full proof of Theorem 7 only in the case of odd prime modulus p, since this case contains most of the essential difficulties. The transition to the general case is relatively easy. At the end of the proof we sketch this transition.

Hence let p be an odd prime and set $\zeta = \cos 2\pi/p + i \sin 2\pi/p$. Since the nonunit quadratic character χ modulo p coincides with the Legendre symbol (x/p) (Problem 4 of Section 2 of Chapter 1), then the normed Gaussian sum $\tau(\chi)$ is given by

$$\tau(\chi) = \sum_x{}' \left(\frac{x}{p}\right) \zeta^x$$

(the prime on the summation sign indicates that x runs through a reduced residue system modulo p). We find the complex conjugate $\overline{\tau(\chi)}$. Since $\bar{\zeta} = \zeta^{-1}$, then

$$\overline{\tau(\chi)} = \sum_x{}' \left(\frac{x}{p}\right)\zeta^{-x} = \sum_x{}' \left(\frac{-x}{p}\right)\zeta^{x} = \left(\frac{-1}{p}\right)\tau(\chi). \tag{4.8}$$

On the other hand, by Theorem 4 of Section 2 of Chapter 1,

$$\overline{\tau(\chi)}\tau(\chi) = p. \tag{4.9}$$

From (4.8) and (4.9) it follows that

$$\tau(\chi)^2 = \left(\frac{-1}{p}\right)p = (-1)^{(p-1)/2}p,$$

and hence

$$\tau(\chi) = \begin{cases} \pm\sqrt{p} & \text{if } p \equiv 1 \pmod 4, \\ \pm i\sqrt{p} & \text{if } p \equiv 3 \pmod 4. \end{cases} \tag{4.10}$$

To complete the proof of Theorem 7 (for $m = p$) we need only determine the signs of $\sqrt{p}$ and $i\sqrt{p}$. But it is precisely here that the principal difficulty of the proof lies.

We represent the sum $\tau(\chi)$ in another form. Let a run through all quadratic residues modulo p, and let b run through all nonresidues. Then

$$\tau(\chi) = \sum_a \zeta^a - \sum_b \zeta^b.$$

But

$$1 + \sum_a \zeta^a + \sum_b \zeta^b = 0,$$

so that

$$\tau(\chi) = 1 + 2\sum_a \zeta^a.$$

If x takes the values $0, 1, \ldots, p-1$, then, modulo p, x^2 takes the value 0 once and each quadratic residue twice. Hence we can write $\tau(\chi)$ in the form

$$\tau(\chi) = \sum_{x=0}^{p-1} \zeta^{x^2}. \tag{4.11}$$

Now consider the matrix

$$A = (\zeta^{xy})_{0 \leqslant x, y \leqslant p-1} = \begin{pmatrix} 1 & 1 & 1 & \cdots & 1 \\ 1 & \zeta & \zeta^2 & \cdots & \zeta^{p-1} \\ 1 & \zeta^2 & \zeta^4 & \cdots & \zeta^{2(p-1)} \\ \cdot & \cdot & \cdot & \cdot & \cdot \\ 1 & \zeta^{p-1} & \zeta^{2(p-1)} & \cdots & \zeta^{(p-1)2} \end{pmatrix}.$$

By (4.11) the Gaussian sum $\tau(\chi)$ coincides with the trace of the matrix A. Hence if we denote the characteristic roots of A by $\lambda_1, \dots, \lambda_p$, then we will have

$$\tau(\chi) = \lambda_1 + \cdots + \lambda_p. \tag{4.12}$$

The computation of $\tau(\chi)$ hence reduces to the finding of the characteristic roots of the matrix A.

We square the matrix A. Since

$$\sum_{t=0}^{p-1} \zeta^{xt}\zeta^{ty} = \sum_{t=0}^{p-1} \zeta^{t(x+y)} = \begin{cases} p & \text{for } x + y \equiv 0 \pmod{p}, \\ 0 & \text{for } x + y \not\equiv 0 \pmod{p}, \end{cases}$$

then

$$A^2 = \begin{pmatrix} p & 0 & \cdots & 0 \\ 0 & 0 & \cdots & p \\ \cdot & \cdot & \cdot & \cdot \\ 0 & p & \cdots & 0 \end{pmatrix}.$$

As is well known, the characteristic roots of the matrix A^2 coincide with the squares

$$\lambda_1^{\,2}, \dots, \lambda_p^{\,2} \tag{4.13}$$

of the characteristic roots of A. But the characteristic polynomial of A^2 is easily computed. It equals

$$(t - p)^{(p+1)/2}(t + p)^{(p-1)/2}.$$

Hence the sequence (4.13) contains $(p + 1)/2$ numbers equal to p, and $(p - 1)/2$ numbers equal to $-p$. Hence each λ_k is one of the numbers $\pm\sqrt{p}$, $\pm i\sqrt{p}$, and if a, b, c, and d denote the multiplicities of the characteristic roots $\sqrt{p}$, $-\sqrt{p}$, $i\sqrt{p}$, and $-i\sqrt{p}$, then

$$a + b = \frac{p+1}{2}, \quad c + d = \frac{p-1}{2}. \tag{4.14}$$

The sum (4.12) can be represented in the form

$$\tau(\chi) = (a - b + (c - d)i)\sqrt{p}. \tag{4.15}$$

Comparing with (4.10), we find that

$$\left.\begin{aligned} a - b = \pm 1, \quad c = d \qquad & \text{for } p \equiv 1 \pmod{4}, \\ a = b, \quad c - d = \pm 1 \qquad & \text{for } p \equiv 3 \pmod{4}. \end{aligned}\right\} \tag{4.16}$$

To determine the multiplicities a, b, c, and d, we must find another relation among them. We compute the determinant of the matrix A. Since $\det(A^2) = p^p(-1)^{p(p-1)/2}$, then

$$\det A = \pm\, i^{p(p-1)/2} p^{p/2}. \tag{4.17}$$

The determinant det A is a Vandermonde determinant. Introducing the notation $\eta = \cos \pi/p + i \sin \pi/p$, we have

$$\begin{aligned}\det A &= \prod_{p-1 \geqslant r > s \geqslant 0} (\zeta^r - \zeta^s) = \prod_{r>s} \eta^{r+s}(\eta^{r-s} - \eta^{-(r-s)}) \\ &= \prod_{r>s} \eta^{r+s} \prod_{r>s} \left(2i \sin \frac{(r-s)\pi}{p}\right) \\ &= i^{p(p-1)/2} 2^{p(p-1)/2} \prod_{r>s} \sin \frac{(r-s)\pi}{p},\end{aligned}$$

since

$$\sum_{r>s}^{p-1} (r+s) = \sum_{r=1}^{p-1} \sum_{s=0}^{r-1} (r+s) = \sum_{r=1}^{p-1} \left(r^2 + \frac{r(r-1)}{2}\right) = 2p\left(\frac{p-1}{2}\right)^2$$

is divisible by $2p$. We compare this expression for det A with (4.17). Since $\sin(r-s)\pi/p > 0$ for $0 \leqslant s < r \leqslant p-1$, we must have the plus sign in (4.17). Thus

$$\det A = i^{p(p-1)/2} p^{p/2}.$$

On the other hand, we have

$$\det A = \prod_{k=1}^{p} \lambda_k = (-1)^b i^c (-i)^d p^{p/2} = i^{2b+c-d} p^{p/2}.$$

This yields the congruence

$$2b + c - d \equiv p\frac{p-1}{2} \pmod 4,$$

from which, in view of (4.14) and (4.16), we deduce

$$\begin{aligned}a - b &= \frac{p+1}{2} - 2b \\ &\equiv \frac{p+1}{2} - \frac{p-1}{2} = 1 \pmod 4 \qquad \text{for } p \equiv 1 \pmod 4, \\ c - d &\equiv -\frac{p-1}{2} + 2b \\ &= -\frac{p-1}{2} + \frac{p+1}{2} = 1 \pmod 4 \qquad \text{for } p \equiv 3 \pmod 4.\end{aligned}$$

These congruences show that the differences $a - b$ and $c - d$ in (4.16) both equal $+1$, and by (4.10) this finally yields

$$\tau(\chi) = \begin{cases} \sqrt{p} & \text{for } p \equiv 1 \pmod 4, \\ i\sqrt{p} & \text{for } p \equiv 3 \pmod 4. \end{cases}$$

This completes the proof of Theorem 7 for the case of a prime modulus $m = p$.

To prove the theorem in the general case, use the result of Problem 4 of Section 2. If χ is a primitive quadratic character modulo m, this problem shows that the normed Gaussian sum $\tau(\chi)$ can easily be expressed in terms of the normed Gaussian sums for the nonunit modulo 4, the two primitive characters modulo 8, and quadratic characters of odd primes p. Since we know all these Gaussian sums (see Problems 10 and 11 for the moduli 4 and 8), then the formula of Problem 4 of Section 2 allows us to give an explicit expression for $\tau(\chi)$. Suppose, for example, that we have the character

$$\chi(x) = (-1)^{[(x^2-1)/8]+[(x-1)/2]}\left(\frac{x}{r}\right), \quad (x, 2r) = 1$$

modulo $m = 8r$, where r is an odd square-free natural number. If $r = p_1 \cdots p_s$, then χ has the representation

$$\chi(x) = (-1)^{[(x^2-1)/8 \;+[(x-1)/2]}\left(\frac{x}{p_1}\right) \cdots \left(\frac{x}{p_s}\right).$$

Let α be the number of primes among $p_1, \ldots, p_s$ which are congruent to 3 modulo 4. Then

$$\begin{aligned} \tau(\chi) &= 2i\sqrt{2}\, i^\alpha \sqrt{r}(-1)^{[(r^2-1)/8]+[(r-1)/2]}\left(\frac{2}{r}\right)\prod_{k \neq j}\left(\frac{p_k}{p_j}\right) \\ &= i^{\alpha+1}\sqrt{m}(-1)^{[(r-1)/2]+C_\alpha{}^2} = \sqrt{m}\, i^{\alpha+1+2\alpha+\alpha(\alpha-1)} \\ &= i^{(\alpha+1)^2}\sqrt{m} = \begin{cases} \sqrt{m} & \text{if } \chi(-1) = (-1)^{\alpha-1} = 1, \\ i\sqrt{m} & \text{if } \chi(-1) = (-1)^{\alpha+1} = -1. \end{cases} \end{aligned}$$

The Gaussian sums for the other primitive quadratic characters are treated analogously.

This proof of Theorem 7 (for prime moduli) is due to Shur. Another proof, found by Kronecker, is given in Problems 13 to 16.

PROBLEMS

1. Knowing that $(1 + \sqrt{5})/2 = 2\cos \pi/5$ is a fundamental unit for the field $R(\sqrt{5})$, use formula (4.2) to compute the number h for this field.

2. Compute the number h for the fields $R(\sqrt{-5})$ and $R(\sqrt{-23})$.

3. Show that a quadratic field with discriminant D is a subfield of the mth cyclotomic field, where $m = |D|$.

4. Let p be an odd prime, and let ζ be a primitive pth root of 1. Show that the cyclotomic field $R(\zeta)$ contains one and only one quadratic subfield. This subfield is $R(\sqrt{p})$ if $p \equiv 1 \pmod 4$, and $R(\sqrt{-p})$ if $p \equiv 3 \pmod 4$. (To solve this and succeeding problems, use the fundamental theorem of Galois theory.)

5. Let $p \equiv 1 \pmod 4$ be a prime, and define the number

$$\prod_b \sin \frac{\pi b}{p} \Big/ \prod_a \sin \frac{\pi a}{p},$$

where a and b run through the quadratic residues and nonresidues modulo p in the interval $(0, p/2)$. Without using Theorem 2 show that this number is a unit of the quadratic field $R(\sqrt{p})$, and that its norm is -1.

6. Using the second assertion of Problem 5, show that the number of divisor classes in the field $R(\sqrt{p})$, p a prime, $p \equiv 1 \pmod 4$, is odd and that the norm of a fundamental unit of this field is -1.

7. Let p be a prime number of the form $8n + 7$. Show that precisely half of the even numbers in the interval $(0, p/2)$ are quadratic residues modulo p.

8. If there is a primitive quadratic character with modulus m, show that m is of the form r, $4r$, or $8r$, where r is an odd square-free natural number. Further, show that every primitive quadratic character is of the form

$$\chi(x) = \left(\frac{x}{r}\right), \ (x, r) = 1 \qquad \text{for the modulus } r,$$

$$\chi(x) = (-1)^{(x-1)/2}\left(\frac{x}{r}\right), \ (x, 2r) = 1 \qquad \text{for the modulus } 4r,$$

$$\left.\begin{aligned} &\chi(x) = (-1)^{(x^2-1)/8}\left(\frac{x}{r}\right), \\ &\chi(x) = (-1)^{[(x^2-1)/84[(x-1)/2]}\left(\frac{x}{r}\right), \end{aligned}\right\} (x, 2r) = 1 \qquad \text{for the modulus } 8r.$$

9. If χ is a primitive quadratic character with even modulus m ($m = 4r$ or $8r$ with odd r), show that

$$\chi\left(x + \frac{m}{2}\right) = -\chi(x).$$

10. Let χ be the character modulo 4 given by $\chi(x) = (-1)^{(x-1)/2}$, $(x, 2) = 1$. Show that the normed Gaussian sum $\tau_1(\chi)$ equals $2i$.

11. Consider the primitive characters

$$\chi'(x) = (-1)^{(x^2-1)/8} \qquad \text{and} \qquad \chi''(x) = (-1)^{[(x^2-1)/8]+[(x-1)/2]} \ (2 \nmid x) \text{ modulo } 8.$$

Verify that the normed Gaussian sums are $\tau_1(\chi') = 2\sqrt{2}$, and $\tau_1(\chi'') = 2i\sqrt{2}$.

12. Give the proof of Theorem 7 for arbitrary moduli.

13. Let p be an odd prime and set $\zeta = \cos 2\pi/p + i \sin 2\pi/p$. Let

$$\delta = \prod_{x=1}^{(p-1)/2} (\zeta^x - \zeta^{-x}).$$

Show that

$$\delta^2 = (-1)^{(p-1)/2} p.$$

Thus δ^2 coincides with the square τ^2 of the Gaussian sum $\tau = \sum_{x=1}^{p-1} (x/p)\zeta^x$.

14. Using the same notations, show that

$$\left(\frac{-2}{p}\right) \delta = \begin{cases} \sqrt{p} & \text{for } p \equiv 1 \pmod 4, \\ i\sqrt{p} & \text{for } p \equiv 3 \pmod 4. \end{cases}$$

Further, setting $\lambda = 1 - \zeta$, show that the congruence

$$\left(\frac{-2}{p}\right) \delta \equiv \left(\frac{p-1}{2}\right)! \, \lambda^{(p-1)/2} \pmod{\lambda^{(p+1)/2}}.$$

holds in the order $Z[\zeta]$.

15. Verify the congruence

$$\sum_{x=1}^{p-1} \left(\frac{x}{p}\right) \zeta^x = \tau \equiv \left(\frac{p-1}{2}\right)! \, \lambda^{(p-1)/2} \pmod{\lambda^{(p+1)/2}}.$$

in the ring $Z[\zeta]$.

Hint: Decompose the sum $\sum_{x=1}^{p-1} x^{(p-1)/2} (1-\lambda)^x$ into powers of λ, using the fact that

$$\sum_{x=1}^{p-1} x^m \equiv \begin{cases} 0 \pmod p & \text{for } 0 < m < p-1, \\ -1 \pmod p & \text{for } m = p-1. \end{cases}$$

16. Use the two preceding problems to show that

$$\tau = \begin{cases} \sqrt{p} & \text{for } p \equiv 1 \pmod 4, \\ i\sqrt{p} & \text{for } p \equiv 3 \pmod 4. \end{cases}$$

5. The Number of Divisor Classes of Prime Cyclotomic Fields

5.1. *The Decomposition of the Number h into Two Factors*

The formulas (2.16) and (2.17) for the number of divisor classes of the mth cyclotomic field do not contain any infinite series or products. But they are somewhat unsatisfactory in that they express the number h of classes, which is of course a natural number, in terms of irrational and complex numbers. In this section we shall put these formulas for h in a more complete form, limiting ourselves to the case of prime cyclotomic fields.

Hence let $l = 2m + 1$ be a prime number, and let $K = K(\zeta)$ be the lth cyclotomic field. For ease of computation we shall assume that K is a subfield of the field of complex numbers, and that $\zeta = \cos 2\pi/l + i \sin 2\pi/l$ (the value of ζ needs to be precisely fixed for analytic computations). We compute for K the terms in the product in (2.16). Since the degree $(K : R)$ is $l - 1$ (corollary of Theorem 1 of Section 2) and all isomorphisms of K into the complex field are complex (they are simply automorphisms of K), then $s = 0$ and $t = (l-1)/2 = m$. By Lemma 3 of Section 1, Chapter 3, the number w, the number of roots in 1 in K, equals $2l$. The norm of the principal divisor $\mathfrak{l} = (1 - \zeta)$ equals $N(\mathfrak{l}) = N(1-\zeta) = l$ [see (1.5) of Chapter 3], so that the divisor $\mathfrak{l}$ is prime, and by Lemma 1 of Section 1, Chapter 3, the number l has the factorization $l = \mathfrak{l}^{l-1}$. Hence the factor $F(s)$ in (2.12) equals

$$F(s) = \left(1 - \frac{1}{N(\mathfrak{l})^s}\right)^{-1}\left(1 - \frac{1}{l^s}\right) = 1.$$

We turn to the computation of the discriminant of the field K.

Theorem 1. The numbers

$$1, \zeta, \ldots, \zeta^{l-2}$$

form a fundamental basis for the lth cyclotomic field $K = R(\zeta)$.

Proof. If $s \not\equiv 0 \pmod{l}$ the characteristic polynomial of the number ζ^s is $X^{l-1} + X^{l-2} + \cdots + X^l + 1$, so that

$$\mathrm{Sp}\, \zeta^s = \begin{cases} -1 & \text{if } s \not\equiv 0 \pmod{l}, \\ l-1 & \text{if } s \equiv 0 \pmod{l}. \end{cases} \tag{5.1}$$

Let

$$\alpha = a_0 + a_1\zeta + \cdots + a_{l-2}\zeta^{l-2} \qquad (a_i \in R)$$

be any integer of the field K. We must show that all the coefficients a_i are rational integers. Since $\alpha\zeta^{-k} - \alpha\zeta$ is an integer, then the trace

$$\mathrm{Sp}(\alpha\zeta^{-k} - \alpha\zeta) = la_k - \sum_{i=0}^{l-2} a_i + \sum_{i=0}^{l-2} a_i = la_k$$

is a rational integer $(0 \leqslant k \leqslant l-2)$. We set $la_k = b_k$, $1 - \zeta = \lambda$ and consider the number

$$l\alpha = b_0 + b_1\zeta + \cdots + b_{l-2}\zeta^{l-2} = c_0 + c_1\lambda + \cdots + c_{l-2}\lambda^{l-2},$$

where the b_k and c_k are all rational integers. We shall show that all the coefficients c_k are divisible by l. Suppose this has been established for $c_1, \ldots, c_{k-1}$ $(0 \leqslant k < l-2)$. Consider the last equation as a congruence modulo λ^{k+1} (in the ring of integers of the field K). Since $l \equiv 0 \pmod{\lambda^{k+1}}$ (Lemma 1 of Section 1, Chapter 3), then this congruence yields

$$c_k\lambda^k \equiv 0 \pmod{\lambda^{k+1}},$$

so that c_k is divisible by λ and hence also divisible by l (Lemma 2 of Section 1, Chapter 3). But then all the coefficients b_k must also be divisible by l, so that all a_k are integers. Theorem 1 is proved.

Corollary. The discriminant of the lth cyclotomic field ($l > 2$) equals $(-1)^{(l-1)/2} l^{l-2}$.

For by formula (5.1) the discriminant of K equals the determinant

$$\det(\operatorname{Sp} \zeta^{i+j})_{1 \leqslant i, j \leqslant l-1} = \begin{vmatrix} -1 & -1 & \cdots & -1 \\ -1 & -1 & \cdots & l-1 \\ \cdot & \cdot & \cdot & \cdot \\ -1 & l-1 & \cdots & -1 \end{vmatrix}$$

(instead of the basis of Theorem 1, we take here the basis $\zeta, \zeta^2, \dots, \zeta^{l-1}$).

Formula (2.16) for the case of the lth cyclotomic field can now be written

$$h = \frac{l^{l/2}}{2^{m-1}\pi^m R} \prod_{\chi \neq \chi_0} L(1, \chi), \tag{5.2}$$

where R is the regulator of the field K, $m = (l-1)/2$ and χ runs through all numerical characters modulo l, except the unit character χ_0.

Since all terms in the formula (5.2) outside the product sign are real and positive, the formula will remain valid if each term $L(1, \chi)$ is replaced by $|L(1, \chi)|$.

For a prime modulus all nonunit numerical characters are primitive. Therefore we can apply Theorem 3 of Section 2. To do this we must separate the even and the odd characters. Let g be a fixed primitive root modulo l (that is, $\bar{g}$ generates the cyclic group G_l of residue classes modulo l), and let θ be a primitive $(l-1)$th root of 1. The group of numerical characters modulo l is cyclic and has order $l-1$. If we denote by χ that character modulo l for which

$$\chi(g) = \theta^{-1},$$

then its powers $\chi, \chi^2, \dots, \chi^{l-1} = \chi_0$ will comprise the entire group of characters modulo l. Since

$$\chi^s(-1) = \chi(g^{(l-1)/2})^s = \theta^{-(l-1)/2s} = (-1)^s$$

then each character of the form χ^{2k-1} will be odd, and each of the form χ^{2k} will be even.

Using formula (2.20) and Theorem 4 of Section 2 of Chapter 1, we obtain for the even characters χ^{2k} $[1 \leqslant k \leqslant (l-3)/2]$:

$$|L(1, \chi^{2k})| = \frac{|\tau(\chi^{2k})|}{l}\left|\sum_{r=0}^{l-2} \bar{\chi}^{2k}(g^r)\ln|1 - \zeta^{g^r}|\right|$$
$$= \frac{1}{\sqrt{l}}\left|\sum_{r=0}^{l-2} \theta^{2kr} \ln|1 - \zeta^{g^r}|\right|.$$

Setting $r = [(l-1)/2] + s$, where $0 \leqslant s < [(l-1)/2] = m$, and using the relation

$$1 - \zeta^{g^{m+s}} = 1 - \zeta^{-g^s} \tag{5.3}$$

we obtain

$$\theta^{2k(m+s)} \ln|1 - \zeta^{g^{m+s}}| = \theta^{2ks} \ln|1 - \zeta^{g^s}|,$$

and thus

$$|L(1, \chi^{2k})| = \frac{2}{\sqrt{l}}\left|\sum_{r=0}^{m-1} \theta^{2kr} \ln|1 - \zeta^{g^r}|\right|.$$

We can apply formula (2.21) to the odd character χ^{2k-1} in an analogous manner. Let g_s denote the smallest positive residue of g^s modulo l. Then

$$\sum_{r=1}^{l-1} \bar{\chi}^{2k-1}(r)r = \sum_{s=0}^{l-2} \chi^{2k-1}(g^s)^{-1} g_s = \sum_{s=0}^{l-2} g_s \theta^{(2k-1)s} = F(\theta^{2k-1}),$$

where F denotes the polynomial

$$F(X) = \sum_{s=0}^{l-2} g_s X^s.$$

Hence

$$|L(1, \chi^{2k-1})| = \frac{\pi\sqrt{l}}{l^2} |F(\theta^{2k-1})|.$$

Substituting these values for $|L(1, \chi^{2k})|$, $1 \leqslant k \leqslant m-1$, and $|L(1, \chi^{2k-1})|$, $1 \leqslant k \leqslant m$, in (5.2), we obtain

$$h = h_0 h^*, \tag{5.4}$$

where

$$h_0 = \frac{2^{m-1}}{R} \prod_{k=1}^{m-1} \left|\sum_{r=0}^{m-1} \theta^{2kr} \ln|1 - \zeta^{g^r}|\right|, \tag{5.5}$$

$$h^* = \frac{1}{(2l)^{m-1}} |F(\theta)F(\theta^3) \cdots F(\theta^{l-2})|. \tag{5.6}$$

In the following sections we shall show that both h_0 and h^* are natural numbers. Hence formula (5.4) gives us a representation of the number h as a product of two natural numbers.

Remark 1. Sometimes h^* is denoted by h_1, and h_0 by h_2, and they are called the *first* and *second factors* of h.

Remark 2. The factor h_0 equals the number of divisor classes of the subfield $R(\zeta + \zeta^{-1})$ of degree $(l-1)/2$, which consists of all real numbers of the field $R(\zeta)$ (see Problems 1 to 4).

5.2. *The Factor* h_0

To shorten our formulas, we set

$$a_r = \ln|1 - \zeta^{g^r}| \qquad (r \leqslant 0).$$

From (5.3) we see that $a_{m+r} = a_r$. This means that the value of a_r depends only the residue of r modulo $m = (l-1)/2$. If we set

$$A = \prod_{k=1}^{m-1} \left(\sum_{r=0}^{m-1} \theta^{2kr} a_r \right),$$

then (5.5) can be written in the form

$$h_0 = \frac{2^{m-1}}{R} |A|. \tag{5.7}$$

We shall show that the product

$$(a_0 + a_1 + \cdots + a_{m-1})A$$

equals, up to sign, the determinant

$$\Delta = \det(a_{i+j})_{0 \leqslant i, j \leqslant m-1} = \begin{vmatrix} a_0 & a_1 & \cdots & a_{m-1} \\ a_1 & a_2 & \cdots & a_0 \\ \cdot & \cdot & \cdot & \cdot \\ a_{m-1} & a_0 & \cdots & a_{m-2} \end{vmatrix}.$$

Consider the cyclic group G or order m, generated by θ^2, which is a primitive mth root of 1. For $0 \leqslant k \leqslant m-1$, set $\chi_k(\theta^{2r}) = \theta^{2rk}$. The function χ_k is clearly a character of the group G. We also define a function f on G, by setting $f(\theta^{2r}) = a_r$. By Problem 13 of Section 5 of the Supplement, our product takes the form

$$\prod_{k=0}^{m-1} \left(\sum_{r=0}^{m-1} \theta^{2kr} a_r \right) = \prod_{k=0}^{m-1} \left(\sum_{r=0}^{m-1} \chi_k(\theta^{2r}) f(\theta^{2r}) \right)$$
$$= \det(f(\theta^{2(i-j)})) = \det(a_{i-j})_{0 \leqslant i, j \leqslant m-1}.$$

Since the matrices (a_{i-j}) and (a_{i+j}) differ only in the order of the columns, we have shown the desired result.

The sum $a_0 + a_1 + \cdots + a_{m-1}$ is nonzero, since

$$a_0 + a_1 + \cdots + a_{m-1} = \ln \left| \prod_{r=0}^{m-1} (1 - \zeta^{g^r}) \right| = \ln \sqrt{l} \tag{5.8}$$

by (1.5) of Chapter 3 and (5.3). Hence we can divide the determinant Δ by (5.8) to obtain a new expression for A. If we add every column in Δ to some fixed column, we obtain a column in which every entry is equal to (5.8). Hence, up to sign, A equals the determinant Δ', obtained from Δ by replacing every element in one column by 1. If we now subtract the first row from all other rows, we see that $|A|$ equals the absolute value of any minor of order $m - 1$ of the matrix

$$(a_{i+j} - a_j)_{\substack{1 \leqslant i \leqslant m-1 \\ 0 \leqslant j \leqslant m-1}}. \tag{5.9}$$

Consider the number

$$\eta = -\zeta^{(l+1)/2} = \cos\frac{\pi}{l} + i \sin\frac{\pi}{l}$$

which is a primitive root of degree $2l$ of 1. Since $\eta^2 = \zeta$, then

$$\frac{1 - \zeta^k}{1 - \zeta} = \eta^{k-1} \frac{\eta^k - \eta^{-k}}{\eta - \eta^{-1}} = \eta^{k-1} \frac{\sin(k\pi/l)}{\sin(\pi/l)}.$$

For $k \not\equiv 0 \pmod{l}$ the number on the left is a unit of the field K (see the proof of Lemma 1 of Section 1, Chapter 3); hence the numbers

$$\theta_k = \frac{\sin(k\pi/l)}{\sin(\pi/l)} \tag{5.10}$$

are also units of the field K for all $k \not\equiv 0 \pmod{l}$. These units clearly are real and positive.

There are $m = (l - 1)/2$ pairs of conjugate isomorphism of the field K into the field of complex numbers. Since the numbers $\zeta, \zeta^g, \ldots, \zeta^{g^{m-1}}$ are all nonconjugate, then the isomorphisms

$$\sigma_j \colon \zeta \to \zeta^{g^j} \qquad (j = 0, 1, \ldots, m - 1)$$

are pairwise-nonconjugate (each σ_j is conjugate to the isomorphism $\zeta \to \zeta^{-g^j} = \zeta^{g^{m+j}}$).

Let $\bar{r}$ denote the absolute value of the smallest residue of g^r in absolute value, modulo l. Then

$$\frac{1 - \zeta^{g^r}}{1 - \zeta} = \pm \eta^{g^r - 1} \theta_{\bar{r}}.$$

Applying the automorphism σ_j to this identity, we obtain

$$\frac{1 - \zeta^{g^{r+j}}}{1 - \zeta^{g^j}} = \pm (\sigma_j \eta)^{g^r - 1} \sigma_j(\theta_{\bar{r}}),$$

and after taking the logarithm of the absolute value we find that

$$a_{r+j} - a_j = \ln|\sigma_j(\theta_{\bar{r}})|. \tag{5.11}$$

We shall show that when r takes the values $1, \dots, m-1$, then $\bar{r}$ runs through $2, \dots, m$. For if $g^i \equiv \pm g^j \pmod{l}$, with $1 \leqslant i \leqslant j \leqslant m-1$, then $g^{j-i} \equiv \pm 1 \pmod{l}$ and $0 \leqslant j - i \leqslant (l-3)/2$, which is possible only for $j - i = 0$. Hence the values of $\bar{r}$ are pairwise-distinct, and since they satisfy $2 \leqslant \bar{r} \leqslant m = (l-1)/2$ and there are $m-1$ of them, then each of the numbers $2, \dots, m$ is some $\bar{r}$.

It follows from (5.11) that the matrix (5.9) differs from the matrix

$$(\ln|\sigma_j(\theta_k)|)_{\substack{2 \leqslant k \leqslant m \\ 0 \leqslant j \leqslant m-1}} \tag{5.12}$$

only in the order of the rows, and hence the absolute value $|A|$ equals the absolute value of any $(m-1)$th minor of the matrix (5.12).

We now turn to a system of fundamental units of the field K. By Lemma 4 of Section 1, Chapter 3, any unit of the field K is the product of a power of ζ with a real unit. Hence the fundamental units $\varepsilon_1, \dots, \varepsilon_{m-1}$ can be chosen real and positive. Then any positive real unit can be represented in the form $\varepsilon_1^{c_1} \cdots \varepsilon_{m-1}^{c_{m-1}}$ with the c_i rational integers. In this case the functions $l_j(\alpha)$, defined in Section 3.3 of Chapter 2, have the form $l_j(\alpha) = \ln|\sigma_j(\alpha)|^2 = 2\ln|\sigma_j(\alpha)|$, $0 \leqslant j \leqslant m-1$. With the fundamental units $\varepsilon_1, \dots, \varepsilon_{m-1}$ we form the matrix

$$(\ln|\sigma_j(\varepsilon_i)|)_{\substack{1 \leqslant i \leqslant m-1 \\ 0 \leqslant j \leqslant m-1}}. \tag{5.13}$$

Since the matrix (4.6) of Chapter 2 is obtained from (5.13) by multiplying all rows by 2, it follows from the definition of the regulator R that the absolute value of any $(m-1)$th minor of the matrix (5.13) equals $R/2^{m-1}$.

The units θ_k of the form (5.10) for $k = 2, \dots, m$ are real and positive, and they can be expressed as

$$\theta_k = \prod_{i=1}^{m-1} \varepsilon_i^{c_{ki}} \qquad (k = 2, \dots, m),$$

with c_{ki} rational integers. Since

$$\ln|\sigma_j(\theta_k)| = \sum_{i=1}^{m-1} c_{ki} \ln|\sigma_j(\varepsilon_i)|$$

the matrix (5.12) is the product of the matrix (c_{ki}) and the matrix (5.13). It follows that each $(m-1)$th minor of the matrix (5.12) equals the product of $\det(c_{ki})$ with the corresponding minor of the matrix (5.13), and this means that

$$|A| = |\det(c_{kj})| \frac{R}{2^{m-1}}.$$

Comparing with (5.7) we obtain

$$h_0 = |\det(c_{kj})|.$$

Since all c_{ki} are rational integers and $h_0 \neq 0$, we have shown that h_0 is a natural number. Further, by Lemma 1 of Section 6 of Chapter 2, we have the following result.

Theorem 2. The factor h_0 in the number of divisor classes of the lth cyclotomic field equals the index $(E : E_0)$ of the group E_0, generated by the units

$$\theta_k = \frac{\sin(k\pi/l)}{\sin(\pi/l)} \qquad \left(k = 2, \ldots, \frac{l-1}{2}\right)$$

of the field K, in the group E of all positive real units of the field K.

In view of Remark 2 at the end of Section 5.1 it is interesting to compare this result with Theorem 2 of Section 4.

5.3. *The Factor h^**

We shall show that the number h^*, defined by (5.6), also is a natural number.

The product

$$B = F(\theta)F(\theta^3) \cdots F(\theta^{l-2})$$

is an integer of the algebraic number field $R(\theta)$, where θ is a primitive $(l-1)$th root of 1. Since the complex conjugate of θ^k is θ^{l-1-k}, then B is left fixed when θ is replaced by $\bar{\theta}$, and hence is a real number. Finally, we note that $|B| = (h/h_0)(2l)^{m-1}$ is a rational number [see (5.4) and (5.6)]. It follows from these three facts that B is a rational integer. We now show that B is divisible by 2^{m-1} and l^{m-1} (here $l \neq 2$).

As in Section 5.1 we let g_s denote the least positive residue of g^s modulo l, where g is a fixed primitive root modulo l. Since

$$g_{m+s} + g_s \equiv g^{m+s} + g^s = g^s(g^{(l-1)/2} + 1) \equiv 0 \pmod{l},$$

then

$$g_{m+s} + g_s = l.$$

It follows that g_{m+s} and g_s differ by an even number. We shall now consider congruences modulo 2 in the ring of integers of the algebraic number field $R(\theta)$. Since $\theta^m = -1$, we have for odd k,

$$\begin{aligned} F(\theta^k) &= \sum_{s=0}^{m-1} (g_s\theta^{ks} + g_{m+s}\theta^{k(m+s)}) \\ &= \sum_{s=0}^{m-1} (g_s - g_{m+s})\theta^{ks} \equiv \sum_{s=0}^{m-1} \theta^{ks} \pmod{2}, \end{aligned}$$

so that

$$F(\theta^k)\,(1-\theta^k) \equiv 0 \pmod{2}.$$

This shows that the product

$$B(1-\theta)\,(1-\theta^3)\cdots(1-\theta^{l-2})$$

is divisible by 2^m. On the other hand, since θ and θ^2 are primitive roots of degrees $l-1$ and $(l-1)/2$, then

$$l-1 = \prod_{k=1}^{l-2}(1-\theta^k), \qquad \frac{l-1}{2} = \prod_{s=1}^{m-1}(1-\theta^{2s}),$$

so that

$$(1-\theta)(1-\theta^3)\cdots(1-\theta^{l-2}) = 2.$$

This shows that B is divisible by 2^{m-1}.

To show that B is divisible by l^{m-1}, we first find the decomposition of the number l into prime divisors in the field $R(\theta)$. Since l is relatively prime to $l-1$ and $l \equiv 1 \pmod{l-1}$, then by Theorem 2 of Section 2, the number l is the product of $\varphi(l-1)$ distinct prime divisors, where the norm of each prime divisor equals l. Let $\mathfrak{q}$ be one of these prime divisors. The numbers $0, 1, \theta, \ldots, \theta^{l-2}$ are pairwise-noncongruent modulo $\mathfrak{q}$ (see the proof of Lemma 3 of Section 2), so they form a complete set of residues modulo $\mathfrak{q}$. Since

$$1-g^{l-1} = \prod_{k=0}^{l-2}(1-\theta^k g) \equiv 0 \pmod{l} \tag{5.14}$$

then $\mathfrak{q}$ must divide one of the differences $1-\theta^k g$. If $1-\theta^k g \equiv 0 \pmod{\mathfrak{q}}$ and $1-\theta^s g \equiv 0 \pmod{\mathfrak{q}}$, then $\theta^k \equiv \theta^s \pmod{\mathfrak{q}}$, and this means that $\theta^k = \theta^s$. Thus $\mathfrak{q}$ divides one and only one of the differences $1-\theta^k g$ in (5.14). We shall show that k is relatively prime to $l-1$. If $(k, l-1) = d$, then by raising the congruence $1 \equiv \theta^k g \pmod{\mathfrak{q}}$ to the power $(l-1)/d$, we see that $g^{(l-1)/d}-1$ is divisible by $\mathfrak{q}$, and hence also divisible by l. But this is possible only if $d = 1$.

Since there are $\varphi(l-1)$ prime divisors $\mathfrak{q}$ of l, and $\varphi(l-1)$ numbers $1-\theta^k g$ in (5.14) with $(k, l-1) = 1$, each of the numbers $1-\theta^k g$ is divisible by one and only one $\mathfrak{q}$. Denoting this prime divisor of l by $\mathfrak{q}_k$, we have

$$1-\theta^k g \equiv 0 \pmod{\mathfrak{q}_k}, \tag{5.15}$$

and we note that if s is not relatively prime to $l-1$, then $1-\theta^s g$ is not divisible by any $\mathfrak{q}_k$. Hence l can be represented as

$$l = \prod_{(k,l-1)=1} \mathfrak{q}_k,$$

where k runs through a reduced system of residues modulo $l-1$.

We shall now show that B is divisible by l^{m-1}. In the ring of integers of the field $R(\theta)$, we have the congruence

$$F(\theta^k)(1-g\theta^k) \equiv \sum_{s=0}^{l-2} (g\theta^k)^s(1-g\theta^k)$$
$$= 1-(g\theta^k)^{l-1} = 1-g^{l-1} \equiv 0 \pmod{l},$$

so $F(\theta^k)(1-g\theta^k)$ is divisible by l. From the above results we see that $F(\theta^k)$ is divisible by l if $(k, l-1) > 1$, and by $l\mathfrak{q}_k^{-1}$ if $(k, l-1) = 1$. If $(k, l-1) > 1$, let $\mathfrak{q}_k$ denote the unit divisor. Then $F(\theta^k)$ is divisible by $l\mathfrak{q}_k^{-1}$ for all k. Hence the product $B = F(\theta)F(\theta^3)\cdots F(\theta^{l-2})$ is divisible by

$$l^m \prod_{k=1,3,\ldots,l-2} \mathfrak{q}_k^{-1} = l^m \prod_{(k,l-1)=1} \mathfrak{q}_k^{-1} = l^{m-1},$$

which completes the proof that h^* is an integer.

5.4. *The Relative Primality of h^* and l*

In Section 7.3 of Chapter 3 we saw how important it is to have criteria for determining whether h and l are relatively prime, that is, for determining whether the prime l is regular. Since $h = h_0 h^m$, then l will be a regular prime if and only if neither of the factors h_0, h^* is divisible by l. In this section we shall find a condition which is necessary and sufficient for l not to divide h^*. In the next section we shall show that if l does not divide h^*, then it also does not divide h_0, so our condition will turn out to be a criterion for the regularity of l.

Preserving the notations of Section 5.3, consider the expression

$$\frac{B}{l^{m-1}} = \prod_{k=1,3,\ldots,l-2} \frac{F(\theta^k)\mathfrak{q}_k}{l} \tag{5.16}$$

[here we identify the principal divisor (α) with the number α]. In view of (5.6) the number h^* is divisible by l if and only if the rational integer (5.16) is divisible by all prime divisors $\mathfrak{q}_s$, with $(s, l-1) = 1$. In particular, the number (5.16) will be divisible by $\mathfrak{q}^l{}_{-2} = \mathfrak{q}_{-1}$, so that at least one of the integral divisors $F(\theta^k)\mathfrak{q}_k l^{-1}$ $(k = 1, 3, \ldots, l-2)$ is divisible by $\mathfrak{q}_{-1}$. For this to happen it is necessary and sufficient that the divisor $F(\theta^k)\mathfrak{q}_k$ be divisible by $\mathfrak{q}^2_{-1}$. We shall show that this cannot happen for $k = l-2 \equiv -1 \pmod{l-1}$. By (5.15), $\theta^{-1}g \equiv 1 \pmod{\mathfrak{q}_{-1}}$, so that

$$F(\theta^{-1}) \equiv \sum_{r=0}^{l-2} (\theta^{-1}g)^r \equiv l-1 \equiv -1 \pmod{\mathfrak{q}_{-1}},$$

that is, $F(\theta^{-1})$ is not divisible by $\mathfrak{q}_{-1}$, and this means that $F(\theta^{-1})\mathfrak{q}_{-1}$ is not divisible by $\mathfrak{q}_{-1}^2$. Thus for h^* to be divisible by l it is necessary and sufficient that $F(\theta^k)$ be divisible by $\mathfrak{q}_{-1}^2$ for some $k = 1, 3, \dots, l-4$.

Up to this time we have not imposed any restrictions on the choice of the primitive root g modulo l. Now we assume that g satisfies the congruence

$$g^{l-1} \equiv 1 \pmod{l^2}$$

(if g does not satisfy this congruence, replace g by $g + xl$ with suitable x). Since the congruence (5.14) will now hold modulo l^2, then $1 - \theta^k g$ will be divisible by $\mathfrak{q}_k^2$ for any k relatively prime to $l - 1$. In particular,

$$\theta \equiv g \pmod{\mathfrak{q}_{-1}^2}.$$

With this choice of g the condition for $\mathfrak{q}_{-1}^2$ to divide $F(\theta^k)$ can easily be found. Indeed, since

$$F(\theta^k) = \sum_{s=0}^{l-2} g_s \theta^{sk} \equiv \sum_{s=0}^{l-2} g_s g^{sk} \pmod{\mathfrak{q}_{-1}^2},$$

then the number $F(\theta^k)$ is divisible by $\mathfrak{q}_{-1}^2$ if and only if

$$\sum_{s=0}^{l-2} g_s g^{sk} \equiv 0 \pmod{l^2}. \tag{5.17}$$

In order to put (5.17) in more convenient form, consider the congruence

$$g_s \equiv g^s + la_s \pmod{l^2} \qquad (0 \leqslant s \leqslant l-2), \tag{5.18}$$

where a_s is an integer. If we raise both sides of (5.18) to the power $k + 1$ ($k = 1, 3, \dots, l-4$), then we find that

$$\begin{aligned} g_s^{k+1} &\equiv g^{s(k+1)} + (k+1)g^{sk}la_s \\ &\equiv g^{s(k+1)} + (k+1)g^{sk}(g_s - g^s) \pmod{l^2}; \end{aligned}$$

that is,

$$g_s^{k+1} \equiv (k+1)g_s g^{sk} - kg^{s(k+1)} \pmod{l^2}. \tag{5.19}$$

Summing (5.19) for $s = 0, 1, \dots, l-2$ and noting that $g^{k+1} \not\equiv 1 \pmod{l}$ for $k + 1 \leqslant l - 3$ and $g^{l-1} \equiv 1 \pmod{l^2}$, we obtain

$$\sum_{s=0}^{l-2} g^{s(k+1)} = \frac{g^{(l-1)(k+1)} - 1}{g^{k+1} - 1} \equiv 0 \pmod{l^2}$$

and hence

$$\sum_{s=0}^{l-2} g_s^{k+1} \equiv (k+1) \sum_{s=0}^{l-2} g_s g^{sk} \pmod{l^2}.$$

But $k + 1 \not\equiv 0 \pmod{l}$, and therefore (5.17) is equivalent to

$$S_{k+1} = \sum_{s=0}^{l-2} g_s^{k+1} = \sum_{n=1}^{l-1} n^{k+1} \equiv 0 \pmod{l^2}.$$

Hence we have proved the following theorem.

Theorem 3. In order that the number h^* not be divisible by l, it is necessary and sufficient that none of the numbers

$$S_k = \sum_{n=1}^{l-1} n^k \qquad (k = 2, 4, \dots, l-3) \tag{5.20}$$

be divisible by l^2.

Note that each of the numbers S_k [$k \not\equiv 0 \pmod{l-1}$] is divisible by l [see (8.10)].

We reformulate Theorem 3 in terms of Bernoulli numbers (Bernoulli numbers will be defined and studied in Section 8). Since the numbers 2, 4, ..., $l-3$ are not divisible by $l-1$, then by the theorem of von Staudt (Theorem 4 of Section 8) the Bernoulli numbers $B_2, B_4, \dots, B_{l-3}$ are l-integers (l does not appear in their denominators). Further, we have the congruence

$$S_k \equiv B_k l \pmod{l^2} \qquad (k = 2, 4, \dots, l-3) \tag{5.21}$$

[in the ring of l-integral numbers; see (8.11)]. Hence the following theorem is valid.

Theorem 4. In order that h^* not be divisible by l, it is necessary and sufficient that the numerators of the Bernoulli numbers $B_2, B_4, \dots, B_{l-3}$ not be divisible by l.

For example, since the numerators of the numbers $B_2, B_4, B_6, B_8, B_{10}, B_{12}, B_{14}$ are not divisible by 17, then $l = 17$ is regular.

Remark. To determine whether h^* and l are relatively prime, it is not necessary to find the precise value of the Bernoulli numbers. It suffices to consider the recurrence relation (8.2) as a congruence modulo l and to use these congruences to compute the sequence $B_2, B_4, \dots, B_{l-3}$. The number h^* will be relatively prime to l if and only if none of these numbers is divisible by l.

PROBLEMS

1. Let K_0 be the subfield of the lth cyclotomic field $R(\zeta)$ which consists of all real numbers in $R(\zeta)$. Show that $K_0 = R(\zeta + \zeta^{-1})$ and is of degree $(l-1)/2$. Further, show that the field K_0 has discriminant $l^{(l-3)/2}$, and that its regulator R_0 is related to the regulator R of the field $R(\zeta)$ by $R = 2^{(l-3)/2} R_0$.

2. Let p be a prime different from l, and let f be the smallest natural number for which $p^f \equiv 1 \pmod{l}$. Show that the number p factors in the field K_0 as the product of $(l-1)/2f$ prime divisors of degree f when f is odd, and as the product of $(l-1)/f$ prime divisors of degree $f/2$ when f is even.

3. Show that the ζ-function $\zeta_{K_0}(s)$ of the field K_0 satisfies

$$\lim_{s \to 1+0} (s-1)\zeta_{K_0}(s) = \prod_{\substack{\chi \neq \chi_0 \\ \chi(-1)=1}} L(1, \chi),$$

where χ runs through all even numerical characters modulo l, except the unit character χ_0.

4. Show that the real subfield $R(\zeta + \zeta^{-1})$ of the lth cyclotomic field has h_0 divisor classes, where h_0 is the factor of the number of divisor classes of the field $R(\zeta)$.

5. Show that h^* is given by

$$h^* = \frac{1}{(2l)^{m-1}} \left| \det(g_{m+i+j} - g_{i+j})_{0 \leqslant i,\, j \leqslant m-1} \right|,$$

where g_s is the smallest positive residue of the number g^s modulo $l = 2m + 1$ (g is a primitive root modulo l).

6. Compute the factor h^* for $l = 7$.

7. Show that the prime number 37 is irregular.

6. A Criterion for Regularity

Our goal in this section is to show that when the factor h^* of the number of divisor classes of the lth cyclotomic field is not divisible by l, then the factor h_0 is also not divisible by l, and hence the prime l is regular. En route we shall also show that when l is regular, every unit of the field $K = R(\zeta)$ which is congruent modulo l to a rational integer is an lth power. On this assertion, known as Kummer's lemma, is based the proof of the second case of Fermat's theorem for regular primes. Both the regularity criterion and Kummer's lemma will be found as simple corollaries of the following result. If $l \nmid h^*$ and $K_\mathfrak{l}$ is the $\mathfrak{l}$-adic completion of the field $K = R(\zeta)$ where $\mathfrak{l} = (1 - \zeta)$, then the numbers $\log \theta_k^{\,l-1}$ $[k = 2, 3, \dots, (l-1)/2]$ form a basis for the set of all "real" $\mathfrak{l}$-adic integers with zero trace [the units θ_k are defined by (5.10)].

6.1. *The Field of* $\mathfrak{l}$*-adic Numbers*

We know that the cyclotomic field $K = R(\zeta)$, $\zeta = \cos 2\pi/l + i \sin 2\pi/l$, $l \geqslant 3$ a prime, has degree $l - 1$ and that the number l has the factorization $l = \mathfrak{l}^{l-1}$, where $\mathfrak{l} = (1 - \zeta)$ is a prime divisor of first degree.

We consider the $\mathfrak{l}$-adic completion $K_\mathfrak{l}$ of the field K. The elements of this completion are called $\mathfrak{l}$-adic numbers. The complete field $K_\mathfrak{l}$ contains a subfield which is canonically isomorphic to the field R_l of l-adic numbers (this subfield coincides with the completion of the field R in $K_\mathfrak{l}$). Using this canonical isomorphism we shall assume that $R_l \subset K_\mathfrak{l}$.

Since $\mathfrak{l}$ is the only prime divisor which divides l, then by Theorem 1 of Section 2, Chapter 4, the degree of the extension $K_\mathfrak{l}/R_l$ equals $l - 1 = (K : R)$. Hence [see (2.6) of Chapter 4] for any $\alpha \in K$ we have

$$N_{K/R}(\alpha) = N_{K_\mathfrak{l}/R_l}(\alpha). \tag{6.1}$$

Lemma 1. There is an element λ in the ring of $\mathfrak{l}$-adic integers such that:

(1) $\lambda^{l-1} + l = 0$,

(2) $\lambda \equiv \zeta - 1 \pmod{\lambda^2}$.

The element λ is uniquely determined by (1) and (2).

In view of (1.5) of Chapter 3 we have

$$\frac{l}{(1-\zeta)^{l-1}} = (1+\zeta)(1+\zeta+\zeta^2)\cdots(1+\zeta+\cdots+\zeta^{l-2}).$$

We now consider congruences modulo the prime element $1-\zeta$ of the field $K_{\mathfrak{l}}$ [recall that $v_{\mathfrak{l}}(1-\zeta)=1$]. Since $\zeta \equiv 1 \pmod{1-\zeta}$ and $(l-1)!+1 \equiv 0 \pmod{l}$ (Wilson's theorem), then

$$\frac{l}{(1-\zeta)^{l-1}} \equiv 2\cdot 3\cdots(l-1) \equiv -1 \pmod{1-\zeta}.$$

We shall show that the $\mathfrak{l}$-adic unit

$$\alpha = \frac{-l}{(1-\zeta)^{l-1}},$$

which is congruent to 1 modulo $1-\zeta$, can be represented in the form $\alpha = \gamma^{l-1}$. Consider the polynomial $F(X) = X^{l-1} - \alpha$. Since $F(1) \equiv 0 \pmod{1-\zeta}$ and $F'(1) \not\equiv 0 \pmod{1-\zeta}$, then there is a unit γ in K for which $F(\gamma)=0$ (see Section 1.2 of Chapter 4). Hence $\alpha = \gamma^{l-1}$, as was claimed. Setting $\lambda = (\zeta-1)\gamma$, we obtain a prime element λ with the desired properties. Any other number λ_1, satisfying the first condition of the lemma, has the form $\lambda\theta$, where θ is a $(l-1)$th root of 1. From $\lambda\theta \equiv \lambda \pmod{\lambda^2}$ it follows that $\theta \equiv 1 \pmod{\lambda}$. If the root θ were different from 1, then $l-1$ would be divisible by λ, which is impossible. Hence $\theta = 1$ and $\lambda_1 = \lambda$. Lemma 1 is proved.

From now on λ will denote that prime element of the field K which is uniquely determined by the conditions of Lemma 1.

For each k which is relatively prime to l the correspondence $\zeta \to \zeta^k$ determines an automorphism σ_k of the extension K/R. If σ is any of these automorphisms, then the function $v'(\alpha) = v_{\mathfrak{l}}(\sigma(\alpha))$, $\alpha \in K$, is a valuation of the field K, and is an extension of the l-adic valuation v_l of the field R. But there is only one extension of v_l to the field K, namely $v_{\mathfrak{l}}$. Hence $v' = v_{\mathfrak{l}}$, and this means that $v_{\mathfrak{l}}(\sigma(\alpha)) = v_{\mathfrak{l}}(\alpha)$ for all $\alpha \in K$. It follows from this that the automorphism σ takes any Cauchy sequence of elements of K (relative to the metric which corresponds to the prime divisor $\mathfrak{l}$) to another Cauchy sequence in K. This allows us to extend the automorphism $\sigma = \sigma_k$ to the field $K_{\mathfrak{l}}$. Namely, if $\xi = \lim_{n\to\infty}\alpha_n$ ($\alpha_n \in K$), then we can set

$$\sigma(\xi) = \lim_{n\to\infty}\sigma(\alpha_n)$$

[it is easily verified that $\sigma(\xi)$ does not depend on the choice of the sequence $\{\alpha_n\}$, and also that the mapping $\xi \to \sigma(\xi)$ is an automorphism of the extension $K_{\mathfrak{l}}/R_l$].

Since the extension $K_{\mathfrak{l}}/R_l$ has degree of inertia 1 and ramification index $l-1$, then by Theorem 4 of Section 1, Chapter 4, all $\mathfrak{l}$-adic integers can be uniquely represented in the form

$$a_0 + a_1\lambda + \cdots + a_{l-2}\lambda^{l-2} \tag{6.2}$$

where the a_i are l-adic integers.

The subfield of real numbers of the field K consists of all $\alpha \in K$ which are left fixed by the automorphism $\sigma_{-1}: \zeta \to \zeta^{-1}$. We shall determine which $\mathfrak{l}$-adic numbers are invariant under the automorphism σ_{-1}. Since $\lambda^{l-1} = -l$, then also $(\sigma_{-1}(\lambda))^{l-1} = -l$, and this means that $\sigma_{-1}(\lambda) = \lambda\theta$, where θ is an $(l-1)$th root of 1. By Problem 4 of Section 3, Chapter 1, the root θ is contained in R_l, so that

$$\sigma_{-1}{}^2(\lambda) = \sigma_{-1}(\sigma_{-1}(\lambda)) = \sigma_{-1}(\theta\lambda) = \theta\sigma_{-1}(\lambda) = \theta^2\lambda,$$

and since also $\sigma_{-1}{}^2(\lambda) = \lambda$, then $\theta = \pm 1$. If $\theta = 1$, then any $\mathfrak{l}$-adic number which can be represented in the form (6.2) with l-adic coefficients a_i, would be left fixed by the automorphism σ_{-1}, and this is not the case. Hence $\theta = -1$, and $\sigma_{-1}(\lambda) = -\lambda$. Hence when the automorphism σ_{-1} acts on the field $K_{\mathfrak{l}}$, the $\mathfrak{l}$-adic numbers which are left fixed are those of the form

$$\sum_{i=0}^{m-1} b_i\lambda^{2i} \qquad \left(b_i \in R_l, \quad m = \frac{l-1}{2}\right). \tag{6.3}$$

The set of all such numbers is a subfield of $K_{\mathfrak{l}}$ of degree $m = (l-1)/2$ over R_l. It will be convenient to call them "real" $\mathfrak{l}$-adic numbers.

We compute the trace of the $\mathfrak{l}$-adic number (6.2) (relative to the extension $K_{\mathfrak{l}}/R_l$). For any $i = 1, \ldots, l-2$, the matrix of the linear transformation $\xi \to \lambda^i\xi$ ($\xi \in K_{\mathfrak{l}}$) with respect to the basis $1, \lambda, \ldots, \lambda^{l-2}$ will have zeros on the main diagonal (since $\lambda^{l-1} = -l$), and therefore $\mathrm{Sp}_{K_{\mathfrak{l}}/R_l}(\lambda^i) = 0$ (for $i = 1, \ldots, l-2$). It follows that the trace of the number (6.2) equals $a_0(l-1)$. The $\mathfrak{l}$-adic numbers with trace (over R_l) equal to zero are thus characterized by having the coefficient a_0 equal to zero in (6.2).

We shall be interested in the set $\mathfrak{M}$ of all "real" $\mathfrak{l}$-adic integers with zero trace. It is clear from the above remarks that $\mathfrak{M}$ coincides with the set of all linear combinations of the form

$$\sum_{i=1}^{m-1} b_i\lambda^{2i} \tag{6.4}$$

where the b_i are l-adic integers.

We consider the functions $\log \varepsilon$ and $\exp \alpha$ over the field $K_{\mathfrak{l}}$, which are defined by power series (see Section 5.2 of Chapter 4). Since the ramification index e of the extension K_{I}/R_l equals $l-1$, then the number $[e/(l-1)]+1$ equals 2, and this means that the series $\exp \alpha$ converges for all integers $\alpha \in K_{\text{I}}$, divisible by λ^2. As we know, the function $\log \varepsilon$ is defined for all principal units of the field K_{I}.

If ε is a principal unit of the field K, that is, $\varepsilon \equiv 1 \pmod{\lambda}$, then for any automorphism σ_k we again have $\sigma_k(\varepsilon) \equiv 1 \pmod{\lambda}$, and this means that $\log \sigma_k(\varepsilon)$ is defined. But then (Corollary 1 of Theorem 11, Section 2 of the Supplement),

$$\mathrm{Sp}_{K_{\text{I}}/R_l} \log \varepsilon = \sum_{k=1}^{l-1} \sigma_k(\log \varepsilon) = \sum_k \log(\sigma_k(\varepsilon))$$
$$= \log\left(\prod_k \sigma_k(\varepsilon)\right) = \log(N_{K_{\text{I}}/R_l}\varepsilon).$$

Now assume that ε is a unit of the field K. It is clear that ε is also a unit in the field K, but $\log \varepsilon$ is not necessarily defined, since ε will not, in general, be a principal unit. But for some rational integer a which is not divisible by l we shall have $\varepsilon \equiv a \pmod{\lambda}$. From $a^{l-1} \equiv 1 \pmod{\lambda}$ it follows that $\varepsilon^{l-1} \equiv 1 \pmod{\lambda}$; that is, ε^{l-1} is a principal unit in K_{I}. The logarithm $\log \varepsilon^{l-1}$ is thus defined, and by formula (6.1)

$$\mathrm{Sp}_{K_{\text{I}}/R_l}(\log \varepsilon^{l-1}) = \log(N_{K_{\text{I}}/R_l}\varepsilon^{l-1}) = \log(N_{K/R}\varepsilon^{l-1}) = 0;$$

that is, the I-adic integer $\log \varepsilon^{l-1}$ has zero trace. If ε is a real unit of the field K, then it is clear that $\log \varepsilon^{l-1}$ will be "real."

Hence for any real unit ε of the field K the I-adic number $\log \varepsilon^{l-1}$ belongs to the set $\mathfrak{M}$; that is, it can be represented in the form (6.4). In particular, this holds for the units θ_k ($k = 2, 3, \ldots, m = (l-1)/2$) defined by (5.10). Thus we have

$$\log \theta_k^{l-1} = \sum_{i=1}^{m-1} b_{ki}\lambda^{2i} \qquad (2 \leqslant k \leqslant m) \tag{6.5}$$

where the coefficients b_{ki} are l-adic integers.

Our problem is now to show that when l does not divide h^* (the factor of the number of divisor classes of the field K), then the I-adic numbers $\log \theta_k^{l-1}$ form a basis for $\mathfrak{M}$ over the ring of l-adic integers in the sense that any $\xi \in \mathfrak{M}$ has a unique representation as a linear combination of the $\log \theta_k^{l-1}$ with l-adic integer coefficients. To do this it clearly suffices to show that $\det(b_{ki})$ is an l-adic unit, that is, that $\det(b_{ki}) \not\equiv 0 \pmod{l}$.

6.2. *Some Congruences*

The series for $\exp x$ in the field K_{I} converges only for those integers x

which are divisible by λ^2. We also consider the polynomial

$$E(x) = 1 + \frac{x}{1!} + \frac{x^2}{2!} + \cdots + \frac{x^{l-1}}{(l-1)!},$$

obtained from the series for $\exp x$ by deleting all terms with degrees $\geqslant l$. Since the coefficients $1/k!$ for $k \leqslant l - 1$ are l-adic integers, then $E(x)$ will be a principal unit of the field K_{I} for all integral $x \equiv 0 \pmod{\lambda}$.

We know that the formal product of the series $\exp x$ and $\exp y$ equals the series $\exp(x + y)$. It follows that

$$E(x)E(y) = E(x + y) + F(x, y), \tag{6.6}$$

where $F(x, y)$ is a polynomial with l-adic integral coefficients in which all terms have degree $\geqslant l$.

Lemma 2. The congruence

$$E(\lambda)^l \equiv 1 \pmod{\lambda^{2l-1}}$$

holds in the ring of I-adic integers.

Set

$$E(x) = 1 + xg(x),$$

where $g(x) = 1 + x/2! + \cdots + x^{l-2}/(l-1)!$ is a polynomial with l-adic integer coefficients. Then

$$\begin{aligned} E(x)^l &= 1 + C_l^{\,1} xg(x) + \cdots + C_l^{\,l-1}(xg(x))^{l-1} + x^l g(x)^l \\ &= 1 + lh(x) + x^l g(x)^l, \end{aligned}$$

where $h(x)$ is also polynomial with l-adic integer coefficients. On the other hand, by (6.6) we see that

$$E(x)^l = E(lx) + x^l M(x),$$

and this means that

$$lh(x) = \frac{lx}{1!} + \frac{(lx)^2}{2!} + \cdots + \frac{(lx)^{l-1}}{(l-1)!} + x^l H(x), \tag{6.7}$$

where $H(x) = M(x) - g(x)^l$. Looking at the coefficients for the various powers of x in this equation, we see that all coefficients of $H(x)$ are l-adic integers divisible by l. Dividing (6.7) by l, we arrive at

$$h(x) = x + \frac{lx^2}{2!} + \cdots + \frac{l^{l-2}x^{l-1}}{(l-1)!} + x^l G(x),$$

where $G(x)$ has l-adic integer coefficients. If we set $x = \lambda$, we obtain the equation

$$h(\lambda) \equiv \lambda \pmod{\lambda^l},$$

and this means that

$$lh(\lambda) \equiv l\lambda \pmod{\lambda^{2l-1}}. \tag{6.8}$$

Further, since $g(\lambda) \equiv 1 \pmod{\lambda}$, then $g(\lambda)^l \equiv 1 \pmod{\lambda^l}$ so that

$$\lambda^l g(\lambda)^l \equiv \lambda^l \pmod{\lambda^{2l}}. \tag{6.9}$$

From (6.8) and (6.9) we obtain

$$E(\lambda)^l = 1 + lh(\lambda) + \lambda^l g(\lambda)^l \equiv 1 + l\lambda + \lambda^l = 1 \pmod{\lambda^{2l-1}}$$

(since $l\lambda + \lambda^l = 0$), which proves the lemma.

Lemma 3. The following congruence holds for any natural number k:

$$E(k\lambda) \equiv \zeta^k \pmod{\lambda^l}.$$

It follows from formula (5.6) that

$$E(k\lambda) \equiv E(\lambda)^k \pmod{\lambda^l},$$

so it suffices to prove the lemma for the case $k = 1$.

By the definition of the prime element λ we have $\zeta \equiv 1 + \lambda \pmod{\lambda^2}$. On the other hand, $E(\lambda) \equiv 1 + \lambda \pmod{\lambda^2}$, and therefore

$$\zeta^{-1}E(\lambda) \equiv 1 \pmod{\lambda^2}.$$

Set

$$\zeta^{-1}E(\lambda) = 1 + \lambda^2\gamma,$$

where γ is an l-adic integer. Raising this equation to the lth power and using Lemma 2, we obtain the congruence

$$\gamma\left(l\lambda^2 + \frac{l(l-1)}{2}\gamma\lambda^4 + \cdots + \gamma^{l-1}\lambda^{2l}\right) \equiv 0 \pmod{\lambda^{2l-1}}.$$

The expression in parentheses is divisible by λ^{l+1} (and by no higher power of λ), so $\gamma \equiv 0 \pmod{\lambda^{l-2}}$, and

$$\zeta^{-1}E(\lambda) \equiv 1 \pmod{\lambda^l},$$

which proves the lemma.

We also consider the polynomial

$$L(1 + x) = x - \frac{x^2}{2} + \cdots + (-1)^{l-2}\frac{x^{l-1}}{l-1}, \tag{6.9'}$$

obtained from the series $\log(1 + x)$ by deleting terms of degree $\geqslant l$.

Lemma 4. If the $\mathfrak{l}$-adic integer α is divisible by λ^2, then

$$L(1 + \alpha) \equiv \log(1 + \alpha) \pmod{\lambda^l}.$$

Indeed, for $n \geqslant l$ we have

$$v_{\mathfrak{l}}\left(\frac{\alpha^n}{n}\right) \geqslant 2n - v_{\mathfrak{l}}(n) \geqslant 2n - (l-1)\frac{\ln n}{\ln l} \geqslant$$

$$\geqslant l + (n - l) + \frac{(l-1)n}{\ln l}\left(\frac{\ln l}{l-1} - \frac{\ln n}{n-1}\right) \geqslant l$$

(see Section 5.2 of Chapter 4).

Lemma 5. Let ε_1 and ε_2 be principal $\mathfrak{l}$-adic units. Then

$$L(\varepsilon_1\varepsilon_2) \equiv L(\varepsilon_1) + L(\varepsilon_2) \pmod{\lambda^l}.$$

Since the series $\log(1 + x + y + xy)$ equals the sum of the series $\log(1 + x)$ and $\log(1 + y)$, then

$$L(1 + x + y + xy) = L(1 + x) + L(1 + y) + G(x, y),$$

where the polynomial $G(x, y)$ contains only terms of degree $\geqslant l$ and has l-adic integer coefficients. The assertion of Lemma 5 follows from the fact that $G(x, y) \equiv 0 \pmod{\lambda^l}$ if x and y are divisible by λ.

Lemma 6. The congruence

$$L(\zeta) \equiv \lambda \pmod{\lambda^l}$$

holds in the ring of $\mathfrak{l}$-adic integers.

To prove this we use the formal equality $\log \exp x = x$. From this it follows easily that

$$L(E(x)) = x + H(x),$$

where $H(x)$ is a polynomial in which all terms have degree $\geqslant l$, and which has l-adic integer coefficients. Setting $x = \lambda$ and using Lemma 3 for $k = 1$ we obtain the desired congruence.

Remark. Let $\mathfrak{U}$ be the multiplicative group of cosets modulo λ^l in the group of all principal $\mathfrak{l}$-adic units, and let $\mathfrak{X}$ be the additive group of cosets modulo λ^l in the group of all l-adic integers which are divisible by λ. It is now easily seen that the mapping $\varepsilon \to L(\varepsilon)$ induces an isomorphism of the group $\mathfrak{U}$ onto the group $\mathfrak{X}$. The inverse isomorphism $\mathfrak{X} \to \mathfrak{U}$ is induced by the mapping $\alpha \to E(\alpha)$ $[\alpha \equiv 0 \pmod{\lambda}]$.

6.3. *A Basis for the Real* $\mathfrak{l}$-*Adic Integers in the Case* $(h^*, l) = 1$

We return to the question which was raised at the end of Section 6.1. To determine whether the determinant $\det(b_{ki})$ is divisible by l, it is only necessary to consider the coefficients b_{ki} modulo l. It is clear that two l-adic integers of the form (6.2) are congruent modulo l if and only if their corresponding coefficients in the expansion (6.2) are congruent modulo l (in the ring of l-adic integers). Hence to find the b_{ki} modulo l we may replace the numbers $\log \theta_k^{\,l-1}$ by any $\mathfrak{l}$-adic integers congruent to them modulo l (that is, modulo λ^{l-1}).

We use the notations of Section 5.2. The principal unit $\theta_k^{\,l-1}$ is real, hence congruent to 1 modulo λ^2, so that, by Lemma 4,

$$\log \theta_k^{\,l-1} \equiv L(\theta_k^{\,l-1}) \pmod{\lambda^l}. \tag{6.10}$$

We now compute $L(\theta_k^{\,l-1})$. Since

$$\theta_k = \frac{\zeta^k - 1}{\zeta - 1}\,\eta^{1-k},$$

then

$$\theta_k^{\,l} = (1 + \zeta + \cdots + \zeta^{k-1})^l(-1)^{1-k}.$$

But $\zeta \equiv 1 \pmod{\lambda}$, so that

$$1 + \zeta + \cdots + \zeta^{k-1} \equiv k \pmod{\lambda},$$

and hence

$$(1 + \zeta + \cdots + \zeta^{k-1})^l \equiv k^l \pmod{\lambda^l}.$$

Since $k^l \equiv k \pmod{\lambda^{l-1}}$, then also

$$(1 + \zeta + \cdots + \zeta^{k-1})^l \equiv k \pmod{\lambda^{l-1}}.$$

Thus

$$\theta_k^{\,l-1} \equiv \theta_k^{\,-1}k(-1)^{1-k} \equiv k\,\frac{\zeta - 1}{\zeta^k - 1}(-\eta)^{k-1} \pmod{\lambda^{l-1}},$$

or

$$\theta_k^{\,l-1} \equiv \frac{\zeta - 1}{\lambda}\left(\frac{\zeta^k - 1}{k\lambda}\right)^{-1}\zeta^{(k-1)[(l+1)/2]} \pmod{\lambda^{l-1}}.$$

By Lemma 5 we have

$$L(\theta_k^{\,l-1}) \equiv L\left(\frac{\zeta - 1}{\lambda}\right) - L\left(\frac{\zeta^k - 1}{k\lambda}\right) + (k-1)\,\frac{l+1}{2}\,L(\zeta) \pmod{\lambda^{l-1}}.$$

But by Lemma 3,

$$\frac{\zeta^k - 1}{k\lambda} \equiv \frac{E(k\lambda) - 1}{k\lambda} \pmod{\lambda^{l-1}},$$

and, therefore, using Lemma 6, we obtain

$$L(\theta_k^{\,l-1}) \equiv L\left(\frac{E(\lambda)-1}{\lambda}\right) - \frac{\lambda}{2} - L\left(\frac{E(k\lambda)-1}{k\lambda}\right) + \frac{k\lambda}{2} \pmod{\lambda^{l-1}}.$$

We now show that

$$L\left(\frac{E(x)-1}{x}\right) - \frac{x}{2} = \sum_{k=1}^{m-1} \frac{B_{2k}x^{2k}}{(2k)!\,2k} + x^{l-1}R(x), \tag{6.11}$$

where the polynomial $R(x)$ has l-adic integer coefficients and B_{2k} are the Bernoulli numbers (see Section 8). We use the identity

$$\frac{x}{e^x-1} = \sum_{n=0}^{\infty} \frac{B_n}{n!} x^n.$$

Since $B_1 = -\frac{1}{2}$, and all remaining Bernoulli numbers with odd index equal zero, then our identity can be written in the form

$$\frac{e^x}{e^x-1} - \frac{1}{2} - \frac{1}{x} = \sum_{k=1}^{\infty} \frac{B_{2k}}{(2k)!} x^{2k-1}.$$

After integrating we obtain

$$\ln \frac{e^x-1}{x} - \frac{x}{2} = \sum_{k=1}^{\infty} \frac{B_{2k}}{(2k)!\,2k} x^{2k} \tag{6.12}$$

(the constant term of the series equals zero, since for $x = 0$ the function on the left vanishes). The formula (6.11) now follows from (6.12). If we substitute the value $k\lambda$ for x in (6.11), we find that

$$L\left(\frac{E(k\lambda)-1}{k\lambda}\right) - \frac{k\lambda}{2} \equiv \sum_{i=1}^{m-1} \frac{B_{2i}k^{2i}\lambda^{2i}}{(2i)!\,2i} \pmod{\lambda^{l-1}},$$

and hence

$$L(\theta_k^{l-1}) \equiv \sum_{i=1}^{m-1} \frac{B_{2i}(1-k^{2i})\lambda^{2i}}{(2i)!\,2i} \pmod{\lambda^{l-1}}. \tag{6.12$'$}$$

This shows that the coefficients b_{ki} in (6.5) satisfy

$$b_{ki} \equiv \frac{B_{2i}(1-k^{2i})}{(2i)!\,2i} \pmod{l} \qquad \left(2 \leqslant k \leqslant m = \frac{l-1}{2}, \quad 1 \leqslant i \leqslant m-1\right).$$

But then $\det(b_{ki})$ is congruent modulo l to the determinant

$$\prod_{i=1}^{m-1} \frac{(-1)^{m-1}B_{2i}}{(2i)!\,2i} \begin{vmatrix} 2^2-1 & 2^4-1 & \cdots & 2^{l-3}-1 \\ 3^2-1 & 3^4-1 & \cdots & 3^{l-3}-1 \\ \cdot & \cdot & \cdot & \cdot \\ m^2-1 & m^4-1 & \cdots & m^{l-3}-1 \end{vmatrix}.$$

We easily evaluate this determinant by reducing it to the Vandermonde determinant. It equals the product

$$\prod_{1 \leqslant s \leqslant r \leqslant m} (r^2 - s^2) = \prod_{s<r} (r + s)(r - s),$$

in which no factor is divisible by l. If $h^* \not\equiv 0 \pmod{l}$, then the numerators of the Bernoulli numbers $B_2, \ldots, B_{l-3}$ are not divisible by l, and we find that

$$\det(b_{ki}) \not\equiv 0 \pmod{l}.$$

We have proved the following theorem.

Theorem 1. If $h^* \not\equiv 0 \pmod{l}$, then the "real" l-adic integers with zero trace are uniquely represented as linear combinations

$$\sum_{k=2}^{m} a_k \log \theta_k^{l-1} \tag{6.13}$$

with l-adic integer coefficients.

6.4. *A Criterion for Regularity and Kummer's Lemma*

Theorem 1 allows us to prove easily the following theorem.

Theorem 2. If the factor h^* of the number of divisor classes of the lth cyclotomic field $R(\zeta)$ is not divisible by l, then the factor h_0 is also not divisible by l.

Proof. Assuming that $h_0 = (E : E_0)$ is divisible by l (see the notations of Theorem 2 of Section 5), we can find a positive real unit $\varepsilon \in E$, which is not contained in E_0, but for which $\varepsilon^l \in E$. Then

$$\varepsilon^l = \prod_{k=2}^{m} \theta_k^{c_k} \tag{6.14}$$

where the rational integers c_k are not all divisible by l (otherwise ε would belong to E_0). Raising (6.14) to the power $l - 1$ and taking the logarithm (in the field $K_\mathfrak{l}$), we obtain

$$l \log \varepsilon^{l-1} = \sum_{k=2}^{m} c_k \log \theta_k^{l-1}. \tag{6.15}$$

Since the number $\log \varepsilon^{l-1}$ belongs to $\mathfrak{M}$, it has a representation in the form (6.13). Comparing this representation with (6.15), we conclude that all the expressions c_k/l are l-adic integers. But this is impossible, since not all c_k are divisible by l. This contradiction proves Theorem 2.

Corollary. The prime number $l \geqslant 3$ is regular if and only if the numerators of the Bernoulli numbers $B_2, B_4, \ldots, B_{l-3}$ are not divisible by l.

Theorem 3 (Kummer's Lemma). Let l be a regular prime number. If a unit of the lth cyclotomic field $R(\zeta)$ is congruent modulo l to a rational integer, then the unit is the lth power of another unit.

Proof. Let $\varepsilon \equiv a \pmod{l}$. We first show that ε is a real unit. If $\varepsilon = \zeta^k \varepsilon_1$, with ε_1 a real unit, then $\varepsilon_1 \equiv b \pmod{\lambda^2}$ with b a rational integer, and $\zeta^k \equiv 1 + k\lambda \pmod{\lambda^2}$. From $a \equiv b(1 + k\lambda) \pmod{\lambda^2}$ it follows that $k \equiv 0 \pmod{l}$, which proves our assertion. Since $-1 = (-1)^l$, then we can assume that $\varepsilon > 0$; that is, $\varepsilon \in E$. From the congruence $\varepsilon^{l-1} \equiv a^{l-1} \equiv 1 \pmod{l}$ it follows that $\log \varepsilon^{l-1} \equiv 0 \pmod{l}$, and therefore by Theorem 1,

$$\log \varepsilon^{l-1} = \sum_{k=2}^{m} lc_k \log \theta_k^{\,l-1}, \tag{6.16}$$

with l-adic integral c_k. On the other hand, since the subgroup E_0 is of finite index in E, then $\varepsilon^a \in E_0$ for some natural number a, and hence

$$\varepsilon^a = \sum_{k=2}^{m} \theta_k^{\,d_k} \tag{6.17}$$

with rational integers d_k. We can assume that the set of numbers $a, d_2, \ldots, d_m$ has greatest common divisor 1 (since the group E has no elements of finite order). Raising (6.17) to the power $l-1$ and taking the logarithm (in the field $K_{\mathfrak{l}}$), we obtain

$$a \log \varepsilon^{l-1} = \sum_{k=2}^{m} d_k \log \theta_k^{\,l-1}.$$

Comparing with (6.16), we arrive at

$$d_k = lac_k \qquad (k = 2, \ldots, m).$$

Since the numbers ac_k are l-adic integers, then it follows that all d_k are divisible by l, and this means that ε^a is an lth power: $\varepsilon^a = \varepsilon_1{}^l$, where $\varepsilon_1 \in E_0$. Since $(a, d_2, \ldots, d_m) = 1$, then a is relatively prime to l, and by picking rational integers u and v such that $1 = au + lv$, we find

$$\varepsilon = (\varepsilon^a)^u (\varepsilon^v)^l = (\varepsilon_1{}^u \varepsilon^v)^l,$$

which proves the theorem.

PROBLEMS

1. Let p be a prime number of the form $4n + 1$, $\zeta = \cos 2\pi/p + i \sin 2\pi/p$, $\lambda = \zeta - 1$, $m = (p-1)/2$. Set

$$\xi = \prod_{k=1}^{p-1} \theta_k^{\,-(k/p)},$$

where $\theta_k = \sin(k\pi/p)[\sin(\pi/p)]^{-1}$, $1 \leqslant k \leqslant p-1$. Show that the congruence

$$L(\xi^{p-1}) \equiv \frac{2B_m}{m!}\lambda^m \equiv -2B_m\sqrt{p} \pmod{\lambda^{m+1}}.$$

holds in the pth cyclotomic field. Here L denotes the function defined by (6.9′) and B_m the Bernoulli number. [Use (6.12′) and Problem 14, Section 4.]

2. Let $\varepsilon = T + U\sqrt{p} > 1$ be a fundamental unit in the quadratic field $R(\sqrt{p})$, where $p \equiv 1 \pmod 4$, and let h be the number of divisor classes of this field. Using the preceding problem and Theorem 2 of Section 4, show that

$$hU \equiv TB_m \pmod p \qquad \left(m = \frac{p-1}{2}\right)$$

(in the ring of p-integral rational numbers).

7. The Second Case of Fermat's Theorem for Regular Exponents

7.1. *Fermat's Theorem*

Theorem 1. If the prime number $l \geqslant 3$ is regular, then the equation

$$x^l + y^l = z^l \tag{7.1}$$

has no solution in nonzero rational numbers x, y, z.

Proof. Assume that x, y, z are relatively prime (nonzero) integers which satisfy (7.1). Since the first case of Fermat's theorem has already been treated in Section 7.3 of Chapter 3, we may assume that one (and only one) of these numbers is divisible by l. We shall let l divide z [if, for example, y is divisible by l, then we can write (7.1) in the form $x^l + (-z)^l = (-y)^l$]. Let $z = l^k z_0$, where $(z_0, l) = 1$, $k \geqslant 1$. In the lth cyclotomic field $R(\zeta)$, the number l has the factorization $l = (1-\zeta)^{l-1}\varepsilon$, where ε is a unit in $R(\zeta)$ (Lemma 1 of Section 1, Chapter 3). Hence we can put (7.1) in the form

$$x^l + y^l = \varepsilon(1-\zeta)^{lm} z_0{}^l, \tag{7.2}$$

where $m = k(l-1) > 0$. To prove the theorem it suffices to show that an equation of the form (7.2) is impossible. We shall actually show somewhat more. Not only will we show that an equation of the form (7.2) is impossible in rational integers x, y, and z_0 relatively prime to l, but even that it is impossible in integers of the field $R(\zeta)$ which are relatively prime to $1-\zeta$. Assuming the converse, we take that solution of (7.2) in which the exponent $m \geqslant 1$ is smallest. To avoid introducing new notation, we shall assume this solution to be given by (7.2). Hence x, y, and z_0 denote integers of $R(\zeta)$ which are relatively prime to $1-\zeta$, and ε is some unit of the field $R(\zeta)$.

As in Section 6, $\mathfrak{l}$ denotes the prime divisor $(1-\zeta)$ of the field $R(\zeta)$. We factor the left side of (7.2) into linear terms and then pass to the corresponding equation in divisors. We obtain

$$\prod_{k=0}^{l-1}(x+\zeta^k y)=\mathfrak{l}^{lm}\mathfrak{a}^l, \tag{7.3}$$

where the divisor $\mathfrak{a}=(z_0)$ is relatively prime to $\mathfrak{l}$. Since $lm \geqslant l > 0$, it follows from (7.3) that at least one of the terms on the left is divisible by $\mathfrak{l}$. But since

$$x+\zeta^i y = x+\zeta^k y - \zeta^k(1-\zeta^{i-k})y,$$

then all the numbers

$$x+\zeta^k y \qquad (0 \leqslant k \leqslant l-1) \tag{7.4}$$

are divisible by $\mathfrak{l}$. If for some $0 \leqslant k < i \leqslant l-1$,

$$x+\zeta^k y \equiv x+\zeta^i y \pmod{\mathfrak{l}^2},$$

then also $\zeta^k y(1-\zeta^{i-k}) \equiv 0 \pmod{\mathfrak{l}^2}$, and this is impossible, since $\zeta^k y$ is relatively prime to $\mathfrak{l}$, and $1-\zeta^{i-k}$ is associate with $1-\zeta$. Hence the numbers (7.4) are pairwise-noncongruent modulo $\mathfrak{l}^2$, so the expressions

$$\frac{x+\zeta^k y}{1-\zeta} \qquad (k=0, 1, \ldots, l-1)$$

are pairwise-noncongruent modulo $\mathfrak{l}$. Since $N(\mathfrak{l})=l$, these expressions form a complete set of residues modulo $\mathfrak{l}$, and hence one of them is divisible by $\mathfrak{l}$. It follows that one (and only one) of the numbers (7.4) is divisible by $\mathfrak{l}^2$. Since we may replace y by any of the numbers $\zeta^k y$ in (7.2), we may assume that $x+y$ is divisible by $\mathfrak{l}^2$, and that all the other numbers $x+\zeta^k y$ are divisible by $\mathfrak{l}$ but not divisible by $\mathfrak{l}^2$. Then the left side of (7.3) is divisible at least by $\mathfrak{l}^{l-1}\mathfrak{l}^2=\mathfrak{l}^{l+1}$, so that $m>1$.

Now let $\mathfrak{m}$ denote the greatest common divisor of the divisors (x) and (y). Since x and y are not divisible by $\mathfrak{l}$, then $\mathfrak{m}$ is not divisible by $\mathfrak{l}$. Then it is clear that $(x+\zeta^k y)$ is divisible by $\mathfrak{l}\mathfrak{m}$, and $(x+y)$ is divisible by $\mathfrak{l}^{l(m-1)+1}\mathfrak{m}$. We set

$$(x+y)=\mathfrak{l}^{l(m-1)+1}\mathfrak{m}\mathfrak{c}_0,$$

$$(x+\zeta^k y)=\mathfrak{l}\mathfrak{m}\mathfrak{c}_k \qquad (k=1, \ldots, l-1),$$

and we shall show that the divisors $\mathfrak{c}_0, \mathfrak{c}_1, \ldots, \mathfrak{c}_{l-1}$ are pairwise relatively prime. Indeed, if $\mathfrak{c}_i$ and $\mathfrak{c}_k$ $(0 \leqslant i < k \leqslant l-1)$ had common divisor $\mathfrak{p}$, then $x+\zeta^i y$ and $x+\zeta^k y$ would be divisible by $\mathfrak{l}\mathfrak{m}\mathfrak{p}$, so that $\zeta^i y(1-\zeta^{k-i})$ and $x(1-\zeta^{k-i})$ would also be divisible by $\mathfrak{l}\mathfrak{m}\mathfrak{p}$. But this would imply that x and y were divisible by $\mathfrak{m}\mathfrak{p}$, contradicting the choice of $\mathfrak{m}$.

Writing (7.3) in the form

$$\mathfrak{m}^l\mathfrak{l}^{lm}\mathfrak{c}_0\mathfrak{c}_1 \cdots \mathfrak{c}_{l-1}=\mathfrak{l}^{lm}\mathfrak{a}^l,$$

we deduce (since the $\mathfrak{c}_k$ are pairwise relatively prime), that

$$\mathfrak{c}_k = \mathfrak{a}_k^{\,l} \qquad (0 \leqslant k \leqslant l-1),$$

and this means that

$$(x+y) = \mathfrak{l}^{l(m-1)+1}\mathfrak{m}_0\mathfrak{a}^l, \tag{7.5}$$

$$(x+\zeta^k y) = \mathfrak{l}\mathfrak{m}\mathfrak{a}_k^{\,l} \qquad (1 \leqslant k \leqslant l-1). \tag{7.6}$$

Solving (7.5) for $\mathfrak{m}$ and substituting in (7.6), we obtain

$$(x+\zeta^k y)\mathfrak{l}^{l(m-1)} = (x+y)(\mathfrak{a}_k\mathfrak{a}_0^{-1})^l, \tag{7.7}$$

from which it follows that the divisors $[(\mathfrak{a}_k\mathfrak{a}_0^{-1})^l$ are principal (since $\mathfrak{l} = (1-\zeta)]$. Now we use the regularity of l. Since the number of classes of divisors of the field $R(\zeta)$ is not divisible by l, then by the corollary to Theorem 3 of Section 7, Chapter 3, the divisors $\mathfrak{a}_k\mathfrak{a}_0^{-1}$ are also principal; that is,

$$\mathfrak{a}_k\mathfrak{a}_0^{-1} = \left(\frac{\alpha_k}{\beta_k}\right) \qquad (1 \leqslant k \leqslant l-1), \tag{7.8}$$

where α_k and β_k are integers of the field $R(\zeta)$. The divisors $\mathfrak{a}_k$ $(1 \leqslant k \leqslant l-1)$ and $\mathfrak{a}_0$ are relatively prime to $\mathfrak{l}$, so we may assume that α_k and β_k are not divisible by $\mathfrak{l}$. Principal divisors are equal if and only if the corresponding numbers differ only by a unit factor. Therefore by (7.7) and (7.8) we have

$$(x+\zeta^k y)(1-\zeta)^{l(m-1)} = (x+y)\left(\frac{\alpha_k}{\beta_k}\right)^l \varepsilon_k \qquad (1 \leqslant k \leqslant l-1), \tag{7.9}$$

where ε_k is a unit of the field $R(\zeta)$.

Now consider the following obvious equation:

$$(x+\zeta y)(1+\zeta) - (x+\zeta^2 y) = \zeta(x+y).$$

If we multiply it by $(1-\zeta)^{l(m-1)}$ and use (7.9) with $k = 1$ and $k = 2$, we obtain

$$(x+y)\left(\frac{\alpha_1}{\beta_1}\right)^l \varepsilon_1(1+\zeta) - (x+y)\left(\frac{\alpha_2}{\beta_2}\right)^l \varepsilon_2 = (x+y)\zeta(1-\zeta)^{l(m-1)},$$

so that

$$(\alpha_1\beta_2)^l - \frac{\varepsilon_2}{\varepsilon_1(1+\zeta)}(\alpha_2\beta_1)^l = \frac{\zeta}{\varepsilon_1(1+\zeta)}(1-\zeta)^{l(m-1)}(\beta_1\beta_2)^l.$$

Hence we have shown that

$$\alpha^l + \varepsilon_0\beta^l = \varepsilon'(1-\zeta)^{l(m-1)}\gamma^l, \tag{7.10}$$

where α, β, and γ are integers of $R(\zeta)$, not divisible by $\mathfrak{l}$, and ε_0 and ε' are units of the field $R(\zeta)$. We shall transform (7.10) to the form (7.2).

We have seen that $m > 1$, so that $m - 1 > 0$ and $l(m-1) \geqslant l$, and this means that

$$\alpha^l + \varepsilon_0 \beta^l \equiv 0 \pmod{\mathfrak{l}^l}.$$

Since β is relatively prime to $\mathfrak{l}$, there is a number β' such that $\beta\beta' \equiv 1 \pmod{\mathfrak{l}^l}$. Multiplying the last congruence by β'^l, we obtain

$$\varepsilon_0 \equiv \omega^l \pmod{\mathfrak{l}^l},$$

where $\omega = -\alpha\beta'$ is an integer of the field $R(\zeta)$. Since $N(\mathfrak{l}) = l$, then any integer of $R(\zeta)$ is congruent modulo $\mathfrak{l}$ to a rational integer. If $\omega \equiv a \pmod{\mathfrak{l}}$, then $\omega^l \equiv a^l \pmod{l\mathfrak{l}}$, and this means that the unit ε_0 is congruent modulo $\mathfrak{l}^l$ to a rational integer. By Kummer's lemma (Theorem 3 of Section 6; here we again use the fact that the prime l is regular) the unit ε_0 is an lth power in $R(\zeta)$, that is, $\varepsilon_0 = \eta^l$, where η is another unit of the field $R(\zeta)$. The equation (7.10) then takes the form

$$\alpha^l + (\eta\beta)^l = \varepsilon'(1-\zeta)^{l(m-1)}\gamma^l.$$

We have obtained an equation of the same type as (7.2), but the exponent m has here been replaced by $m - 1$. But this is impossible, since we chose m to be as small as possible. This contradiction shows that the equation (7.1) has no solution in nonzero rational integers x, y, and z, one of which is divisible by l, that is, that the second case of Fermat's theorem holds for regular primes. Theorem 1 is proved.

7.2. *The Infinitude of the Number of Irregular Primes*

In all existing tables the irregular primes are outnumbered by the regular ones. However, it is not known whether this is true for all intervals $(1, N)$. Further, the infinitude of the set of regular primes is still an open question. Hence the following theorem is of considerable interest.

Theorem 2. There are infinitely many irregular prime numbers.

The proof of Theorem 2 is based on some properties of Bernoulli numbers. These properties are formulated and proved in the following section.

Let $p_1, \ldots, p_s$ be any finite set of irregular prime numbers. Theorem 2 will be proved if we can find an irregular prime number p, different from $p_1, \ldots, p_s$. Set

$$n = r(p_1 - 1) \cdots (p_s - 1).$$

Since the Bernoulli numbers B_{2k} have the property

$$\left|\frac{B_{2k}}{2k}\right| \to \infty \qquad \text{as } k \to \infty$$

(see the end of Section 8), then we can choose r large enough so that B_n/n has absolute value greater than 1. Let p be any prime number which divides the numerator of this number (in lowest terms). If $(p-1)\,|\,n$, then by von Staudt's theorem (Theorem 4 of Section 8), p would divide the denominator of B_n, and this is not the case by choice of p. Hence $(p-1)\,|\,n$, and p is different from $p_1, \dots, p_s$ (and different from 2). Let $n = m + a(p-1)$, with $2 \leqslant m \leqslant p-3$ (note that m is even). We now use Kummer's congruence (Theorem 5 of Section 8), and obtain the congruence

$$\frac{B_m}{m} \equiv \frac{B_n}{n} \pmod{p}$$

in the ring of p-integral rational numbers. But $B_n/n \equiv 0 \pmod{p}$, so $B_m/m \equiv 0 \pmod{p}$ and $B_m \equiv 0 \pmod{p}$. Since m is one of the numbers $2, 4, \dots, p-3$, it follows from the corollary of Theorem 2 of Section 6 that the number p is irregular. Theorem 2 is proved.

PROBLEMS

1. Show that the equation $x^3 + y^3 = 5z^3$ has no solution in rational integers with $z \neq 0$.

2. Show that there are infinitely many irregular primes of the form $4n + 3$ (use Problems 9 and 10 of Section 8).

8. Bernoulli Numbers

In this section we shall prove those properties of Bernoulli numbers which have been used in the preceding sections.

All the power series to be considered converge in some neighborhood of the origin, but their radii of convergence could easily be computed. But we shall not worry about questions of convergence, since for our purposes it suffices to consider all series formally (except in the proof of Theorem 6).

Definition. The rational numbers B_m $(m \geqslant 1)$, defined by

$$\frac{t}{e^t - 1} = 1 + \sum_{m=1}^{\infty} \frac{B_m}{m!} t^m, \tag{8.1}$$

are called *Bernoulli numbers*.

We use the following notations. If $f(x) = a + ax + \cdots + ax$ is a polynomial, then by $f(B)$ we mean the number $a_0 + a_1 B_1 + \cdots + a_n B_n$. Analogously, if (x, t) is a power series of the form $\sum_{n=0}^{\infty} f_n(x) t^n$, where $f_n(x)$ is a polynomial, then by $f(B, t)$ we mean the series $\sum_{n=0}^{\infty} f_n(B)t^n$. Using this notation, the expansion (8.1), which defines the Bernoulli numbers, can be written in the form

$$\frac{t}{e^t - 1} = e^{Bt}.$$

It is easily seen that for any number a

$$e^{at}e^{Bt} = e^{(a+B)t}$$

(to prove this it suffices to multiply the series on the left).

Theorem 1. The Bernoulli numbers satisfy the recurrence relation

$$(1 + B)^m - B^m = 0 \qquad \text{for } m \geqslant 2, \tag{8.2}$$

which in expanded form becomes

$$1 + \sum_{k=1}^{m-1} C_m{}^k B_k = 0 \qquad (m \geqslant 2)$$

(the $C_m{}^k$ are the binomial coefficients).

To prove this theorem we write (8.1) in the form

$$t = e^{(1+B)t} - e^{Bt}.$$

Comparing the coefficients of the terms $t^m/m!$ $(m \geqslant 2)$, we obtain the relation (8.2).

For $m = 2$ formula (8.2) gives us $1 + 2B_1 = 0$, and this means that

$$B_1 = -\tfrac{1}{2}.$$

Theorem 2. All Bernoulli numbers with odd index, except B_1, equal zero:

$$B_{2m+1} = 0 \qquad \text{for } m \geqslant 1. \tag{8.3}$$

The equality (8.3) is clearly equivalent to the fact that the function

$$\frac{t}{e^t - 1} + \frac{t}{2} = 1 + \sum_{m=2}^{\infty} \frac{B_m}{m!} t^m$$

is even, and this is easily verified.

We give the values of the first 12 Bernoulli numbers with even index:

$$B_2 = \frac{1}{6}, \quad B_4 = -\frac{1}{30}, \quad B_6 = \frac{1}{42}, \quad B_8 = -\frac{1}{30}, \quad B_{10} = \frac{5}{66},$$

$$B_{12} = -\frac{691}{2730}, \quad B_{14} = \frac{7}{6}, \quad B_{16} = -\frac{3617}{510}, \quad B_{18} = \frac{43\,867}{798},$$

$$B_{20} = -\frac{174\,611}{330}, \quad B_{22} = \frac{854\,513}{138}, \quad B_{24} = -\frac{236\,364\,091}{2730}.$$

Bernoulli numbers are connected with the sums of series of natural numbers. Set

$$S_k(n) = 1^k + 2^k + \cdots + (n-1)^k.$$

Theorem 3. The sums $S_k(n)$ satisfy the formula

$$(m+1)S_m(n) = (n+B)^{m+1} - B^{m+1}, \quad m \geqslant 1, \tag{8.4}$$

or in expanded form

$$(m+1)S_m(n) = \sum_{k=0}^{m} C_{m+1}^k B_k n^{m+1-k}, \quad m \geqslant 1 \qquad (B_0 = 1). \tag{8.5}$$

In fact, the expression on the right in (8.4) equals the coefficient of $t^{m+1}/(m+1)!$ in the series $e^{(n+B)t} - e^{Bt}$. On the other hand,

$$e^{(n+B)t} - e^{Bt} = e^{Bt}(e^{nt}-1) = t\,\frac{e^{nt}-1}{e^t-1} = t\sum_{r=0}^{n-1} e^{rt}$$

$$= nt + \sum_{m=1}^{\infty}\left(\sum_{r=1}^{n-1} r^m\right)\frac{t^{m+1}}{m!} = nt + \sum_{m=1}^{\infty}\frac{(m+1)S_m(n)t^{m+1}}{(m+1)!},$$

which proves the formula (8.4).

Note that for $n = 1$ the formula (8.4) coincides with (8.2).

Theorem 4 (von Staudt's Theorem). Let p be a prime and m and even integer. If $(p-1) \nmid m$, then B_m is p-integral (that is, p does not appear in the denominator of B_m). If $(p-1) \mid m$, then pB_m is p-integral, and

$$pB_m \equiv 1 \pmod{p}.$$

We prove Theorem 4 by induction on m, using the relation

$$(m+1)S_m(p) = (m+1)B_m p + \sum_{k=0}^{m-1} C_{m+1}^k B_k p^{m+1-k},$$

which is obtained from (8.5) by substituting p for n. We write this in the form

$$pB_m = S_m(p) - \sum_{k=0}^{m-1} \frac{1}{m+1} C_{m+1}^k p^{m-k} pB_k, \tag{8.6}$$

and we shall show that all terms under the summation sign are p-integers which are divisible by p (in the ring of p-integral numbers). The term pB_k for $k < m$ is a p-integer by the induction assumption. We consider the terms

$$\frac{1}{m+1} C_{m+1}^k p^{m-k}. \tag{8.7}$$

If $p = 2$, then since $m + 1$ is odd, this number is a 2-integer and is divisible by 2 (since $k < m$). If $p \neq 2$, we write (8.7) in the form

$$\frac{1}{m+1} C_{m+1}^{m+1-k} p^{m-k} = \frac{m(m-1)\cdots(k+1)}{(m-k+1)!} p^{m-k}.$$

The number p occurs in $(m - k + 1)! = r!$ with exponent

$$\left[\frac{r}{p}\right] + \left[\frac{r}{p^2}\right] + \cdots < \frac{r}{p} + \frac{r}{p^2} + \cdots = \frac{r}{p-1} \leqslant \frac{r}{2} \leqslant r - 1 = m - k,$$

and hence $[1/(m - k + 1)!]p^{m-k}$ is a p-integer and is divisible by p.

Hence we have shown that pB_m is p-integral and that

$$pB_m \equiv S_m(p) \pmod{p} \tag{8.8}$$

in the ring of p-integral numbers.

On the other hand, we have the congruences

$$S_m(p) \equiv -1 \pmod{p} \qquad \text{if } (p-1) \mid m, \tag{8.9}$$

$$S_m(p) \equiv 0 \pmod{p} \qquad \text{if } (p-1) \nmid m. \tag{8.10}$$

Indeed, if $(p - 1) \mid m$, then $x^m \equiv 1 \pmod{p}$, for $1 \leqslant x \leqslant p - 1$, and hence

$$S_m(p) = \sum_{x=1}^{p-1} x^m \equiv \sum_{x=1}^{p-1} 1 = p - 1 \equiv -1 \pmod{p}.$$

If $(p - 1) \nmid m$, then, taking g to be a primitive root modulo p, we shall have

$$S_m(p) = \sum_{x=1}^{p-1} x^m \equiv \sum_{r=0}^{p-2} g^{mr} = \frac{g^{(p-1)m} - 1}{g^m - 1} \equiv 0 \pmod{p},$$

since $g^{p-1} \equiv 1 \pmod{p}$ and $g^m \not\equiv 1 \pmod{p}$.

Comparing (8.8) and (8.10), we see that if $(p - 1) \nmid m$, then $pB_m \equiv 0 \pmod{0}$, and this means that B_m is p-integral. The second assertion of Theorem 4 follows from (8.8) and (8.9).

In the case $m \leqslant p - 1$ the number $p - 1$ does not divide any number $k < m$, so that all B_k for $k < m$ are p-integral. Hence every term on the right in (8.6) is divisible by p^2, and we have the following assertion.

Corollary. If $p \neq 2$ and $m \leqslant p - 1$ (m even), then

$$pB_m \equiv S_m(p) \pmod{p^2}. \tag{8.11}$$

Theorem 5 (Kummer's Congruence). If p is prime and $(p - 1) \nmid m$ (m positive and even), then the number B_m/m is a p-integer, and

$$\frac{B_{m+p-1}}{m+p-1} \equiv \frac{B_m}{m} \pmod{p}. \tag{8.12}$$

In other words, the expression B_m/m (for $(p-1) \nmid m$) has period $p-1$ modulo p.

Proof. Consider the function

$$F(t) = \frac{gt}{e^{gt}-1} - \frac{t}{e^t-1} = \sum_{m=1}^{\infty} \frac{B_m(g^m-1)}{m!} t^m, \tag{8.13}$$

where g is a primitive root modulo p, $1 < g < p$. We set $e^t - 1 = u$. Then

$$F(t) = \frac{gt}{(1+u)^g - 1} - \frac{t}{u} = tG(u),$$

where

$$G(u) = \frac{g}{(1+u)^g - 1} - \frac{1}{u} = \frac{g}{gu + \cdots + u^g} - \frac{1}{u} = \sum_{k=0}^{\infty} c_k u^k.$$

It is clear that the numbers c_k are p-integral.

We shall show that in the expansion of the function $G(u)$ in powers of t:

$$G(u) = G(e^t - 1) = \sum_{k=0}^{\infty} c_k(e^t-1)^k = \sum_{m=0}^{\infty} \frac{A_m}{m!} t^m, \tag{8.14}$$

all the coefficients A_m are p-integral, and that they have period $p-1$ modulo p (for $m > 0$). It is clear that if this latter property holds for some collection of series, then it also holds for any linear combination of these series with p-integral coefficients. Hence it suffices to verify it for the functions $(e^t-1)^k$. But these functions are in turn linear combinations of the functions e^{rt} with $r \geqslant 0$, and

$$e^{rt} = \sum_{n=0}^{\infty} \frac{r^n}{n!} t^n,$$

so by the small Fermat theorem

$$r^{n+p-1} \equiv r^n \pmod{p} \qquad (n > 0).$$

Hence the functions e^{rt} have the desired property, and our assertion about the coefficients A_m is proved.

Comparing the coefficients in (8.13) and (8.14), we see that

$$\frac{B_m(g^m-1)}{m!} = \frac{A_{m-1}}{(m-1)!},$$

so that

$$\frac{B_m}{m}(g^m - 1) = A_{m-1}.$$

Since $g^m - 1 \not\equiv 0 \pmod{p}$ (because $(p-1) \nmid m$), then the sequence of numbers $g^m - 1$ also has period $p-1$ modulo p, by the small Fermat theorem.

It now follows from what we have proved about the numbers A_m that the numbers B_m/m, when $(p-1) \nmid m$, are p-integral and have period $p-1$ modulo p. Theorem 5 is proved.

Theorem 6. The Bernoulli number B_{2m} is given by the formula

$$B_{2m} = (-1)^{m-1} \frac{2(2m)!}{(2\pi)^{2m}} \zeta(2m), \tag{8.15}$$

where $\zeta(2m)$ is the value of the Riemann ζ-function $\zeta(s)$ for $s = 2m$.

To prove this we use the expansion of the function $1/(e^t - 1)$ into partial fractions

$$\begin{aligned} \frac{1}{e^t - 1} &= -\frac{1}{2} + \sum_{n=-\infty}^{+\infty} \frac{1}{t - 2\pi i n} \\ &= -\frac{1}{2} + \frac{1}{t} + \sum_{n=1}^{\infty} \frac{2t}{t^2 + (2\pi n)^2}. \end{aligned} \tag{8.16}$$

This expansion can be derived from the familiar expansion for the cotangent

$$\cot z = \frac{1}{z} + \sum_{n=1}^{\infty} \frac{2z}{z^2 - (\pi n)^2},$$

by using the fact that

$$\cot z = i \frac{e^{iz} + e^{-iz}}{e^{iz} - e^{-iz}} = i + \frac{2i}{e^{2iz} - 1}.$$

It follows from (8.16) that

$$\frac{t}{e^t - 1} = 1 - \frac{t}{2} + 2 \sum_{n=1}^{\infty} \frac{t^2}{t^2 + (2\pi n)^2},$$

and since

$$\frac{t^2}{t^2 + (2\pi n)^2} = \sum_{m=1}^{\infty} (-1)^{m-1} \left(\frac{t}{2\pi n}\right)^{2m},$$

then

$$\begin{aligned} \frac{t}{e^t - 1} &= 1 - \frac{t}{2} + 2 \sum_{n=1}^{\infty} \sum_{m=1}^{\infty} (-1)^{m-1} \frac{t^{2m}}{(2\pi n)^{2m}} \\ &= 1 - \frac{t}{2} + \sum_{m=1}^{\infty} (-1)^{m-1} \frac{2\zeta(2m)}{(2\pi)^{2m}} t^{2m}. \end{aligned}$$

Comparing this equation with (8.1) and equating the coefficients of the various powers of t, we obtain (8.15).

From the formula (8.15) we obtain an estimate for the growth of the numbers $|B_{2m}|$ with increasing m. Since $\zeta(2m) > 1$ and $(2m)! > (2m/e)^{2m}$ (by Stirling's formula), then

$$|B_{2m}| > 2\left(\frac{m}{\pi e}\right)^{2m}.$$

In particular, we find that

$$\left|\frac{B_{2m}}{2m}\right| \to \infty \qquad \text{as } m \to \infty.$$

PROBLEMS

1. Show that

$$(x+B)^m = (x-1-B)^m, \quad m \geqslant 1.$$

2. Show that

$$\left(\frac{1}{2}+B\right)^m = \left(\frac{1}{2^{m-1}}-1\right)B^m.$$

3. Let p be an odd prime number. Show that

$$\sum_{x=1}^{(p-1)/2} x^{(p-1)/2} \equiv 2\left(\left(\frac{2}{p}\right)-2\right)B_{(p+1)/2} \pmod p.$$

4. Let $p > 3$ be a prime number of the form $4k+3$. If h denotes the number of divisor classes of the imaginary quadratic field $R(\sqrt{-p})$, show that h satisfies the congruence

$$h \equiv -2B_{(p+1)/2} \pmod p.$$

5. If $p > 3$ is a prime number, show that

$$1+\frac{1}{2}+\frac{1}{3}+\cdots+\frac{1}{p-1} \equiv 0 \pmod{p^2}.$$

6. Prove the formula

$$(kx+B)^m = k^{m-1}\sum_{s=0}^{k-1}\left(x+\frac{s}{k}+B\right)^m$$

(k and m are natural numbers).

7. The function $\tan x$ has the expansion

$$\tan x = \sum_{n=1}^{\infty} T_n \frac{x^{2n-1}}{(2n-1)!},$$

where

$$T_n = 2^{2n}(2^{2n}-1)\frac{|B_{2n}|}{2n}.$$

Show that all the coefficients T_n are natural numbers.

8. If $m > 1$, show that

$$2B_{2m} \equiv 1 \pmod{4}.$$

9. Let q be a prime number such that $2q + 1$ is composite [for instance, $q \equiv 1 \pmod{3}$]. Show that the numerator of the Bernoulli number B_{2q} is divisible by a prime of the form $4n + 3$.

10. Let $p_1, \ldots, p_s$ be prime numbers greater than 3, and let q be a natural number such that $q \equiv 1 \pmod{M}$, where $M = (p_1 - 1) \cdots (p_s - 1)$. Show that none of the prime numbers $p_1, \ldots, p_s$ divide the numerator of the fraction $B_{2q}/2q$.

Algebraic Supplement

1. Quadratic Forms over Arbitrary Fields of Characteristic $\neq 2$

In this section we describe some of the general properties of quadratic forms over arbitrary fields. We shall state some well-known results without proof. Throughout, K will denote an arbitrary field whose characteristic is not 2. For any matrix A, we shall denote the transpose by A'.

1.1. *Equivalence of Quadratic Forms*

By a quadratic form over the field K we mean a homogeneous polynomial of degree 2 with coefficients in K. Any quadratic form f can be written

$$f = \sum_{i,j=1}^{n} a_{ij} x_i x_j ,$$

where $a_{ij} = a_{ji}$. The symmetric matrix

$$A = (a_{ij})$$

is called the *matrix* of the quadratic form f. If the matrix is given, the quadratic form is completely determined (except for the names of the variables). The determinant $d = \det A$ is called the *determinant* of the quadratic form f. If $d = 0$, the form f is called *singular*, and otherwise it is called *nonsingular*. If we let X denote the column vector of the variables $x_1, x_2, \ldots, x_n$, then the quadratic form can be written

$$f = X'AX.$$

Suppose we replace the variables $x_1, \dots, x_n$ by the new variables $y_1, \dots, y_n$ according to the formula

$$x_i = \sum_{j=1}^{n} c_{ij} y_j \qquad (1 \leqslant i \leqslant n,\ c_{ij} \in K).$$

In matrix form this linear substitution becomes

$$X = CY,$$

where Y is the column vector of the variables $y_1, \dots, y_n$, and C is the matrix (c_{ij}). If we replace the variables $x_1, \dots, x_n$ in f by the corresponding expressions in $y_1, \dots, y_n$, then (after carrying out the indicated operations) we shall obtain a quadratic form g (also over the field K) in the variables $y_1, \dots, y_n$. The matrix A_1 of the quadratic form g equals

$$A_1 = C'AC. \tag{1.1}$$

Two quadratic forms f and g are called *equivalent*, and we write $f \sim g$, if there is a nonsingular change of variables which takes one form to the other. From formula (1.1) we obtain

Theorem 1. If two quadratic forms are equivalent, then their determinants differ by a nonzero factor which is a square in K.

Let γ be an element of K. If there exist elements $\alpha_1, \dots, \alpha_n$ in K for which

$$f(\alpha_1, \dots, \alpha_n) = \gamma,$$

then we say that the form f represents γ. In other words, a number is represented by a quadratic form if it is the value of the form for some values of the variables. It is easily seen that equivalent quadratic forms represent the same elements of the field K.

We shall further say that the form f represents zero in the field K if there exist values $\alpha_i \in K$, not all zero, such that $f(\alpha_1, \dots, \alpha_n) = 0$. The property of representing zero is clearly preserved if we pass to an equivalent form.

Theorem 2. If a quadratic form f in n variables represents an element $\alpha \neq 0$, then it is equivalent to a form of the type

$$\alpha {x_1}^2 + g(x_2, \dots, x_n),$$

where g is a quadratic form in $n - 1$ variables.

Regarding the proof of this theorem we note only the following. If $f(\alpha_1, \dots, \alpha_n) = \alpha$, then not all α_i are equal to zero, so we can find a nonsingular matrix C, whose first row is $\alpha_1, \dots, \alpha_n$. If we apply to f the linear substitution whose matrix is C, we obtain a form in which the coefficient of the square of the first variable is α. The rest of the proof is carried out as usual.

If the matrix of a quadratic form is diagonal (that is, if the coefficient of every product of distinct variables equals zero), then we say that the form is *diagonal*. Theorem 2 now implies

Theorem 3. Any quadratic form over K can be put in diagonal form by some nonsingular linear substitution. In other words, every form is equivalent to a diagonal form.

In terms of matrices, Theorem 3 shows that for any symmetric matrix A there exists a nonsingular matrix C such that the matrix $C'AC$ is diagonal.

1.2. *The Direct Sum of Quadratic Forms*

Since the names of the variables are not of any significance, we can assume that two given quadratic forms f and g have different variables. In this case the form $f + g$ is called the *direct sum* of f and g, and is denoted by $f \dotplus g$ (this must not be confused with the usual addition of quadratic forms when the forms have the same variables). It is clear that if $g \sim h$, then $f \dotplus g \sim f \dotplus h$. We shall now show that the following converse holds.

Theorem 4 (Witt's Theorem). Let f, g, and h be nonsingular quadratic forms over the field K. If the forms $f \dotplus g$ and $f \dotplus h$ are equivalent, then the forms g and h are also equivalent.

Proof. Let f_0 be a diagonal form equivalent to f. Then, as noted above, $f \dotplus g \sim f_0 \dotplus g$ and $f \dotplus h \sim f_0 \dotplus h$, so that $f_0 \dotplus g \sim f_0 \dotplus h$. Hence we may assume that f is a diagonal form. It is now easily seen that to prove the theorem it suffices to consider the case $f = ax^2$, $a \neq 0$. Let A and B denote the matrices of g and h. Since the forms $ax^2 \dotplus g$ and $ax^2 \dotplus h$ are equivalent, there exists a matrix

$$C = \begin{pmatrix} \gamma & S \\ T & Q \end{pmatrix}$$

such that

$$\begin{pmatrix} \gamma & T' \\ S' & Q' \end{pmatrix} \begin{pmatrix} a & 0 \\ 0 & A \end{pmatrix} \begin{pmatrix} \gamma & S \\ T & Q \end{pmatrix} = \begin{pmatrix} a & 0 \\ 0 & B \end{pmatrix}.$$

(Here S is a row matrix and T is a column matrix.) From this equation we obtain

$$\gamma^2 a + T'AT = a, \tag{1.2}$$

$$\gamma a S + T'AQ = 0, \tag{1.3}$$

$$S'aS + Q'AQ = B. \tag{1.4}$$

We must show that there exists a nonsingular matrix C_0 such that $C_0'AC_0 = B$. The matrix C_0 will be found in the form

$$C_0 = Q + \xi TS,$$

where the element ξ must be suitably chosen. By (1.2) and (1.3) we have

$$\begin{aligned} C_0'AC_0 &= (Q' + \xi S'T')A(Q + \xi TS) \\ &= Q'AQ + \xi S'T'AQ + \xi Q'ATS + \xi^2 S'T'ATS \\ &= Q'AQ + a[(1 - \gamma^2)\xi^2 - 2\gamma\xi]S'S. \end{aligned}$$

In view of (1.4) the last expression will equal the matrix B, provided that $(1 - \gamma^2)\xi^2 - 2\gamma\xi = 1$. This equation, which can also be written in the form $\xi^2 - (\gamma\xi + 1)^2 = 0$, always has a solution $\xi_0 \in K$ for any $\gamma \in K$ (recall that the characteristic of K is not 2). Hence we have found a matrix $C_0 = Q + \xi_0 TS$, for which $C_0'AC_0 = B$. Since the matrix B is nonsingular, then C_0 is also nonsingular. Theorem 4 is proved.

1.3. *Representation of Field Elements*

Theorem 5. If a nonsingular quadratic form represents zero in the field K, then it also represents all elements of K.

Proof. Since equivalent forms represent the same field elements, it suffices to prove the theorem for a diagonal form $f = a_1x_1^2 + \cdots + a_nx_n^2$. Let $a_1\alpha_1^2 + \cdots + a_n\alpha_n^2 = 0$ be a representation of zero, and let γ be any element of K. We can assume that $\alpha_1 \neq 0$. We express the variables $x_1, \ldots, x_n$ in terms of a new variable t:

$$x_1 = \alpha_1(1 + t), \quad x_k = \alpha_k(1 - t) \qquad (k = 2, \ldots, n).$$

Substituting in the form f we obtain

$$f^* = f^*(t) = 2a_1\alpha_1^2 t - 2a_2\alpha_2^2 t - \cdots - 2a_n\alpha_n^2 t = 4a_1\alpha_1^2 t.$$

If we now set $t = \gamma/4a_1\alpha_1^2$, we obtain $f^* = \gamma$.

Theorem 6. A nonsingular quadratic form f represents the element $\gamma \neq 0$ in K if and only if the form $-\gamma x_0^2 \dotplus f$ represents zero.

Proof. The necessity of the condition is clear. Assume that

$$-\gamma\alpha_0^2 + f(\alpha_1, \ldots, \alpha_n) = 0,$$

where not all α_i equal zero. If $\alpha_0 \neq 0$, then $\gamma = f(\alpha_1/\alpha_0, \ldots, \alpha_n/\alpha_0)$. If $\alpha_0 = 0$, then the form f represents zero, and hence by Theorem 5 it represents all elements of the field K.

Remark. From the proof of Theorem 6 it is clear that if we determine all representations of zero by the form $-\gamma x_0^2 + f$ (only those in which $x_0 \neq 0$ are relevant), then we have also determined all representations of γ by the form f. Hence the question of the representability of an element of the field K by a nonsingular form can be reduced to the question of the representability of zero by a nonsingular form in one more variable.

Theorem 7. If a nonsingular form f represents zero, then it is equivalent to a form of the following type:

$$y_1 y_2 + g(y_3, \ldots, y_n).$$

Proof. Using Theorem 5, we first find $\alpha_1, \ldots, \alpha_n$ such that $f(\alpha_1, \ldots, \alpha_n) = 1$. By Theorem 2 we can now put f in the form $x_1^2 + f_1(x_2, \ldots, x_n)$. Since the form $x_1^2 + f_1$ represents zero, we can find $\beta_2, \ldots, \beta_n$ such that $f_1(\beta_2, \ldots \beta_n) = -1$. Again applying Theorem 2, we can put f_1 in the form $-x_2^2 + g(y_3, \ldots, y_n)$. Setting $x_1 - x_2 = y_1$, and $x_1 + x_2 = y_2$, we obtain the desired result.

Remark. If we know some representation of zero by the form f, then all the operations described in the proof of Theorem 7 can be carried out explicitly, and the form $g(y_3, \ldots, y_n)$ can be determined. Now assume that for any quadratic form which represents zero over the field K, an actual representation of zero can be found. Then any nonsingular form can be transformed to a form of the type

$$y_1 y_2 + \cdots + y_{2s-1} y_{2s} + h(y_{2s+1}, \ldots, y_n), \tag{1.5}$$

where the form h does not represent zero. In any representation of zero by the form (1.5), at least one of the variables $y_1, y_2, \ldots, y_{2s+1}, y_{2s}$ must be nonzero. To determine all representations of zero in which, say, $y_1 = \alpha_1 \neq 0$, we note that we can give $y_3, \ldots, y_n$ arbitrary values $\alpha_3, \ldots, \alpha_n$ and then determine y_2 by the condition

$$\alpha_1 y_2 + \alpha_3 \alpha_4 + \cdots + g(\alpha_{2s+1}, \ldots, \alpha_n) = 0.$$

This gives us an effective method for finding all representations of zero by a nonsingular quadratic form over the field K, provided that we have a method for determining whether or not a given form represents zero, and, in case it does, an algorithm for finding some specific representation of zero.

Theorem 8. Let the field K contain more than five elements. If the diagonal form

$$a_1 x_1^2 + \cdots + a_n x_n^2 \qquad (a_i \in K)$$

represents zero in the field K, then there is a representation of zero in which all the variables take nonzero values.

Proof. We first show that if $a\xi^2 = \lambda \neq 0$, then for any $b \neq 0$ there exist nonzero elements α and β such that $a\alpha^2 + b\beta^2 = \lambda$. To prove this fact we consider the identity

$$\frac{(t-1)^2}{(t+1)^2} + \frac{4t}{(t+1)^2} = 1.$$

Multiplying this identity by $a\xi^2 = \lambda$, we obtain

$$a\left(\xi\frac{t-1}{t+1}\right)^2 + at\left(\frac{2\xi}{t+1}\right)^2 = \lambda. \tag{1.6}$$

Choose a nonzero γ in K so that the value of $t = t_0 = b\gamma^2/a$ is not ± 1. This can be done because each of the equations $bx^2 - a = 0$ and $bx^2 + a = 0$ has at most two solutions for x in K, and the field K has more than five elements. Setting $t = t_0$ in (1.6), we obtain

$$a\left(\xi\frac{t_0-1}{t_0+1}\right)^2 + b\left(\frac{2\xi\gamma}{t_0+1}\right)^2 = \lambda,$$

and our assertion is proved. We can now easily complete the proof of the theorem. If the representation $a_1\xi_1{}^2 + \cdots + a_n\xi_n{}^2 = 0$ is such that $\xi_1 \neq 0, \ldots,$ $\xi_r \neq 0$, $\xi_{r+1} = \cdots = \xi_n = 0$, where $r \geqslant 2$, then we have shown that we can find $\alpha \neq 0$ and $\beta \neq 0$ such that $a_r\xi_r{}^2 = a_r\alpha^2 + a_{r+1}\beta^2$, and this yields a representation of zero in which the number of nonzero variables is increased by one. Repeating this process, we arrive at a representation in which all the variables have nonzero value.

1.4. *Binary Quadratic Forms*

A quadratic form in two variables is called a *binary quadratic form.*

Theorem 9. All nonsingular binary quadratic forms which represent zero in K are equivalent.

Indeed, by Theorem 7, any such form is equivalent to the form y_1y_2.

Theorem 10. In order that the binary quadratic form f with determinant $d \neq 0$ represents zero in K, it is necessary and sufficient that the element $-d$ be a square in K (that is, $-d = \alpha^2$, $\alpha \in K$).

Proof. The necessity of the condition follows from Theorems 1 and 7. Conversely, if $f = ax^2 + by^2$ and $-d = -ab = \alpha^2$, then $f(\alpha, a) = a\alpha^2 + ba^2 = 0$.

Theorem 11. Let f and g be two nonsingular binary quadratic forms over the field K. In order that f and g be equivalent, it is necessary and sufficient that their determinants differ by a factor which is a square in K, and that there exist some nonzero element of K which is represented by both f and g.

Proof. Both conditions are clearly necessary. To prove sufficiency, let $\alpha \neq 0$ be an element of K which is represented by both f and g. By Theorem 2 f and g are equivalent to the forms $f_1 = \alpha x^2 + \beta y^2$ and $g_1 = \alpha x^2 + \beta' y^2$. Since $\alpha\beta$ and $\alpha\beta'$ differ by a square factor, then $\beta' = \beta\gamma^2$, $\gamma \in K$, and this means that $f_1 \sim g_1$ and $f \sim g$.

PROBLEMS

1. Show that a singular quadratic form always represents zero.

2. Show that Theorem 5 does not hold for singular quadratic forms.

3. If the binary form $x^2 - \alpha y^2$ represents the elements γ_1 and γ_2 of K, show that it also represents their product.

4. Show that Theorem 8 is not valid for fields with not more than five elements.

5. We shall decompose the set of all nonsingular quadratic forms over K in $n = 0, 1, 2, \ldots$ variables into the so-called *Witt classes.* (We treat the zero form as a nonsingular form on the empty set of variables, and consider it to represent zero.) Two forms f_1 and f_2 belong to the same Witt class $[f_1] = [f_2]$, if the corresponding forms h in (1.5) have the same number of variables and are equivalent. We add Witt classes by the formula $[f_1] + [f_2] = [f_1 \dotplus f_2]$. Show that these definitions make sense and that under this operation the set of Witt classes becomes a group.

6. Determine the group of Witt classes for the real and complex fields.

7. Show that a quadratic form over a finite field (characteristic $\neq 2$) in three or more variables represents zero.

2. Algebraic Extensions

Many theorems of this section are given without proof. For the proofs the reader may consult, for example, "Modern Algebra," by B. L. van der Waerden, Vol. 1, Chap. 5, Ungar, New York, 1950.

2.1. *Finite Extensions*

If the field Ω contains the field k as a subfield, then we say that Ω is an extension of the field k. To denote that Ω is being considered as an extension field of k, we write Ω/k. If K is a subfield of Ω which contains k, that is, $k \subset K \subset \Omega$, then K is called an *intermediate field* for the extension Ω/k.

For any extension Ω/k, we may consider Ω as a vector space over k.

Definition. The extension K/k is called *finite* if K, considered as a vector space over k, is finite-dimensional. This dimension is called the *degree* of the extension and is denoted by $(K:k)$. Any basis for K as a vector space over k is called a basis for the extension K/k.

If the extension K/k is finite, then for any intermediate field K_0, the extensions K_0/k and K/K_0 are clearly both finite. The following converse also holds.

Theorem 1. Let K_0 be an intermediate field for the extension K/k. If the extensions K/K_0 and K_0/k are finite, then K/k is also finite, and its degree equals the products of the degrees of the extensions K/K_0 and K_0/k:

$$(K:k) = (K:K_0)(K_0:k).$$

Proof. Let $\theta_1, \ldots \theta_m$, be a basis for K/K_0, and $\omega_1, \ldots, \omega_n$ be a basis for K_0/k. Then every element of K can be represented as a linear combination (over k) of the products $\omega_i\theta_j$, so the extension K/k is finite. Further, it is easily checked that these products are linearly independent over k, so $(K:k) = mn$.

For any field k we denote the ring of polynomials in the variable t with coefficients in k by $k[t]$.

Let Ω/k be an extension of the field k. An element $\alpha \in \Omega$ is called *algebraic* over k; it is the root of some nonzero polynomial $f(t)$ of $k[t]$. Among all such polynomials we take that polynomial $\varphi(t) \neq 0$ which is of lowest degree and has leading coefficient 1. Since all polynomials $f(t)$ which have α as a root are divisible by $\varphi(t)$ (otherwise the remainder after division of f by φ would be a polynomial of lower degree with α as a root), then the polynomial $\varphi(t)$ is uniquely determined. It is called the *minimum polynomial* of the algebraic element α over the field k. The minimum polynomial $\varphi \in k[t]$ is always irreducible, since from $\varphi = gh$ it follows that α is either a root of $g(t)$ or of $h(t)$. Any element $a \in k$ is algebraic over k, and its minimum polynomial is $t - a$. An element $\xi \in \Omega$ which is not algebraic over k is called *transcendental* over k.

The extension Ω/k is called algebraic if every $\alpha \in \Omega$ is algebraic over k.

Theorem 2. Any finite extension is algebraic.

Theorem 3. Let the element α of the extension Ω/k be algebraic over k, and let its minimum polynomial $\varphi(t) \in k[t]$ have degree m. Then the elements $1, \alpha, \ldots, \alpha^{m-1}$ are linearly independent over k and the set of all linear combinations

$$a_0 + a_1\alpha + \cdots + a_{m+1}\alpha^{m-1} \tag{2.1}$$

with coefficients a_i in k is an intermediate field, denoted by $k(\alpha)$. The extension $k(\alpha)/k$ is finite of degree m.

To add two elements of the field $k(\alpha)$, written in the form (2.1), we simply add the corresponding coefficients. To put the product of the elements

$\xi = g(\alpha)$ and $\eta = h(\alpha)$ (g and h are polynomials in $k[t]$ of degree $\leqslant m - 1$) in the form (2.1), we must divide gh by φ with remainder:

$$g(t)h(t) = \varphi(t)q(t) + r(t),$$

where the degree of $r(t)$ does not exceed $m - 1$; since $\varphi(\alpha) = 0$, then $\xi\eta = r(\alpha)$. Hence the operation of multiplication in the field $k(\alpha)$ is determined by the minimum polynomial $\varphi(t)$ of the element α.

Let $\alpha_1, \ldots, \alpha_s$ be a finite set of elements of Ω, which are algebraic over the field k, and let $m_1, \ldots, m_s$ be the degrees of their minimum polynomials over k. The set of all linear combinations of the elements

$$\alpha_1^{k_1} \cdots \alpha_s^{k_s} \qquad (0 \leqslant k_1 < m_1, \ldots, 0 \leqslant k_s < m_s)$$

with coefficients in k is an intermediate field. It is denoted by $k(\alpha_1, \ldots, \alpha_s)$ and is called the *field* generated by the elements $\alpha_1, \ldots, \alpha_s$. Its degree over k does not exceed the product $m_1 \cdots m_s$.

Any finite extension K/k, contained in Ω, can be represented in the form $k(\alpha_1, \ldots, \alpha_s)$ for some $\alpha_1, \ldots, \alpha_s$.

Definition. A finite extension K/k is called *simple* if there is an element θ such that $K = k(\theta)$. Any element $\theta \in K$, for which $K = k(\theta)$, is called a *primitive element* of the extension K/k.

The primitive elements of K over k are those elements whose minimum polynomial has degree equal to the degree of K/k.

Theorem 4. Let Ω/k and Ω'/k be two extensions of the field k, and let $\theta \in \Omega$ and $\theta' \in \Omega'$ be algebraic elements over k with the same minimum polynomial $\varphi(t)$. Then there is a unique isomorphism of the field $k(\theta)$ onto the field $k(\theta')$ for which $\theta \to \theta'$ and $a \to a$ for all $a \in k$.

Let m be the degree of the polynomial $\varphi(t)$. The isomorphism $k(\theta) \to k(\theta')$ of Theorem 4 coincides with the mapping

$$a_0 + a_1\theta + \cdots + a_{m+1}\theta^{m-1} \to a_0 + a_1\theta' + \cdots + a_{m+1}\theta'^{m-1} \tag{2.2}$$

($a_1, \ldots, a_{m-1}$ are arbitrary elements of the field k).

So far we have considered finite extensions K/k which are contained in a given extension Ω/k. We now turn to the question of the construction of finite extensions over a fixed field k.

Theorem 5. Let k be a field, and let $\varphi(t)$ be an irreducible polynomial of $k[t]$ of degree n. Then there exists a finite extension K/k of degree n in which the polynomial φ has a root. The extension K/k is unique (up to an isomorphism which is the identity map on k). If $\varphi(\theta) = 0$, $\theta \in K$, then $K = k(\theta)$.

The field K (in the case $n > 1$) is constructed in the following manner. We choose some new object θ and consider the set K of all formal linear combinations

$$a_0 + a_1\theta + \cdots + a_{n+1}\theta^{n-1} \tag{2.3}$$

with coefficients in k. If we denote the polynomial $a_0 + a_1t + \cdots + a_{n-1}t^{n-1}$ by $g(t)$, then the expression (2.3) can be written $g(\theta)$. Let $\xi = g(\theta)$ and $\eta = h(\theta)$ be two linear combinations of the type (2.3) (g and h are polynomials in $k[t]$ of degree $\leqslant n - 1$). Let $s(t)$ denote the sum $g(t) + h(t)$ and let $r(t)$ denote the remainder after division of the product $g(t)h(t)$ by $\varphi(t)$. Set

$$\xi + \eta = s(\theta),$$

$$\xi\eta = r(\theta).$$

It is easily verified that under these operations K is a field with the desired properties.

Corollary. For any polynomial $f(t) \in k[t]$ there is a finite extension K/k in which $f(t)$ factors into linear factors.

If k is a field such that the only algebraic extension of k is k itself, then k is called *algebraically closed*. It is clear that k is algebraically closed if and only if every polynomial in $k[t]$ factors into linear factors.

2.2. *Norm and Trace*

Let K/k be a finite extension of degree n. For any $\alpha \in K$ the mapping $\xi \to \alpha\xi$ ($\xi \in K$) is a linear transformation of K (considered as a vector space over k). The characteristic polynomial $f_\alpha(t)$ of this transformation is also called the *characteristic polynomial* of the element $\alpha \in K$, relative to the extension K/k. If $\omega_1, \ldots, \omega_n$ is a basis for the extension K/k and

$$\alpha\omega_i = \sum_{j=1}^{n} a_{ij}\omega_j \qquad (a_{ij} \in k), \tag{2.4}$$

then

$$f_\alpha(t) = \det(tE - (a_{ij})),$$

where E is the n by n identity matrix.

Theorem 6. The characteristic polynomial $f_\alpha(t)$ of an element $\alpha \in K$ relative to the extension K/k is a power of its minimum polynomial $\varphi_\alpha(t)$ over k.

Proof. Let

$$\phi_\alpha(t) = t^m + c_1t^{m-1} + \cdots + c_m .$$

By Theorem 3 the powers, $1, \alpha, \ldots, \alpha^{m-1}$ form a basis for the extension $k(\alpha)/k$. If $\theta_1, \ldots, \theta_s$ is a basis for $K/k(\alpha)$, then we can take for a basis of K/k the products

$$\theta_1, \alpha\theta_1, \ldots, \alpha^{m-1}\theta_1; \ldots; \theta_s, \alpha\theta_s, \ldots, \alpha^{m-1}\theta_s.$$

The matrix of the linear transformation $\xi \to \alpha\xi$ in this basis will clearly be a block-diagonal matrix, with s blocks down the main diagonal, each block being equal to

$$\begin{pmatrix} 0 & 1 & 0 & \cdots & 0 & 0 \\ 0 & 0 & 1 & \cdots & 0 & 0 \\ \cdot & \cdot & \cdot & \cdot & \cdot & \cdot \\ 0 & 0 & 0 & \cdots & 0 & 1 \\ -c_m & -c_{m-1} & -c_{m-2} & \cdots & -c_2 & -c_1 \end{pmatrix}$$

The characteristic polynomial of each block is easily computed to be $t^m + c_1 t^{m-1} + \cdots + c_m = \varphi_\alpha(t)$. Hence $f_\alpha = \varphi_\alpha{}^s$, and Theorem 6 is proved.

Since when we pass from one basis to another the matrix of a linear transformation is replaced by a similar matrix, then the determinant and trace of the matrix (a_{ij}), defined by (2.4), do not depend on the choice of the basis $\omega_1, \ldots, \omega_n$.

Definition. The determinant $\det(a_{ij})$ of the matrix (a_{ij}) of (2.4) is called the *norm*, and its trace $\mathrm{Sp}(a_{ij}) = \sum_{i=1}^n a_{ii}$ is called the *trace* of the element $\alpha \in K$ relative to the extension K/k. The norm and trace are denoted by $N_{K/k}(\alpha)$ and $\mathrm{Sp}_{K/k}(\alpha)$, or, more briefly, by $N(\alpha)$ and $\mathrm{Sp}(\alpha)$.

If $a \in k$ the matrix of the linear transformation $\xi \to a\xi$ $(\xi \in K)$ will be the diagonal matrix aE. Therefore for every element of k we have

$$N_{K/k}(a) = a^n,$$

$$\mathrm{Sp}_{K/k}(a) = na.$$

When linear transformations are added or composed, their matrices are added or multiplied (for a fixed basis), and hence for any elements α and β of K we have the formulas

$$N_{K/k}(\alpha\beta) = N_{K/k}(\alpha)N_{K/k}(\beta), \tag{2.5}$$

$$\mathrm{Sp}_{K/k}(\alpha + \beta) = \mathrm{Sp}_{K/k}(\alpha) + \mathrm{Sp}_{K/k}(\beta). \tag{2.6}$$

The matrix of the linear transformation $\xi \to a\alpha\xi$ $(a \in k, \xi \in K)$ is obtained from the matrix of the transformation $\xi \to \alpha\xi$ by multiplying all entries by a. Hence we also have the formula

$$\mathrm{Sp}_{K/k}(a\alpha) = a\,\mathrm{Sp}_{K/k}(\alpha) \qquad (a \in k, \alpha \in K). \tag{2.7}$$

If $\alpha \neq 0$, then the transformation $\xi \to \alpha\xi$ is nonsingular, and hence the norm $N_{K/k}(\alpha)$ is also nonzero. Hence we see by (2.5) that the mapping $\alpha \to N_{K/k}(\alpha)$ is a homomorphism of the multiplicative group K^* of the field K to the multiplicative group k^* of the field k. As for the mapping $\alpha \to \mathrm{Sp}_{K/k}(\alpha)$, by (8.6) and (8.7) it is a linear function on K with values in the field k.

Theorem 7. Let $\alpha \in K$ have characteristic polynomial $f_\alpha(t)$ relative to the extension K/k, and let Ω/k be an extension in which $f_\alpha(t)$ factors into linear factors:

$$f_\alpha(t) = (t - \alpha_1) \cdots (t - \alpha_n).$$

Then

$$N_{K/k}(\alpha) = \alpha_1 \alpha_2 \cdots \alpha_n,$$

$$\mathrm{Sp}_{K/k}(\alpha) = \alpha_1 + \alpha_2 + \cdots + \alpha_n.$$

Proof. If

$$f_\alpha(t) = \det(tE - (a_{ij})) = t^n + a_1 t^{n-1} + \cdots + a_n,$$

then

$$a_1 = -\mathrm{Sp}(a_{ij}), \quad a_n = (-1)^n \det(a_{ij}).$$

On the other hand, it is easily checked that

$$\alpha_1 + \alpha_2 + \cdots + \alpha_n = -\alpha_1, \quad \alpha_1 \alpha_2 \cdots \alpha_n = (-1)^n a_n,$$

and this proves the theorem.

Theorem 8. We keep the notations of Theorem 7. If $\gamma = g(\alpha) \in K$ $(g(t) \in [t])$, then the characteristic polynomial $f_\gamma(t)$ has the following factorization in

$$(t - g(\alpha_1))(t - g(\alpha_2)) \cdots (t - g(\alpha_n)). \tag{2.8}$$

Proof. We first note that the coefficients of the polynomial (2.8), being symmetric expressions in $\alpha_1, \ldots, \alpha_n$, belong to the field k. Let $\varphi_\gamma(t)$ be the minimum polynomial of γ over k. If we apply the isomorphism $k(\alpha) \to k(\alpha_i)$ (in which $\alpha \to \alpha_i$ and $a \to a$ for $a \in k$) to the equation $\varphi(g(\alpha)) = 0$, we obtain $\varphi(g(\alpha_i)) = 0$. Hence every root of the polynomial (2.8) is a root of the polynomial $\varphi_\gamma(t)$, which is irreducible over k. This is possible only if the polynomial (2.8) is a power of $\varphi_\gamma(t)$. Thus, we can see that the theorem now follows from Theorem 6.

Let $k \subset K \subset L$ be a tower of finite extensions. We choose bases $\omega_1, \ldots, \omega_n$

and $\theta_1, \ldots, \theta_m$ for the extensions K/k and L/K. For any $\gamma \in L$ set

$$\gamma\theta_j = \sum_{s=1}^{m} \alpha_{js}\theta_s \qquad (\alpha_{js} \in K),$$

$$\alpha_{js}\omega_i = \sum_{r=1}^{n} a_{jsir}\omega_r \qquad (a_{jsir} \in k).$$

Since

$$\gamma\omega_i\theta_j = \sum_{s,r} a_{jsir}\omega_r\theta_s,$$

then $\mathrm{Sp}_{L/K}(\gamma) = \sum_{i,j} a_{iijj}$. On the other hand, we also have

$$\mathrm{Sp}_{K/k}(\mathrm{Sp}_{L/K}(\gamma)) = \mathrm{Sp}_{K/k}\left(\sum_j \alpha_{jj}\right) = \sum_{i,j} a_{jjii}.$$

Hence for any $\gamma \in L$,

$$\mathrm{Sp}_{L/k}(\gamma) = \mathrm{Sp}_{K/k}(\mathrm{Sp}_{L/K}(\gamma)). \tag{2.9}$$

An analogous formula holds for the norm (Problem 2).

2.3. *Separable Extensions*

Definition. A finite extension K/k is called separable if the linear mapping $\xi \to \mathrm{Sp}_{K/k}(\xi)$, $\xi \in K$, is not identically zero.

Since $\mathrm{Sp}_{K/k}(1) = n = (K:k)$, every finite extension over a field of characteristic zero is separable. The same holds for every extension over a field of characteristic p for which the degree of the extension is not divisible by p.

For a finite separable extension K/k we choose a basis $\omega_1, \ldots, \omega_n$ and consider the matrix

$$(\mathrm{Sp}(\omega_i\omega_j))_{1 \leqslant i,j \leqslant n}. \tag{2.10}$$

If the determinant of this matrix were zero, then we could find elements $c_1, \ldots, c_n$ in k, not all zero, for which

$$\sum_{j=1}^{n} c_j \,\mathrm{Sp}(\omega_i\omega_j) = 0 \qquad (i = 1, \ldots, n).$$

Setting $\gamma = c_1\omega_1 + \cdots + c_n\omega_n$, we can write this last equation

$$\mathrm{Sp}(\omega_i\gamma) = 0 \qquad (i = 1, \ldots, n). \tag{2.11}$$

Let ξ be any element of K. Since $\gamma \neq 0$, then ξ can be represented in the form $\xi = a_1\omega_1\gamma + \cdots + a_n\omega_n\gamma$, $a_i \in k$, and by (2.6), (2.7), and (2.11) we have $\mathrm{Sp}\,\xi = 0$. But this would contradict the separability of K/k. Thus for separable extensions the matrix (2.10) is always nonsingular.

Definition. The determinant $\det(\mathrm{Sp}(\omega_i\omega_j))$ is called the *discriminant* of the basis $\omega_1, \ldots, \omega_n$ of the finite separable extension K/k and is denoted by $D(\omega_1, \ldots, \omega_n)$.

We have shown that the discriminant of any basis of a finite separable extension is a nonzero element of the ground field.

Let $\omega_1', \ldots, \omega_n'$ be any other basis of the extension K/k, and let

$$\omega_i' = \sum_{j=1}^{n} c_{ij}\omega_j \qquad (i = 1, \ldots, n).$$

Since the matrix $(\mathrm{Sp}(\omega_i'\omega_j'))$ equals the product $(c_{ij})(\mathrm{Sp}(\omega_i\omega_j))(c_{ij})'$, then

$$D(\omega_1', \ldots, \omega_n') = (\det(c_{ij}))^2 D(\omega_1, \ldots, \omega_n). \tag{2.12}$$

Thus the discriminants of two different bases differ by a factor which is a square in the ground field.

We fix a basis $\omega_1, \ldots, \omega_n$ for the extension K/k. Then for any elements $c_1, \ldots, c_n$ of k there exists a unique element $\alpha \in K$ such that

$$\mathrm{Sp}(\omega_i\alpha) = c_i \qquad (i = 1, \ldots, n). \tag{2.13}$$

(indeed, representing α in the form $\alpha = x_1\omega_1 + \cdots + x_n\omega_1$ $(x_i \in k)$ and substituting in (2.13), we obtain a system of n linear equations in the n unknowns x_i with nonzero determinant. In particular, we can find elements $\omega_1^*, \ldots, \omega_n^*$ in the field K such that

$$\mathrm{Sp}(\omega_i\omega_j^*) = \begin{cases} 1 & \text{for } i = j, \\ 0 & \text{for } i \neq j. \end{cases} \tag{2.14}$$

These n elements are linearly independent over k, since if $c_1\omega_1^* + \cdots + c_n\omega_n^* = 0$ $(c_i \in k)$, then, multiplying by ω_i and taking the trace we obtain $c_i = 0$.

Definition. The basis $\omega_1^*, \ldots, \omega_n^*$ of the separable extension K/k, which is determined by (2.14), is called the *dual basis* to the basis $\omega_1, \ldots, \omega_n$.

The dual basis allows us to express the coefficients in

$$\alpha = a_1\omega_1 + \cdots + a_n\omega_n$$

explicitly in terms of α. Indeed, taking the trace of the product $\alpha\omega_i^*$, we obtain

$$a_i = \mathrm{Sp}(\alpha\omega_i^*) \qquad (i = 1, \ldots, n).$$

Assume that the minimum polynomial $\varphi(t)$ of the element α of the separable extension K/k factors completely into linear factors in the extension Ω/k:

$$\varphi(t) = (t - \alpha_1) \cdots (t - \alpha_m).$$

It follows easily from (2.9) that the extension $k(\alpha)/k$ is also separable. Since the minimum polynomial φ is also the characteristic polynomial for α relative to the extension $k(\alpha)/k$, then by Theorems 7 and 8

$$\mathrm{Sp}_{k(\alpha)/k}\alpha^k = \sum_{s=1}^{m} \alpha_s{}^k,$$

and hence we have the following expression for the discriminant $D(1, \alpha, \ldots, \alpha^{m-1}) = D$ of the basis $1, \alpha, \ldots, \alpha^{m-1}$ of the extension $k(\alpha)/k$:

$$\begin{aligned} D &= \det\left(\sum_{s=1}^{m} \alpha_s{}^{i+j}\right)_{0 \leqslant i,j \leqslant m-1} \\ &= \det(\alpha_s{}^i) \cdot \det(\alpha_s{}^j) = \prod_{0 \leqslant i < j \leqslant m-1} (\alpha_i - \alpha_j)^2. \end{aligned}$$

Since $D \neq 0$, $\alpha_i \neq \alpha_j$, and we have proved the following fact.

Theorem 9. The minimum polynomial of any element of a separable extension has no multiple roots (in that field in which it factors into linear factors).

Theorem 10. Any finite separable extension K/k is simple; that is, there exists an element α such that $K = k(\alpha)$.

Theorem 11. Let K/k be a finite separable extension of degree n. There is an extension Ω/k such that there are precisely n isomorphisms of K into Ω which are the identity map on k. Denote these isomorphisms by $\sigma_1, \ldots, \sigma_n$. If α is any element of K, then the characteristic polynomial $f_\alpha(t)$ has the factorization

$$f_\alpha(t) = (t - \sigma_1(\alpha))(t - \sigma_2(\alpha) \cdots (t - \sigma_n(\alpha))$$

in the field Ω.

The elements $\sigma_1(\alpha), \ldots, \sigma_n(\alpha)$ (which lie in the field Ω) are called the *conjugates* of the element $\alpha \in K$. The images $\sigma_1(K), \ldots, \sigma_n(K)$ of the field K under the isomorphisms σ_i are called the *conjugate fields* of the field K. If θ is a primitive element of the field K over k, then it is clear that $\sigma_i(K) = k(\sigma_i(\theta))$.

Corollary 1. Using the above notations we have

$$N_{K/k}(\alpha) = \sigma_1(\alpha)\sigma_2(\alpha) \cdots \sigma_n(\alpha),$$

$$\mathrm{Sp}_{K/k}(\alpha) = \sigma_1(\alpha) + \sigma_2(\alpha) + \cdots + \sigma_n(\alpha).$$

Corollary 2. There are precisely n isomorphisms of any finite extension of the rational numbers of degree n into the field of complex numbers.

Let $\omega_1, \ldots, \omega_n$ be a basis for K/k. Since $\mathrm{Sp}(\omega_i\omega_j) = \sum_{s=1}^{n} \sigma_s(\omega_i)\sigma_s(\omega_j)$, then the matrix $(\mathrm{Sp}(\omega_i\omega_j))$ is the product of the matrices $(\sigma_i(\omega_j))'$ and $(\sigma_i(\omega_j))$. Hence we have the following formula for the discriminant of the basis ω_i:

$$D(\omega_1, \ldots, \omega_n) = (\det(\sigma_i(\omega_j)))^2. \tag{2.15}$$

PROBLEMS

1. Let $\Omega = k(x)$ be the field of rational functions in x with coefficients from k. Show that any element of Ω which does not lie in k is transcendental over k.

2. Let $k \subset K \subset L$ be a tower of finite extensions. If θ is any element of L, prove the formula

$$N_{K/k}(N_{L/K}(\theta)) = N_{L/K}(\theta).$$

[First assume that $L = K(\theta)$, and consider the basis $\omega_i\theta^j$ for L/k, where ω_i is a basis for K/k.]

3. Find a primitive element for the extension $R(\sqrt{2}, \sqrt{3})$ of the field R of rational numbers, and express it in terms of $\sqrt{2}$ and $\sqrt{3}$.

4. Show that a finite extension K/k is simple if and only if there are only a finite number of intermediate fields.

5. Let k be any field of characteristic $p \neq 0$. Show that the polynomial $f(t) = t^p - t - a$ $(a \in k)$ is either irreducible or factors completely into linear factors in $k[t]$. Further, in the former case show that the extension $k(\theta)/k$, where $f(\theta) = 0$, is separable.

6. Let k_0 be a field of characteristic $p \neq 0$, and let $k = k_0(x)$ be the field of rational functions in x with coefficients from k_0. Show that the polynomial $f(t) = t^p - x$ is irreducible in $k[t]$. Further, show that the extension $k(\theta)/k$, where $f(\theta) = 0$, is inseparable.

7. Let K/k be a finite extension of degree n. If there exists some extension Ω/k for which there are n isomorphisms of K into Ω which leave every element of k fixed, show that the extension K/k is separable.

8. Let k be a field of characteristic $\neq p$ which contains a primitive pth root of 1 (that is, an element ε with $\varepsilon^p = 1$ and $\varepsilon^k \neq 1$ for $0 < k < p$). If the element $\alpha \in k$ is not the pth power of some element of k, show that $(k(\sqrt[p]{\alpha}) : k) = p$.

9. Let K/k be a finite separable extension and let φ be a linear mapping from K (considered as a vector space over k) to k. Show that there is a unique element α in the field K such that

$$\varphi(\xi) = \mathrm{Sp}_{K/k}(\alpha\xi) \qquad (\xi \in K).$$

3. Finite Fields

A field Σ is called *finite* if it has only a finite number of elements. The field Z_p of residue classes in the ring Z of integers modulo a prime number p is an example of a finite field. Every finite field is of finite characteristic and, if the characteristic of the finite field Σ equals p, then this field contains a prime subfield (a subfield not containing any proper subfield) which is isomorphic to Z_p. Hence we may assume that $Z_p \subset \Sigma$. The extension Σ/Z_p is clearly finite. If it is of degree m and if $\omega_1, \ldots, \omega_m$ is a basis for Σ/Z, then every element $\xi \in \Sigma$ has a unique representation in the form $\xi = c_1\omega_1 + \cdots + c_m\omega_m$, where $c_i \in Z_p$. Since there are p^m such linear combinations, we have proved that the number of elements of any finite field is a power of its characteristic.

The multiplicative group Σ^* of a finite field Σ is a finite Abelian group. We consider its structure.

Lemma 1. A finite subgroup G of the multiplicative group K^* of any field K is always cyclic.

Proof. We first show that if an Abelian group G contains elements of orders m and n, then it contains an element whose order equals the least common multiple r of m and n. Let the elements x and y of G have orders m and n, respectively. If $(m, n) = 1$, then it is easily seen that the product xy has order $r = mn$. In general, by considering the prime factorizations of the numbers m and n, we can find factorizations

$$m = m_0 m_1, \quad n = n_0 n_1,$$

so that $(m_0, n_0) = 1$ and $r = m_0 n_0$. The elements x^{m_1} and y^{n_1} have orders m_0 and n_0, and their product $x^{m_1}y^{n_1}$ has order $r = m_0 n_0$.

Now let G be a finite subgroup of order g of the multiplicative group of the field K. If m is the maximum of the orders of the elements of G, then clearly $m \leqslant g$. On the other hand, it follows from what we have just shown that the order of every element divides m; that is, every element of the group G is a root of the polynomial $t^m - 1$. Since a polynomial of degree m can have at most m roots, $g \leqslant m$. Hence $g = m$, and this means that G is cyclic.

Applying this lemma to the case of finite fields, we obtain the following fact.

Theorem 1. The multiplicative group of a finite field which has p^m elements is a cyclic group of order $p^m - 1$.

Corollary. Any finite extension of a finite field is simple.

Indeed, if θ is a generating element of the group Σ^*, then it is clear that $Z_p(\theta) = \Sigma$. Hence for any intermediate field Σ_0 we have $\Sigma_0(\theta) = \Sigma$.

It also follows from Theorem 1 that all elements of Σ are roots of the polynomial $f^{p^m} - t$, and since the degree of this polynomial equals the number of elements in Σ, then in the ring $\Sigma[t]$ we have the factorization

$$t^{p^m} - t = \prod_{\xi \in \Sigma} (t - \xi)$$

(ξ runs through all elements of the field Σ).

Theorem 2. For any prime number p and any natural number m there exists one and only one (up to isomorphism) finite field with p^m elements.

Proof. By the corollary to Theorem 5 of Section 2 there is an extension Ω/Z_p in which the polynomial $t^{p^m} - t$ factors into linear factors. Let Σ denote the set of all roots of this polynomial (in Ω). Since in any field of characteristic p the formula

$$(x \pm y)^{p^m} = x^{p^m} \pm y^{p^m}$$

holds, then the sum and difference of any two elements of Σ also belongs to Σ. It is clear that the set Σ is closed under the operations of multiplication and division (except division by zero). Hence Σ is a subfield of the field Ω. The polynomial $t^{p^m} - t$ has no multiple roots (since its derivative $p^m t^{p^m - 1} - 1 = -1$ never vanishes) and hence Σ consists of p^m elements. The existence of a finite field with p^m elements is proved.

Let Σ and Σ' be two extensions of Z_p of degree m. Choose a primitive element θ in Σ (corollary of Theorem 1) and denote its minimum polynomial by $\varphi(t)$. Since $\varphi(t)$ divides the polynomial $t^{p^m} - t$, and the latter polynomial splits into linear factors over Σ', then $\varphi(t)$ has a root $\theta' \in \Sigma'$. The degree of the extension $Z_p(\theta')/Z_p$ equals the degree of the polynomial $\varphi(t)$, that is, m, and therefore $Z_p(\theta') = \Sigma'$. The existence of an isomorphism of Σ onto Σ' now follows from Theorem 4 of Section 2.

The finite field with p^m elements is often denoted by $GF(p^m)$ (and finite fields are often called *Galois fields*).

Corollary. For every natural number n there is an irreducible polynomial of degree n over the finite field $\Sigma_0 = GF(p^r)$.

Indeed, $p^r - 1$ divides $p^{nr} - 1$, so that the set of all roots of the polynomial $t - t$ in the field $\Sigma = GF(p^{rn})$ forms a subfield which is isomorphic to the field Σ_0. Hence we can assume that $\Sigma_0 \subset \Sigma$. If $\theta \in \Sigma$ is a primitive element for the extension Σ/Σ_0, then the minimum polynomial of θ will be an irreducible polynomial in $\Sigma_0[t]$ of degree n, since

$$(\Sigma : \Sigma_0) = \frac{(\Sigma : Z_p)}{(\Sigma_0 : Z_p)} = \frac{rn}{r} = n.$$

In conclusion we note that in order to show that a given finite commutative ring is a field, it suffices to show that it has no divisors of zero. Indeed, let $\mathfrak{O}$ be a finite ring without zero divisors, and let a be a nonzero element of $\mathfrak{O}$. If $ax_1 = ax_2$, then $a(x_1 - x_2) = 0$, and $x_1 = x_2$. Thus as x runs through all elements of the (finite) ring $\mathfrak{O}$, ax also takes on every value in $\mathfrak{O}$. Then for any b in $\mathfrak{O}$ the equation $ax = b$ is solvable in $\mathfrak{O}$, and this means that $\mathfrak{O}$ is a field.

PROBLEMS

1. Let $r(m)$ denote the number of distinct irreducible polynomials of degree m with leading coefficient 1 in the ring $Z_p[t]$. Show that

$$r(m) = \frac{1}{m} \sum_{d|m} \mu\left(\frac{m}{d}\right) p^d$$

(d runs through all divisors of m, and μ denotes the Möbius function).

2. Find all irreducible polynomials of degree 2 over the field $Z_5 = GF(5)$.

3. Show that the field $GF(p^m)$ is contained in the field $GF(p^n)$ (up to isomorphism) if and only if $m|n$.

4. What is the degree over Z_p of the splitting field of the polynomial $t^n - 1$?

5. Let $\Sigma = GF(p^m)$. Show that each mapping $\sigma_i : \xi \to \xi^{p^i}$, $\xi \in \Sigma$ $(i = 0, 1, \ldots, m-1)$ is an automorphism of the field Σ and show that every automorphism of Σ coincides with one and only one of the σ_i.

6. Let $\Sigma_0 = GF(p^r)$ and let Σ be a finite extension of Σ_0 of degree n. Show that each mapping $\xi \to \xi^{p^{ri}}$ $(i = 0, 1, \ldots, n-1)$ is an automorphism of the field Σ which leaves every element of Σ_0 fixed. Further, show that these n automorphisms are distinct and that every automorphism of Σ which is the identity on Σ_0 coincides with one of them. Let $f_\xi(t)$ be the characteristic polynomial of the element $\xi \in \Sigma$ relative to the extension Σ/Σ_0. Show that we have the factorization

$$f_\xi(t) = (t - \xi)(t - \xi^q) \cdots (t - \xi^{q^{n-1}}),$$

in the field Σ, where $q = p^r$ (use Theorem 8 of Section 2). Deduce that

$$\mathrm{Sp}_{\Sigma/\Sigma_0}(\xi) = \xi + \xi^q + \cdots + \xi^{q^{n-1}}, \quad N_{\Sigma/\Sigma_0}(\xi) = \xi^{1+q+\cdots+q^{n-1}}.$$

7. Show that a finite extension of a finite field is always separable.

8. Using the above notations, show that every element of the field Σ_0 is the norm of some element of Σ.

9. Let $\Sigma = GF(q^r)$, where $p^m = q$, and let $\alpha \in \Sigma$. Show that the equation $\xi^q - \xi = \alpha$ is solvable in Σ if and only if $\alpha + \alpha^q + \cdots + \alpha^{q^{r-1}} = 0$.

10. Let ε be a primitive pth root of 1 (over the field of rational numbers). Let $\Sigma_0 = GF(p)$ and $\Sigma = GF(p^m)$. Since the elements of the field Σ_0 can be considered as residue classes of integers modulo the prime p, then the expression $\varepsilon^{\mathrm{Sp}(\gamma)}$ makes sense for any $\gamma \in \Sigma$ (the trace is taken relative to the extension Σ/Σ_0). Show that

$$\sum_{\xi \in \Sigma} \varepsilon^{\mathrm{Sp}\, \xi\alpha} = \begin{cases} 0 & \text{for } \alpha \neq 0, \\ p^m & \text{for } \alpha = 0. \end{cases}$$

11. Let χ be a character of the multiplicative group of the field $\Sigma = GF(p^m)$, and set $p^m = q$ (for the definition of a character, see Section 5). Extend χ to the whole field Σ by setting $\chi(0) = 0$. The expression

$$\tau_\alpha(\chi) = \sum_{\xi \in \Sigma} \chi(\xi) \varepsilon^{\mathrm{Sp}\, \alpha\xi} \qquad (\alpha \in \Sigma),$$

which is a complex number, is called the *Gaussian sum* of the finite field Σ. Assuming that the character χ is not the unit character, prove the formulas

$$\tau_\alpha(\chi) = \chi(\alpha)^{-1}\tau_1(\chi) \qquad \alpha \neq 0,$$

$$|\tau_\alpha(\chi)| = \sqrt{q} \qquad \alpha \neq 0,$$

$$\sum_{\alpha \neq 0} \tau_\alpha(\chi) = 0.$$

12. Let $p \neq 2$. Then the set of all squares in the multiplicative group Σ^* of the field $\Sigma = GF(p^m)$ is a subgroup of index 2. If we set $\psi(\alpha) = +1$ if $\alpha \neq 0$ is a square, and $\psi(\alpha) = -1$ otherwise, we obtain a character of the group Σ^*. Show that if $\alpha\beta \neq 0$, then

$$\tau_\alpha(\psi)\tau_\beta(\psi) = \psi(-\alpha\beta)p^m.$$

13. Show that for $\alpha \neq 0$

$$\sum_{\xi \in \Sigma} \psi(\xi^2 - \alpha) = -1.$$

14. Let $f(x_1, \ldots, x_n)$ be a nonsingular quadratic form with determinant δ and with coefficients in $\Sigma = GF(p^m)$, where $p \neq 2$ and we set $p^m = q$. Let α be any element of Σ and let N denote the number of solutions in Σ of the equation

$$f(x_1, \ldots, x_n) = \alpha.$$

Show that N satisfies the formulas

$$N = q^{2r} + q^r\psi((-1)^r\alpha\delta) \qquad \text{if } n = 2r + 1,$$

$$N = q^{2r-1} + \omega q^{r-1}\psi((-1)^r\delta \qquad \text{if } n = 2r,$$

where $\omega = -1$ if $\alpha \neq 0$ and $\omega = q - 1$ if $\alpha = 0$.

15. Let p and q be distinct odd primes. If x is an integer, we shall also use the letter x to denote the corresponding residue classes in the fields $GF(p)$ and $GF(q)$. Let Δ be an extension of $GF(q)$ in which the polynomial $t^p - 1$ splits into linear factors, and let ε denote a primitive pth root of 1 contained in Δ. The Legendre symbol (x/p) clearly coincides with the character $\psi(x)$ of the field $GF(p)$ which was defined in Problem 12. Since it takes the values ± 1, we may assume that $(x/p) \in \Delta$. Show that the "Gaussian sum"

$$\tau = \sum_{x \in GF(p)} \left(\frac{x}{p}\right) \varepsilon x \in \Delta$$

of the field $GF(p)$ satisfies the equations

$$\tau^2 = (-1)^{(p-1)/2}p, \tag{1}$$

$$\tau q = \left(\frac{q}{p}\right)\tau. \tag{2}$$

16. Use the representation of the Legendre symbol $(p/q) = p^{(q-1)/2}$ in the field $GF(q)$ and formulas (1) and (2) to prove the Gaussian reciprocity law:

$$(-1)^{-[(p-1)/2]\,[(q-1)/2]}\left(\frac{p}{q}\right) = \left(\frac{q}{p}\right).$$

4. Some Results on Commutative Rings

Throughout this section the word ring will mean a commutative ring with unit element 1 and without divisors of zero (that is, an integral domain).

4.1. *Divisibility in Rings*

Let $\mathfrak{O}$ be a ring, and let α and $\beta \neq 0$ be two elements of $\mathfrak{O}$. If there exists an element $\xi \in \mathfrak{O}$ such that $\beta\xi = \alpha$, then we say that α is divisible by β (or that β divides α), and we write $\beta|\alpha$. Since $\mathfrak{O}$ contains no divisors of zero, there is at most one element ξ such that $\alpha = \beta\xi$. The concept of divisibility in an arbitrary ring clearly possesses all the usual properties of divisibility in the ring of rational integers. For example, if $\gamma|\beta$ and $\beta|\alpha$, then $\gamma|\alpha$.

An element $\varepsilon \in \mathfrak{O}$ which divides the unit element 1, is called a *unit* of the ring $\mathfrak{O}$ (or an *invertible element*).

Theorem 1. The units of the ring $\mathfrak{O}$ form a group under multiplication.

Proof. Let E be the set of all units of the ring $\mathfrak{O}$. If $\varepsilon \in E$ and $\eta \in E$, then $\varepsilon\varepsilon' = 1$ and $\varepsilon\eta = 1$ for some ε' and η' of $\mathfrak{O}$. But then $\varepsilon\eta(\varepsilon'\eta') = 1$, and this means that $\varepsilon\eta \in E$. Since $1 \in E$, and if $\varepsilon \in E$, then $\varepsilon' \in E$, where $\varepsilon\varepsilon' = 1$, we have verified that E is a group under multiplication, and this proves the theorem.

Elements $\alpha \neq 0$ and $\beta \neq 0$ of the ring $\mathfrak{O}$ are called *associate* if they divide each other. From $\alpha = \beta\xi$ and $\beta = \alpha\eta$ ($\xi \in \mathfrak{O}$, $\eta \in \mathfrak{O}$) it follows that $\alpha = \alpha\xi\eta$ and hence $1 = \xi\eta$. Thus two nonzero elements of $\mathfrak{O}$ are associate if one is a unit multiple of the other.

Let μ be a nonzero element of the ring $\mathfrak{O}$ which is not a unit. We shall say that the elements α and β of $\mathfrak{O}$ are congruent modulo μ and write $\alpha \equiv \beta \pmod{\mu}$, if the difference $\alpha - \beta$ is divisible by μ. All the usual properties of congruences in the ring of integers also hold for congruences in the ring $\mathfrak{O}$. For any $\alpha \in \mathfrak{O}$ we denote by $\bar{\alpha}$ the set of all elements of $\mathfrak{O}$ which are congruent to α modulo μ. The set $\bar{\alpha}$ is called a *residue class modulo* μ. We clearly have $\bar{\alpha} = \bar{\beta}$ if and only if $\alpha \equiv \beta \pmod{\mu}$. We can define the sum and product of two residue classes modulo μ by setting

$$\bar{\alpha} + \bar{\beta} = \overline{\alpha + \beta}, \quad \bar{\alpha}\bar{\beta} = \overline{\alpha\beta}.$$

It is easily checked that these definitions do not depend on the choice of the representatives (residues) α and β. It is also easily verified that under these operations the set of all residue classes modulo μ becomes a commutative ring with unit element 1 (but possibly with divisors of zero). It is called the *ring* of residue classes modulo μ.

If in each residue class modulo μ we choose a representative, then the set S of all representatives is called a *complete system* of residues modulo μ. A complete system S of residues is clearly characterized by the property that every element of $\mathfrak{O}$ is congruent modulo μ to one and only one element of S.

4.2. *Ideals*

A subset A of the ring $\mathfrak{O}$ is called an *ideal* if it is a subgroup of the additive group of $\mathfrak{O}$, and if for any $\alpha \in A$ and any $\xi \in \mathfrak{O}$ the product $\xi\alpha$ lies in A. The subset consisting only of zero and the entire ring are trivially examples of ideals. The first of these ideals is called the *zero ideal*, the second the *unit ideal.*

Let $\alpha_1, \ldots, \alpha_m$ be any elements of the ring $\mathfrak{O}$. It is clear that the set A of all linear combinations $\xi_1\alpha_1 + \cdots + \xi_m\alpha_m$ of these elements with coefficients ξ_i in $\mathfrak{O}$ is an ideal in $\mathfrak{O}$. It is called the *ideal generated by the elements* $\alpha_1, \ldots, \alpha_m$, and is denoted by $A = (\alpha_1, \ldots, \alpha_m)$. The elements $\alpha_1, \ldots, \alpha_m$ are called *generators* of the ideal A. In general, not every ideal has a finite system of generators. An ideal A is called *principal* if it has a system of generators consisting of a single element, that is, if it has the form $A = (\alpha)$. A nonzero principal ideal (α) consists of all elements of A which are divisible by α. The zero and unit ideals are both principal. The zero ideal is generated by the zero element, and the unit ideal is generated by any unit ε of the ring $\mathfrak{O}$. Two principal ideals (α) and (β) are equal if and only if the elements α and β are associate.

Let A and B be two ideals of the ring $\mathfrak{O}$. The set of all elements $\xi \in \mathfrak{O}$ which can be represented in the form

$$\xi = \alpha_1\beta_1 + \cdots + \alpha_s\beta_s,$$

where $\alpha_i \in A$ and $\beta_i \in \beta$ $(s \geqslant 1)$, is also an ideal in $\mathfrak{O}$. This ideal is called the *product* of the ideals A and B, and is denoted by AB. Since multiplication of ideals is commutative and associative, then the set of all ideals of the ring $\mathfrak{O}$ is a commutative semigroup under the operation of multiplication.

Two elements α and β of $\mathfrak{O}$ are said to be *congruent* modulo the ideal A, and we write $\alpha \equiv \beta \pmod{A}$, if their difference $\alpha - \beta$ lies in A, that is, if α and β belong to the same coset of the additive group A. If $\bar{\gamma}$ denotes the coset of A which contains γ, then we have $\bar{\alpha} = \bar{\beta}$ if and only if $\alpha \equiv \beta \pmod{A}$. For the principal ideal (μ), the concept of congruence modulo (μ) coincides with that modulo the element μ. Consider the factor group $\mathfrak{O}/A$ of the additive group of the ring $\mathfrak{O}$. When the subgroup A is an ideal, then we can define multiplication in $\mathfrak{O}/A$. Namely, for $\bar{\alpha}$ and $\bar{\beta}$ in $\mathfrak{O}/A$ we set

$$\bar{\alpha}\bar{\beta} = \overline{\alpha\beta}.$$

If $\bar{\alpha} = \bar{\alpha}_1$ and $\bar{\beta} = \bar{\beta}_1$, then since $\alpha_1\beta_1 - \alpha\beta = \alpha_1(\beta_1 - \beta) + \beta(\alpha_1 - \alpha)$, we have $\alpha\beta \equiv \alpha_1\beta_1 \pmod{A}$. This means that the product $\bar{\alpha}\bar{\beta}$ does not depend on the choice of the representatives α and β (it is essential here that A is an ideal). It is easily verified that under this definition the factor group $\mathfrak{O}/A$ becomes a ring. The ring $\mathfrak{O}/A$ is called the *factor ring* of the ring $\mathfrak{O}$ by the ideal A. For a principal ideal (μ) the factor ring $\mathfrak{O}/(\mu)$ coincides with the ring of residue classes modulo μ.

4.3. *Integral Elements*

Any ring $\mathfrak{o}$ (commutative and without zero divisors) can be embedded in a field. To show this we consider the set of all formal fractions a/b, where a and b are elements of $\mathfrak{o}$ and $b \neq 0$. Two fractions a/b and c/d are called equal if and only if $ad = bc$. Addition and multiplication are defined by the formulas

$$\frac{a}{b} + \frac{c}{d} = \frac{ad + bc}{bd},$$

$$\frac{a}{b} \cdot \frac{c}{d} = \frac{ac}{bd}.$$

It is easily verified that these operations are compatible with the notion of equality, and that with these operations the set of all fractions a/b becomes a field. We denote this field by k_0. If we identify each fraction $a/1 = ac/c$ $(c \neq 0)$ with the element $a \in \mathfrak{o}$, then $\mathfrak{o}$ will be a subring of the field k_0. Hence every element of k_0 is the quotient of two elements of $\mathfrak{o}$.

Now let Ω be any field which contains $\mathfrak{o}$ as a subring. Let k be the set of all quotients a/b, where a and b lie in $\mathfrak{o}$ $(b \neq 0)$. Clearly, k is a subfield of the field Ω. This subfield is called the *quotient field* of $\mathfrak{o}$. It is easily checked that the field k is isomorphic to the field k_0 constructed above, and hence that it is uniquely determined by the ring $\mathfrak{o}$ (up to isomorphism).

Definition. Let the ring $\mathfrak{o}$ be contained in the field Ω. An element $\alpha \in \Omega$ is called integral over $\mathfrak{o}$, if it is the root of a polynomial with coefficients in $\mathfrak{o}$ and with leading coefficient 1.

Since any element $a \in \mathfrak{o}$ is the root of the polynomial $t - a$, then every element of $\mathfrak{o}$ is integral over $\mathfrak{o}$.

Let $\omega_1, \ldots, \omega_m$ be arbitrary element of Ω. The set M of all linear combinations $a_1\omega_1 + \cdots + a_m\omega_m$ with coefficients $a_i \in \mathfrak{o}$ is called a finitely generated $\mathfrak{o}$-module in Ω, and the elements $\omega_1, \ldots, \omega_m$ are called generators of the $\mathfrak{o}$-module M. Since $1 \in \mathfrak{o}$, then all the ω_i are contained in M.

Lemma 1. If the finitely generated $\mathfrak{o}$-module M is a ring, then all its elements are integral over $\mathfrak{o}$.

Proof. We can of course assume that not all ω_i are zero. Let α be any element of M. Since for any i the product $\alpha\omega_i$ belongs to M, then

$$\alpha\omega_i = \sum_{j=1}^{m} a_{ij}\omega_j, \quad a_{ij} \in \mathfrak{o} \qquad (i = 1, \dots, m).$$

It follows that $\det(\alpha E - (a_{ij})) = 0$ (E is the unit matrix). Hence the element α is a root of the polynomial $f(t) = \det(tE - (a_{ij}))$ which has all coefficients in $\mathfrak{o}$ and has leading coefficient 1, and this proves the lemma.

Theorem 2. The set of all elements of Ω which are integral over $\mathfrak{o}$ is a ring.

Proof. We must verify that the sum, difference, and product of two integral elements of the field Ω are again integral over $\mathfrak{o}$. If α and β are the roots of the polynomials

$$t^m - a_m t^{m-1} - \cdots - a_1, \quad t^n - b_n t^{n-1} - \cdots - b_1,$$

where a_i and b_j are elements of $\mathfrak{o}$, then

$$\alpha^m = a_1 + a_2\alpha + \cdots + a_m\alpha^{m-1}, \quad \beta^n = b_1 = b_2\beta + \cdots + b_n\beta^{n-1}.$$

It easily follows that the $\mathfrak{o}$-module which consists of all linear combinations of the products

$$\alpha^i\beta^j \qquad (0 \leqslant i < m, 0 \leqslant j < n) \tag{4.1}$$

with coefficients in $\mathfrak{o}$, is a ring (since any product with $k \geqslant 0$ and $l \geqslant 0$ can be expressed as a linear combination of the elements $\alpha^k\beta^l$ with coefficients in $\mathfrak{o}$). By Lemma 1 all elements of this ring are integral over $\mathfrak{o}$; in particular, this will hold for $\alpha \pm \beta$ and $\alpha\beta$. Theorem 2 is proved.

Definition. Let $\mathfrak{o}$ be a subring of the field Ω, and let $\mathfrak{O}$ be the set (which is a ring by Theorem 2) of all elements of Ω which are integral over $\mathfrak{o}$. The ring $\mathfrak{O}$ is called the *integral closure* of the ring $\mathfrak{o}$ in the field Ω.

Definition. A subring $\mathfrak{O}_0$ of a field K is called *integrally closed* in K if its integral closure in K coincides with $\mathfrak{O}_0$.

Definition. A ring $\mathfrak{O}$ is called *integrally closed* if it is integrally closed in its quotient field k.

Theorem 3. Let $\mathfrak{o}$ be a subring of the field Ω, and let $\mathfrak{O}$ be the integral closure of $\mathfrak{o}$ in Ω. Then the ring $\mathfrak{O}$ is integrally closed in Ω.

Proof. Let θ be any element of Ω which is integral over $\mathfrak{O}$, so that

$$\theta^n = \alpha_1 + \alpha_2\theta + \cdots + \alpha_n\theta^{n-1}, \tag{4.2}$$

where all α_i lie in $\mathfrak{O}$. We must show that $\theta \in \mathfrak{O}$. For each $i = 1, \ldots, n$ there is an integer m for which

$$\alpha_i^{m_i} = \sum_{j=1}^{m_i} a_{ij}\alpha_i^{j-1} \qquad (a_{ij} \in \mathfrak{o}) \tag{4.3}$$

(since α_i is integral over $\mathfrak{o}$). Consider the $\mathfrak{o}$-module M which is generated by the products

$$\alpha_i^{k_1} \cdots \alpha_n^{k_n}\theta^k \qquad (0 \leqslant k_i < m_i,\ 0 \leqslant k < n). \tag{4.4}$$

It easily follows from (4.2) and (4.3) that any product $\alpha_1^{l_1} \cdots \alpha_n^{l_n}\theta^l$ with non-negative exponents can be expressed as a linear combination of the elements (4.4) with coefficients in $\mathfrak{o}$, and this means that the module M is a ring. By Lemma 1 every element of M is integral over $\mathfrak{o}$. In particular, θ is integral over $\mathfrak{o}$ and this proves the theorem.

Lemma 2. Let $\mathfrak{o}$ be an integrally closed ring with quotient field k, and let $f(t) \in \mathfrak{o}[t]$ be a polynomial with leading coefficient 1. If the polynomial $\varphi(t) \in \mathfrak{o}[t]$ divides $f(t)$ and has leading coefficient 1, then $\varphi(t) \in \theta[t]$.

Proof. Let Ω/k be an extension in which the polynomial $f(t)$ factors into linear factors. If $\mathfrak{O}$ is the integral closure of $\mathfrak{o}$ in Ω, then every root of $f(t)$ clearly lies in $\mathfrak{O}$, and this will also be true for all roots of $\varphi(t)$. From the identity $\varphi(t) = (t - \gamma_1) \cdots (t - \gamma_s)$, it follows that all coefficients of $\varphi(t)$ also lie in $\mathfrak{O}$, and since $\mathfrak{O} \cap k = \mathfrak{o}$ (as $\mathfrak{o}$ is integrally closed), then these coefficients lie in $\mathfrak{o}$, which proves the lemma.

The following fact is an obvious consequence of Lemma 2.

Theorem 4. Let $\mathfrak{o}$ be an integrally closed ring with quotient field k, and let Ω/k be an algebraic extension of the field k. In order that the element $\alpha \in \Omega$ be integral over $\mathfrak{o}$, it is necessary and sufficient that all coefficients of its minimum polynomial lie in $\mathfrak{o}$.

PROBLEMS

1. An ideal A of the ring $\mathfrak{O}$ is called *maximal* if $A \neq \mathfrak{O}$ and the only ideal of $\mathfrak{O}$ which properly contains A is the unit ideal $\mathfrak{O}$. Show that the ideal A is maximal if and only if the factor ring $\mathfrak{O}/A$ is a field.

2. If $\mathfrak{o}$ is an integrally closed ring, show that the polynomial ring $\mathfrak{o}[t]$ is also integrally closed.

5. Characters

In this section we describe some properties of characters of finite Abelian groups and numerical characters.

5.1. *The Structure of Finite Abelian Groups*

The structure of finite Abelian groups is determined by the following theorem (see, for instance, M. Hall, "The Theory of Groups," Macmillan, New York, (1959)).

Theorem 1. Every finite Abelian group can be represented as the direct product of cyclic subgroups.

By Problems 1 and 2 a finite cyclic group cannot be represented as the direct product of proper subgroups if and only if its order is a power of a prime. Therefore a finite Abelian group G can be represented as a direct product $G = A_1 \times \cdots \times A_s$ of cyclic subgroups A_i of prime power order. This representation is not, in general, unique. But the orders of the cyclic subgroups of prime power order are uniquely determined by G. These orders (which are powers of prime numbers) are called the *invariants* of the finite Abelian group G. The product of all invariants of G clearly equals the order of G.

5.2. *Characters of Finite Abelian Groups*

Definition. A homomorphism of the finite Abelian group G into the multiplicative group of the field of complex numbers is called a *character* of the group G.

In other words, a character of G is a nonzero complex-valued function χ on G for which

$$\chi(xy) = \chi(x)\chi(y) \tag{5.1}$$

for any x and y of G.

Since any homomorphism of groups takes the unit onto the unit, then $\chi(1) = 1$. If the element $x \in G$ has order k, then

$$(\chi(x))^k = \chi(x^k) = \chi(1) = 1; \tag{5.2}$$

that is, $\chi(x)$ is a kth root of 1. If m is the maximum of the orders of the elements of G, then by Problem 3 the order of every element of G divides m. Hence $\chi(x)$ is an mth root of 1 for all $x \in g$, and this means that a character could also be defined as a homomorphism from G to the group of all mth roots of 1.

We represent G as a direct product of cyclic groups:

$$G = \{a_1\} \times \cdots \times \{a_s\}.$$

Since every element $x \in G$ can be represented in the form

$$x = a_1^{k_1} \cdots a_s^{k_s}, \tag{5.3}$$

then by (5.1),

$$\chi(x) = \chi(a_1)^{k_1} \cdots \chi(a_s)^{k_s},$$

so the character χ is completely determined by the values $\chi(a_1), \ldots, \chi(a_s)$. If a_i has order m, then by (5.2) $\chi(a_i)$ is an m_ith root of 1. Conversely, for $i = 1, \ldots, s$ let ε_i be any m_ith root of 1, and for any $x \in G$ of the form (5.3) set

$$\chi(x) = \varepsilon_1^{k_1} \cdots \varepsilon_s^{k_s}. \tag{5.4}$$

It is easily seen that the value of (5.4) is independent of the choice of the exponents k_i (which are only defined modulo m_i), and that this function is a character of the group G. Each root ε_i can be chosen in m_i ways, so there are $m_1 \cdots m_s$ distinct functions of the type (5.4). Hence we have the following theorem.

Theorem 2. The number of characters of a finite Abelian group equals the order of the group.

We shall define a multiplication for characters. If χ and χ' are characters of the group G, set

$$(\chi\chi')(x) = \chi(x)\chi'(x) \qquad (x \in G).$$

It is clear that the function χ also is a character of the group G. The character χ_0, for which $\chi_0(x) = 1$ for all $x \in G$, is called the *unit character*. It is clear that $\chi\chi_0 = \chi$ for any character χ. If for a character χ of the group G we set

$$\bar{\chi}(x) = \overline{\chi(x)} \qquad (x \in G),$$

where $\overline{\chi(x)}$ is the complex conjugate of the number $\chi(x)$, then the function $\bar{\chi}$ also will be a character of the group G, and $\chi\bar{\chi} = \chi_0$. Since multiplication of characters is clearly associative, then the set of all characters of a finite Abelian group is a group under the operation of multiplication.

Let $G = \{a\}$ be a cyclic group of order m and let ε be a fixed primitive mth root of 1. Let χ be that character of the group G for which $\chi(a) = \varepsilon$ (and hence $\chi(a^k) = \varepsilon^k$). Since $\chi^r(a) = \varepsilon^r$, then the characters $\chi_0 = \chi^m, \chi, \chi^2, \ldots, \chi^{m-1}$ are pairwise-distinct, and hence exhaust the characters of the group G. Hence we have shown that the character group of a cyclic group is cyclic. In the general case it is easy to prove the following theorem: Any finite Abelian group is isomorphic to its character group.

Let G be an Abelian group of order n and let H be a subgroup of order m. If we restrict a character of G to H, we clearly obtain a character of the group H. We denote this character by χ. It is clear that the mapping $\chi \to \hat{\chi}$ is a homomorphism from the character group X of G to the character group Y of H. Let A be the kernel of this map. The characters of A are characterized by having $\chi(z) = 1$ for all $z \in H$. If $\chi \in A$ and x and x' lie in the same coset of H in G, then $\chi(x) = \chi(x')$. Setting $\bar{\chi}(\bar{x}) = \chi(x)$, where $\chi \in A$ and $\bar{x}$ is the coset of H in G which contains x, we obtain a function $\bar{\chi}$ on the factor group G/H, which is a character of this group. Conversely, if ψ is any character of the factor group G/H, then by setting

$$\chi(x) = \psi(\bar{x}) \qquad (x \in G),$$

we obtain a character $\chi \in A$, for which $\hat{\chi} = \psi$. Since under the mapping $\chi \to \bar{\chi}$ distinct characters in A go to distinct characters of the group G/H, we have shown that the number of characters contained in A equals the number of characters of the factor group G/H, that is, equals n/m (Theorem 2). Hence the image of the group X under the homomorphism $\chi \to \hat{\chi}$ must have order $n/(n/m) = m$, and since by Theorem 2 the group Y has order m, then the image coincides with Y. This means that every character of the group H is of the form $\hat{\chi}$ for some character χ of the group G. It is clear that the number of characters $\chi \in X$ which induce a given character of H equals $n/m = (G:H)$.

We have proved the following theorem.

Theorem 3. If G is a finite Abelian group and H is a subgroup, then any character of the group H can be extended to a character of the group G, and the number of such extensions equals the index $(G : H)$.

Corollary 1. If x is any nonunit element of the group G, then there exists a character χ of the group G for which $\chi(x) \neq 1$.

Indeed, consider the cyclic group $\{x\} = H$. Since H has order greater than 1, there is a nonunit character χ' of H, and $\chi'(x) \neq 1$. Extending χ' to a character of the group G, we obtain the desired character χ.

Corollary 2. If the element x of the group G is not contained in the subgroup H, then there is a character χ of the group G such that $\chi(x) \neq 1$ and $\chi(z) = 1$ for all $z \in H$.

Indeed, the unit character of the group H can be extended to a nonunit character of the subgroup $\{x, H\}$, which in turn can be extended to a character of the group G.

We now consider some relations between the values of characters. If χ_0 is the unit character, then $\chi_0(x) = 1$ for all $x \in G$, and hence $\sum_{x \in G} \chi_0(x) = n$, where n is the order of the group G. Assume that the character χ is not the unit character, so that $\chi(z) \neq 1$ for some $z \in G$. If x runs through all elements of the group G, then zx also runs through all elements of G. Setting $s = \sum_{x \in G} \chi(x)$, we have

$$S = \sum_{x \in G} \chi(zx) = \chi(z)S.$$

Since $\chi(z) \neq 1$, we must have $S = 0$. Thus we have the formula

$$\sum_{x \in G} \chi(x) = \begin{cases} n & \text{if } \chi = \chi_0, \\ 0 & \text{if } \chi \neq \chi_0. \end{cases} \tag{5.5}$$

The value of any character χ on the unit element of the group equals 1, hence $\sum_\chi \chi(1) = n$ (here χ runs through all characters of the group G). We set $T = \sum_\chi \chi(x)$. By the first corollary to Theorem 3 there is a character χ' for which $\chi'(x) \neq 1$ (if $x \neq 1$). As χ runs through the character group of G, so does $\chi\chi'$. Therefore,

$$T = \sum_\chi (\chi'\chi)(x) = \sum_\chi \chi'(x)\chi(x) = \chi'(x)T,$$

and since $\chi'(x) \neq 1$, then $T = 0$. We have proved the formula

$$\sum_\chi \chi(x) = \begin{cases} n & \text{if } x = 1, \\ 0 & \text{if } x \neq 1. \end{cases} \tag{5.6}$$

5.3. *Numerical Characters*

For any natural number m let G_m denote the group under multiplication of all residue classes modulo m which consists of all residue classes of numbers relatively prime to m. The residue class modulo m which contains a will be denoted by $\bar{a}$.

Every character χ of the group G_m can be associated to a function χ^* on rational integers relatively prime to m by setting

$$\chi^*(a) = \chi(\bar{a}).$$

We extend this function to all rational integers by setting $\chi^*(a) = 0$ if a and m are not relatively prime. Such a function χ^* (defined on all rational integers) is called a *numerical character* modulo m. In the future we shall denote χ^* by the same symbol, χ, as used for the corresponding character of the group G_m. It is clear that distinct characters of the group G_m correspond to distinct numerical characters modulo m, and that the number of numerical characters modulo m equal $\varphi(m)$.

The following properties of numerical characters easily follow from the definition.

(1) For any rational integer a the value of $\chi(a)$ is a complex number which is zero if and only if a and m are not relatively prime.

(2) If $a \equiv a' \pmod{m}$, then $\chi(a) = \chi(a')$.

(3) For any rational integers a and b we have $\chi(ab) = \chi(a)\chi(b)$.

We shall show that numerical characters are completely characterized by these three properties. Let η be a function which satisfies (1) to (3). For any class $\bar{a} \in G_m$, $(a, m) = 1$, we set $\chi(\bar{a}) = \eta(a)$. By (2) the value of $\chi(\bar{a})$ does not depend on the choice of a, and by (1) it is nonzero. Also, if $(a, m) = 1$ and $(b, m) = 1$, then by condition (3)

$$\chi(\bar{a}\,\bar{b}) = \chi(\overline{ab}) = \eta(ab) = \eta(a)\eta(b) = \chi(\bar{a})\chi(\bar{b}).$$

Hence χ is a character of the group G_m, and the corresponding numerical character coincides with the function η.

Let m' be a natural number which is divisible by m. From any character χ modulo m we can form a character χ' modulo m'. Namely, if a is relatively prime to m' (and hence also to m), set $\chi'(a) = \chi(a)$; if $(a, m') > 1$, set $\chi'(a) = 0$. The function χ' satisfies conditions (1) to (3), and hence is a numerical character modulo m'. We shall say that χ' is *induced* by the character χ.

Definition. Let χ be a numerical character modulo m. If there is a proper divisor d of the number m and a character χ_1 modulo d such that χ_1 induces χ, then the character is called *nonprimitive*; otherwise it is called *primitive.*

Theorem 4. In order that the character χ modulo m be primitive, it is necessary and sufficient that for any proper divisor d of the number m, there be a number x which is congruent to 1 modulo d and relatively prime to m, such that $\chi(x) \neq 1$.

Proof. If the character χ is nonprimitive, then it is induced by some character χ_1 modulo d, where d is a proper divisor of m. This means that for any x which is relatively prime to m we have $\chi(x) = \chi_1(x)$. If $x \equiv 1 \pmod{d}$, then $\chi(x) = \chi_1(x) = \chi_1(1) = 1$. Conversely, assume that for some proper divisor d of the number m we have $\chi(x) = 1$ for any x which is relatively prime to m and congruent to 1 modulo d. For any a which is relatively prime to d, we can find a number a', for which $(a', m) = 1$ and $a' \equiv a \pmod{d}$. Set

$$\chi_1(a) = \chi(a').$$

We claim that the value of $\chi_1(a)$ does not depend on the choice of a'. Indeed, if $a' \equiv a'' \pmod{d}$, where a'' is also relatively prime to m, then $a'' \equiv xa' \pmod{m}$ for some x relatively prime to m (since a' and a'' are both relatively prime to m). Since $x \equiv 1 \pmod{d}$, by the conditions of the theorem we have $\chi(x) = 1$, and then $\chi(a'') = \chi(x)\chi(a') = \chi(a')$. By setting $\chi_1(a) = 0$ when $(a, d) \neq 1$, we obtain a function χ_1 which is easily seen to be a numerical character modulo d. Since $\chi(a) = \chi_1(a)$ for $(a, m) = 1$, then χ is induced by the character χ_1. The theorem is proved.

PROBLEMS

1. If a finite cyclic group has prime power order, show that it is not the direct product of proper subgroups.

2. Let G be a finite cyclic group whose order is the product of the relatively prime numbers k and l. Show that G is the direct product of two subgroups of orders k and l.

3. Let a be an element of maximum order in the finite Abelian group G. Show that the cyclic subgroup $\{a\}$ is a direct factor of G.

4. Let k be a natural number. Show that the element x of the finite Abelian group G is a kth power in G if and only if $\chi(x) = 1$ for all characters χ of the group G for which $\chi^k = \chi_0$.

5. Let G be a finite Abelian group of order n. Write in any order the elements $x_1, \ldots, x_n$ and the characters $\chi_1, \ldots, \chi_n$. Show that the matrix

$$\left(\frac{1}{n}\,\chi_i(x_j)\right)_{i,j}$$

is unitary.

6. Let $m_1, \ldots, m_k$ be pairwise relatively prime natural numbers and let $m = m_1 \cdots m_k$. Show that any numerical character modulo m has a unique representation as a product of characters χ_i modulo m_i ($i = 1, \ldots, k$), that is, that

$$\chi(a) = \chi_1(a) \cdots \chi_k(a)$$

for any rational integer a. [For any i the character χ_i is defined by $\chi_i(a) = \chi(a')$, where a' is determined by the congruences $a' \equiv a \pmod{m_i}$, $a' \equiv 1 \pmod{m/m_i}$.]

7. If the character χ of Problem 6 is primitive, show that the characters $\chi_1, \ldots, \chi_k$ are also primitive.

8. Let d_1 and d_2 be divisors of the natural number m and let $d = (d_1, d_2)$. If the character χ modulo m is induced by some character modulo d and also induced by some character modulo d, show that it is induced by some character modulo d.

9. Show that every character modulo m is induced by a unique primitive character. The modulus f of this primitive character is called the *fundamental modulus* of the character χ.

10. Show that the number of primitive characters modulo m equals

$$\sum_{d \mid m} \mu(d)\varphi\left(\frac{m}{d}\right)$$

(d runs through all divisors of the number m; μ is the Möbius function; φ is Euler's function).

11. Show that there exist primitive characters modulo m if and only if m is either odd or is divisible by 4.

12. Let $\mathfrak{F}$ be the vector space over the complex numbers which consists of all complex-valued functions on the finite Abelian group G. For each element $\omega \in G$ let T_ω denote the shift operator, defined by the formula $(T_\omega f)(\sigma) = f(\omega\sigma)$. Show that every character of the group G is an eigenvector of the operator T_ω. What are the corresponding eigenvalues?

13. We keep the notations of the preceding problem and consider, for some fixed $f \in \mathfrak{F}$, the square matrix

$$A = (f(\sigma\tau^{-1}))_{\sigma,\tau},$$

where σ and τ run through all elements of the group G, arranged in some order. Show that the determinant of this matrix is

$$\prod_\chi \left(\sum_\sigma f(\sigma)\chi(\sigma) \right)$$

(σ runs through all elements and χ through all characters of the group G).

[*Hint:* The matrix A is the matrix of the operator $T = \sum_\omega f(\omega)T_\omega$ in the basis which consists of the functions l_σ, where

$$l_\sigma(\tau) = \begin{cases} 1 & \text{for } \sigma = \tau, \\ 0 & \text{for } \sigma \neq \tau. \end{cases}$$

Find the eigenvalues of the operator T.]

14. Prove the assertion of Problem 13 by considering the determinant of the product of the matrices $(\chi(\sigma))_{\chi,\sigma}$ and A.

Tables

TABLE 1

h, THE NUMBER OF DIVISOR CLASSES, AND ε, THE FUNDAMENTAL UNIT GREATER THAN 1, FOR THE REAL QUADRATIC FIELDS $R(\sqrt{d})$, WHERE d IS A SQUARE-FREE INTEGER, $2 \leqslant d \leqslant 101$. THE NORM $N(\varepsilon)$ IS ALSO GIVEN. HERE $\omega = (1+\sqrt{d})/2$ WHEN $d \equiv 1 \pmod 4$, AND $\omega = \sqrt{d}$ WHEN $d \equiv 2, 3 \pmod 4$

d	h	ε	$N(\varepsilon)$	d	h	ε	$N(\varepsilon)$
2	1	$1+\omega$	-1	53	1	$3+\omega$	-1
3	1	$2+\omega$	$+1$	55	2	$89+12\omega$	$+1$
5	1	ω	-1	57	1	$131+40\omega$	$+1$
6	1	$5+2\omega$	$+1$	58	2	$99+13\omega$	-1
7	1	$8+3\omega$	$+1$	59	1	$530+69\omega$	$+1$
10	2	$3+\omega$	-1	61	1	$17+5\omega$	-1
11	1	$10+3\omega$	$+1$	62	1	$63+8\omega$	$+1$
13	1	$1+\omega$	-1	65	2	$7+2\omega$	-1
14	1	$15+4\omega$	$+1$	66	2	$65+8\omega$	$+1$
15	2	$4+\omega$	$+1$	67	1	$48\,842+5\,967\omega$	$+1$
17	1	$3+2\omega$	-1	69	1	$11+3\omega$	$+1$
19	1	$170+39\omega$	$+1$	70	2	$251+30\omega$	$+1$
21	1	$2+\omega$	$+1$	71	1	$3\,480+413\omega$	$+1$
22	1	$197+42\omega$	$+1$	73	1	$943+250\omega$	-1
23	1	$24+5\omega$	$+1$	74	2	$43+5\omega$	-1
26	2	$5+\omega$	-1	77	1	$4+\omega$	$+1$
29	1	$2+\omega$	-1	78	2	$53+6\omega$	$+1$
30	2	$11+2\omega$	$+1$	79	3	$80+9\omega$	$+1$
31	1	$1\,520+273\omega$	$+1$	82	4	$9+\omega$	-1
33	1	$19+8\omega$	$+1$	83	1	$82+9\omega$	$+1$
34	2	$35+6\omega$	$+1$	85	2	$4+\omega$	-1
35	2	$6+\omega$	$+1$	86	1	$10\,405+1\,122\omega$	$+1$
37	1	$5+2\omega$	-1	87	2	$28+3\omega$	$+1$
38	1	$37+6\omega$	$+1$	89	1	$447+106\omega$	-1
39	2	$25+4\omega$	$+1$	91	2	$1\,574+165\omega$	$+1$
41	1	$27+10\omega$	-1	93	1	$13+3\omega$	$+1$
42	2	$13+2\omega$	$+1$	94	1	$2\,143\,295+221\,064\omega$	$+1$
43	1	$3\,482+531\omega$	$+1$	95	2	$39+4\omega$	$+1$
46	1	$24\,335+3\,588\omega$	$+1$	97	1	$5\,035+1\,138\omega$	-1
47	1	$48+7\omega$	$+1$	101	1	$9+2\omega$	-1
51	2	$50+7\omega$	$+1$				

TABLE 2

h, THE NUMBER OF DIVISOR CLASSES, AND $N(\varepsilon)$, THE NORM OF A FUNDAMENTAL UNIT, FOR THE REAL QUADRATIC FIELDS $R(\sqrt{d})$, WHERE d IS A SQUARE-FREE INTEGER, $101 \leqslant d < 500$

d	h	$N(\varepsilon)$	d	h	$N(\varepsilon)$	d	h	$N(\varepsilon)$	d	h	$N(\varepsilon)$
101	1	−1	166	1	+1	229	3	−1	298	2	−1
102	2	+1	167	1	+1	230	2	+1	299	2	+1
103	1	+1	170	4	−1	231	4	+1	301	1	+1
105	2	+1	173	1	−1	233	1	−1	302	1	+1
106	2	−1	174	2	+1	235	6	+1	303	2	+1
107	1	+1	177	1	+1	237	1	+1	305	2	+1
109	1	−1	178	2	+1	238	2	+1	307	1	+1
110	2	+1	179	1	+1	239	1	+1	309	1	+1
111	2	+1	181	1	−1	241	1	−1	310	2	+1
113	1	−1	182	2	+1	246	2	+1	311	1	+1
114	2	+1	183	2	+1	247	2	+1	313	1	−1
115	2	+1	185	2	−1	249	1	+1	314	2	−1
118	1	+1	186	2	+1	251	1	+1	317	1	−1
119	2	+1	187	2	+1	253	1	+1	318	2	+1
122	2	−1	190	2	+1	254	3	+1	319	2	+1
123	2	+1	191	1	+1	255	4	+1	321	3	+1
127	1	+1	193	1	−1	257	3	−1	322	4	+1
129	1	+1	194	2	+1	258	2	+1	323	4	+1
130	4	−1	195	4	+1	259	2	+1	326	3	+1
131	1	+1	197	1	−1	262	1	+1	327	2	+1
133	1	+1	199	1	+1	263	1	+1	329	1	+1
134	1	+1	201	1	+1	265	2	−1	330	4	+1
137	1	−1	202	2	−1	266	2	+1	331	1	+1
138	2	+1	203	2	+1	267	2	+1	334	1	+1
139	1	+1	205	2	+1	269	1	−1	335	2	+1
141	1	+1	206	1	+1	271	1	+1	337	1	−1
142	3	+1	209	1	+1	273	2	+1	339	2	+1
143	2	+1	210	4	+1	274	4	−1	341	1	+1
145	4	−1	211	1	+1	277	1	−1	345	2	+1
146	2	+1	213	1	+1	278	1	+1	346	6	−1
149	1	−1	214	1	+1	281	1	−1	347	1	+1
151	1	+1	215	2	+1	282	2	+1	349	1	−1
154	2	+1	217	1	+1	283	1	+1	353	1	−1
155	2	+1	218	2	−1	285	2	+1	354	2	+1
157	1	−1	219	4	+1	286	2	+1	355	2	+1
158	1	+1	221	2	+1	287	2	+1	357	2	+1
159	2	+1	222	2	+1	290	4	−1	358	1	+1
161	1	+1	223	3	+1	291	4	+1	359	3	+1
163	1	+1	226	8	−1	293	1	−1	362	2	−1
165	2	+1	227	1	+1	295	2	+1	365	2	−1

TABLE 3

h, THE NUMBER OF DIVISOR CLASSES, FOR THE REAL QUADRATIC FIELD $R(\sqrt{p})$, WHERE p IS A PRIME AND $p < 2000$[a,b]

d	h	$N(\varepsilon)$	d	h	$N(\varepsilon)$	d	h	$N(\varepsilon)$	d	h	$N(\varepsilon)$
366	2	+1	401	5	−1	435	4	+1	469	3	+1
367	1	+1	402	2	+1	437	1	+1	470	2	+1
370	4	−1	403	2	+1	438	4	+1	471	2	+1
371	2	+1	406	2	+1	439	5	+1	473	3	+1
373	1	−1	407	2	+1	442	8	−1	474	2	+1
374	2	+1	409	1	−1	443	3	+1	478	1	+1
377	2	+1	410	4	+1	445	4	−1	479	1	+1
379	1	+1	411	2	+1	446	1	+1	481	2	−1
381	1	+1	413	1	+1	447	2	+1	482	2	+1
382	1	+1	415	2	+1	449	1	−1	483	4	+1
383	1	+1	417	1	+1	451	2	+1	485	2	−1
385	2	+1	418	2	+1	453	1	+1	487	1	+1
386	2	+1	419	1	+1	454	1	+1	489	1	+1
389	1	−1	421	1	−1	455	4	+1	491	1	+1
390	4	+1	422	1	+1	457	1	−1	493	2	−1
391	2	+1	426	2	+1	458	2	−1	494	2	+1
393	1	+1	427	6	+1	461	1	−1	497	1	+1
394	2	−1	429	2	+1	462	4	+1	498	2	+1
395	2	+1	430	2	+1	463	1	+1	499	5	+1
397	1	−1	431	1	+1	465	2	+1			
398	1	+1	433	1	−1	466	2	+1			
399	8	+1	434	4	+1	467	1	+1			

[a] E. L. Ince, Cycles of reduced ideals in quadratic fields, *in* "Mathematical Tables," Vol. IV, British Association for the Advancement of Science, London 1934.

[b] There are 303 prime numbers p less than 2000 (including $p = 2$).

For 26 of these primes $h = 3$ for $R(\sqrt{p})$. They are: $p = 79$, 223, 229, 257, 359, 443, 659, 733, 761, 839, 1091, 1171, 1223, 1229, 1367, 1373, 1489, 1523, 1567, 1627, 1787, 1811, 1847, 1901, 1907, 1987. For 7 primes $h = 5$. They are: $p = 401$, 439, 499, 727, 1093, 1327, 1429. For 4 primes $h = 7$. They are: $p = 577$, 1009, 1087, 1601. For $p = 1129$ we have $h = 9$ (and the group of divisor classes is cyclic), and for $p = 1297$ we have $h = 11$. For the remaining 264 prime numbers $p < 2000$, the field $R(\sqrt{p})$ has only one divisor class (that is, every divisor is principal).

TABLE 4

h, THE NUMBER OF DIVISOR CLASSES, FOR THE IMAGINARY QUADRATIC FIELDS $R(\sqrt{-a})$, WHERE *a* IS SQUARE-FREE AND $1 \leqslant a < 500$

a	*h*	*a*	*h*	*a*	*h*	*a*	*h*	*a*	*h*
1	1	67	1	134	14	202	6	267	2
2	1	69	8	137	8	203	4	269	22
3	1	70	4	138	8	205	8	271	11
5	2	71	7	139	3	206	20	273	8
6	2	73	4	141	8	209	20	274	12
7	1	74	10	142	4	210	8	277	6
10	2	77	8	143	10	211	3	278	14
11	1	78	4	145	8	213	8	281	20
13	2	79	5	146	16	214	6	282	8
14	4	82	4	149	14	215	14	283	3
15	2	83	3	151	7	217	8	285	16
17	4	85	4	154	8	218	10	286	12
19	1	86	10	155	4	219	4	287	14
21	4	87	6	157	6	221	16	290	20
22	2	89	12	158	8	222	12	291	4
23	3	91	2	159	10	223	7	293	18
26	6	93	4	161	16	226	8	295	8
29	6	94	8	163	1	227	5	298	6
30	4	95	8	165	8	229	10	299	8
31	3	97	4	166	10	230	20	301	8
33	4	101	14	167	11	231	12	302	12
34	4	102	4	170	12	233	12	303	10
35	2	103	5	173	14	235	2	305	16
37	2	105	8	174	12	237	12	307	3
38	6	106	6	177	4	238	8	309	12
39	4	107	3	178	8	239	15	310	8
41	8	109	6	179	5	241	12	311	19
42	4	110	12	181	10	246	12	313	8
43	1	111	8	182	12	247	6	314	26
46	4	113	8	183	8	249	12	317	10
47	5	114	8	185	16	251	7	318	12
51	2	115	2	186	12	253	4	319	10
53	6	118	6	187	2	254	16	321	20
55	4	119	10	190	4	255	12	322	8
57	4	122	10	191	13	257	16	323	4
58	2	123	2	193	4	258	8	326	22
59	3	127	5	194	20	259	4	327	12
61	6	129	12	195	4	262	6	329	24
62	8	130	4	197	10	263	13	330	8
65	8	131	5	199	9	265	8	331	3
66	8	133	4	201	12	266	20	334	12

TABLE 5

THE DISCRIMINANTS OF ALL KNOWN ORDERS OF IMAGINARY QUADRATIC FIELDS FOR WHICH EVERY GENUS OF MODULES BELONGING TO THE ORDER CONSISTS OF A SINGLE CLASS[a,b]

a	*h*	*a*	*h*	*a*	*h*	*a*	*h*	*a*	*h*
335	18	374	28	406	16	437	20	467	7
337	8	377	16	407	16	438	8	469	16
339	6	379	3	409	16	439	15	470	20
341	28	381	20	410	16	442	8	471	16
345	8	382	8	411	6	443	5	473	12
346	10	383	17	413	20	445	8	474	20
347	5	385	8	415	10	446	32	478	8
349	14	386	20	417	12	447	14	479	25
353	16	389	22	418	8	449	20	481	16
354	16	390	16	419	9	451	6	482	20
355	4	391	14	421	10	453	12	483	4
357	8	393	12	422	10	454	14	485	16
358	6	394	10	426	24	455	20	487	7
359	19	395	8	427	2	457	8	489	20
362	18	397	6	429	16	458	26	491	9
365	20	398	20	430	12	461	30	493	12
366	12	399	16	431	21	462	8	494	28
367	9	401	20	433	12	463	7	497	24
370	12	402	16	434	24	465	16	498	8
371	8	403	2	435	4	466	8	499	3
373	10								

[a] L. E. Dickson, "Introduction to the Theory of Numbers," Dover, New York, 1929.

[b]1. The discriminants of maximal orders (65 values):

−3	−43	−148	−340	−595	−1320
−4	−51	−163	−372	−627	−1380
−7	−52	−168	−403	−660	−1428
−8	−67	−187	−408	−708	−1435
−11	−84	−195	−420	−715	−1540
−15	−88	−228	−427	−760	−1848
−19	−91	−232	−435	−795	−1995
−20	−115	−235	−483	−840	−3003
−24	−120	−267	−520	−1012	−3315
−35	−123	−280	−532	−1092	−5460
−40	−132	−312	−555	−1155	

2. The discriminants of nonmaximal orders (36 values):

$-3\cdot 2^2$	$-4\cdot 2^2$	$-7\cdot 8^2$	$-15\cdot 4^2$	$-88\cdot 2^2$	$-408\cdot 2^2$
$-3\cdot 3^2$	$-4\cdot 3^2$	$-8\cdot 2^2$	$-15\cdot 8^2$	$-120\cdot 2^2$	$-520\cdot 2^2$
$-3\cdot 4^2$	$-4\cdot 4^2$	$-8\cdot 3^2$	$-20\cdot 3^2$	$-168\cdot 2^2$	$-760\cdot 2^2$
$-3\cdot 5^2$	$-4\cdot 5^2$	$-8\cdot 6^2$	$-24\cdot 2^2$	$-232\cdot 2^2$	$-840\cdot 2^2$
$-3\cdot 7^2$	$-7\cdot 2^2$	$-11\cdot 3^2$	$-35\cdot 3^2$	$-280\cdot 2^2$	$-1320\cdot 2^2$
$-3\cdot 8^2$	$-7\cdot 4^2$	$-15\cdot 2^2$	$-40\cdot 2^2$	$-312\cdot 2^2$	$-1848\cdot 2^2$

The suitable numbers of Euler: 1, 2, 3, 4, 5, 6, 7, 8, 9, 10, 12, 13, 15, 16, 18, 21, 22, 24, 25, 28, 30, 33, 37, 40, 42, 45, 48, 57, 58, 60, 70, 72, 78, 85, 88, 93, 102, 105, 112, 120, 130, 133, 165, 168, 177, 190, 210, 232, 240, 253, 273, 280, 312, 330, 345, 357, 385, 408, 462, 520, 760, 840, 1320, 1365, 1848.

TABLE 6

h, THE NUMBER OF DIVISOR CLASSES, FOR CERTAIN CUBIC FIELDS $R(\sqrt[3]{m})$

m	n	m	n	m	n
2	1	22	3	43	12
3	1	23	1	44	1
5	1	26	3	45	1
6	1	28	3	46	1
7	3	29	1	47	2
10	1	30	3	63	6
11	2	31	3	65	18
12	1	33	1	91	9
13	3	34	3	124	9
14	3	35	3	126	9
15	2	37	3	182	27
17	1	38	3	215	21
19	3	39	6	217	27
20	3	41	1	342	27
21	3	42	3	422	21

TABLE 7

h, THE NUMBER OF DIVISOR CLASSES, FOR ALL TOTALLY REAL CUBIC FIELDS WITH DISCRIMINANT $< 20{,}000$[a,b]

Bounds for the discriminants	*Number of fields*	*Bounds for the discriminants*	*Number of fields*
1– 1,000	22	11,001–12,000	52
1,001– 2,000	32	12,001–13,000	37
2,001– 3,000	35	13,001–14,000	43
3,001– 4,000	39	14,001–15,000	42
4,001– 5,000	34	15,001–16,000	46
5,001– 6,000	41	16,001–17,000	52
6,001– 7,000	37	17,001–18,000	39
7,001– 8,000	47	18,001–19,000	39
8,001– 9,000	40	19,001–20,000	48
9,001–10,000	39		
10,001–11,000	42	Total	806

[a] H. J. Godwin and P. A. Samet, *J. London Math. Soc.* **34**, 108–110 (1959); H. J. Godwin, *Proc. Cambridge Phil. Soc.* **57**, 728–730 (1961).

[b] The cubic field $R(\theta)$ is called totally real if $s = 3$, $t = 0$, that is, if all its isomorphisms into the complex numbers are real. If the minimum polynomial of θ factors into linear factors in $R(\theta)$, then the field $R(\theta)$ is called *cyclic*. Cyclic cubic fields are characterized by the fact that their discriminant is the square of a rational integer.

There are 830 totally real cubic fields with discriminant $< 20{,}000$. Of these 24 are cyclic. For 16 cyclic cubic fields $h=1$. These fields have discriminants 7^2, 9^2, 13^2, 19^2, 31^2, 37^2, 43^2, 61^2, 67^2, 73^2, 79^2, 97^2, 103^2, 109^2, 127^2, 139^2.

For each of the discriminants 63^2, 91^2, 117^2, 133^2, there are precisely two cyclic cubic fields, and for all of these fields, $h=3$.

The noncyclic totally real cubic fields with discriminant $< 20{,}000$ are distributed as follows (for each value of the discriminant there is only one field:

Among these fields there are 748 with $h=1$. There are 29 fields with $h=2$. Their discriminants are 1,957, 2,777, 3,981, 6,809, 7,053, 7,537, 8,468, 8,789, 9,301, 10,273, 10,889, 11,197, 1,324, 11,348, 12,197, 13,676, 13,768, 14,013, 14,197, 15,188, 15,529, 16,609, 16,997, 17,417, 17,428, 17,609, 17,989, 18,097, 19,429. Fields with $h=3$ (there are 26 of them), have discriminants 2,597, 4,212, 4,312, 5,684, 6,885, 7,220, 8,829, 9,653, 9,800, 9,996, 10,309, 11,417, 13,916, 13,932, 14,661, 14,945, 15,141, 15,884, 16,660, 16,905, 18,228, 18,252, 18,792, 19,220, 19,604, 19,764. The three fields with discriminants 8,069, 16,357, 19,821 have $h=4$. There are no fields with $h \geqslant 5$ (among the totally real cubic fields with discriminant $< 20{,}000$).

Remark: Within the limits of the table there is one and only one noncyclic totally real cubic field for each value of the discriminant. However, this is not always true. Thus, for example, there are at least three fields with discriminant 22,356 (Problem 21 of Section 2, Chapter 2).

TABLE 8

THE FACTOR $h^* = h^*(l)$ OF THE NUMBER OF DIVISOR CLASSES FOR THE lth CYCLOTOMIC FIELD FOR PRIME NUMBERS $l < 100$

l	h^*	l	h^*
3	1	43	211
5	1	47	$5 \cdot 139$
7	1	53	4,889
11	1	59	$3 \cdot 59 \cdot 233$
13	1	61	$41 \cdot 1{,}861$
17	1	67	$67 \cdot 12{,}739$
19	1	71	$7^2 \cdot 79{,}241$
23	3	73	$89 \cdot 134{,}353$
29	2^3	79	$5 \cdot 53 \cdot 377{,}911$
31	9	83	$3 \cdot 279{,}405{,}653$
37	37	89	$113 \cdot 118{,}401{,}449$
41	11^2	97	$577 \cdot 3{,}457 \cdot 206{,}209$

TABLE 9

ALL IRREGULAR PRIME NUMBERS $\leqslant$ 4001. ALONG WITH THE PRIME l ARE LISTED THOSE NUMBERS $2a$ ($2 \leqslant 2a \leqslant l-3$) FOR WHICH THE NUMERATOR OF THE BERNOULLI NUMBER B_{2a} IS DIVISIBLE BY l. THERE ARE 219 IRREGULAR PRIME NUMBERS $\leqslant$ 4001. ALL ODD PRIMES $<$ 4000 WHICH DO NOT APPEAR IN THE TABLE ARE REGULAR (THERE ARE 331 OF THEM)[ab]

l	$2a$	l	$2a$	l	$2a$
37	32	647	236, 242, 554	1307	382, 852
59	44	653	48	1319	304
67	58	659	224	1327	466
101	68	673	408, 502	1367	234
103	24	677	628	1381	266
131	22	683	32	1409	358
149	130	691	12, 200	1429	996
157	62, 110	727	378	1439	574
233	84	751	290	1483	224
257	164	757	514	1499	94
263	100	761	260	1523	1310
271	84	773	732	1559	862
283	20	797	220	1597	842
293	156	809	330, 628	1609	1356
307	88	811	544	1613	172
311	292	821	744	1619	560
347	280	827	102	1621	980
353	186, 300	839	66	1637	718
379	100, 174	877	868	1663	270, 1508
389	200	881	162	1669	388, 1086
401	382	887	418	1721	30
409	126	929	520, 820	1733	810, 942
421	240	953	156	1753	712
433	366	971	166	1759	1520
461	196	1061	474	1777	1192
463	130	1091	888	1787	1606
467	94, 194	1117	794	1789	848, 1442
491	292, 336, 338	1129	348	1811	550, 698, 1520
523	400	1151	534, 784, 968	1831	1274
541	86	1153	802	1847	954, 1016, 1558
547	270, 486	1193	262	1871	1794
557	222	1201	676	1877	1026
577	52	1217	784, 866, 1118	1879	1260
587	90, 92	1229	784	1889	242
593	22	1237	874	1901	1722
607	592	1279	518	1933	1058, 1320
613	522	1283	510	1951	1656
617	20, 174, 338	1291	206, 824	1979	148
619	428	1297	202, 220	1987	510
631	80, 226	1301	176	1993	912

l	$2a$	l	$2a$	l	$2a$
1997	772, 1888	2663	1244	3491	2544
2003	60, 600	2671	404, 2394	3511	1416, 1724
2017	1204	2689	926	3517	1836, 2586
2039	1300	2753	482	3529	3490
2053	1932	2767	2528	3533	2314, 3136
2087	376, 1298	2777	1600	3539	2082, 2130
2099	1230	2789	1984, 2154	3559	344, 1592
2111	1038	2791	2554	3581	1466
2137	1624	2833	1832	3583	1922
2143	1916	2857	98	3593	360, 642
2153	1832	2861	352	3607	1976
2213	154	2909	400, 950	3613	2082
2239	1826	2927	242	3617	16, 2856
2267	2234	2939	332, 1102, 2748	3631	1104
2273	876, 2166	2957	138, 788	3637	2526, 3202
2293	2040	2999	776	3671	1580
2309	1660, 1772	3011	1496	3677	2238
2357	2204	3023	2020	3697	1884
2371	242, 2274	3049	700	3779	2362
2377	1226	3061	2522	3797	1256
2381	2060	3083	1450	3821	3296
2383	842, 2278	3089	1706	3833	1840, 1998, 3286
2389	776	3119	1704	3851	216, 404
2411	2126	3181	3142	3853	748
2423	290, 884	3203	2368	3881	1686, 2138
2441	366, 1750	3221	98	3917	1490
2503	1044	3229	1634	3967	106
2543	2374	3257	922	3989	1936
2557	1464	3313	2222	4001	534
2579	1730	3323	3292		
2591	854, 2574	3329	1378		
2621	1772	3391	2232, 2534		
2633	1416	3407	2076, 2558		
2647	1172	3433	1300		
2657	710	3469	1174		

[a] D. H. Lehmer, Emma Lehmer, H. S. Vandiver, J. L. Selfridge, and C. A. Nicol, *Proc. Natl. Acad. Sci. U.S.* **40,** No. 1, 25–33. (1954); No. 8, 732–735 (1954); **41,** No. 11. 970–973 (1955).

[b] Table 9 was corrected in the second printing to include four additional pairs as noted in Wells Johnson, *Math. Comp.* **27** (1973), 387-396 and in V.V. Kobelev, *Soviet Math. Dokl.* **11** (1970), 188-189. See also J.L. Selfridge and R. Pollack, *Notices Amer. Math. Soc.* **11** (1964), 97.

Index

R

S

T

U

V

Z

Pure and Applied Mathematics

A Series of Monographs and Textbooks

Editors **Samuel Eilenberg and Hyman Bass**

Columbia University, New York

RECENT TITLES

E. R. Kolchin. Differential Algebra and Algebraic Groups

Gerald J. Janusz. Algebraic Number Fields

A. S. B. Holland. Introduction to the Theory of Entire Functions

Wayne Roberts and Dale Varberg. Convex Functions

A. M. Ostrowski. Solution of Equations in Euclidean and Banach Spaces, Third Edition of Solution of Equations and Systems of Equations

H. M. Edwards. Riemann's Zeta Function

Samuel Eilenberg. Automata, Languages, and Machines: Volumes A and B

Morris Hirsch and Stephen Smale. Differential Equations, Dynamical Systems, and Linear Algebra

Wilhelm Magnus. Noneuclidean Tesselations and Their Groups

François Treves. Basic Linear Partial Differential Equations

William M. Boothby. An Introduction to Differentiable Manifolds and Riemannian Geometry

Brayton Gray. Homotopy Theory: An Introduction to Algebraic Topology

Robert A. Adams. Sobolev Spaces

John J. Benedetto. Spectral Synthesis

D. V. Widder. The Heat Equation

Irving Ezra Segal. Mathematical Cosmology and Extragalactic Astronomy

J. Dieudonné. Treatise on Analysis: Volume II, enlarged and corrected printing; Volume IV; Volume V. *In preparation*

Werner Greub, Stephen Halperin, and Ray Vanstone. Connections, Curvature, and Cohomology: Volume III, Cohomology of Principal Bundles and Homogeneous Spaces

I. Martin Isaacs. Character Theory of Finite Groups

James R. Brown. Ergodic Theory and Topological Dynamics

C. Truesdell. A First Course in Rational Continuum Mechanics: Volume 1, General Concepts

George Gratzer. Lattice Theory

K. D. Stroyan and W. A. J. Luxemburg. Introduction to the Theory of Infinitesimals

B. M. Puttaswamaiah and John D. Dixon. Modular Representations of Finite Groups

Melvyn Berger. Nonlinearity and Functional Analysis: Lectures on Nonlinear Problems in Mathematical Analysis

Jan Mikusinski. The Bochner Integral

7 8 9 E 0 F 1 G 2 H 3 I 4 J 5